MEMPHIS STATE UNIVERSITY
3 2109 00448 7675

D1068352

Atlas of Creep and Stress-Rupture Curves

Edited by
Howard E. Boyer

ASM INTERNATIONAL™
Metals Park, Ohio 44073

Library of Congress Catalog Card Number: 88-070471
ISNB: 0-87170-322-X

PRINTED IN THE UNITED STATES OF AMERICA

PREFACE

Currently, creep and stress-rupture curves represent an important aspect of the materials engineering area. The need to understand just how a given metal or metal alloy will perform at elevated temperatures has greatly increased in recent years, and there are two reasons for this. First, design requirements for performance at higher and higher temperatures is a constant objective of contemporary industry. Secondly, engineers have become more deeply involved with materials under dynamic--as opposed to static--operating conditions. Thus, *easily accessed* high-temperature, material-performance data generally has great value.

Earlier industrial requirements for materials which withstand high elevated temperatures are well known. Such items as parts and fixtures used in heat-processing furnaces and kilns have long taken up the attention of metallurgists. The demands created by these items, simply because they usually did not involve dynamic applications, were not as rigorous and stringent as those expected of items used in rotating equipment. The gas turbine best exemplifies the operating environment in which excessively high temperatures and exaggerated dynamic motions over long time periods exist, yet may not affect strength--unless degradation occurs within predictable and tolerable limits. In fact, the major limitation on gas turbine efficiency, especially in aero-space applications, has been the high strength at high temperature design ratio.

Until now, there has not been a concise consolidation of creep and stress-rupture curves into a single volume. Time-consuming research has been the common methodology for both designer and application engineer. With the appearance of the *Atlas of Creep and Stress-Rupture Curves* well over 600 curves and tables are now available in convenient reference form.

The *Atlas...* includes all metal and metal alloy categories used, to any appreciable extent, for high-temperature service. Both ferrous and nonferrous types are covered. Product forms include, when available, wrought, cast, and P/M.

This reference work is, to an extent, international in character. While a majority of curves and associated data reflect sources from within the United States, a number of other countries are represented.

Chagrin Falls, OH
1988

Howard E. Boyer

INTRODUCTION

The *Atlas of Creep and Stress-Rupture Curves* is divided into 22 sections plus an appendix. Sections 1, 2, and 3 contain technical discussions of various aspects of the central topic. For those not familiar with the concept of creep, its measurement, and the interpretation of test results, this introductory material will be very helpful. The appendix, placed at the end of the book, is a glossary of terms relevant to creep.

The remaining sections are:

* Iron-Base Superalloys
* Nickel-Base Superalloys
* Cobalt-Base Superalloys
* Superalloy Comparisons
* Refractory Metals
* Refractory Metals Comparisons
* ACI Casting Alloys
* Austenitic Stainless Steels
* Ferritic Stainless Steels
* Martensitic Stainless Steels
* Precipitation Hardening Stainless Steels
* Higher-Nickel Austenitic Alloys
* Stainless Steels Comparisons
* Nickel-Base Alloys
* Cast Irons
* Carbon and Alloy Steels
* Copper and Copper Alloys
* Magnesium and Magnesium Alloys
* Titanium and Titanium Alloys

It will become immediately obvious that there is a certain unevenness in the number of items contained in the various sections. This is caused by two circumstances. First, some materials, by their very nature, are regularly applied in high-temperature operating conditions. For this reason Section 5, Nickel-Base Superalloys, is one of the largest of the groupings, possibly because of its extensive usage in jet engines. Therefore extensive research has been carried on for many years. Secondly, certain metals and alloys do not, historically speaking, lend themselves to extensive research in high-temperature operating environments. Frequently room temperature applications predominate the use patterns, and little high-temperature creep work has been performed.

Although producers and fabricators have done extensive research on the subjects of creep and stress-rupture, the editor avoided, as much as possible, using company-provided literature.

High-temperature is not clearly specified here, but it is generally understood that the lower limits of the range begin at about 425 °C (800 °F). This measurement is purely arbitrary; however, it is the point at which plain carbon steel begins to form scale. Obviously, there are exceptions to this generality--magnesium for example. A temperature of 205 °C (400 °F) is generally considered high temperature for magnesium alloys.

In most cases a single curve appears on a given page. There may in some instances be more, and some designations or product forms may even cascade to a second page. There was an underlying attempt to choose the most clear and simple curves available so that little explanatory information was required. At times the curves would be unnecessarily complex or even meaningless without a paragraph or two of text, and so explanations appear when required. In all cases data is fully referenced in order to support a reader's desire to obtain more information. All of this has the objective of making each curve as useful as possible.

In addition to the vast array of compositions and variation of product form, a wide variety of data resulting from testing in differing environments is an essential part of the presentation.

HOW TO USE THIS BOOK

In order to find an individual metal or alloy, consult the Table of Contents. There, each major category may be found, and items contained within each may be noted. Page numbers are to be used to locate a specific designation.

This *Atlas...* is divided into sections. Pages within a section are numbered according to the section number. Thus page 5-2 represents page 2 of Nickel-Base Superalloys.

Generally all items within a section are ordered alphabetically. The alphabetizing is based *only* on the designation or identification appearing in the topmost line of the page.

Comparisons within a metal and/or alloy grouping are always placed after all other items at the end of the section. (In a sense, alphabetizing begins again for comparisons.) The word *comparison* always appears so that there is no ambiguity.

Regarding superalloys, all iron-base, or nickel-base, or cobalt-base comparisons are placed at the end of the respective sections. However, cross-material references appear in Section 7, Superalloy Comparisons. In much the same way, cross-stainless steel comparisons appear in Section 16, Stainless Steel Comparisons, although any comparison within an individual type of stainless appears at the end of the particular section.

Section 8, Refractory Metals, is subdivided into four groupings: niobium, molybdenum, tantalum, and tungsten. (Each of these is separated by a divider page.) Cross-material comparisons appear in Section 9, Refractory Metals Comparisons.

Section 19, Carbon and Alloy Steels, are combined in one section, but they are subdivided and separated physically by a divider page.

The Table of Contents is the key for quick and efficient use of this reference volume.

CONTENTS

5. NICKEL-BASE SUPERALLOYS

9. REFRACTORY METALS COMPARISONS

10. ACI CASTING METALS

11. AUSTENITIC STAINLESS STEELS

12. FERRITIC STAINLESS STEELS

13. MARTENSITIC STAINLESS STEELS

14. PRECIPITATION HARDENING STAINLESS STEELS

15. HIGHER-NICKEL AUSTENITIC ALLOYS

16. STAINLESS STEELS COMPARISONS

17. NICKEL-BASE ALLOYS

20. COPPER AND COPPER ALLOYS

21. MAGNESIUM AND MAGNESIUM ALLOYS

22. TITANIUM AND TITANIUM ALLOYS

1

General Introduction to Creep

Introduction to Creep

CREEP

Creep is the slow deformation of a material under a stress that results in a permanent change in shape. Generally, creep pertains to rates of deformation less than 1.0%/min; faster rates are usually associated with mechanical working (processes such as forging and rolling). Shape changes arising from creep generally are undesirable and can be the limiting factor in the life of a part. For example, blades on the spinning rotors in turbine engines slowly grow in length during operation and must be replaced before they touch the housing.

Although creep can occur at any temperature, only at temperatures exceeding about 0.4 of the melting point of the material is the full range of effects visible ($T \geq 0.4\ T_M$, where T is temperature and T_M is the melting point of the material). At lower temperatures, creep is generally characterized by an ever-decreasing strain rate, while at elevated temperature, creep usually proceeds through three distinct stages and ultimately results in failure.

A schematic representation of creep in both temperature regimes is shown in Fig. 1.1. At time = 0, the load is applied, which produces an immediate elastic extension that is greater for high-temperature tests due to the lower modulus. Once loaded, the material initially deforms at a very rapid rate, but, as time proceeds, the rate of deformation progressively decreases.

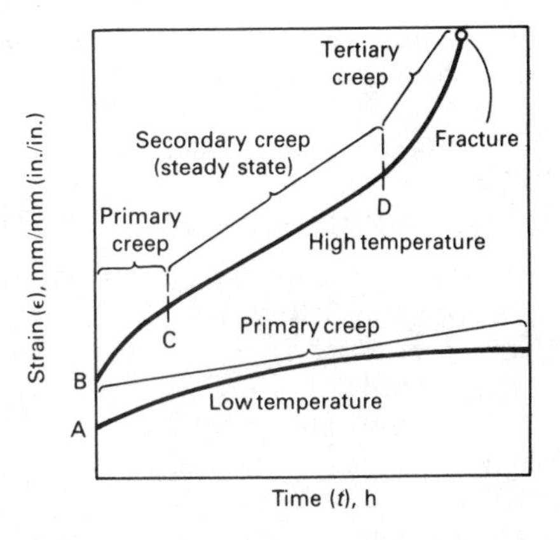

A and B denote the elastic strain on loading; C denotes transition from primary (first-stage) to steady-state (second-stage) creep; D denotes transition from steady-state to tertiary (third-stage) creep.

Fig. 1.1 Low-temperature and high-temperature creep of a material under a constant engineering stress

For low temperatures, this type of behavior can continue indefinitely. At high temperatures, however, the regime of constantly decreasing strain rate (primary, or first-stage, creep) leads to conditions where the rate of deformation becomes independent of time and strain. When this occurs, creep is in its second stage, or steady-state regime.

Although considerable deformation can occur under these steady-state conditions, eventually the strain rate begins to accelerate with time, and the material enters tertiary or third-stage creep. Deformation then proceeds at

Table 1.1 General behavior during creep

Stage	Temperature	Characteristic
First (primary)	$T > 0.4\ T_M$ or $T \leq 0.4\ T_M$	$\dot{\epsilon}$ decreases as t and ϵ increase
Secondary (steady state)	$T \geq 0.4\ T_M$	$\dot{\epsilon}$ is constant ($\dot{\epsilon}_{ss}$)
Third (tertiary)	$T \geq 0.4\ T_M$	$\dot{\epsilon}$ increases as t and ϵ increase

an ever-faster rate until the material can no longer support the applied stress and fracture occurs. With ϵ, t, and $\dot{\epsilon}$ representing strain, time, and strain rate, respectively, creep consists of the components listed in Table 1.1.

In addition to temperature, stress also affects creep, as shown in Fig. 1.2. In both temperature regimes, the elastic strain on loading increases with increasing applied stress. At low temperatures (as shown in Fig. 1.2a), very high stresses (σ_4) near or above the ultimate tensile stress result in rapid deformation and fracture at time t_4. A somewhat lesser stress (σ_3) can result in a long period of constantly decreasing strain rate, followed by a short transition to an accelerating rate and failure at t_3. Finally, lowered stresses (σ_2 and σ_1) exhibit ever-decreasing creep rates, where σ_2 produces more elastic and plastic strain than σ_1 in the same period. The stress range over which behavior changes from σ_4 to σ_2 is small, and fracture under stress σ_3 is likely to be the result of microstructural and/or mechanical instabilities.

At elevated temperatures (Fig. 1.2b), increasing the initial stress usually shortens the period of time spent in each stage of creep. Hence, the

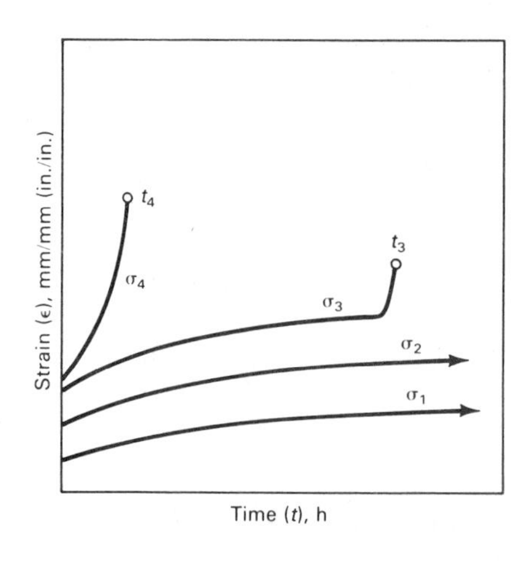

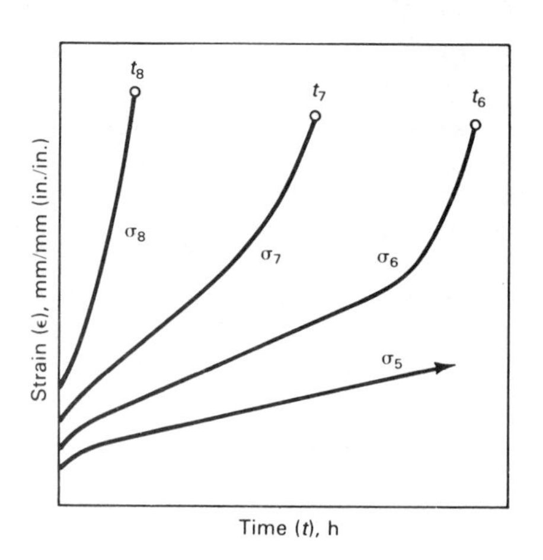

(a) Low-temperature creep, in which $\sigma_4 > \sigma_3 > \sigma_2 > \sigma_1$. (b) High-temperature creep, in which $\sigma_8 > \sigma_7 > \sigma_6 > \sigma_5$.

Fig. 1.2 Elevated-temperature creep in a material as a function of stress where the time-to-rupture is t_i for stress σ_i

time-to-rupture (t_6, t_7, and t_8) decreases as stress is increased. Additionally, the steady-state creep rate decreases as the applied stress is decreased. Also, the stress range over which behavior changes from that exhibited by σ_8 and σ_5 (Fig. 1.2b) is much broader than the range necessary to yield similar behavior at low temperatures (Fig. 1.2b).

Most of the behavior shown in Fig. 1.1 and 1.2 can be understood in terms of the Bailey-Orowan model, which views creep as the result of competition between recovery and work-hardening processes. Recovery is the mechanism(s) through which a material becomes softer and regains its ability to undergo additional deformation. In general, exposure to high temperature (stress relieving after cold working, for example) is necessary for recovery processes to be activated. Work-hardening processes make a material increasingly more difficult to deform as it is strained. The increasing load required to continue deformation between the yield stress and the ultimate tensile stress during a short-term tensile test is an example of work hardening.

After the load is applied, fast deformation begins, but this is not maintained as the material work hardens and becomes increasingly more resistant to further deformation. At low temperatures, recovery cannot occur; hence, the creep rate is always decreasing. However, at elevated temperatures, softening can occur, which leads to the steady state in which recovery and hardening processes balance one another. As the temperature increases, recovery becomes easier to activate and overcomes hardening. Thus, the transition from primary to secondary creep generally occurs at lower strains as temperature increases.

Third-stage creep cannot be rationalized in terms of the Bailey-Orowan model. Instead, tertiary creep is the result of microstructural and/or mechanical instabilities. For instance, defects in the microstructure, such as cavities, grain-boundary separations, and cracks, develop. These result in a local decrease in cross-sectional area that corresponds to a slightly higher stress in this region.

Because creep rate is dependent on stress, the strain and strain rate in the vicinity of a defect will increase. This then leads to an increase in the number and size of microstructural faults, which in turn further decreases the local cross-sectional area and increases the strain rate. Additionally, the microstructural defects, as well as other heterogeneities, can act as sites for necking. Once formed, deformation tends to increase in this region, because local stress is higher than in other parts of the specimen. The neck continues to grow, because more local deformation yields higher stresses.

CREEP EXPERIMENTS

The creep behavior of a material is generally determined by uniaxial loading of test specimens heated to temperature in some environment. Creep-rupture experiments measure the deformation as a function of time to failure. If strain-time behavior is measured, but the test is stopped before failure, this is termed an interrupted creep experiment. Finally, if an inadequate strain-measuring system or no attempt to determine length is employed, and the test is run to fracture, a stress-rupture experiment results.

In terms of data that characterize creep, the stress-rupture test provides the least amount, because only the time-to-rupture and strain-at-rupture data are available for correlation with temperature and stress. These data and other information, however, can be obtained from creep-rupture experiments. Such additional measurements can include elastic strain on loading, amount of primary creep strain, time to onset of secondary creep, steady-state creep rate, amount of secondary creep, time to onset of tertiary creep, time to 0.5% strain, time to 1.0% strain, etc. All of these data can be fitted to equations involving temperature and stress. An interrupted creep test provides much the same data as a creep-rupture experiment within the imposed strain-time limitations.

Direction of Loading

Most creep-rupture tests of metallic materials are conducted in uniaxial tension. Although this method is suitable for ductile metals, compressive testing is more appropriate for brittle, flaw-sensitive materials. In compression, cracks perpendicular to the applied stress do not propagate as they would in tension; thus, a better measure of the inherent plastic properties of a brittle material can be obtained.

In general, loading direction has little influence on many creep properties, for example, steady-state creep rate in ductile materials. However, even in these materials, the onset of third-stage creep and fracture is usually delayed in compression compared to tension. This delay is due to the minimized effect of microstructural flaws and the inability to form a "neck-like" mechanical instability. For brittle materials, the difference in behavior between tension and compression can be extreme, primarily due to the response to flaws. Consequently, care must be exercised when using compressive creep properties of a brittle material to estimate tensile behavior.

Test Specimens

Test specimens for uniaxial tensile creep-rupture tests are the same as those used in short-term tensile tests. Solid round bars with threaded or tapered grip ends or thin sheet specimens with pin and clevis grip ends are typical. However, many other types and sizes of specimens have been used successfully where the choice of geometry was dictated by the available materials. For example, small threaded round bars with a 12-mm (0.47-in.) overall length and a 1.52-mm (0.06-in.) diam by 5-mm (0.2-in.) long reduced section have been used to measure transverse stress-rupture properties of a 13-mm (0.51-in.) diam directionally solidified eutectic alloy bar.

In the case of uniaxial compression testing, specimen design can be simple small-diameter right cylinders or parallelepipeds with length-to-diameter ratios ranging from approximately 2 to 4. Larger ratios tend to enhance elastic buckling, and smaller ratios magnify the effects of friction between the test specimen and the load-transmitting member. These specimen geometries are well suited for creep testing when only a small amount of material is available, or when the material is difficult to machine.

Environment

The optimum conditions for a creep-rupture test are those in which the specimen is influenced only by the applied stress and temperature. This rarely occurs, particularly at elevated temperatures, and these conditions do not exist for real structures and equipment operating under creep conditions. For example, turbine blades are continuously exposed to hot, reactive gases that cause corrosion and oxidation.

Reactions between the test environment and material vary greatly, ranging from no visible effect to large-scale attack. For example, creep-rupture testing of aluminum, iron-chromium-aluminum, nickel-chromium, and nickel-based superalloys at elevated temperatures in air can generally be accomplished without problems because these materials form thin, stable, protective oxide films. This is not the case for refractory metals (molybdenum, niobium, tantalum, and tungsten) and their alloys, due to their strong reaction with oxygen, which leads to the formation of porous, nonprotective, and, in some cases, volatile oxides. Environmental effects such as oxidation and corrosion reduce the load-bearing cross-sectional area and can also facilitate the formation and growth of cracks.

Reactions are also possible in inert atmospheres (such as vacuum) and in reducing gas environments. Elevated-temperature testing in vacuum can result in the loss of volatile alloying elements and subsequent loss of strength. Exposure to reducing gases can result in the absorption of interstitial atoms (carbon, hydrogen, and nitrogen), which may increase strength, but also induce brittleness.

A "perfect" environment does not exist for all creep-rupture testing. The appropriate choice depends on the material, its intended use, and the available environmental protection methods. If creep mechanisms are being determined, then the atmosphere should be as inert, or nonreactive, as possible. However, if the material is to be used in an unprotected state in a reactive atmosphere, then creep-rupture testing should reflect these conditions.

Creep-rupture data from inert atmosphere tests cannot be used for design purposes when the material will be exposed to conditions of severe oxidation. However, if environmental protection methods, such as oxidation- or corrosion-resistant coatings, are available, then testing in inert gas is acceptable, and the resulting data can be used for design.

If reactions occur between the test environment and the specimen, the resultant creep-rupture data will not reflect the true creep properties of the material. Rather, the measured data are indicative of a complex interaction between creep and environmental attack, where the effects of environmental attack become more important in long-term exposure.

Strain Measurement

Care must be taken to ensure that the measured deformation occurs only in the gage section. Thus, measurements based on the relative motion of parts of the gripping system above and below the test specimen are generally

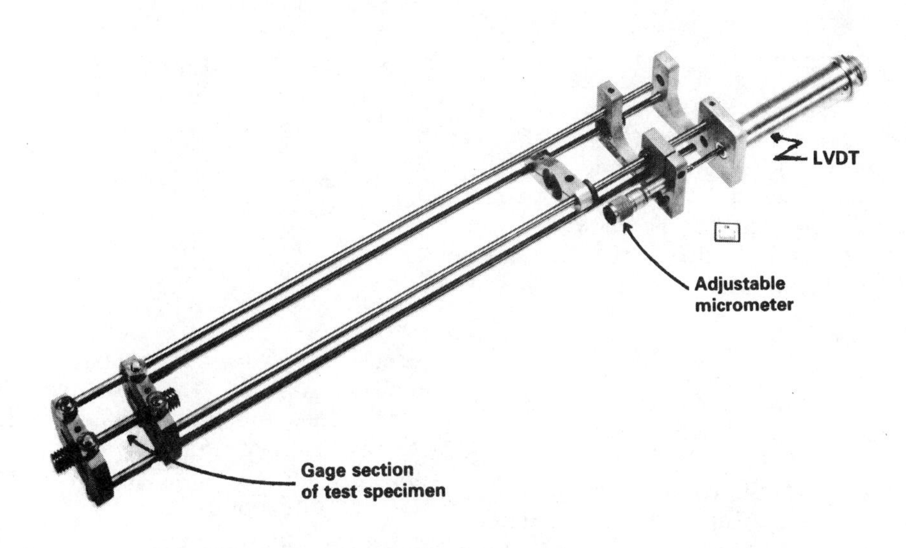

Extensometer is clamped to grooves machined in the shoulders of the test specimen.

Fig. 1.3 Typical rod-and-tube-type extensometer for elevated-temperature creep testing

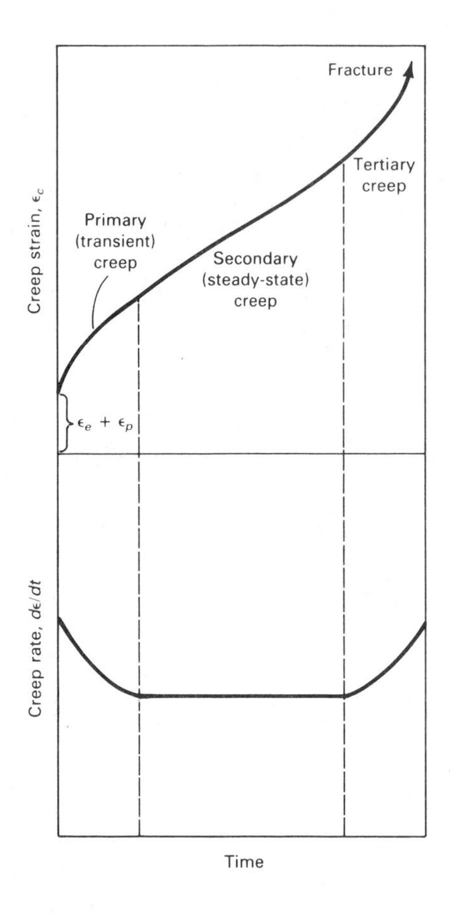

Fig. 1.4 Variation of creep and creep rate with time

inaccurate, because the site of deformation is unknown. Extensometry systems are currently available that attach directly to the specimen (shoulders, special ridges machined on the reduced section, or the gage section itself) and transmit the relative motion of the top and bottom of the gage section via tubes and rods to a sensing device such as a linear variable differential transformer (LVDT). Figure 1.3 illustrates such a system. These systems are quite accurate and stable over long periods of time.

Other methods of direct strain measurement exist and, under certain circumstances, are suitable. At low temperatures, strain gages can be directly bonded to the gage section and can be used to follow deformation over the range of extension for which the strain gage is valid. For specimens that will undergo reasonable deformations ($\epsilon > 1.0\%$), the distance between two gage marks can be optically tracked with a cathetometer as a function of time. While the location of strain is known, use of this technique is operator dependent and is generally limited to tests of less than 8 h or greater than 100 h in duration in order to permit sufficient readings to properly define the creep curve.

The dependence of creep rate on stress and temperature has been well characterized. Recently, however, insight on the atomic processes that produce creep deformation and that are responsible for the observed stress and temperature dependence of creep deformation has been gained. Knowledge of the important creep mechanisms for a particular alloy or alloy system aids the metallurgist in varying composition and microstructure to improve creep resistance. Thus, creep life prediction methods, particularly for alloys that undergo microstructural changes during creep, can be improved significantly.

CREEP CURVES

If a stress is suddenly applied to pure metals, some solid solutions, and most engineering alloys at a temperature near or greater than $0.5T_M$ (where T_M is the absolute melting point of the metal or alloy), deformation proceeds as shown in Fig. 1.4. The initial application of stress causes an instantaneous elastic strain, ϵ_e, to occur. If the stress is sufficiently high, an initial plastic deformation, ϵ_p, also occurs. At low temperatures, significant deformation ceases after the initial application of stress, and an increase in stress is required to cause further deformation.

At elevated temperatures ($T \geq 0.5T_M$), deformation under a constant applied load continues with time. The early stage of such deformation, called primary creep, is characterized by an initially high creep rate, $d\epsilon/dt$, which gradually decreases with time. Eventually, a linear variation of creep strain accumulation with time is observed.

This steady-state creep region is characterized by a constant, minimum creep rate. Steady-state creep rates depend significantly on stress and temperature and are used frequently to compare the creep resistance among alloys.

After significant deformation in steady-state creep, necking occurs, or sufficient internal damage in the form of voids or cavities accumulates to reduce the cross-sectional area, resulting in an increase in stress and creep rate. The process accelerates rapidly and failure occurs. This region of the creep curve is called tertiary creep.

Figure 1.4 also shows the derivative of the creep curve, or the creep rate curve. Although Fig. 1.4 represents the most common creep strain/time behavior, other behavior modes also occur. Figure 1.5 illustrates the strain rate/time behavior at temperatures below approximately $0.3T_M$. In Fig. 1.5, only transient creep is observed, which is characterized by a continuously decreasing creep rate that approaches zero as the inverse of time. Such low-temperature creep behavior is called logarithmic creep.

Figure 1.5 also illustrates the creep rate/time behavior of alloys that exhibit a continuously increasing creep rate in the early stages of creep. Such alloys have been designated Class I alloys. In Class I alloys – for example, Al–3% Mg – creep may occur by viscous glide of dislocations. In pure metals, dislocation climb is thought to be the dominant creep mechanism.

Creep by viscous dislocation glide results when dislocations glide, or move, in slip planes under the action of an applied stress. These dislocations drag along solute atoms attracted to the strain fields of the dislocations.

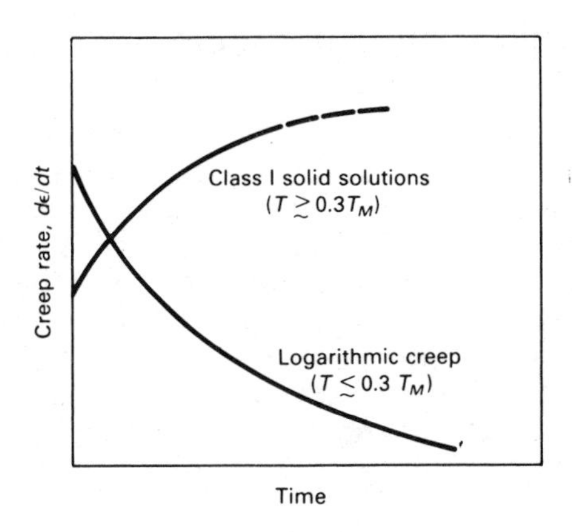

Fig. 1.5 Creep rate versus time for logarithmic creep and creep in Class I solid solutions

In order for the dislocations to move, the solute atmospheres must diffuse in the direction of dislocation motion. As a result, viscous dislocation glide is quite slow and is the rate-controlling creep process in Class I alloys.

In pure metals where no such atmospheres surround dislocations, dislocation glide can occur very rapidly until dislocation motion is stopped by barriers to motion, such as other dislocations or obstacles. The dislocations may surmount the barriers by a process called climb. In dislocation climb, vacancies diffuse to or away from the dislocation, and the dislocation moves perpendicular to its glide plane. When sufficient clearance of the obstacle is attained, the dislocation may combine with another dislocation or may glide to the next barrier.

At lower temperatures ($\sim 0.3T_M$) or at higher stress levels, creep occurs by thermally activated dislocation glide. Under these conditions, dislocations can overcome barriers to motion without dislocation climb. The localized motion of small dislocations is important in overcoming barriers. Because of the larger contribution of stress in thermally activated dislocation glide, a temperature and stress dependence for creep different from that for dislocation climb or viscous glide will be observed.

NONCLASSICAL CREEP BEHAVIOR

The curve of creep deformation versus time traditionally displays three consecutive stages. (See Fig. 1.1.) The longest period of substantially constant creep rate is preceded by a primary stage during which the rate declines from an initial high value and is followed by a tertiary stage of rising creep rate as rupture is approached.

Although this classical pattern can be made to fit many materials and test conditions, the relative duration of the three periods differs widely with materials and conditions. For example, in many superalloys and other materials in which a strengthening precipitate continues to age at creep temperatures, brief primary creep often shows transition to a long upward sweep of creep rate, with only a point of inflection for the secondary period.

Aging of normalized and tempered 0.5Cr-0.5Mo-0.25V steel during creep under 80 MPa (11.6 ksi) stress at about 565 °C (1050 °F) has been reported to cause the creep curve to effectively exhibit only a continuously increasing creep rate to fracture. For twice the amount of stress, the creep curve in this case followed the classical trends. In other alloys, such as titanium alloys, with limited elongation before fracture, the tertiary stage may be brief and may show little increase in creep rate before rupture occurs.

A more obvious departure from classical behavior develops during the early portion of many tests when precise creep measurements are taken. When 34 ferritic steels were studied for as long as 100,000 h at temperatures ranging from 450 to 600 °C (842 to 1112 °F), step-form irregularities were observed, with an extended period of secondary creep preceded by a lower creep rate of shorter duration during primary creep.

Negative Creep

Because a variety of metallurgical processes can be involved and because the rates and direction of these processes can vary with time and temperature, departures from classical creep curves can take many forms and can be overlooked, unless accurate creep readings are taken at sufficiently close intervals, particularly during early stages of the test. For 2.25Cr-1Mo, Cr-Ni-Mo, and Cr-Mo-V steels, some tests have demonstrated an abrupt drop to negative creep rate (contraction) after a brief beginning period of positive primary creep. Once this contraction ceased, the remaining portion of the test displayed the classical succession of declining, steady, and then rising creep rates.

Definite negative creep was noted in at least one test each at temperatures of 482, 704, 816, and 871 °C (900, 1300, 1500, and 1600 °F) for cast CF8 austenitic stainless steel in tests of boiler and pressure vessel materials. Rupture times for these tests ranged from 1000 h to longer than 30,000 h. For some combinations of material lots and test temperatures, nearly all creep curves of these tests showed an early "false" minimum rate during part of the primary stage. Structural changes responsible for the measured contraction were undetermined.

Short-term negative creep also was observed in tests on quenched and tempered 2.25Cr-1Mo steel at 482 °C (900 °F) and at 482 and 538 °C (900 and 1000 °F) for the same steel in the normalized and tempered condition. Two steady-state creep stages for annealed 2.25Cr-1Mo steel have been reported which were due to the interaction of molybdenum and carbon atoms with dislocations and the subsequent decrease in the number of these atoms as Mo_2C precipitated. A volume decrease associated with the precipitation process also could account for the observed creep curve trends.

Interstitial diffusion of carbon and hydrogen into dislocations has been observed, and alloy strain-aging effects have been found to cause creep rate transitions noted for carbon steels and normalized 0.5% Mo steel. Negative creep in Nimonic 80A appears related to an ordering reaction in the Ni-Cr matrix and possible formation of Ni_3Cr.

Oxide, Nitride Strengthening

An entirely different source of variation from classical patterns occurs in creep tests at high temperature due to reaction with the air that forms the environment. Tests longer than 50 h with 80Ni-20Cr alloys at temperatures of 816 and 982 °C (1500 and 1800 °F) showed a deceleration of creep after the normal tertiary stage was reached, resulting in a second period of steady-state creep and later another period of last-stage creep.

This behavior, which prolonged rupture life and caused a slope decrease in curves of log stress versus log rupture life, was due to oxide and nitride formation on surfaces of the intercrystalline cracks that form extensively during tertiary creep. Observed interconnection of the bulk of these cracks added substantially to strengthening against creep deformation in the late stages of the tests.

This effect also was observed in 99.8% Ni tested at 816 °C (1500 °F) under 20.7 MPa (3000 psi) stress. Fracture after a prolonged time occurred in a lower stressed section of the fillet. Fewer intergranular cracks in this region resulted in less oxidation strengthening than in the gage section.

STRESS DEPENDENCE OF STEADY-STATE CREEP

At intermediate stress levels and at temperatures above $0.5T_M$, the steady-state creep rate for most metals and alloys varies with stress according to:

$$\dot{\epsilon}_s = A\sigma^n$$

where A and n are constants. For pure metals, n generally varies from 4 to 5, and for solid-solution alloys, n has a value of approximately 3. For precipitation- or dispersion-strengthened alloys, the reported values of n can range as high as 30 to 40. Such high values for precipitation- and dispersion-strengthened alloys can be explained in terms of the interaction stresses between dislocations and the barriers to dislocation motion during creep.

At stress levels higher than approximately $5 \times 10^{-4}G$, where G is the shear modulus, the use of a power-law creep equation underestimates the creep rate. At the higher stresses, creep rate varies exponentially with applied stress. The breakdown in power-law creep behavior apparently results from a transition from diffusion-controlled dislocation climb creep to thermally activated dislocation glide processes similar to deformation found at lower temperatures.

TEMPERATURE DEPENDENCE OF STEADY-STATE CREEP

Although creep is known to be a thermally activated process, it is less generally understood that the measured activation energy varies significantly with temperature at low temperatures and becomes independent of temperature only above approximately $0.6T_M$.

The low activation energy for creep at low temperatures indicates that the deformation processes are only weakly temperature dependent at low temperatures, which is anticipated if dislocation glide processes are the dominant deformation mechanism in this temperature regime. Above $0.6T_M$, the activation energy for creep deformation is independent of temperature and is equal to the activation energy for self-diffusion. For many pure metals, an excellent correlation exists between the activation energy for self-diffusion and the activation energy for creep at high temperatures.

DORN EQUATION

In the temperature/stress regime in which power-law stress dependence exists and in which dislocation climb determines the temperature dependence of creep, the temperature/stress dependence of steady-state creep may be expressed by the Dorn equation:

$$\dot{\epsilon}_s = AD_L \frac{G\mathbf{b}}{kT}\left(\frac{\sigma}{G}\right)^n$$

where A and n are constants; $\mathbf{b}$ is the burgers vector, representative of the discontinuity in the crystal caused by the dislocation; D_L is the self-diffusion coefficient; k is Boltzman's constant; T is the absolute temperature; σ is the applied stress; and G is the shear modulus. The constant A incorporates material properties that cannot be explicitly determined and is determined empirically.

The equivalence of the activation energy for self-diffusion and creep indicates that dislocation climb is an integral part of creep at high temperatures. However, several different dislocation theories of power-law creep involve dislocation climb and therefore identical activation energies. The stress exponent, n, can be determined by the dominant dislocation mechanisms involved in the creep process.

DISLOCATION CREEP MECHANISMS

Creep has been considered to result from the competing processes of work hardening and thermal recovery. The Bailey-Orowan equation defines this concept mathematically:

$$d\sigma = \left(\frac{d\sigma}{d\epsilon}\right)d\epsilon + \left(\frac{d\sigma}{dt}\right)dt$$

where $d\sigma$ represents the change in flow stress, $d\sigma/d\epsilon$ represents the hardening that results from an increment of plastic strain $d\epsilon$, and $d\sigma/dt$ represents the softening due to recovery in a time increment dt. For steady-state creep, the rate of work hardening equals the rate of recovery and $d\sigma = 0$. Although this approach is simplistic, the concept of work hardening balanced by recovery forms the basis for most dislocation creep theories.

The rate of creep deformation initiated by dislocation motion can be expressed as:

$$\dot{\gamma} = \rho v \mathbf{b}$$

where $\dot{\gamma}$ is the shear strain rate, ρ is the mobile dislocation density and defines the number of dislocations free to move under an applied stress with a mean velocity v, and $\mathbf{b}$ is the burgers vector. If the stress dependence of ρ and v can be determined, the stress dependence of the steady-state creep rate can also be determined. As described earlier, the dislocation processes of thermally activated dislocation glide, viscous dislocation glide, and dislocation climb contribute to creep strain.

Most current dislocation creep models assume that immediately after application of stress all dislocations are mobile, and dislocation multiplication and glide can occur readily. This accounts for the initially high creep rate in primary creep. With continued straining, however, the dislocation configuration changes from a random distribution to a more orderly distribution, in which three-dimensional cellular networks of dislocations are formed.

Cell formation results in a decrease in the mobile dislocation density and a corresponding decrease in creep rate. The long-range stresses existing within the dislocation cells cause dislocation climb recovery. When the dislocation density in the cell is reduced by such recovery, new mobile dislocations are created, and an increment of strain is produced.

A recent model illustrates that power-law creep occurs by dislocation glide in the cell interiors and dislocation climb in the cell boundaries. Such a model, for which the steady-state creep rate is the sum of the creep rates produced by each mechanism, accounts for power-law breakdown by presuming that, as stress level increases, the contribution to creep from dislocation glide is much greater than the contribution from dislocation climb.

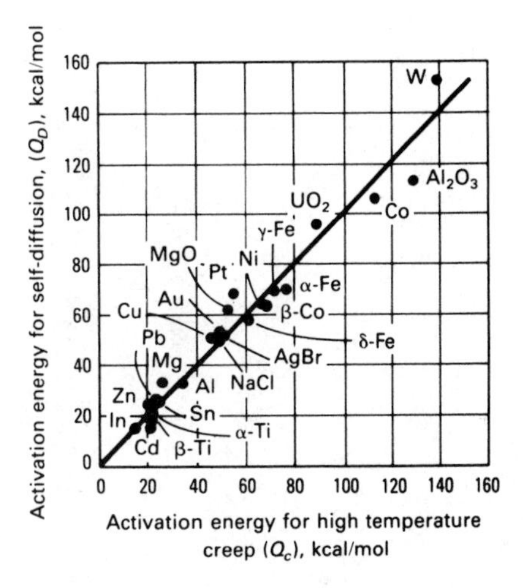

Fig. 1.6 Relationship between the activation energy for creep and the activation energy for self-diffusion for several metals and compounds

DEFORMATION MECHANISM MAPS

In addition to thermally activated glide, viscous glide, and dislocation climb, creep may also occur by strictly diffusional mass transport through grain interiors and along grain boundaries. The relative contribution of a particular mechanism depends on its stress and temperature sensitivity, as illustrated by the development of deformation mechanism maps.

Figure 1.6 is a deformation mechanism map of aluminum. The coordinates of the map are stress, normalized by shear modulus, and temperature, normalized by absolute melting temperature. Each field on the map represents the ranges of stresses and temperature at which a particular mechanism dominates the creep behavior. Thus, at low stresses and high temperatures, diffusional processes dominate. At lower temperatures and higher stresses, dislocation creep dominates. Specific knowledge of stress and temperature dependence of creep mechanisms thus allows the metallurgist to determine more efficiently alloy modifications for improved creep resistance.

2

Test Methods and Equipment

Test Methods and Equipment

GENERAL

Use of tests that measure creep, stress-rupture, and stress-relaxation properties has grown due to the design and application of metal parts that must withstand high loads at high temperatures for long periods of time. Many parts are designed for a given expected life span. To determine that life span, accurate data are needed to predict expected deformation under the conditions of stress and temperature to be encountered in service. These data can be obtained under tension, compression, combined tension and compression (bending), or torsion.

Creep Test

The creep test measures the deformation of a metal as a function of time at constant temperature. In an engineering creep test, the load is usually maintained constant throughout the test. Thus, as the specimen elongates and decreases in cross-sectional area, the axial stress increases. The initial stress that was applied to the specimen is usually the reported value of stress.

Methods of compensating for the change in dimensions of the specimen so as to carry out the creep test under constant-stress conditions have been developed. When constant-stress tests are made, no region of accelerated creep rate occurs.

Stress-Rupture Test

The stress-rupture test determines the tendencies of materials that may break under an overload. It is used widely in the selection of materials for applications in which dimensioned tolerances are not critical, but in which rupture cannot be tolerated. The stress-rupture test is similar to a constant-load creep test, using the same type of specimen and apparatus. However, no strain measurements are made during the test. The specimen is stressed under a constant load at constant temperature, as in the creep test, and the time to fracture is measured. If, however, an adequate strain-measuring system is employed and the test is run to fracture, a creep-rupture experiment results. Because of their similarities with regard to test specimens and test apparatus, constant-load creep and stress-rupture test methods will be discussed together in this article.

Stress Relaxation Test

The stress-relaxation test is somewhat similar to the creep test, but the load continually decreases instead of remaining constant. In a stress-relaxation test, the load is reduced at intervals to maintain a constant strain. The y axis in a stress-relation curve is stress or load rather than strain (elongation) as in the creep curve.

Exposure Test

Long-term elevated temperature tests, such as creep, creep-rupture, and stress-rupture tests, may be combined with exposure to an aggressive environment to evaluate the effect of the environment on creep behavior.

An increase in either stress or temperature accelerates the creep process; thus, the minimum creep rate is increased. If either stress or temperature is increased beyond the design levels for boiler tubes or heat exchangers, for example, the increased deformation can alter the temperature distribution in the system, which may cause local zones to be overheated, eventually resulting in failure.

For those alloys in which failure occurs before a well-defined start of tertiary creep, it is useful to use notched specimens or specimens with both smooth and notched test sections (with the cross-sectional area of the notch equal to that of the smooth test section). If the material is notch sensitive, the specimen will fail in the notch before failure occurs in the smooth section. The purpose is to identify notch-sensitive materials and conditions that cause notch sensitivity. Limited published data on notched stress-rupture properties of low-alloy constructional steels indicate that these steels generally are not notch sensitive.

In some procedures for design for elevated temperature service, the critical materials properties are time to reach a specified strain or time prior to onset of tertiary creep. Because the end of secondary creep often blends smoothly into the beginning of tertiary creep, an offset method is sometimes used to define the onset of tertiary creep.

Relaxation is another elevated temperature property that is important to the design of bolts or other devices intended to hold components in contact under pressure. If the service temperature is high enough, the extended-time stress on the bolt causes a minute amount of creep, which results in a major reduction in the restraining force.

Determination of creep characteristics of metals at high temperatures requires the use of a loading device or test stand, an electric furnace with suitable temperature control, and an extensometer. Equipment discussed in this section is for uniaxially loaded specimens in tension.

TEST STANDS

The test stand is designed to apply static stress to a test specimen for an extended period of time at a constant elevated temperature. Typical test stands have a balance beam that connects the test specimen to a weight pan, as shown in Fig. 2.1. Ratios of 3-to-1 up to 20-to-1 are commonly used between the weight pan and the specimen. The lower ratios are used to provide optimum accuracy at lower loads. The weight pan is part of the overall weights and frequently is suspended with a chain to prevent bending moments on the load train.

On the specimen side of the machine, a balance beam leveling motor is recommended to compensate for elongation of the test specimen. If this is not available, the balance beam may become unlevel, thus changing the calibration of the weight system. However, properly designed creep testing machines will maintain load accuracy well within ASTM requirements, even when out of level by as much as ±10°.

The test specimen is connected to the balance beam through the load train, a system of pull rods and couplings manufactured from high-temperature alloys that are capable of maintaining strength and corrosion resistance at the test temperatures encountered. The load train should be machined and assembled such that minimum bending moments are imposed on the test specimen. ASTM recommends a maximum of 10% bending strain, compared to the axial strain due to misalignment of the load train. To overcome this problem, alignment couplings (such as ball and socket or knife-edge systems) are used in the load train to facilitate self-alignment.

Vibration and shock loads can have a significant effect on the end results in

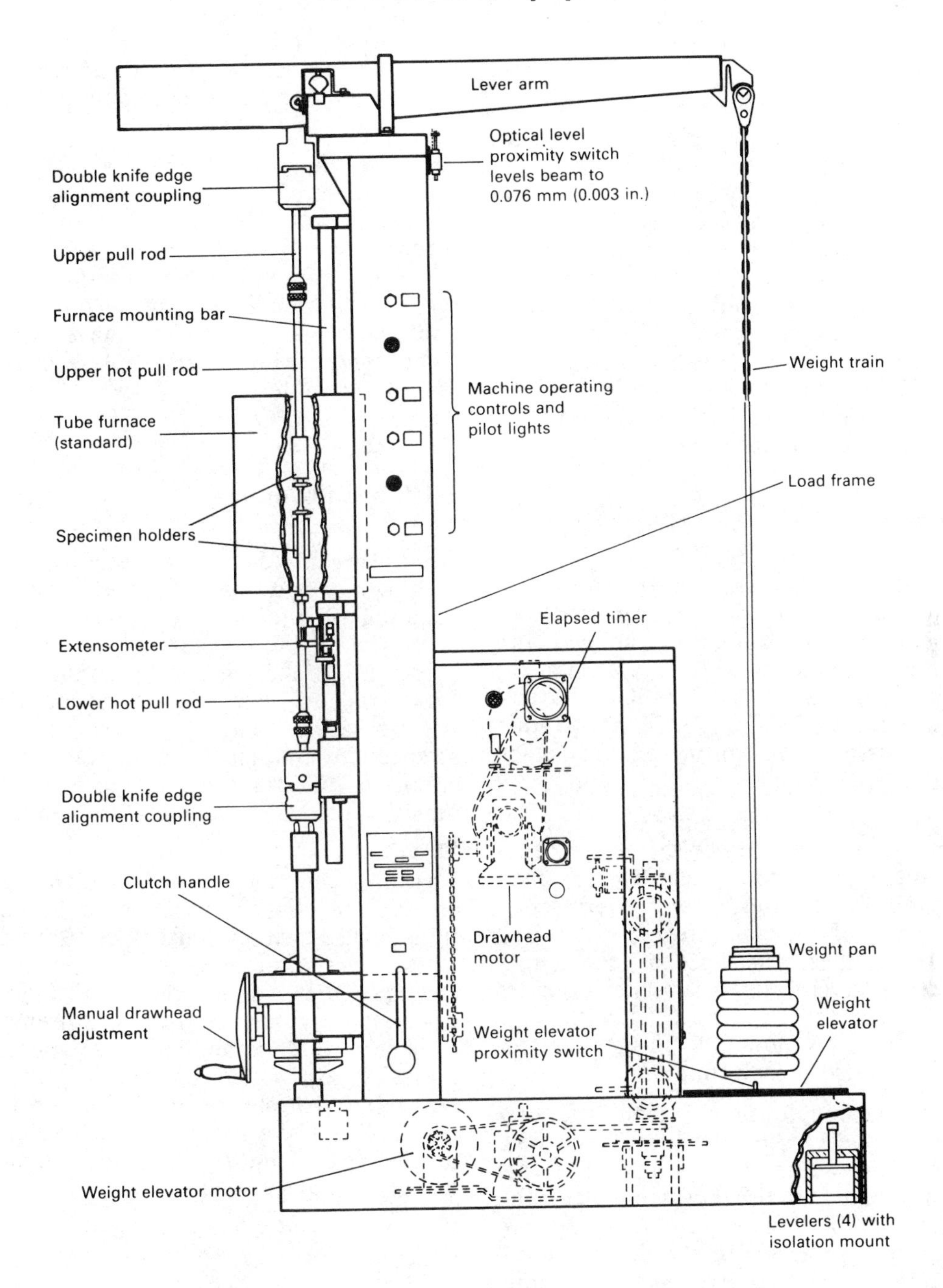

Fig. 2.1 Schematic of a test stand used for creep and stress-rupture testing

creep and stress-rupture testing. Care must be taken in selecting the test site to ensure that vibration or shock is minimal. Additionally, the test stand should be isolated from the floor with a vibration-damping material such as cork or rubber. The leveling motor can introduce vibration that may affect long-term creep tests, or shorter tests if the vibration is significant.

A timer is also included on most test stands that automatically records the time to rupture. During creep tests, the time must be recorded with the creep values.

FURNACES

Furnaces used in creep and stress-rupture testing generally are tubular, with an electrical-resistance winding that heats the test specimen through radiation in an air atmosphere. These furnaces can have single or multiple heating zones. The tube is located in a vertical position, with the pull rods connected to the specimen. Care must be taken to seal the opening of the furnace without interfering with the alignment of the load train or the action of the linkage for creep tests.

Temperature Control, Measurement

Material properties frequently are affected by temperature. The requirement for temperature control of creep and stress-rupture tests is ±1.7 °C (±3 °F) when testing at 982 °C (1800 °F) and below, and ±2.8 °C (±5 °F) above that value. Maintaining control requires practice.

Temperature measurement systems require a transducer to convert a temperature differential to an electrical signal. The transducer typically is a thermocouple that is attached directly to the specimen. Specimens with a gage length of 25.4 mm (1.0 in.) or greater require two thermocouples; specimens with gage lengths of 50.8 mm (2.0 in.) or greater require a third thermocouple.

The thermocouple should maintain intimate contact with the test specimen during the entire test. Inherent errors are associated with thermocouples, including calibration error, drift due to metallurgical changes to the thermocouple junction during the test, lead wire error, attachment gap error, radiation error, and conduction error.

The present trend in temperature control incorporates computers with two basic control schemes. Stand-alone control systems are available that control only one test stand. This type of control was used for many years with analog controllers and is now used with digital controllers. One advantage of a stand-alone controller is that if it fails only one test is lost. One disadvantage is the cost of dedicating a computer to each test stand.

The second control scheme utilizes a micro- or minicomputer to control multiple test stands. The loss of multiple tests through computer failure can be overcome by using a "host" computer that provides automatic backup. Control systems currently exist that have a host computer backing up multiple front-end control computers.

EXTENSOMETERS

When creep data are required, the specimen strain must be measured as a function of time. With most metals this is difficult, because use of strain-measuring transducers generally is not practical at the test temperatures encountered. A mechanical linkage must then be attached to the specimen to transmit the strain to the strain-measuring equipment outside the high-temperature environment.

The most commonly used strain transducer is the linear variable differential transformer, which consists of a movable metal core that changes the electrical characteristics with the small motion associated with strain measurements. Linkages that attach to the specimen typically are made of alloys that can withstand the test temperatures encountered. The linkage can be attached either to the gage length or on the shoulders outside the gage length.

Problems encountered with the use of extensometers include:

* Error due to strain in the fillet of the test specimen when the extensometer linkage is attached to the shoulder
* Fracture where the extensometer linkage attaches to the specimen gage length
* Error in strain measurement when the extensometer linkage is attached such that it measures the strain only on one side of the specimen

* Error due to slippage of extensometer linkage on the gage section of ductile specimens
* Damage to linkages and extensometers when the specimen fails
* Bending strain introduced to smaller test specimens, particularly strip specimens, due to the weight of the linkage and extensometer

Most uniaxial creep and stress-rupture tests are conducted under constant-load conditions. Typical procedures for constant-load tests and methods of data presentation are discussed below. In addition, procedures for determining the rupture life of notched specimens are presented.

SPECIMEN PREPARATION

Test specimens for constant-load tests are prepared to meet material specifications and end-use design parameters; thus, a wide variety of configurations are available.

Test specimens for creep and stress-rupture tests in tension are similar to those used for short-time tensile tests.

Specimens with round cross sections have threaded ends, except those used at very high temperatures or those of materials that are difficult to machine or display high sensitivity to notches. Specimens with shouldered ends (no threads) are called "buttonhead" specimens and are used for tests conducted at very high temperatures, particularly those in vacuum. At high temperatures, threaded specimens tend to seize in the adapters after testing due to oxidation or diffusion. The method of gripping buttonhead specimens provides self-alignment, which is advantageous with brittle materials.

For sheet or plate specimens, the load is often applied via pins placed in through holes in the shoulder ends of the specimens. If the specimen is sufficiently large, it may have elongated shoulder ends extending outside the furnace. Extension tabs sometimes are welded or brazed to the specimen shoulder and extend outside the furnace. Specially designed grips that fit the fillets at the end of the gage length are also used to apply load.

The gage length of a specimen should be uniform. The diameter or width at the ends of the reduced section should not be less than the diameter or width at the center. It is sometimes desirable to have the diameter or width of the reduced section slightly smaller at its center. The difference should range from maximums of 0.5% for a 2.54-mm (0.100-in.) diam or width specimen to 0.2% for a 12.8-mm (0.505-in.) diam or width specimen. Specimens should be smooth and free from scratches or other stress raisers and should be machined to minimize cold working or surface distortion. In computing the stress or the load required to provide a certain stress, the smallest original cross-sectional area should be used.

Misalignment can cause high local stresses and premature failure. If threaded specimens are used, the threaded adapters that the specimens fit into should be inspected frequently to ensure proper alignment. These devices may creep during a test and after several tests may undergo appreciable misalignment.

Buttonhead specimens and adapters tend to be self-aligning and pose fewer alignment problems. With sheet specimens, it is important that any brazing or welding of extension tabs be done in an alignment fixture. For sheet specimens using a pin in each tab to apply the load, the pin holes must be centered on a line running through the center of the reduced section rather than centered on the tabs. Brittle materials are more sensitive to misalignment than ductile materials.

SPECIMEN LOADING

Care is required to avoid straining the specimen when mounting it in the adapters and load train, particularly if the specimen is small or brittle. With the specimen in place, the load train (specimen adapters or grips, pull rods, etc.) should be examined carefully for any misalignment that may cause bending of the specimen under load.

The upper load train should be suspended from the lever arm, and the compensating weight adjusted so that the lever arm balances. Strain-measuring clamps and an extensometer or platinum strips are attached to the specimen, and the load train is inserted into the furnace with the specimen centered. The specimen must be stabilized at temperature before loading. Also, the extensometer should be adjusted and zeroed.

Loading the weight pan should be done smoothly and without excessive shock. If the specimen is to be step-loaded, the weight is placed on the weight pan in measured increments, and the strain corresponding to each step of loading is recorded. The loading curve thus obtained is used in determining the elastic modulus and plastic strain from load application. If step-loading is not used, a method of smoothly applying the load must be used. This can be done by placing a support such as a scissors jack under the load pan during loading. When all weights are in place, the supporting jack is lowered smoothly from under the weight pan.

TEMPERATURE CONTROL

The specimen should not be overheated while brought to temperature. A common practice is first to bring the specimen to about 10 °C (18 °F) below the desired temperature in about 1 to 4 h. Then, over a longer period, the specimen is brought to the desired temperature.

A period of time above the desired temperature is not "cancelled out" by an equal period at a temperature the same amount below the desired temperature. If the temperature rises above the desired temperature by more than a small amount, the test should be rejected. Specified limits are ±1.7 °C (±3 °F) up to 982 °C (1800 °F) and ±2.8 °C (±5 °F) above 982 °C (1800 °F). At temperatures significantly above 1093 °C (2000 °F), the limits are broadened. Variations of temperature along the specimen from the nominal test temperature should vary no more than these limits at these temperatures. These limits refer to indicated variations in temperature according to the temperature recorder.

The indicated temperature must be as close to the true temperature as possible to prevent thermocouple error or instrument error. Thermocouples, particularly base-metal thermocouples, drift in calibration with use or when contaminated. Other sources of error are incorrect lead wires, lead wires that are connected incorrectly, and direct radiation on the thermocouple bead.

Representative thermocouples should be calibrated from each lot of wires used for base-metal thermocouples. At high temperatures, base-metal thermocouples should not be reused without first removing the wire exposed to high temperature and rewelding. Nobel-metal thermocouples generally are more stable. However, they are also subject to error due to contamination and must be annealed periodically by connecting a variable transformer and passing sufficient current through the wires to make them incandescent.

When attaching the thermocouple to the specimen, the junction must be kept in intimate contact with the specimen. The bead at the junction should be as small as possible, and there must be no twisting of the thermocouple that could cause shorting. Any other metal contact across the two wires will cause shorting and erroneous readings. Shielding of the thermocouple junction from radiant heating is also recommended.

Temperature measuring, controlling, and recording instruments must be calibrated periodically against a standard. This usually is done by connecting a precision potentiometer to the thermocouple terminals on the instrument and applying potential corresponding to the output of the thermocouple at each of several temperatures. Tables of millivolt output as a function of temperature for various types of thermocouples are available in ASTM Standard E 230, "Temperature Electromotive Force (EMF) Tables for Standardized Thermocouples," and from manufacturers of precision potentiometers.

Most creep and stress-rupture machines are equipped with a switch that automatically shuts off a timer when the specimen breaks. In creep tests, the load usually is low enough that rupture does not occur. The microswitch that shuts off the timer often also shuts off or lowers the temperature of the furnace. In some furnaces, the life of the heating element is reduced significantly if the furnace is shut off after each test; instead, the temperature is merely lowered.

NOTCHED-SPECIMEN TESTING

Notched specimens are used principally as a qualitative alloy selection tool for comparing the suitability of materials for components that may contain deliberate or accidental stress concentrations. The rupture life of notched specimens is an indication of the ability of a material to deform locally without cracking under multiaxial stresses. Because this behavior is typical of superalloys, the majority of notched-specimen testing is performed on superalloys.

The most common practice is to use a circumferential 60° V-notch in round specimens, with a cross-sectional area at the base of the notch one half that of the unnotched section. However, size and shape of test specimens should be based on requirements necessary for obtaining representative samples of the material being investigated.

In a notch test, the material being tested most severely is the small volume at the root of the notch. Therefore, surface effects and residual stresses can be very influential. The notch radius must be carefully machined or ground, because it can have a pronounced effect on test results. The root radius is generally 0.13 mm (0.005 in.) or less and should be measured using an optical comparator or other equally accurate means. Size effects, stress-concentration factors introduced by notches, notch preparation, grain size, and hardness are all known to affect notch-rupture life.

Notch-rupture properties can be obtained by using individual notched and unnotched specimens, or by using a specimen with a combined notched and unnotched test section. The ratio of rupture strength of notched specimens to that of unnotched specimens varies with (1) notch shape and acuity, (2) specimen size, (3) rupture life (and therefore stress level), (4) testing temperature, and (5) heat treatment and processing history.

To avoid introducing large experimental errors, notched and unnotched specimens must be machined from adjacent sections of the same piece of material, and the gage sections must be machined to very accurate dimensions. For the combination specimen, the diameter of the unnotched section and the diameter at the root of the notch should be the same within ±0.025 mm (±0.001 in.).

Notch Sensitivity

Notch sensitivity in creep rupture is influenced by various factors, including material and test conditions. The presence of a notch may increase life, decrease life, or have no effect. When the presence of a notch increases life over the entire range of rupture time, as shown in Fig. 2.2(a), the alloy is said to be notch strengthened; that is, the notched specimen can withstand higher nominal stresses than the unnotched specimen. Conversely, when the notch-rupture strength is consistently below the unnotched-rupture strength, as in Fig. 2.2(c), the alloy is said to be notch sensitive, or notch weakened. Many investigators have defined a notch-sensitive condition as one for which the notch strength ratio is below unity. However, this ratio is unreliable and can vary according to class of alloy and rupture time.

Certain alloys and test conditions show notch strengthening at high nominal stresses (short rupture times) and notch weakening at lower nominal stresses (longer rupture times), with the result that the stress-rupture curve for notched specimens crosses the curve for unnotched specimens as nominal stress is

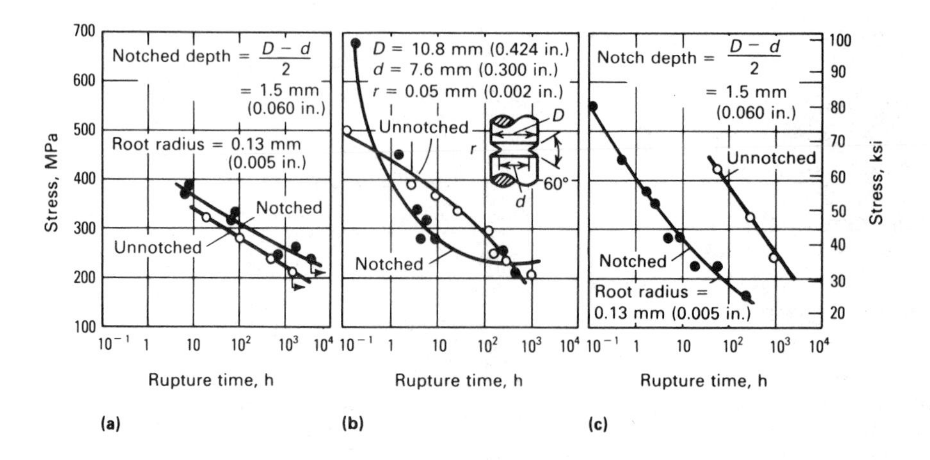

(a) Notch strengthening in 19-9 DL heat treated 50 h at 650 °C (1200 °F) and air cooled. (b) Mixed behavior in Haynes 88 heat treated 1 h at 1150 °C (2100 °F), air cooled, and worked 40% at 760 °C (1400 °F). (c) Notch weakening in K-42-B heat treated 1 h at 955 °C (1750 °F), water quenched, reheated 24 h at 650 °C, and air cooled.

Fig. 2.2 Three general types of notch effects in stress-rupture tests

reduced. Figure 2.2(b) shows that Haynes 88 becomes notch sensitive under high nominal stresses in a rupture time of about 2 h and that the material becomes notch strengthened again at lower nominal stresses at a rupture time of approximately 400 h. This same phenomenon has been observed in many superalloys and is illustrated in a different manner in Fig. 2.3. The "notch ductility trough" varies with alloy composition. For example, A-286 is notch sensitive at 540 °C (1000 °F), whereas Inconel X-750 is notch sensitive at 650 °C (1200 °F). A given alloy may show notch weakening at some temperatures and notch strengthening at others. Generally, notch sensitivity appears to increase as temperature is reduced.

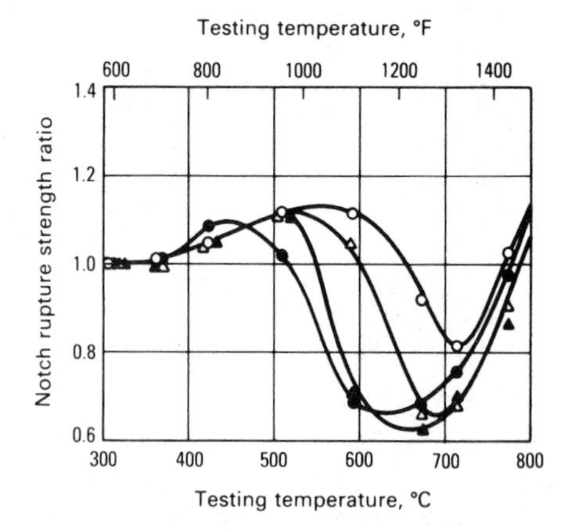

Fig. 2.3 Notch-rupture strength ratio versus temperature at four different rupture times for Inconel X-750

Changes in heat treatment of some alloys may alter notch sensitivity significantly. For example, single low-temperature aging of some alloys may produce very low rupture ductilities, because the structure is not sufficiently stabilized. Consequently, exposure of such materials for prolonged rupture times will further reduce rupture ductility because of continued precipitation of particles that enhance notch sensitivity. On the other hand, multiple aging usually stabilizes the structure and thus reduces notch sensitivity.

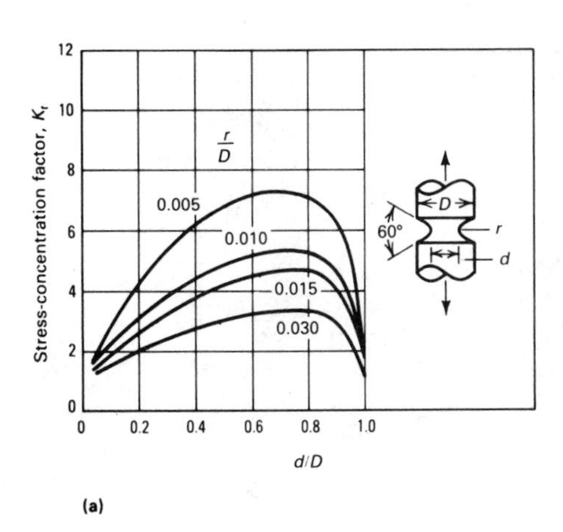

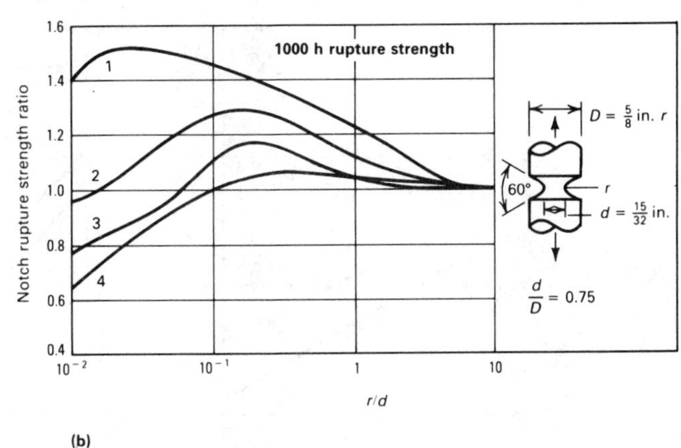

(a) Variation of stress-concentration factor with ratio of minor to major diameter and with ratio of root radius to major diameter for notched bar stressed in tension within the elastic range. (b) Variation of notch-rupture strength ratio for 1000 h life with ratio of root radius to minor diameter. Curve 1 is for 12Cr-3W steel heated 3 h at 900 °C (1650 °F) and air cooled. Grain size, ASTM No. 12; hardness, 215 HV; unnotched rupture ductility, 40%; test temperature, 540 °C (1000 °F). Curve 2 is for Refractaloy 26 oil quenched from 1010 °C (1850 °F); reheated 20 h at 815 °C (1500 °F) and air cooled; reheated 20 h at 650 °C (1200 °F) and air cooled; reheated 20 h at 815 °C and air cooled; and finally reheated 20 h at 65 0 °C and air cooled. Grain size, ASTM No. 7 to 8; hardness 330 HV; unnotched rupture ductility, 7%; test temperature 650 °C. Curve 3 is for Refractaloy 26 oil quenched from 1175 °C (2150 °F); reheated 20 h at 815 °C and air cooled; reheated 20 h at 730 °C (1350 °F) and air cooled; and finally reheated 20 h at 650 °C and air cooled. Grain size, ASTM No. 2 to 3; hardness, 325 HV; unnotched rupture ductility, 10%; test temperature, 650 °C. Curve 4 is for Refractaloy 26 oil quenched from 980 °C (1800 °F); reheated 44 h at 730 °C and air cooled; and finally reheated 20 h at 650 °C and air cooled. Grain size, ASTM 7 to 8; hardness, 375 HV; unnotched rupture ductility, 3%; test temperature, 650 °C.

Fig. 2.4 Effect of notch dimensions on stress concentration and notch-rupture strength ratio

Notch Configuration

Notch configuration can have a profound effect on test results, particularly in notch-sensitive alloys. Most studies on notch configuration present results in terms of the elastic stress-concentration factor. The design criterion for the weakening effect of notches at normal and low temperatures is that of complete elasticity. The design stress is the yield stress divided by the elastic stress-concentration factor K_t (Fig. 2.4a). The value of the peak axial (design) stress depends on the configuration of the notch.

There is no simple relationship for the effect of notches at elevated temperatures. The metallurgical effects that influence the behavior of notched material are complex and include composition, fabrication history, and heat treatment.

For ductile metals, the ratio of rupture strength of notched specimens to that of unnotched specimens usually increases to some maximum as the stress-concentration factor is increased. For very insensitive alloys, there may be little further change. Metals that are more notch sensitive may undergo a reduction in ratio as the notch sharpness (stress-concentration factor) is increased beyond the maximum and may show notch weakening for even sharper notches. Very notch-sensitive alloys may undergo little or no notch strengthening, even for very

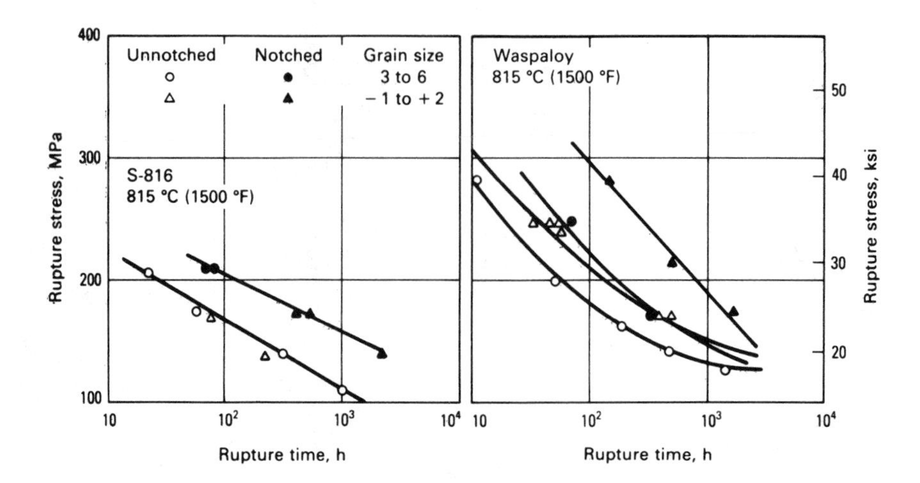

S-816 was heated to 1175 °C (2150 °F) and water quenched, reheated to 760 °C (1400 °F), held 12 h and air cooled. Waspaloy was heated to 1080 °C (1975 °F), held 4 h, and air cooled; reheated to 840 °C (1550 °F), held 4 h, and air cooled; and finally reheated to 760 °C (1400 °F), held 16 h, and air cooled. Smaller grain sizes were produced by cold reducing the S-816 1%, and the Waspaloy 1.25%, by cold rolling at 24 °C (75 °F), and then heat treating. Diameter of specimens was 12.7 mm (0.5 in.), diameter at base of notch was 8.9 mm (0.35 in.), root radius was 0.1 mm (0.004 in.), and notch angle was 60°. Data are a composite of results from two laboratories.

Fig. 2.5 Rupture strength as a function of time for notched and unnotched bars of different grain size

blunt notches (low stress-concentration factor) and may undergo progressive weakening as notch sharpness increases.

Relationships between notch configuration and the ratio of rupture strengths of notched and unnotched specimens are shown in Fig. 2.4(b). In curve 1, for an alloy with an unnotched rupture ductility of 40%, the notch-strengthening factor decreases as the notch is decreased in sharpness (increase in ratio r/d). In curve 2, for an alloy with unnotched rupture ductility of 7%, the notch-strength factor increases with increasing notch sharpness, reaches a peak, and then drops to a notch-strength reduction factor of less than unity. For an alloy with a still lower unnotched rupture ductility of 3% (curve 4), the notch-strength factor is only slightly greater than unity for large radii of curvature and becomes less than unity. It continues to decrease as the notches become sharper.

Effect of Grain Size and Other Variables

The effects of grain size on notched and unnotched rupture strength are shown in Fig. 2.5 The coarse grain sizes (for example, ASTM -1 to +2) were obtained by reheating bars in which small strains had been introduced by cold reducing them 1 to 1.25%. Notches had a strengthening effect on S-816 and Waspaloy when tested at 815 °C (1500 °F). There was no measured effect of grain size on either the notched or unnotched specimens of S-816. On the other hand, the coarse-grained Waspaloy specimens showed a longer rupture time at the same rupture stress for both notched and unnotched specimens.

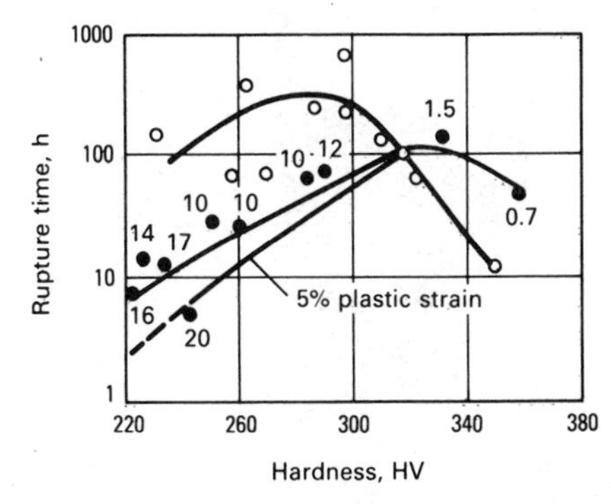

Open symbols indicate notched-bar tests (K_t = 3.9); solid symbols indicate smooth-bar tests. Numbers adjacent to points are total elongations for these tests.

Fig. 2.6 Variation of rupture time at 650 °C with initial hardness for Discaloy

The rupture time for Discaloy at 650 °C (1200 °F) increases with increasing hardness up to about 290 HV for notched specimens (K_t = 3.9) and up to 330 HV for unnotched specimens, as shown in Fig. 2.6. Ductility, as measured by elongation values for unnotched bars, decreases with increasing hardness.

The peak in rupture time at 650 °C (1200 °F) corresponds to a rupture elongation of 1.5%. The continual reduction in rupture elongation with increasing hardness indicates that the alloy exhibits time-dependent notch sensitivity. Notched bars exhibit a strengthening effect at lower hardnesses and higher ductilities; for specimens of higher hardness and lower ductility, rapid notch weakening is apparent.

For this particular alloy at this temperature, 5% rupture elongation indicates the point at which no notch strengthening or weakening occurs; this point is also indicated by the crossover of the two curves in Fig. 2.6 at about 318 HV. For other alloys, this crossover may occur at rupture ductilities as low as 3% or as high as 25%. Alloys with lower rupture ductilities are more notch sensitive.

INTERRUPTED TESTS

Power failure or some other problem may make it necessary to interrupt a test, during which time the specimen cools and must then be reheated. For many materials, this appears to have little effect on either creep properties or time to rupture if cooling and heating times are not too long. However, such treatment may affect the test material. Any interruption of a test should be reported.

DATA PRESENTATION

Readings of strain should be made frequently enough to produce a well-defined curve. This necessitates more frequent readings during the early part of the test than during later stages. The elastic portion of the stress-strain curve can be obtained from the step-loading curve or estimated by measurement of the instantaneous contraction when the load is removed at the end of the test, if the specimen has not broken.

Creep data usually are presented in the form of a curve showing percent creep strain as the vertical axis and time as the horizontal axis. Time usually is plotted on a log scale to illustrate the early part of the curve in detail. Sometimes, a family of curves is plotted on the same coordinates to show the effect of different temperatures or different stresses on one material.

Other methods for plotting data include time to reach a given percent of creep versus load at a constant temperature, and time to reach a given percent of creep versus temperature at constant load. The loading curve, showing the strain versus load as the specimen is loaded, is plotted separately and is used in computing the elastic modulus of the material at temperature.

Rupture data are presented in several types of graphs. One common format has stress as the ordinate versus log of time to rupture (at constant temperature) on the abscissa. Usually, stress-rupture data are presented by means of a parameter plot; i.e., stress is plotted against a parameter value that relates it to both time and temperature. Several different parameters have been used.

3

Manipulation and Interpretation of Data

Manipulation and Interpretation of Data

GENERAL

Use of creep-rupture properties to determine allowable stresses for service parts has evolved with experience, although guidelines for use differ among specifications. For temperatures in the creep range, the *ASME Boiler and Pressure Vessel Code* uses the following criteria as a lower limit in determining creep values: 100% of the stress to produce a creep rate of 0.01%/1000 h, which is based on a conservative average of reported tests, as evaluated by an ASME committee (in assessing data, greater weight is given to those tests run for longer times); 67% of the average stress, or 80% of the minimum stress, required to produce rupture at the end of 100,000 h, as determined from available extrapolated data.

For most commercial steels and alloys, the available raw data are obtained from many tests of durations ranging from a few hundred to a few thousand hours. Tests seldom run longer than 10,000 h, and durations of 100,000 h are rare. Measured secondary creep rates as low as 0.01%/1000 h are also unusual. Allowable stresses recommended in existing specifications usually are derived from extrapolations. Considerable scatter in test results may be observed even for a given heat of an alloy, so interpolated creep and rupture strengths are not precisely known.

Although measurement and application of creep-rupture properties is often imprecise, general trends are evident. Methods for assessing creep-rupture properties (including nonclassical creep behavior), common interpolation and extrapolation procedures, and properties estimation based on insufficient data are discussed in this article.

TEST DATA SCATTER

Reliable creep-rupture property measurements require that the test specimens be representative of the material to be used in service, preferably in the product form and condition of intended service. Testing results may vary significantly with sampling procedures. For example, heavy sections may exhibit variations in strength level with depth after normalizing or quenching and tempering treatments. A common practice is to take samples midway between the center and the surface of the specimen.

Similar variations in strength can occur for fully annealed plate or bar that has been cold straightened. Re-annealing may be required before valid testing results can be obtained. For some materials (e.g., stainless steel castings several inches thick, or alloys subjected to elevated temperature after critical prior plastic deformation), the grains may become so coarse that a specimen cross section contains only one or several grains. Local creep rates may then vary considerably along the specimen gage length according to the orientation of the individual large grains with respect to the loading direction.

Such effects caused cross sections of specimens from 127-mm (5-in.) thick castings of ACI CF-8M to distort from circular to oval, or even to nearly flat, as creep progressed in tests whose rupture life at 595 to 870 °C (1100 to 1600 °F) ranged between a few hours and 40,000 h. In this example, time to rupture was not markedly affected by gage section diameter of 8.9 versus 12.7 mm (0.35 versus 0.5 in.) or by the nonuniform deformation of the gage section. Measured creep tended to be more variable with the smaller specimens.

For materials in which grain-boundary material and that in the body of the grains differ markedly in either composition or strength, the effect of number and orientation of grains in the cross section would be expected to result in greater scatter than was noted in the above example.

Sampling direction seldom affects creep and rupture properties for materials with uniform single-phase structures of small equiaxed grains. For such materials, specimen size also exhibits minimal influence, except for the greater relative importance of surface oxidation and similar effects in small specimens. Due to solidification and processing conditions, preferred orientation of secondary phases or grains can alter test results. In critical situations, the loading direction in the test specimen should parallel the major stress direction expected under service loading. For rolled materials, ASTM Standard E 139 recommends that specimens be taken in the direction of rolling, unless otherwise specified.

The presence of large discrete particles of lower ductility and strength (e.g., graphite flakes in gray cast iron, or large glassy inclusions in steel) may significantly lower the sound cross section of a small specimen, but have a lesser influence in a larger specimen. Specimen size may have the opposite effect if only a few scattered low-strength particles are present; consequently, a specimen small enough could be free of these large-scale areas of weakness. More subtle variations in local creep and rupture strengths arise at grain boundaries, precipitates, voids, composition gradients, and other regions of microscale nonuniformity.

Specimen Size

For many steels, the influence of specimen size is minimal compared to other variables. During testing of a low-alloy steel (ASTM A193, grade B16) and a high-alloy steel (ASTM A453, grade 660) using five different specimen diameters ranging from 5 to 84 mm (0.2 to 3.3 in.), rupture strength appeared to be independent of size for unnotched specimens. Variation in rupture time with size was erratic for notched specimens, but the largest size had about three quarters of the strength of the smallest geometrically similar specimen.

Scatter From Heterogeneities

Rupture properties were not uniform with position in the original bar, as well as among bars from the same heat that received the same treatment. Fracture in notched specimens sampled from the mid-radius location of the A193 alloy originated consistently in the area of the cross section nearest the outside of the original bar.

Three tests at 227.5 MPa/538 °C (33 ksi/1000 °F) produced rupture times from 10,330 to 12,000+ h (test incomplete) for specimens from one bar, whereas three specimens from the other two bars of the A193 steel lasted only 7198 to 7820 h under similar conditions. These six tests included four different specimen diameters, but scatter among bars significantly exceeded that due to test specimen size.

Heterogeneities in a low-alloy steel of commercial quality have been found to cause much of the typical 30% scatter encountered in determining creep rates under strict testing conditions. Scatter for rupture times is approximately half that magnitude with materials that are specially prepared to be uniform.

Temperature and Other Testing Variables

Even with the use of precise temperature controllers and high-quality pyrometric practice, care is required to hold the average indicated specimen temperature within usual specifications of ±2 °C (±3.6 °F) with time. Variation from one location to another along the specimen gage length typically approaches this magnitude, even in furnaces with independent adjustment of power input within zones.

These variations in indicated temperature do not include errors in initial thermocouple and pyrometer calibrations, lead-wire mismatch, or drift in thermocouple output with time. Actual temperature can differ from the reported value by at least 5 °C (9 °F) during some portion of a representative creep or rupture test.

For steel at 450 °C (842 °F), a temperature disparity of 5 °C (9 °F) corresponds approximately to 20% variation in rupture time. A 10% change in creep rate results from a load change of 1.5%, or a temperature change of 2.5 °C (4.5 °F). An error of more than 1% in the applied load is unlikely under typical conditions.

Typical Data Scatter

Rolled bars from a single heat of 2.25Cr-1Mo steel were tested by 21 laboratories in eight countries. The largest deviations in average results for one laboratory from the arithmetic mean of values from all laboratories for the stress to cause rupture in 1000, 3000, and 10,000 h are noted in Table 3.1.

Table 3.1 Average scatter results

Temperature	Deviation
550 °C (1022 °F)	+14.3 and −16.6%
600 °C (1112 °F)	+21.2 and −12.1%

Another group of 18 laboratories in seven countries cooperated in tests of Nimonic 105 alloy bars at 900 °C (1652 °F). Mean rupture times adjusted to a common stress for individual laboratories (four tests each) ranged from a high of 1491.4 h to a low of 1090.2 h (15.5% deviation). Time to 0.5% total deformation varied from 75.4 to 182.3 h for 16 of these laboratories; deviation from the mean value thus increased to 41.5%

Temperature control was found to be the most serious source of variation in rupture measurement. In particular, calibration drift of Chromel-Alumel thermocouples at this high test temperature was responsible for the long mean rupture times found by five of the laboratories. Not more than about 15% scatter generally can be attributed to testing variables, if a laboratory has followed standard procedures.

Multiple Heats, Product Forms

Scatter bands become much broader when data originate from tests on numerous heats, particularly when data include a broad range of product forms and sizes. Elevated-temperature properties for steel and superalloys are available. Typically, data for each grade of steel include plots of the stress for rupture in 1000, 10,000, or 100,000 h versus the test temperature. Where appropriate, distinctive shapes of data points identify different product forms or heat treatments.

When data at a given temperature are extensive, the reported range of derived stresses to produce rupture in a given time typically consists of a two-fold ratio of the highest to the lowest stress level. The corresponding spread of rupture times for a given test stress is on the order of 100-fold.

If a new set of test results does not fall within the broad scatter bands, careful review and confirmation is required before the new data are accepted as valid. Bias of new data tends toward the upper half of the scatter bands compiled for older data, because the increase of residual alloying elements from scrap recycling, higher nitrogen content, and improved alignment and temperature control in modern testers tend to raise indicated creep and rupture properties.

At a test temperature of 593 °C (1100 °F) and at a stress of 207 MPa (30 ksi), rupture life ranged from 84 to 2580 h and secondary creep rate ranged from 0.16 to 0.00077%/h for 20 heats of type 304 stainless steel in the re-annealed condition. Corresponding large variations were observed at all test temperatures and for tests on seven heats of type 316 stainless steel. Re-annealing of as-received material lowered time to rupture in some cases, but the degree of variation persisted in properties among heats.

Good linear correlation was obtained when the logarithm of rupture strength was plotted against ultimate tensile strength at the same temperature for various types of austenitic stainless steels in the annealed and cold worked conditions at temperatures ranging from 538 to 816 °C (1000 to 1500 °F) and for test times approaching 10,000 h. Tensile strength, in turn, was reported to be essentially proportional to $(C + N)d^{-1/2}$, where C and N represent the weight percentage of carbon and nitrogen content, respectively, and d is average grain diameter.

Although long-time performance generally cannot be accurately predicted from short-time data, the location of a particular set of rupture data within a published scatter band should agree with the relative tensile strength of the material being tested in the range of tensile strengths spanned by all heats and product forms. Tests to measure low rates of secondary creep frequently are terminated after the creep rate appears to have become reasonably constant. If the test duration had been prolonged substantially, a continued slow decline in creep rate may have been observed.

A "false" minimum rate may occur in some tests before the classical secondary creep period has been reached. Observations of many comparative creep and rupture strengths for steels has established the following general relationships:

$$\frac{0.0001\%/\text{h creep strength}}{10\,000 \text{ h rupture strength}} = 0.7 \text{ to } 0.8$$

$$\frac{0.00001\%/\text{h creep strength}}{100\,000 \text{ h rupture strength}} = 0.5 \text{ to } 0.6$$

If a new set of test results differs from these patterns, verification is suggested before the new results are accepted.

EXTRAPOLATION, INTERPOLATION PROCEDURES

The determination of creep-rupture behavior under the conditions of intended service requires extrapolation and/or interpolation of raw data. No single method for determination of properties exists; however, a variety of techniques have evolved for data handling of most materials and applications of engineering interest. These techniques include graphical methods, time-temperature parameters, and methods used for estimations when data are sparse or hard to obtain.

Graphical Methods

Test results frequently are displayed as plots of log stress versus log rupture time and log stress versus log secondary creep rate, with a separate curve (isotherm) for each test temperature. For limited ranges of test variables, test points frequently fall in a straight line for each temperature. Nonlinearity of isotherms with broader ranges of test parameters has been treated variously, but common practice is to represent the data by two or more intersecting straight line segments. Figure 3.1 illustrates such treatment for an aluminum alloy.

Isotherms for lower temperatures characteristically display a flatter slope than those for higher temperatures. At a given temperature, when the test stresses drop below a given level that varies with alloy composition and metallurgical condition, the slope of the isotherm usually steepens. This steeper slope often approximates the slope for early times at the next higher test temperature.

Early investigators of engineering creep behavior introduced a "conservative" practice of using the slope from the next higher temperature when an isotherm had to be extended to longer times. Use of this method is limited to the specific

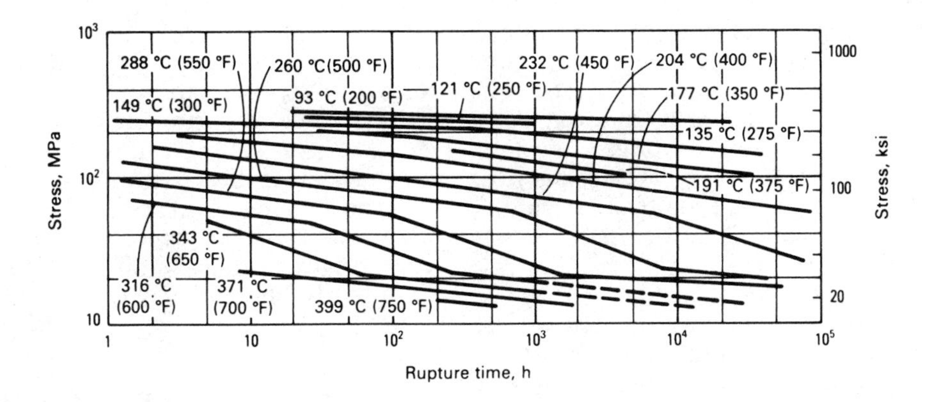

Fig. 3.1 Logarithmic plot of stress versus rupture life for aluminum alloy 6061-T651

temperatures of the test runs. Even under these conditions, extrapolations should be only in the direction of longer times for the lower range of test temperatures.

Because the change in slope of log stress versus log time isotherms historically appeared to be associated with a gradual change in fracture mode from transgranular at lower temperatures and higher stress to intergranular at relatively high temperatures and low stress, the belief developed that once the slope of the longer time portion became established, further slope change would not occur. Experimental data available at that time provided no indication that these linear plots could not be extrapolated to long times with confidence. Subsequent long-time data demonstrate that such extrapolations may lead to erroneous results.

Upward Inflection of Log-Log Rupture Plots at Long Times

Review of 52 heats from 31 wrought and cast steels, each with test times longer than 50,000 h, indicated that some portion of the log stress versus log rupture time curves for all ferritic steels showed an increase in slope when tests were of sufficiently long durations. This upward inflection was pronounced depending on composition, heat treatment, and particularly test temperature.

A sharp inflection at one temperature (500 °C, or 930 °F, for example) was usually accompanied by a less distinct inflection covering a broader time range at a higher test temperature (e.g., 550 °C, or 1022 °F). Generally, these inflections shifted to shorter times and lower stresses with increasing test temperature. Existence of inflections appeared to be related to precipitation phenomena.

For the heat-treatable aluminum alloy 6061-T651, test stresses between about 20 and 50 MPa (2900 and 7250 psi) for temperatures ranging from 260 to 343 °C (500 to 650 °F) exhibited nearly the same slope on a plot of log stress versus log rupture time, which was steeper than for either higher or lower stress levels. The long-time rupture results obtained had been predicted by separate graphical extrapolation of each of three regimes of rather constant slope (see Fig. 3.1).

In this instance, the curves that were actually extended were lines for fixed stress levels (isostress lines) on plots of log rupture time versus temperature, or the reciprocal of absolute temperature. However, extrapolation could have been carried out on the usual log stress versus log rupture time plot by treating the data as a family of curves, with different portions of each curve falling into different slope regimes. Direct graphical extension of isostress lines appeared to provide better extrapolation of rupture data than other common methods.

Curves of log stress versus log rupture life for two chromium-molybdenum steels (ASTM A387, grades 22 and 11) typically show an increase followed by a decrease in steepness for tests at 538 to 566 °C (1000 to 1050 °F). Consequently, correct prediction of 100,000-h strengths requires that these changes in slope be incorporated into the analysis. This requirement applies to all evaluation methods. Unless the input data include results that encompass structural changes of the type expected under intended service conditions, accurate extrapolation cannot be expected.

Some metallurgists prefer a semilogarithmic plot of stress versus log rupture time. The sigmoidal shape of isotherms is thus more evident, but extrapolation difficulties remain. The double inflections (or sigmoidal shape) for rupture curves can be greatly accentuated when notched specimens are tested. In the intermediate stress regime, rupture life can actually decrease as the level of test stress is lowered.

Time-Temperature Parameters

Use of time-temperature parameters (TTP's) for presenting and extrapolating high-temperature creep-rupture data has been practiced for many years. For reasons that will be evident, time-temperature parameters have not become the ultimate tools for reliable long-time data prediction, but in the course of their development they have provided much useful information and have aided in practical decision making. Furthermore, because of continuing interest in their development, they have the best potential among all available techniques for improving the technology of reliable prediction of creep-rupture data for superalloys.

A TTP is basically a function correlating the creep-rupture test variables of stress (load), temperature and time, which normally are recorded for every standard uniaxial, isothermal, constant-load test performed throughout the world. When properly developed, these correlations can be used (*a*) to represent creep-rupture data in a compact form, allowing for analytical representation and interpolation of data not experimentally determined; (*b*) to provide a simple means of comparing the behavior of materials and of rating them in a relative manner; or (*c*) to extrapolate experimental data to time ranges ordinarily difficult to evaluate directly because of test limitations.

Conceptually, the TTP has a physical basis in chemical rate theory in that the Arrhenius equation, with suitable but not rigorous assumptions, is used to derive a number of useful expressions. Grounes has reviewed this thoroughly, showing over thirty existing equations that have been proposed. By far the most common and illustrative of these are the four linear parameters described in Fig. 3.2. Here they are shown by name, form of equation, type of graphical plot they suggest and finally a schematic of that plot. The reasons for the graphs will become apparent, but first it will help to give an actual example to illustrate the technique and its uses. In Fig. 3.3(a), an actual set of stress-rupture data (log stress vs log time to rupture) is shown for the nickel-base alloy Inconel 718. Next, the data are replotted in the form shown in Fig. 3.2 for the Larson-Miller parameter. This is done by first using constant-time intercepts, shown in Fig. 3.3(b) by the dashed lines, to arrive at the constant-time curves shown in Fig. 3.3(b). Next, constant-stress curves are plotted as shown in Fig. 3.3(c) using again the intercepts shown by the dashed lines in Fig. 3.3(b). By extending the data in Fig. 3.3(c), a plausible set of convergent isostress lines meeting on the ordinate at a value of log t = −25 can be obtained. Of course, such data could have been obtained by direct experiment also. Referring again to Fig. 3.2, it can be established that the Larson-Miller equation for this set of data is:

$$P = f(\sigma) = (T + 460)(\log t_R + 25)$$

where T is temperature in °F and t_R is time to rupture, in hours. At this stage, for each data point in the original set of

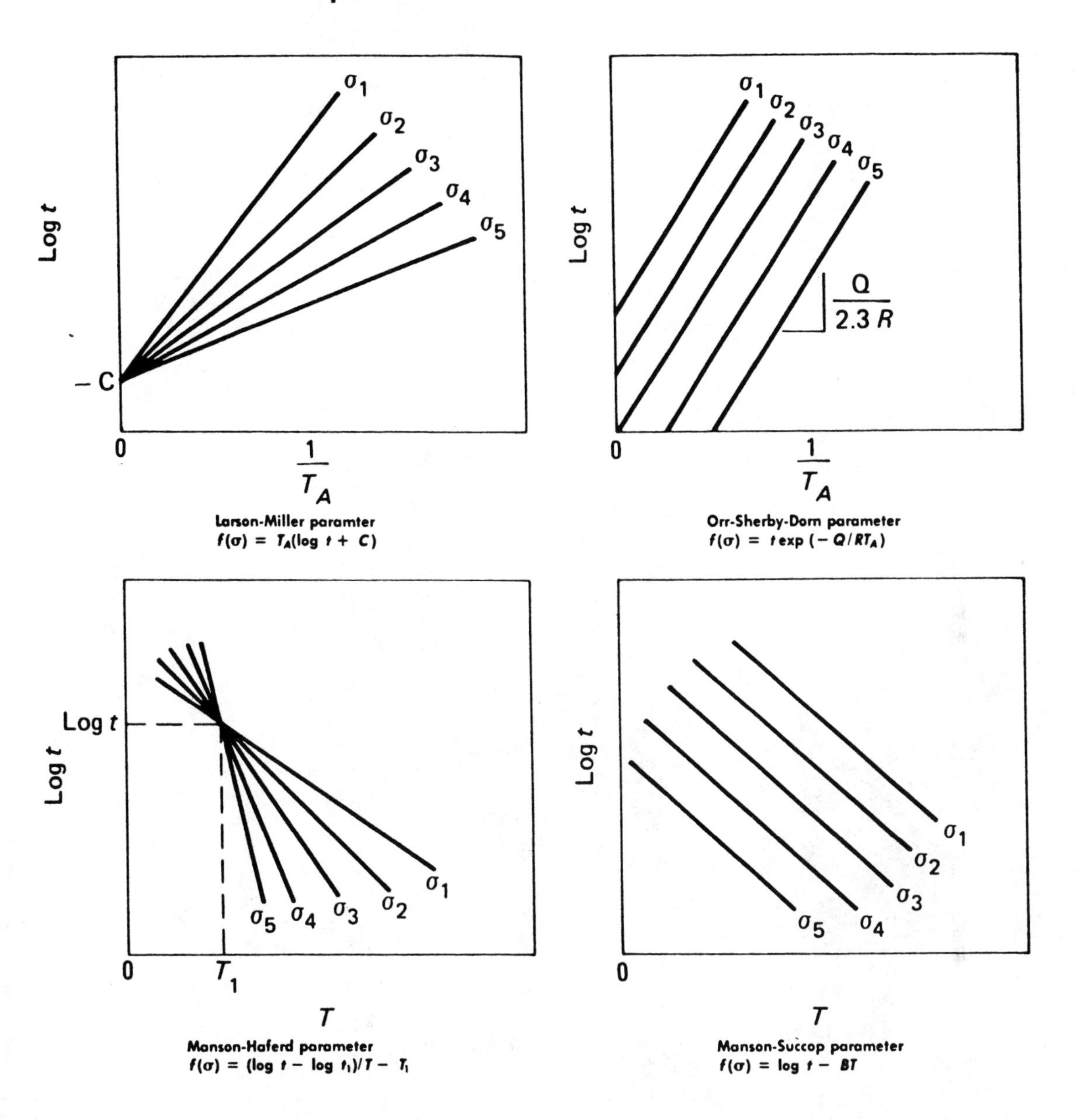

In the equations given above for each parameter, σ is applied stress, t is time, T is temperature in °C or °F, T_A is temperature in K or °R, Q is the activation energy, R is the gas constant, and B and C are numerical constants characteristic of the material and its metallurgical condition.

Fig. 3.2 Characteristic appearance of creep-rupture correlations using four different time-temperature parameters

stress, time and temperature data, the proper value can be substituted in Eq 1 and plotted as shown in Fig. 3.3(d). This is the compact parameter form of graphical representation known as the "master curve". It is possible, of course, to find the optimum constant(s) for any of the parameter models by computational data-fitting techniques. In either event it is now possible to see that a TTP

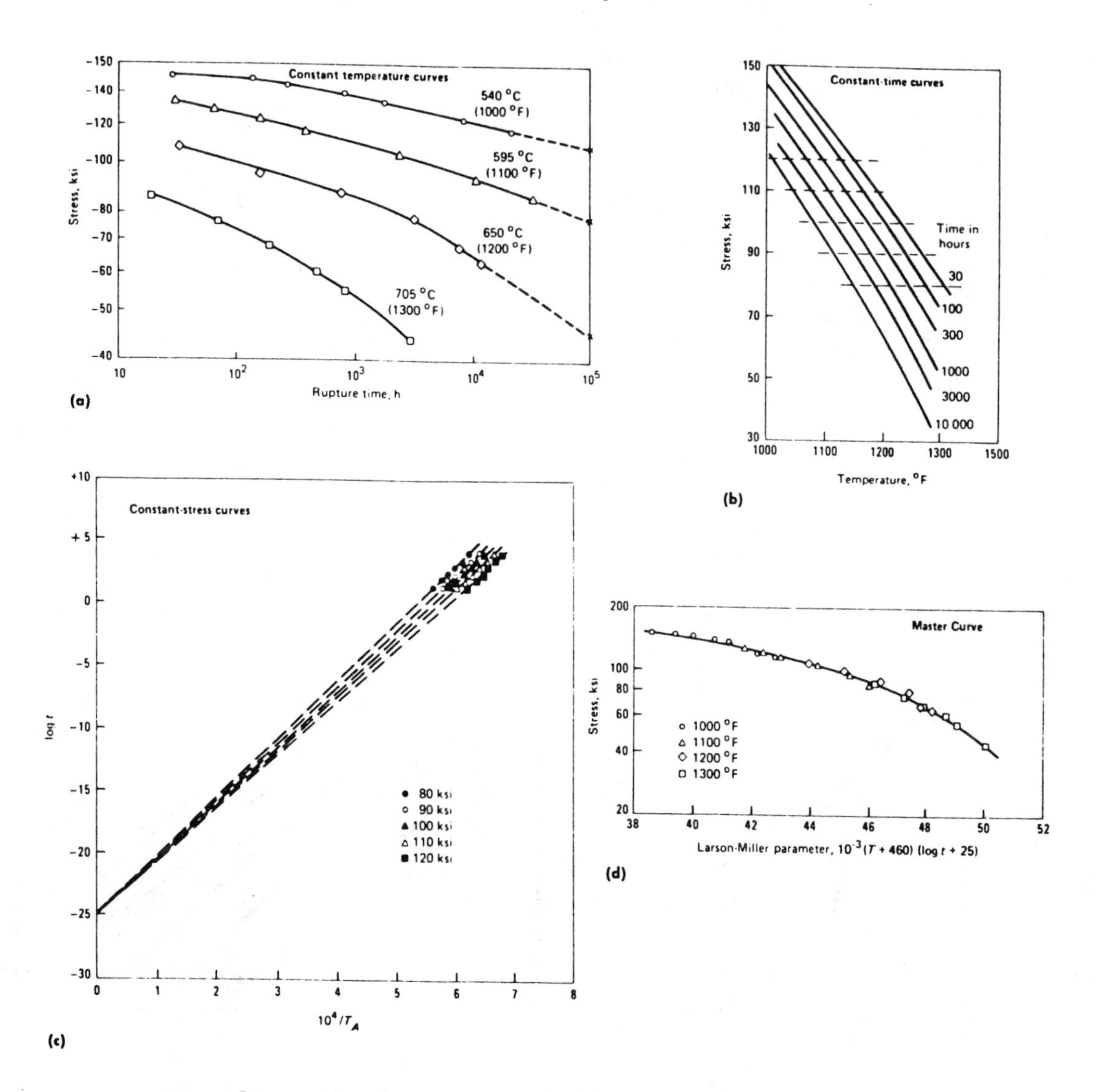

Conventional constant-temperature rupture curves shown at upper left. See text for discussion of correlating method.

Fig. 3.3 Method of creating master Larson-Miller curve for Inconel 718 from experimental stress-rupture curves

is an analytical expression containing both time and temperature such that any combination of these two variables that produces the same numerical value of parameter number will cause failure at the same stress level. Thus the parameter is a means of interchanging time and temperature in the analysis for the creep-rupture process. Note then that the expression for time in the parameter in Eq 1 may also be the time to reach a fixed amount of creep strain (such as 1%), and indeed such expressions will accommodate the normal minimum creep-rate data from evaluation of engineering creep curves when such values are substituted for time in a pertinent equation.

Returning to description of the uses of parameters and referring to Fig. 3.3(d), it is immediately recognized that the master curve can be described analytically by some relation between stress and the chosen parameter. It is most useful in this form for engineers who need to interpolate creep-rupture data. For example, assuming the validity of the master curve, isothermal rupture curves between 540 to 700 °C (1000 to 1300 °F) can be reproduced for parameter values included in the span of the master curve. Again viewing the master curve in Fig. 3.3(d), it is evident that the line drawn could have been determined by fewer data points - in fact, as few as three: the points on either extreme and one midway on the curve. In practice, for comparing materials for acceptance or in an alloy development program, three or four well-chosen points at various stresses and temperatures will produce such a curve in only several hundred hours of testing. This is an excellent method for comparing heats within a specification over a period of time or alternatively as a first cut in selecting the few best performing materials out of many prepared in a new-material development program.

Although these are useful procedures, by far the most needed and controversial employment of parameters is in extrapolation. To identify the stress required for failure of Inconel 718 in 10^5 h at temperatures of 1000, 1100 and 1200 °F, simply substitute in Eq 1 the value 10^5 for t_R and, consecutively, the values 1000, 1100, and 1200 for T and solve for the parameter numbers, which are approximately 43,800, 46,900 and 49,900. Now proceed from the abscissa of the graph in Fig. 3.3(d) at each of these values to the master curve and thence to the ordinate to arrive at approximate stresses of 110, 78, and 46 ksi. These values are shown as dashed extensions on the curves in Fig. 3.3(a). Obviously, this is a very simple way of arriving at necessary design inputs without having to wait for eleven years of testing time to elapse. But we must consider the assumptions that are made and ask how sure we can be that these answers are correct.

First consider Fig. 3.2, in which common forms of linear parameters are shown. This suggests, and correctly so, that among many available data sets one or more of the types shown will better fit

Table 3.2 Statistically useful linear parameter constants

Parameter	Constant(s)
Larson-Miller	$C = 20$
Manson-Haferd	$t_1 = 10$; $T = 600$
Manson-Succop	$B = 0.01$

the data and hence should be used. Note that the word "linear" is used to describe these common forms because linear isostress lines are used to construct the graphical presentations. Practically, of course, inherent scatter in the test data, metallurgical characteristics of the material (discussed later) and extent of the experimental data may have to be taken into account when deciding which method to use. The most important factors affecting this decision are the degree of accuracy needed and the end use for which the procedure is being employed. Most complete data sets do not conform to the simple requirement of linearity. That is to say, the techniques described up to this point may not produce the linear isostress lines and convergence or parallelism so neatly shown in Fig. 3.2. This then requires much more sophisticated mathematical treatment and perhaps even less certainty in the results.

Up to now it has been assumed that large quantities of data, such as shown in Fig. 3.2, are available to be treated as described, but this is most emphatically not the case. Because of this and the pressing need for a parameter extrapolation estimate, the universalized constant parameter has come into use. In these cases, experience on many materials over extended periods has indicated that on the average certain constants in the linear parameters are most statistically useful. For example (refer to Fig. 3.2), the values of constants shown in Table 3.2 have been suggested.

Thus, it is reasonable in certain instances to make comparisons and extrapolations for incomplete sets of data by assuming the validity of one of these universalized parameter methods. In so doing, the analyst is well advised of the probability of inaccuracy. Further to this point, because of observations that Larson-Miller evaluations were too liberal and Manson-Haferd evaluations too conservative, the following "compromise" parameter was devised by Manson:

$$f(\sigma) = \log t + 1/40 (\log t)^2 - \frac{40\,000}{T + 460}$$

This, too, is a universal parameter that may provide the analyst with a better fit for the data at hand.

By now it is clear that the TTP concept of interchanging time and temperature is not without problems and in certain cases is suspect for some materials, at least in some ranges of time and temperature. The most serious lack is the inability to predict metallurgical phase instabilities, which may cause, for example, curvature in isothermal stress-rupture lines beyond the point at which data are available. Naturally, the parameter cannot represent effects other than those apparent in the short-time data involved in its derivation. Note, too, that optimizing a constant in a parametric method ensures only that a best fit of short-time data is obtained; it does not guarantee the reliability of extrapolation beyond that range. Of the types of instabilities occurring in metallurgical systems, those in which precipitates nucleate and grow in a discontinuous way are the most disturbing. The occurrence of "sigma" phase is perhaps the best known of these. Such precipitations usually are accompanied by changes in fracture mode occurring discontinuously over certain time and temperature ranges, which disturbs the "family" relationship of curves needed to justify the parameter usage. On the other hand, some metallurgical changes (such as grain growth and overaging [softening]) will occur after short times at higher temperatures much as they occur after longer times at lower temperatures. Here the parameter methods remain tractable. The obvious point to be made is that all possible information concerning the high-temperature physical metallurgy of the alloy in question should be sought out ahead of analysis and judgment used in parameter application. All sets of data should be further examined to ensure that testing stresses above the yield point are not included, that temperatures inducing such surface problems as excessive oxidation have not been used (such data should be discarded) and that sufficient data are available to justify parameter use (master curves should not be extrapolated).

It has become obvious that the metallurgical complexity of engineering materials precludes a strict theoretical basis for the derivation of correlating parameters. The tendency now is to seek expressions in which material performance itself reveals the best form of parameter to use. Despite the inherent problems in existing TTP's and the pressures for development of new ones, current technology dictates continued use of TTP's. What is required is complete cooperation between the metallurgist and the analyst in making objective decisions on how and where to use the techniques. Finally, it is more than clear that the parameter does not obviate the need for generating data, rather it justifies the continued accumulation of long-time data for analysis of these complex engineering materials.

ESTIMATION OF REQUIRED PROPERTIES BASED ON INSUFFICIENT DATA

Complete independent evaluation of creep-rupture properties for a new lot of material, whether by graphical, parametric, or minimum commitment methods, requires numerous test points, covering an extensive range of test variables. Frequently, the amount of data available is too limited for full treatment by usual procedures. Experimental difficulties often limit obtaining accurate test results at conditions of interest, such as evaluation of creep-rupture properties near the low end of the temperature range in which time-dependent effects are significant.

Tests of short or moderate duration at these temperatures frequently require use of such high stress levels that the immediate high plastic strains at load

Table 3.3 Comparative extrapolation abilities

Parameter or method	Ratio: predicted life/actual life Minimum	Average	Maximum
Larson-Miller	0.34	1.57	5.64
Manson-Haferd	0.44	1.51	6.30
Goldhoff-Sherby	0.36	1.64	8.85
Manson-Succop	0.39	1.53	4.96
Orr-Sherby-Dorn	0.11	1.09	4.01
Monkman-Grant	0.33	0.93	1.82

application alter the nature of the material from that which exists during service under lower stresses. Testing at or near a stress of intended application often requires more time and/or expense than is feasible before the material is to be put into use. Approximate methods permit such difficulties to be treated in a generally satisfactory manner. Established correlations also permit estimation of some unmeasured properties from other types of available results.

Monkman-Grant Relationship

Analysis of data for a variety of aluminum-, iron-, nickel-, titanium-, cobalt-, and copper-based alloys led Monkman and Grant to the following empirical relationship:

$$\log t_r + m \log (\text{mcr}) = C$$

where t_r is time to rupture; mcr is minimum creep rate; and m and C are constants that differ significantly among alloy groups, but exhibit nearly fixed values for a given heat of material, or for different lots within the same alloy group.

Eq 3 enables assessment of the reliability of each individual test by examining its fit within the scatter band for all tests. Once a minimum creep rate has been determined in a long test, rupture life can be estimated without running the test to failure. Although Monkman and Grant stated that this relationship was not intended for extrapolation, it can be used for that purpose, particularly when only low-stress tests are acceptable to prevent large initial plastic strains.

Table 3.3 includes the results obtained when the Monkman-Grant relationship is applied to data obtained on seven materials. The prediction of rupture life for these 46 extrapolations using this technique was more accurate than that provided by any of the five time-temperature parameters.

For additional materials where good fit is obtained to a single linear plot on the coordinates of log time versus log secondary creep rate, extrapolation of a known secondary creep rate to the corresponding rupture life appears to be as good or better than by other extrapolation methods. One advantage of Eq 3 is that it can be applied successfully to as few as five or six data points, in contrast to the approximately 30 tests needed to establish the entire Manson-Haferd master curve. For the minimum commitment method, even more data points are usually required.

One advantage of this correlation, particularly with materials that exhibit structural instability under testing, is that the specimens used to determine the input data for secondary creep rates experience the same history of structural change that exists during the corresponding period of a test carried to rupture. Best predictions result by concentrating on tests encompassing a limited range of stresses and temperatures, thus discounting results obtained at considerably higher combined temperatures and stresses.

Reduced scatter was noted for eight nonferrous alloys and two superalloys when the term $\log t_r$ in Eq 3 was replaced by log (t_r/ϵ_c), where ϵ_c is the total creep deformation at fracture. This trend was confirmed in tests on a 2.25Cr-1Mo steel.

Although deformation-modified rupture time may improve correlation in some instances, other cases exist where use of the original relationship is sufficient or better. Data for 17 test points for 4% cold worked type 304 stainless steel exhibited a spread in creep elongation from 1.5 to 24%. Goodness of fit was identical (the coefficient of determination was r_2 = 0.86) for linear regression of the data treated by the original versus the modified log-log relationships.

Extrapolation is fast and direct when using the Monkman-Grant coordinates, but with the modified relationship, creep elongation at the given temperature and corresponding to the rupture time sought must first be estimated. This usually requires subjective extrapolation of only a few elongation values displaying wide scatter and with no evident single trend. Introduction of a creep elongation factor may have value when only correlation or interpolation of test results is desired, but it is not recommended for extrapolation.

One occasional problem in estimation of rupture life from creep data is uncertainty whether secondary creep has truly been established. Changes in creep rate with continuing test time are often sufficiently gradual and so close to the sensitivity of measurement that what appears to be a steady-rate condition may in fact still be a late portion of primary creep.

A distinctive slope change in a plot of log creep rate versus time or log time often provides better assurance that the secondary creep period has been entered than study of the deformation-time curve itself. Although an equation expressing true strain in terms of elapsed time, secondary creep rate, and three constants deviates markedly from actual behavior during the early portion of primary creep, a statistical analysis may predict acceptable values of secondary creep rate from transient data.

For type 316 stainless steel tested at 704 to 830 °C (1300 to 1525 °F), the initial transient rate at t = 0 was found to be almost equal to 3.3 times the secondary creep rate in the same test. A significantly different magnitude (near 1000) for this ratio of initial and secondary creep rates has also been found for a high-temperature alloy. A simple proportionality of this type and the more general analysis cited above are tempting alternatives for shortened test durations, but both suffer from the need for creep measurements that are more precise than those commonly obtained. Currently, neither method is capable of replacing long-time testing.

Gill-Goldhoff Method

Many designs for elevated-temperature service require that deformation not exceed some maximum value; in these cases, creep strain rather than rupture life becomes the focus. Although published compilations and computer banks of data include rupture properties for most materials of engineering interest, corresponding information on the time-dependency of strain is frequently sparse or nonexistent. Many early studies did not include strain measurement during rupture tests. When creep data were obtained, accuracy was sometimes questionable due to inadequate control of temperature or low precision of strain measurements. Typically, the only listed creep data are minimum creep rates. Most of these results were obtained from tests that were terminated after a few thousand hours, or even less, and true secondary creep rate may not have been established.

Studies by the Metal Properties Council and similar groups attempt to report both the total strain on loading and the times to various levels of creep strain. Until such results are more universally available, estimates may still be required of the creep strain to be expected in given design situations.

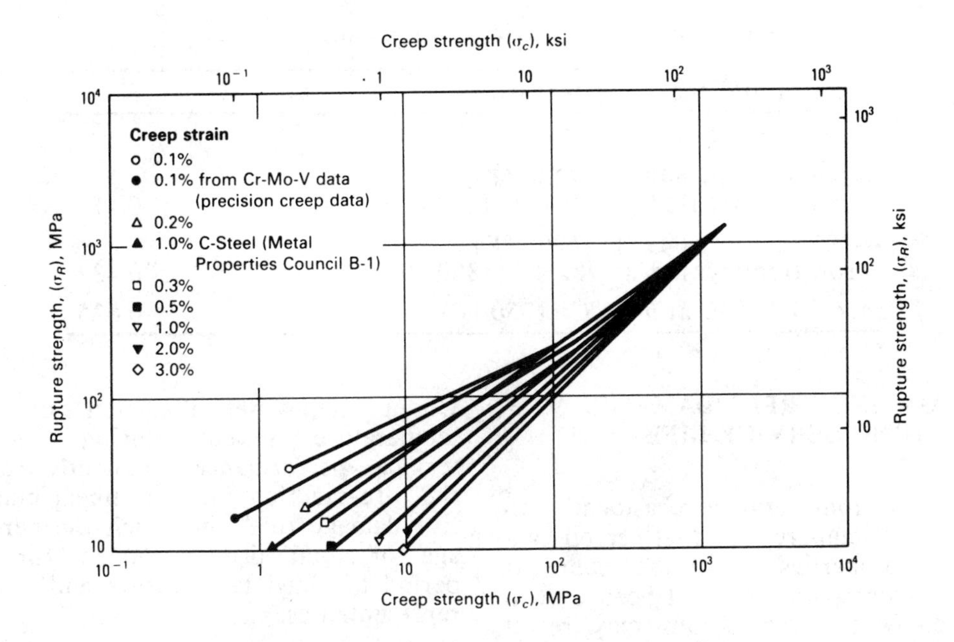

Fig. 3.4 Composite graph for the Gill-Goldhoff correlation

Gill and Goldhoff found a log-log correlation between stress to cause rupture and stress for a given creep strain for the same time and temperature. Figure 3.4 shows their composite plot for aluminum-based alloys and stainless steels, including several superalloys.

To obtain this correlation, tests in which 0.1% creep occurred in less than 100 h were rejected to prevent intolerable data scatter. Despite this, the "universal" curves of Fig. 3.4 can be associated with fairly wide data scatter, particularly at low creep strains. Some deviations from the correlations were related to microstructural instabilities, which produce differing proportions of primary, secondary, and tertiary creep between alloys and for varying test conditions.

Despite occasional anomalies, the Gill-Goldhoff correlation meets some preliminary design needs, particularly if the technique is tailored to grades of alloys similar to those of immediate concern. In principle, this technique can also be used to predict rupture properties from early creep measurements from tests that are not continued to rupture. This use is limited by the short rupture times that are derived from tests terminated at creep strains of 1% or less. If these tests were continued to higher levels of creep, improved predictions of rupture could be obtained by determining the secondary creep rate and then applying the Monkman-Grant relationship.

Treatment of Isolated Test Points

Particularly at the start of a testing program, the need may arise to extract information from a single available test. The form of most parameters limits their use to situations in which multiple test results are available. The Larson-Miller parameter is an exception if the generalized constant $C = 20$ is used.

For the stress of the test, longer rupture times within a factor of ten from the test duration frequently can be estimated satisfactorily for temperatures below that of the test. If a master curve or graph of isothermals is available for another lot of like or similar alloy, a parallel curve passed through the coordinates of the test point for the new lot can serve as an approximate representation of expected behavior for limited ranges of variables from the test conditions.

Table 3.4 Typical material constant (K) values

Material and test temperature	K
Copper at 250 °C (482 °F)	0.3
A-286 alloy at 649 °C (1200 °F), solution treated at 1204 °C (2200 °F)	0.47
A-286 alloy at 649 °C (1200 °F), solution treated at 982 °C (1800 °F)	0.43
Inconel X-550 at 732 °C (1350 °F)	0.625

EVALUATING CREEP DAMAGE, REMAINING SERVICE LIFE

Specimens from service occasionally are tested to compare residual creep and rupture properties with the same or similar material that has not been used in service. Diverse evaluations may result, depending on the conditions selected for the tests and on the criteria used to define damage. As shown below, a false prediction of drastic drop in rupture strength can result if tests are of rather short duration and if a simple direct ratio is taken of rupture life after service versus before service for similar test conditions.

Discrete Changes in Test Temperature or Load

When a creep-rupture test is interrupted by cooling and reheating at a moderate rate at constant load, and if the time under changing temperature is brief compared to the original test duration, the effect of the interruption on either the creep curve or rupture life usually cannot be readily detected, unless thermal gradients cause gross plastic deformation or spalling of surface layers.

Similar results can be expected when temperature and stress rise and fall in unison, as during start-up and shutdown of a steam boiler. For the alternate situation in which unloading occurs while the creep temperature is maintained, significant recovery of primary creep can ensue; reapplying the load results in a period of primary creep.

Under step-wise alteration of load, temperature, or both during a test or service, performance frequently follows the "life-fraction rule" or "linear cumulative damage rule", in which the percentage of total life consumed for each period of fixed temperature and stress is represented as:

$$\% \text{ total life} = \frac{\text{actual time at the given conditions}}{\text{rupture life at those conditions without alteration}} \times 100$$

Accuracy of this rule ranges from excellent to rather poor, with best results for multiple small excursions. Although solid-state reactions, which can reverse at different exposure temperatures, may introduce complications under some conditions, investigators have found the life-fraction rule more appropriate for steels under temperature changes than under stress changes.

Life-fraction summations at failure as low as 0.36 and as high as 2.43 have been reported. The spread was 0.75 to 1.50 for the same tests using damage fractions defined by $K(t/t_r) + (1 - K)(\epsilon/\epsilon_r)$, where t and ϵ are the time and strain under a period of fixed conditions, for which the rupture time and fracture strain are t_r and ϵ_r. K is a material constant ranging from zero to 1; the zero limit applies to materials that develop cracks early in life, and K approaches 1 for materials that exhibit no cracking until rupture is imminent. Typical values of K are shown in Table 3.4.

When data are insufficient for determination of K, acceptable results frequently can be obtained with an empirical rule, by which the life-fractions added are defined by the square root of $(t/t_r)(\epsilon/\epsilon_r)$. With any of these cumulative damage rules, comparison usually is against rupture life from constant-load tests – i.e., with actual stress rising as creep reduces the cross section. When the same specimen undergoes load changes for different periods, respective stress levels have been based on the initial cross section. This corresponds to using the same original load if an interrupted test must be restarted. The actual stress at the time of test restart is $(\sigma_n)(A_o/A_c)$, where σ_n is the present nominal stress, A_o is the initial specimen cross section, and A_c is the specimen cross section after creep deformation up to the time of the test interruption.

When the specimen for a test in a later portion of creep has been machined from a part that has already undergone considerable reduction in cross section by prior creep, a corresponding area-modified stress must be employed for consistent interpretation of the results. The load applied to produce the desired nominal stress σ_n related to the virgin material is calculated to make the actual stress a value of $\sigma_a = \sigma_n(A_o/A_c)$, where the latter term relates cross-sectional areas of the original part before and after creep. Use of such an area-modified stress has been reported to improve prediction of remaining life from post-service rupture tests.

An approximation of remaining rupture life for components that have undergone prolonged service can be calculated by introducing best estimates for operating temperatures and stresses into the above damage rules. More exact evaluation, however, can be obtained by testing representative samples removed from service.

Measurement of Rupture Properties After Service

Direct measurement of remaining life of a part at conditions near the service stress and temperature generally requires impractically lengthy testing times. Studies on material after creep service have involved increased temperature, stress, or both, with subsequent extrapolation of results to nominal service conditions.

Assessments made in this manner generally conclude that carbon, carbon-molybdenum, and chromium-molybdenum steels operated at or below allowable stresses (recommended by the *ASME Boiler and Pressure Vessel Code*) experienced negligible creep damage from service exposures up to 200,000 h. Significantly different, and presumably erroneous, conclusions can result from cursory studies or from poor selection of post-service tests.

Operating conditions for 1Cr-0.5Mo steel tubing removed from a superheater after 33,000 h of service were sufficiently well known to permit true residual life to be determined by continuing samples to rupture under the same stress and temperature (66.26 MPa/557 °C, or 9610 psi/1035 °F). These tests indicted service to have accounted for 78% of the total life of the tubes.

In that study a series of accelerated tests at 66.26 MPa (9610 psi) stress, but with test temperatures ranging from 561 to 610 °C (1042 to 1130 °F), yielded service lives ranging from 6535 to 432 h. A plot of these results on coordinates of test temperature versus log test life was linear and extrapolated rather well to the total life established by the samples continued to rupture at service stress and temperature. The simple life-fraction rule yielded values ranging from 0.93 to 1.05 for the sum of test and service fractions when area-modified stresses were used and 0.96 to 1.11 without the area correction.

Acceleration of tests by using stress levels ranging from 77.15 to 141.75 MPa (11.2 to 20.6 ksi) resulted in test durations between 2057 and 61 h at temperatures of 557 °C (1035 °F). Departure from linearity of the plot of test stress versus log test life as the service stress was approached made extrapolation to expected total life at service conditions difficult. Deviation trends between mean ISO data and properties of this tubing before service may bias the life-fraction comparisons for increased temperature versus increased stress tests after service. However, the sum of test and service fractions was found to be only 0.83 for the highest test stress and was found to increase only marginally to 0.89 for the lowest accelerated stress (77.85 MPa, or 11.3 ksi).

Life assessment based on extrapolation of temperature at the service stress is generally preferred to life assessment based on extrapolation of stress at the service temperature. However, many apparently successful evaluations of materials after creep service have used isothermal tests similar to those used to establish the original allowable stresses of the *ASME Boiler and Pressure Vessel Code* for creep temperatures.

A common finding for steels used to construct boilers and heated vessels has been that materials taken from service show lower rupture strengths than the same original material or other typical new material made to the same specifications. However, the slope of the log stress versus log rupture time curve is flatter than for virgin samples, and each isotherm frequently approximates a single straight line. Extrapolation of that line characteristically predicts negligible change from original long-time strength values for specimens whose actual service stress was low enough to preclude extensive creep damage after prolonged operation.

Subjecting such essentially undamaged service specimens to a heat treatment similar to that originally given the component restores short-time rupture curves to approximately the level and form reported for the steel before service. Note that sigmoidal isotherms have been reported for these steels, with a final lower slope linear portion after time-dependent structural alterations produced a slope increase encompassing tests with intermediate durations to rupture.

Dependable application of these techniques to estimate remaining life requires that the loading direction for the final test corresponds to the largest principal stress of service and that the specimen is representative of surface deterioration or other damage present in the part. Possible temperature and loading gradients in service must be kept in mind when selecting a sampling site and when applying test results to predictions of further serviceability. Despite these possible additional variables, published assessments of post-service rupture properties require only about the same number and duration of tests as conventional evaluation of any new lot of familiar material.

Creep-Rate Measurements After Service

An alternative testing approach has been to perform a creep test on the specimen from service and to compare the measured rate against creep for the same or like material without service. An ideal test would be conducted at the operating temperature and stress, but more extreme test conditions are common. Although simple in principle, this test can entail several difficulties.

If the specimen of previously crept material is heated to the test temperature before stressing, recovery effects are likely to produce a higher creep rate at the start of the test than the rate that existed when service was interrupted. Even if the test is brought to temperature with the load applied, an imprecise match of test and service stresses commonly leads to early nonsteady creep rates, even when a steady condition was previously established in service.

Reference creep rates, whether from tests on identical new material or not, usually are limited to reported extrapolated stresses for 0.0001%/h or 0.00001%/h secondary creep rate at several temperatures. To fix the creep rate to be expected for virgin material at the service stress and temperature requires nonlinear interpolation of a few tabulated data points.

The considerable scatter typical in creep measurement poses a particular problem, because under the specified stresses of most elevated-temperature designs, creep deformation obtained in a reasonable time can approach the sensitivity limit of most extensometer systems. A large increase in that sensitivity limit is impractical, because movements caused by thermal expansion due to unavoidable temperature fluctuations necessitate averaging of the resulting unsteady strain indications.

Even determination of test duration can be an important test decision. If the component in service was well within an extended period of secondary creep, a test conducted for about a thousand hours at near-service conditions should display a creep rate that is close to the latest period of service during the entire test. However, if secondary creep is not distinct and prolonged for the alloy under study, or if service exposure has reached a stage significantly earlier or later than the minimum rate, extended testing deviates increasingly from the last rate of service. Greatest concern arises when post-service test conditions are made more severe than those of service, either to shorten the study or to yield rates that are more convenient to measure.

Procedure modifications minimize these difficulties, but their use has not been documented. The "ideal" test does not require a virgin specimen or knowledge of its initial properties; it can be used with a single representative specimen and permits evaluation in no longer than a few months at conditions reasonably near those of service.

Assuming that prior operation typifies specified design, so that rupture life at service temperature T_s is 100,000 h or more, a suitable test procedure may be:

1. Load a cold post-service specimen to the estimated service stress (or, if this is unknown, to the applicable specified allowable stress).

2. Increase the specimen temperature to T_s + 50 °C (T_s + 90 °F) within 3 h. Immediately on reaching that temperature, gradually reduce furnace power to obtain a uniform specimen temperature of T_s + 20 °C (T_s + 36 °F) at the end of one additional hour.

3. Take creep measurements until a rate is established with a precision of two significant digits.

4. Increase the specimen temperature to T_s + 50 °C (T_s + 90 °F) (with the load maintained) and hold until about 0.5% additional creep strain occurs at that temperature.

5. Reduce the temperature to T_s + 20 °C (T_s + 36 °F) as in Step 2 and obtain a second creep rate as in Step 3.

This procedure minimizes anomalous recovery effects at the start of each creep-rate determination, permitting valid direct comparison of rates for the material as removed from service and after an additional exposure, which should account for another significant fraction of total rupture life.

A measured creep rate from Step 5 that is lower than the rate determined in Step 3 under identical conditions indicates that service has been so mild that steady secondary creep has not yet become fully established. The more common situation, in which the two rates are approximately the same or the second is faster than the first, would indicate that the component is within or past the secondary creep stage.

Creep and rupture tests appear to offer confident determination of remaining life of elevated-temperature materials; however, more rapid and less expensive methods of assessing creep damage continue to be sought.

Microstructural Evaluation of Creep Damage

Numerous studies beyond the realm of creep-rupture testing have sought immediate warning of creep damage, while monitoring its development, without the need for destructive sampling. Pending improvement and acceptance of such methods, it has been reported that remaining lifetime can be estimated from quantitative measurements of the microstructure. The procedure involves comparisons between a specimen taken from service and other specimens from accelerated tests at the service stress, but at increased temperatures.

4

Iron-Base Superalloys

16-25-6: Solution Treated

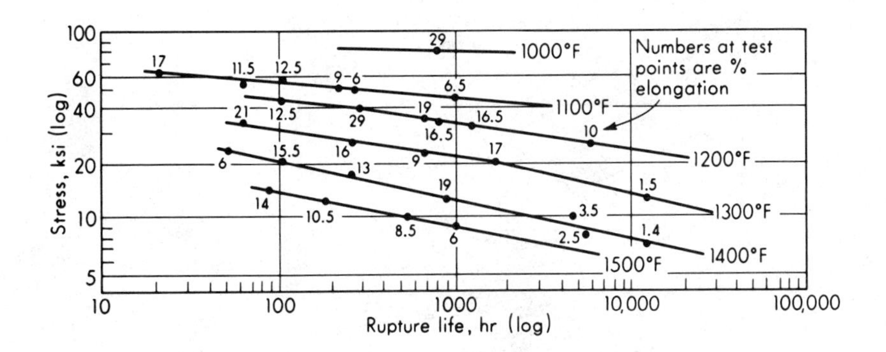

Stress vs rupture life for solution treated 16-25-6 tested at various temperatures.

Source: R.C. Juvinall, Stress, Strain, and Strength, McGraw-Hill, New York, 1967, p 416

16-25-6: Total Deformation-Time Curves

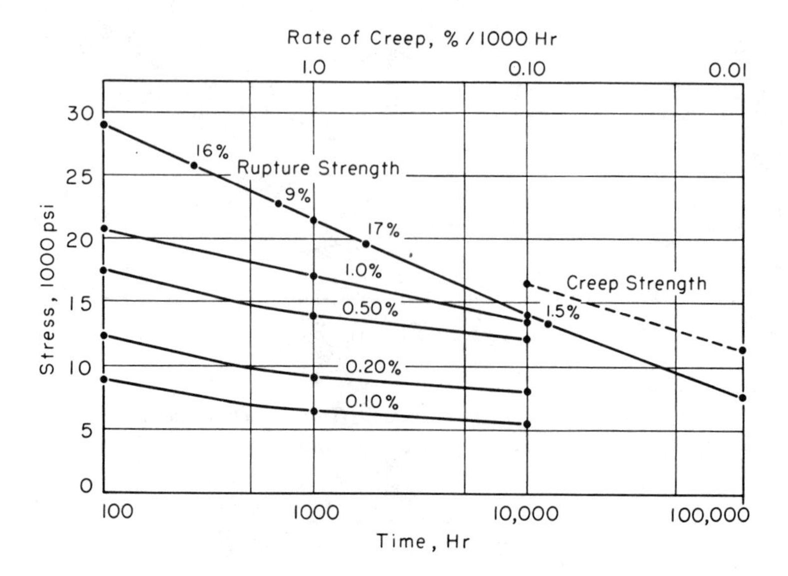

Total deformation-time curves at 1300 °F for 16-25-6 water quenched at 2150 °F.

At times, for shorter time applications, even though the design is based on a rather short-time rupture-strength value, it is necessary to know and to restrict the total amount of deformation during the expected service life. In these cases, a relationship can be developed, as shown above. Stresses for total plastic deformations of 0.10 to 1.0%, in periods extending from 100 to 10,000 h, are given, as well as the usual rupture and creep values. Thus, the designer can select the stress that will produce the limiting total deformation in the desired period.

Source: C.L. Clark, "Factors Involved in Using High Temperature Test Data for Selecting Materials and Proportioning Parts," in Utilization of Heat Resistant Alloys, American Society for Metals, Metals Park, OH, 1954, p 39-40

19-9DL: Forged and Aged

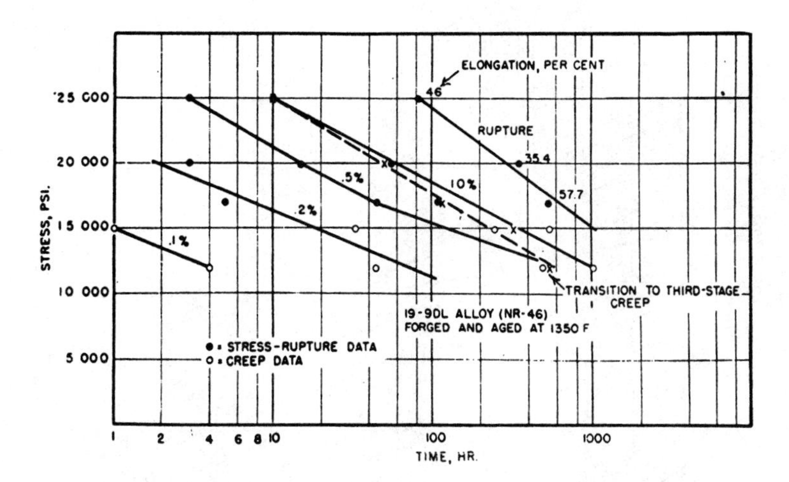

Design curves for 19-9DL at 732 °C (1350 °F). Forged and aged at 732 °C (1350 °F).

Source: F.H. Clark, Metals at High Temperatures, Reinhold, New York, 1950, p 161

19-9DL: Heat Treated and Aged

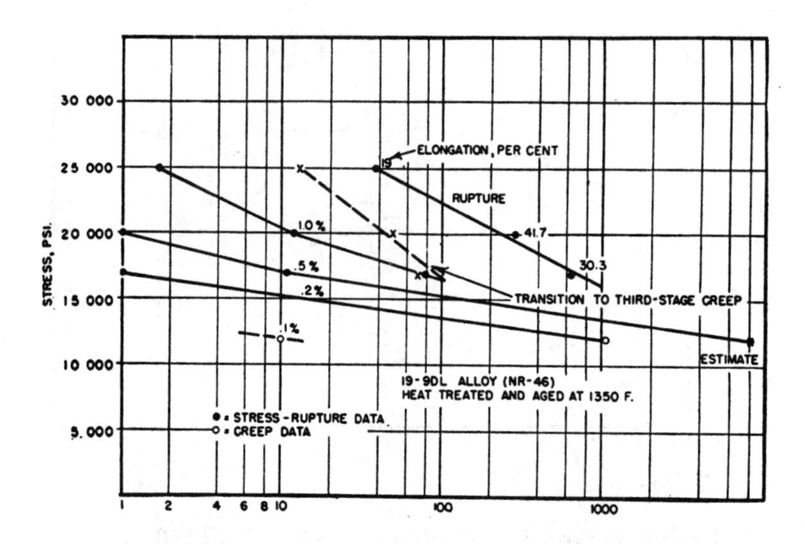

Design curves for 19-9DL at 732 °C (1350 °F). Heat treated and aged at 732 °C (1350 °F).

Source: F.H. Clark, Metals at High Temperatures, Reinhold, New York, 1950, p 162

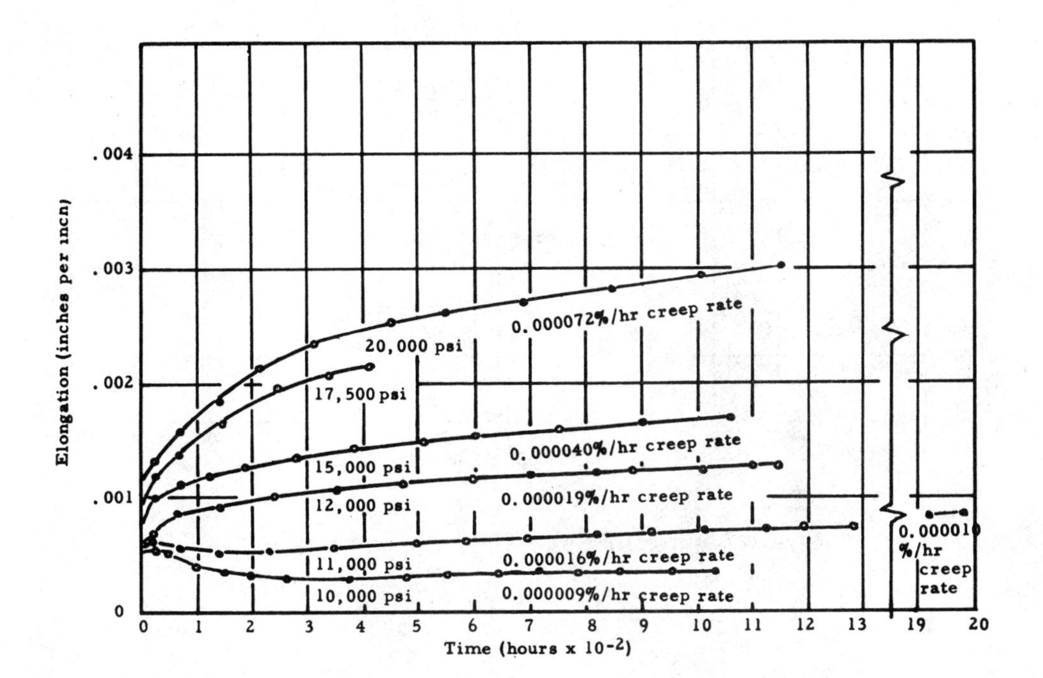

Time-elongation curves showing negative creep at 649 °C (1200 °F) of hot rolled 19-9DL bar stock.

Source: J.W. Freeman and C.L. Corey, "Microstructure and Creep," in Creep and Fracture of Metals at High Temperature, Philosophical Library, New York, 1957, p 169

A286: Bar and Forgings

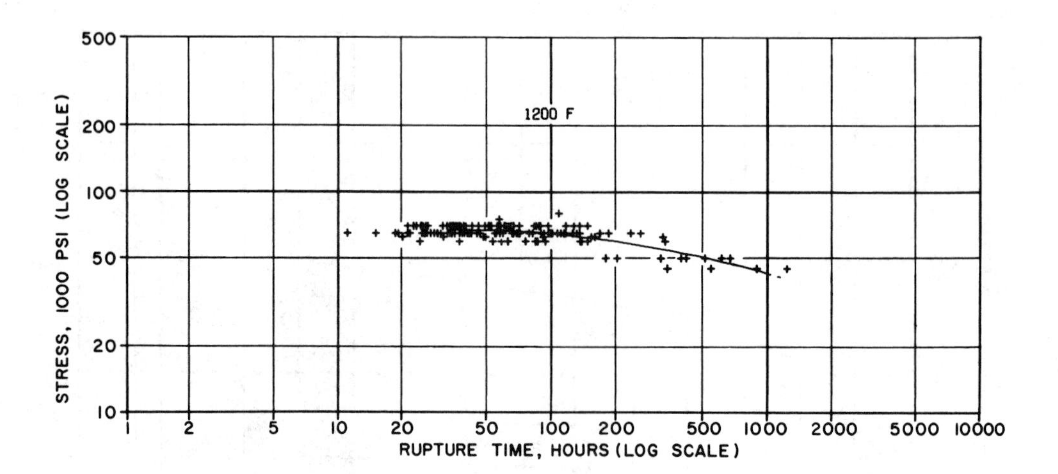

Stress-rupture time for A286 bar and forgings.

The Elevated-Temperature Properties of Selected Superalloys, ASTM Data Series DS 7-S1, American Society for Testing and Materials, Philadelphia, 1968, p 117

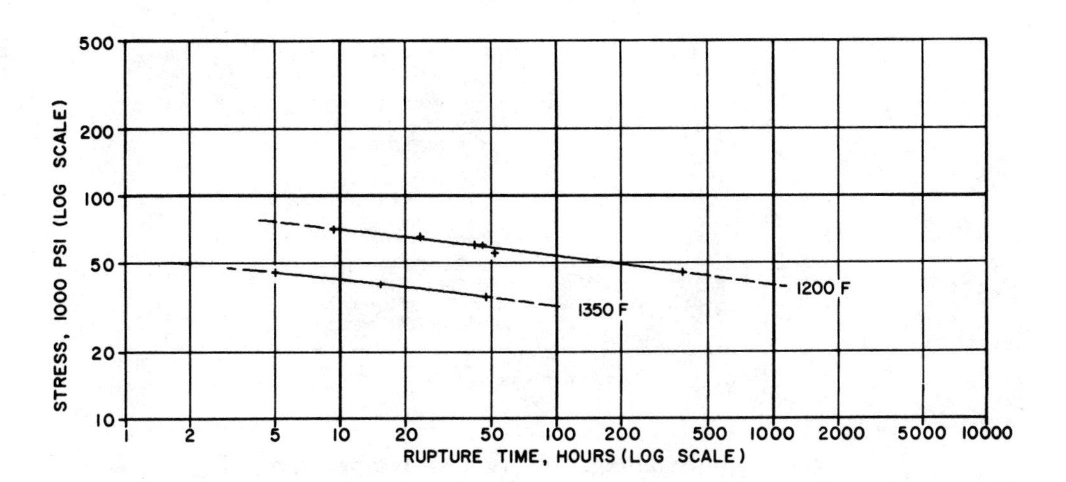

Stress-rupture time for A286 sheet.

Source: The Elevated-Temperature Properties of Selected Superalloys, ASTM Data Series DS 7-S1, American Society for Testing and Materials, Philadelphia, 1968, p 112

A286: Stress-Rupture Elongation vs Test Temperature

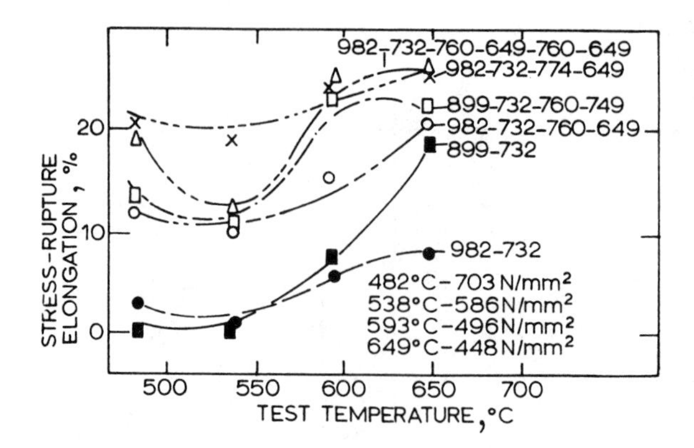

Stress-rupture elongation vs test temperature of A286. Numbers on curves refer to solution and subsequent aging temperatures in °C.

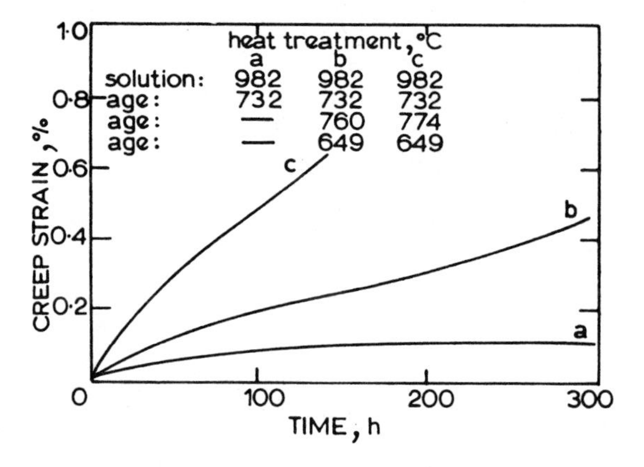

Creep curves for A286 at 482 °C and 703 N/mm².

The relationship between stress-rupture elongation and test temperature for A286 is shown in the top figure, where it can be seen that there is minimum ductility at about 538 °C. The figure shows the influence of a variety of heat treatments, solution treating at 899 and 982 °C, and single and multiple aging up to 774 °C. Material heat treated for 1 h at 899 °C and then for 16 h at 732 °C exhibited notch brittleness both at 482 and 538 °C. After re-aging at 760 and 649 °C, notch brittleness was eliminated and rupture ductility improved considerably, but with some sacrifice in rupture life and creep strength. Some typical effects of this treatment are shown in the bottom figure.

Source: N.A. Wilkinson, "Technological Considerations in the Forging of Superalloy Rotor Parts," in Superalloys: Source Book, M.J. Donachie, Jr., Ed., American Society for Metals, Metals Park, OH, 1984, p 239-240

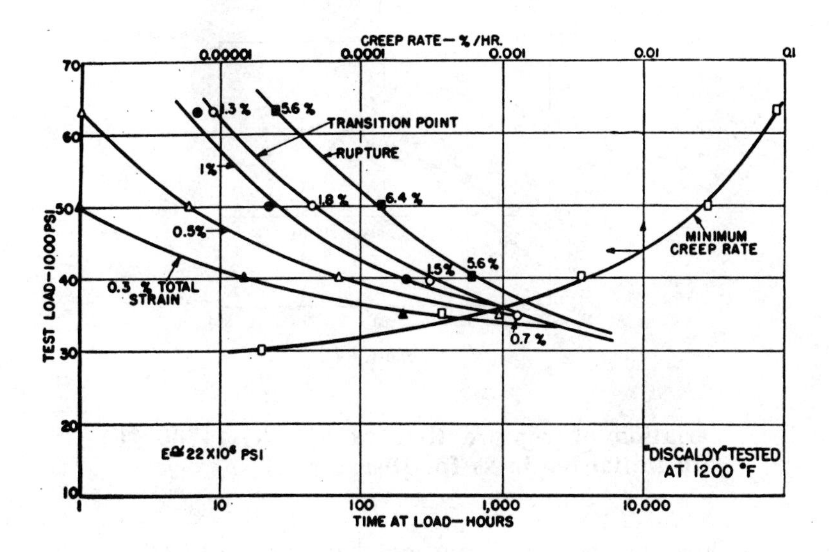

Design curves for Discaloy at 650 °C (1200 °F). Specimens cut from a forged rotor. Heat treatment: solution treated at 1065 °C (1950 °F) 1 h, furnace cooled, aged at 732 °C (1350 °F) 20 h.

Source: F.H. Clark, Metals at High Temperatures, Reinhold, New York, 1950, p 188

Discaloy: Effect of Initial Hardness

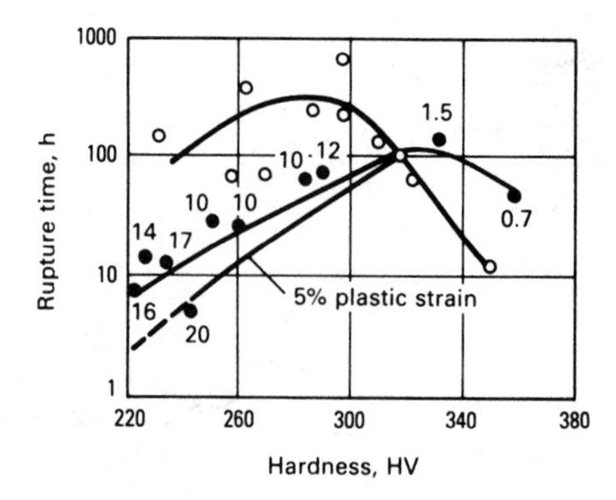

Variation of rupture time at 650 °C (1200 °F) with initial hardness for Discaloy.

Open symbols indicate notched bar tests (K_t = 3.9); solid symbols indicate smooth bar tests. Numbers adjacent to points are total elongations for those tests.

Source: Metals Handbook, Ninth Edition, Volume 3, Properties and Selection: Stainless Steels, Tool Materials and Special-Purpose Metals, American Society for Metals, Metals Park, OH, 1980, p 233

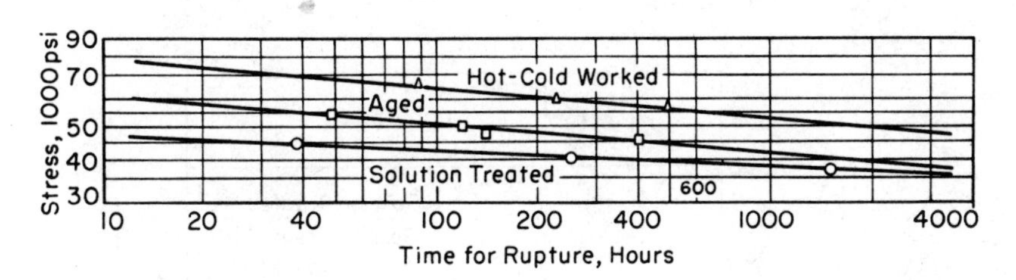

Effect of aging and hot/cold working on the stress-rupture strength of N-155 at 1200 °F.

Source: J.W. Freeman, C.L. Corey, and A.I. Rush, "Metallurgical Variables Influencing Properties of Heat Resistant Alloys," in Utilization of Heat Resistant Alloys, American Society for Metals, Metals Park, OH, 1954, p 251

N-155: Effect of Solution Temperature and Solution Treatment Prior to Hot/Cold Working

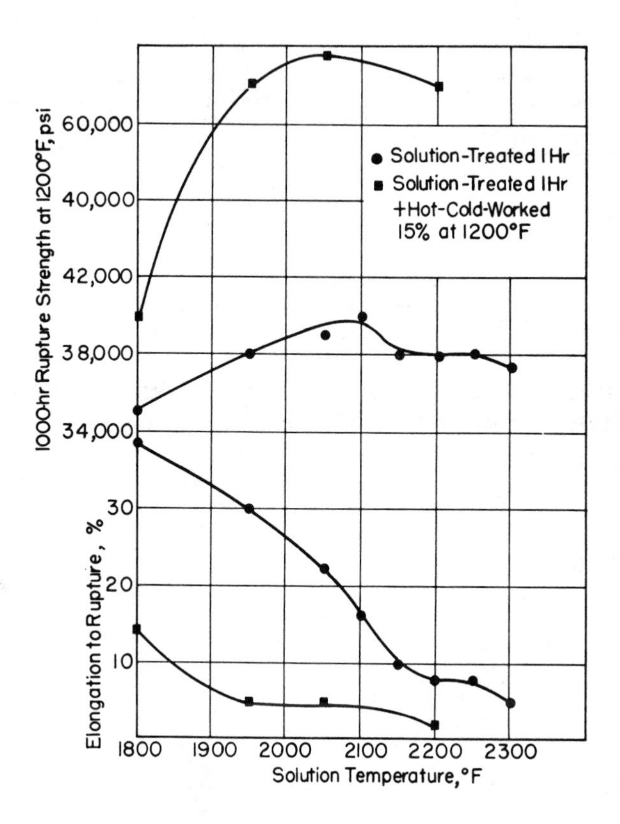

Effect of solution temperature and solution treatment prior to hot/cold working on the 1000-h rupture strength and rupture elongation of N-155.

Source: J.W. Freeman, C.L. Corey, and A.I. Rush, "Metallurgical Variables Influencing Properties of Heat Resistant Alloys," in Utilization of Heat Resistant Alloys, American Society for Metals, Metals Park, OH, 1954, p 249

5

Nickel-Base Superalloys

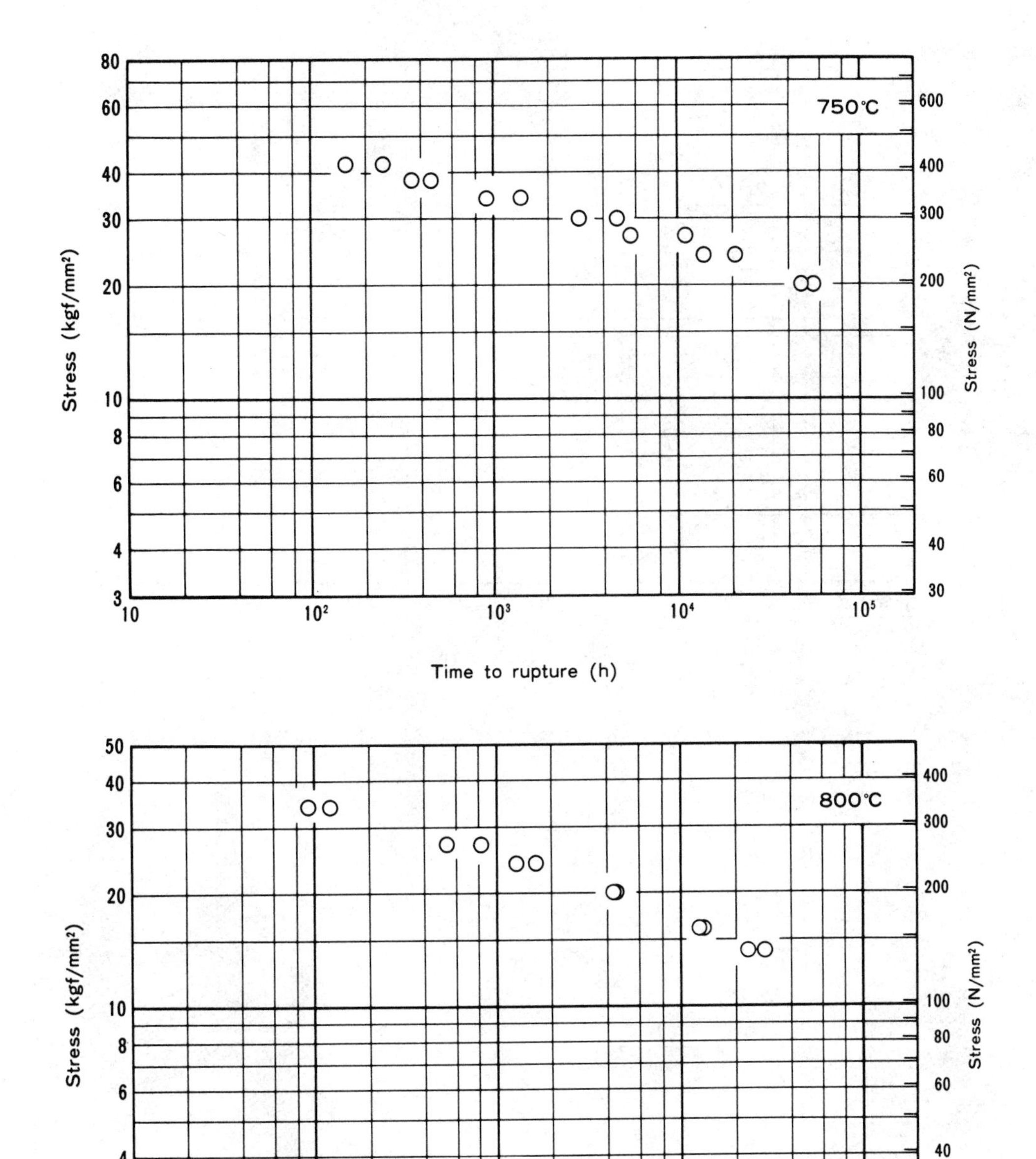

Stress vs time to rupture for 15Cr-28Co-4Mo-2.5Ti-3Al bar at various temperatures.

15Cr-28Co-4Mo-2.5Ti-3Al: Bar (Continued)

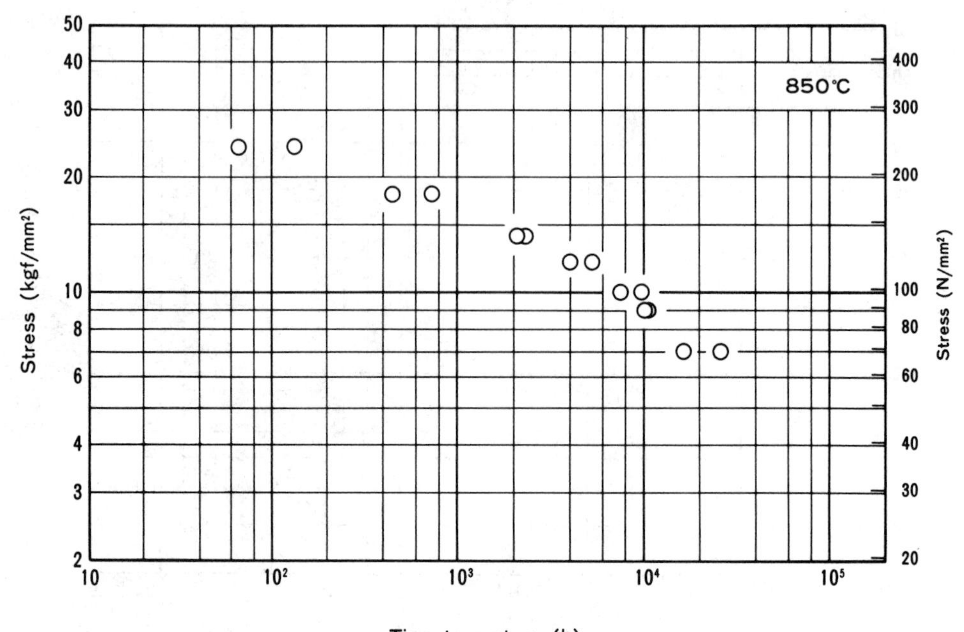

Stress vs time to rupture for 15Cr-28Co-4Mo-2.5Ti-3Al bar at various temperatures.

Source: Data Sheets on the Elevated-Temperature Properties of Nickel Base 15Cr-28Co-4Mo-2.5Ti-3Al Alloy Bars for Gas Turbine Blades, NRIM Creep Data Sheet No. 24A, 1982, p 6

15Cr-28Co-4Mo-2.5Ti-3Al: Temperature Dependence of 0.2% Proof Stress, Tensile Strength, and Rupture Strength

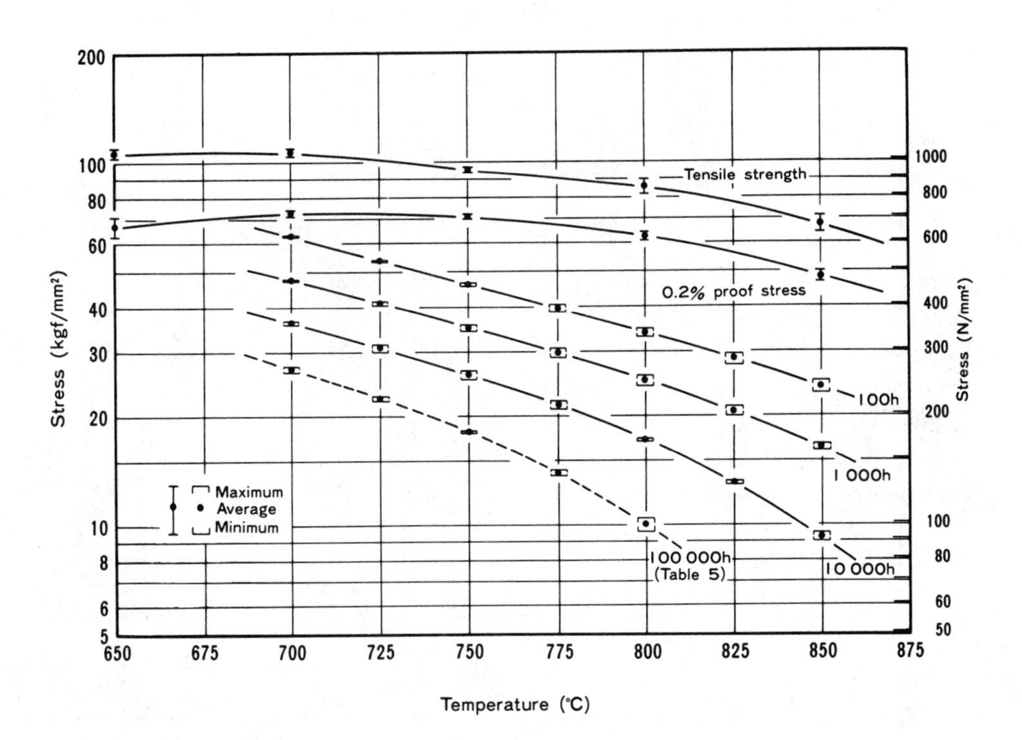

Temperature dependence of 0.2% proof stress, tensile strength, and rupture strength of 15Cr-28Co-4Mo-2.5Ti-3Al bar.

Rupture-strength curves have been drawn from the results of curvilinear regression using the Manson-Succop parameter method for individual heats.

Source: Data Sheets on the Elevated-Temperature Properties of Nickel Base 15Cr-28Co-4Mo-2.5Ti-3Al Alloy Bars for Gas Turbine Blades, NRIM Creep Data Sheet No. 24A, 1982, p 9

5.4 Alloy 800H: Tensile, Creep-Rupture, and Creep Strain-Time Data

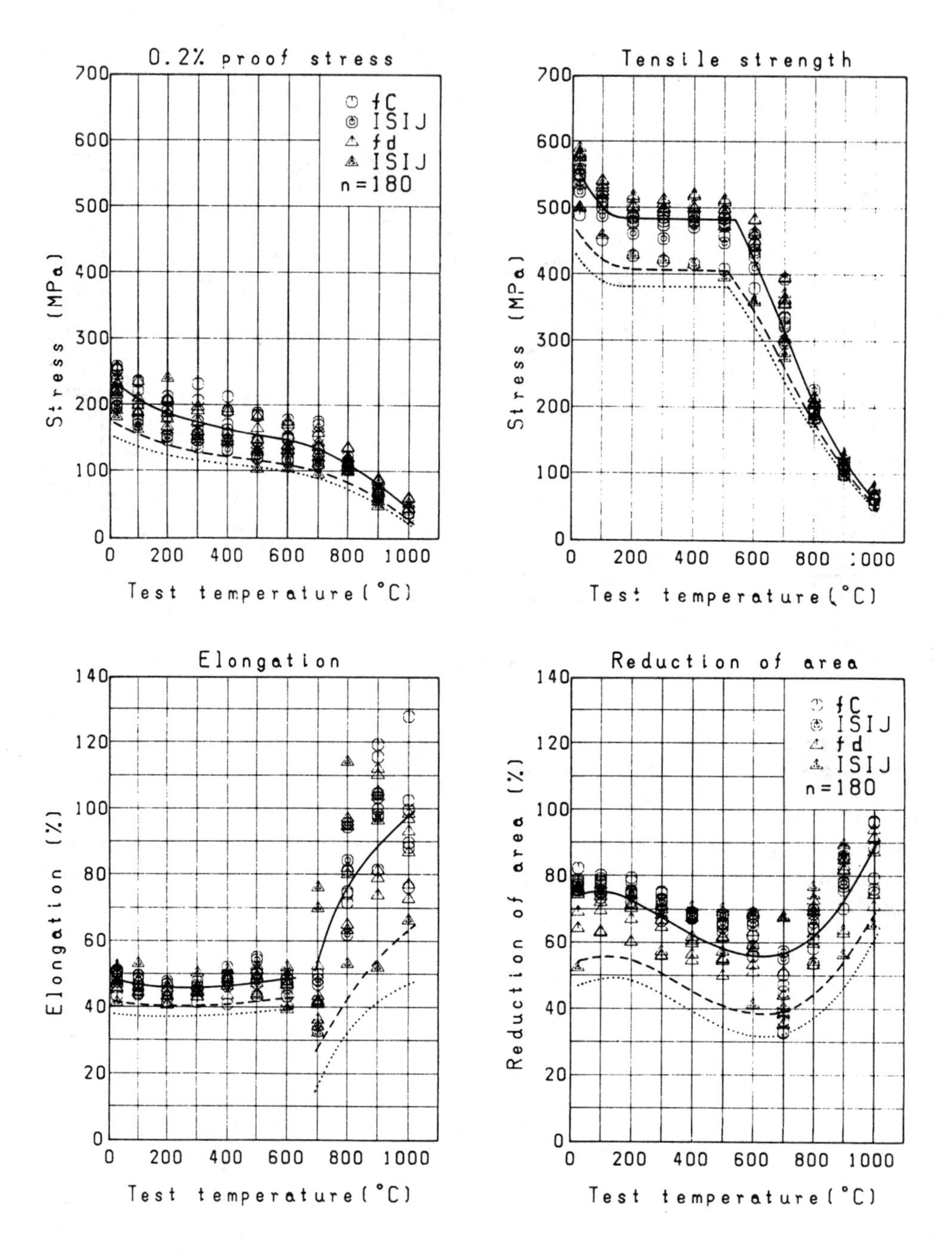

Comparison of experimental data and regression curves with lower 95% limits of the prediction intervals (PI) and the simultaneous tolerance intervals (STI) for tensile properties of alloy 800H. Circles and triangles indicate tubes and plates, respectively. Solid lines are the average; broken and dotted lines are lower limits of PI and STI, respectively.

Source: Y. Monma, M. Sakamoto, A. Miyazaki, H. Nagai, and S. Yokoi, "Assessment of Elevated-Temperature Property Data for Alloy 800H," Trans. Natl. Res. Inst. Met. (Jpn.), Vol. 26, Sept. 1984, p 35

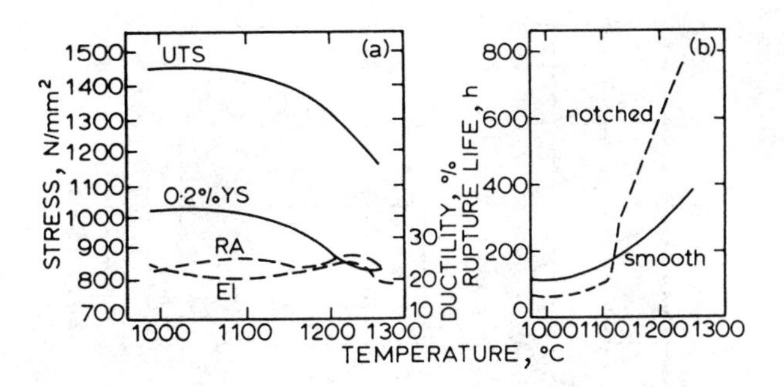

Variation of (a) 538 °C (1000 °F) tensile and (b) 732 °C (1350 °F), 560 N/mm² stress-rupture properties of low-carbon Astroloy with pressing temperature.

Phase I alloy screening: alloy compositions

Base alloy	Modifications
Astroloy	W, Ta, Hf, Nb, 0·03–0·07C
IN 100	No V, Nb, Hf, 0·02–0·07C
MAR–M–432	No W, high Al : Ti, Hf, 0·03C
PA101	Low W, Nb, 0·05C
AF2–1DA	Nb, no Ta, 0·10C
René 95	No modifications
AF115	No modifications

Variation in strength and ductility with HIP temperature, after a constant heat treatment, is shown above. Note that yield-strength level can be changed by 70-100 N/mm² with little change in tensile ductility (Fig. a). However, the stress-rupture characteristics change dramatically, as shown in Fig. b, a discontinuity occurring near the γ solvus temperature of ≈1120 °C (2192 °F) in this alloy.

Source: M.J. Blackburn and R.A. Sprague, "Production of Components by Hot Isostatic Pressing of Nickel-Base Superalloy Powders," in Superalloys: Source Book, M.J. Donachie, Jr., Ed., American Society for Metals, Metals Park, OH, 1984, p 273

B-1900: Effect of Gamma Prime Coarsening

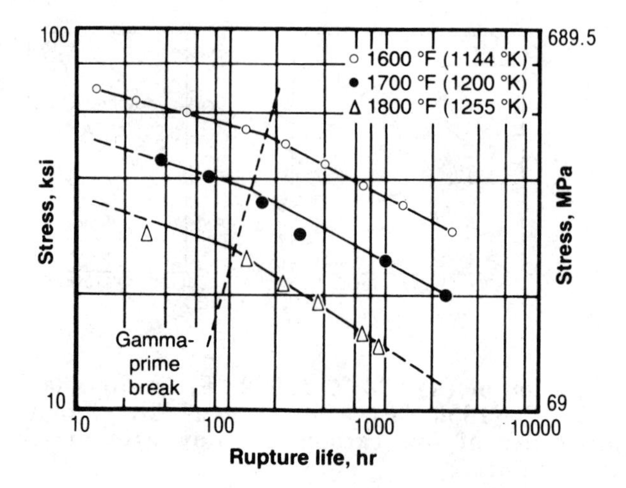

Rupture behavior of B-1900 showing break in slope believed to be caused by γ' coarsening.

The coarsening of γ' can cause reduced rupture life. Overheating can result in accelerated coarsening as well as solutioning of some fine γ'. Properties may be reduced, but reprecipitation of fine γ' occurs in the case of mild overheating, and property losses may be hard to detect.

Source: M.J. Donachie, Jr., "Relationship of Properties to Microstructure in Superalloys," in Superalloys: Source Book, M.J. Donachie, Jr., Ed., American Society for Metals, Metals Park, OH, 1984, p 110

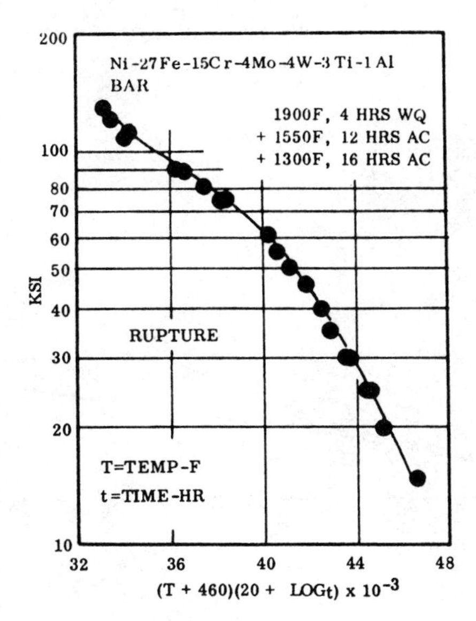

Stress-rupture parameter master curve for bar stock.

Source: Structural Alloys Handbook, Volume 4, Mechanical Properties Data Center, Battelle Columbus Laboratories, revised 1980, p 4109-5

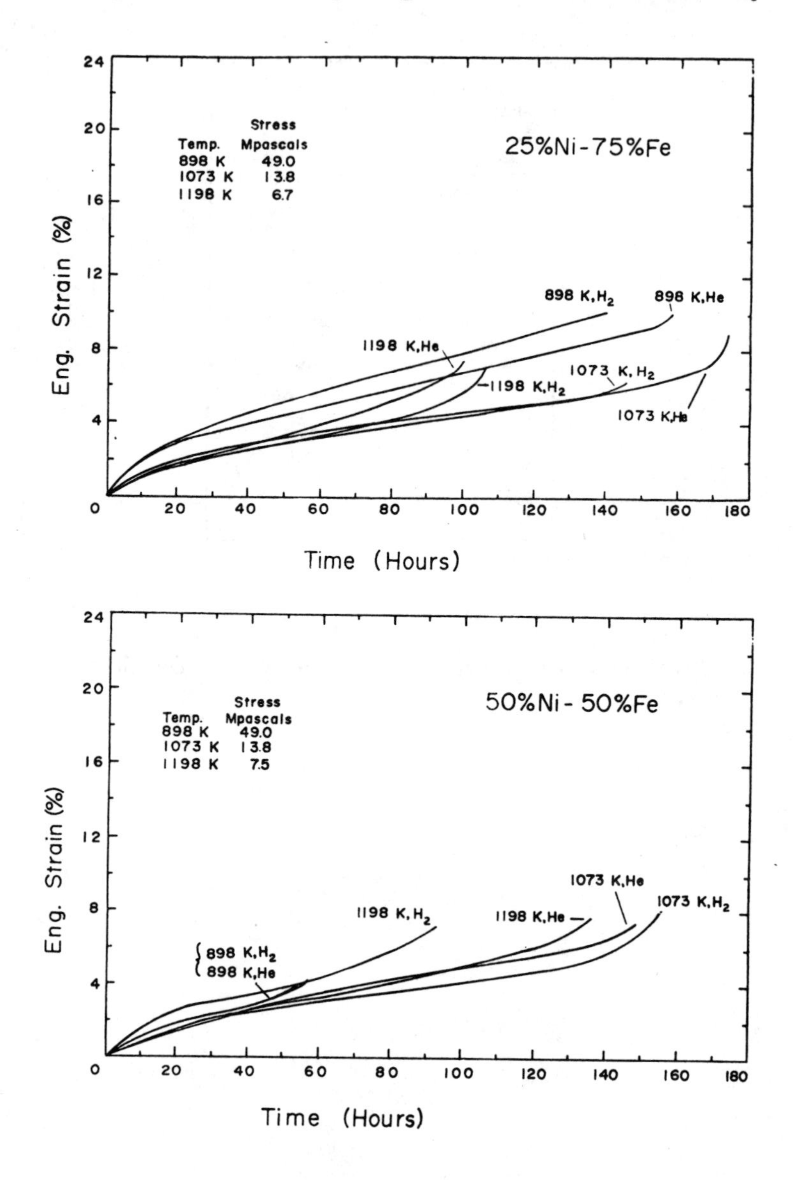

25Ni-75Fe (top) and 50Ni-50Fe (bottom) tested at several temperatures and stresses in hydrogen and in helium.

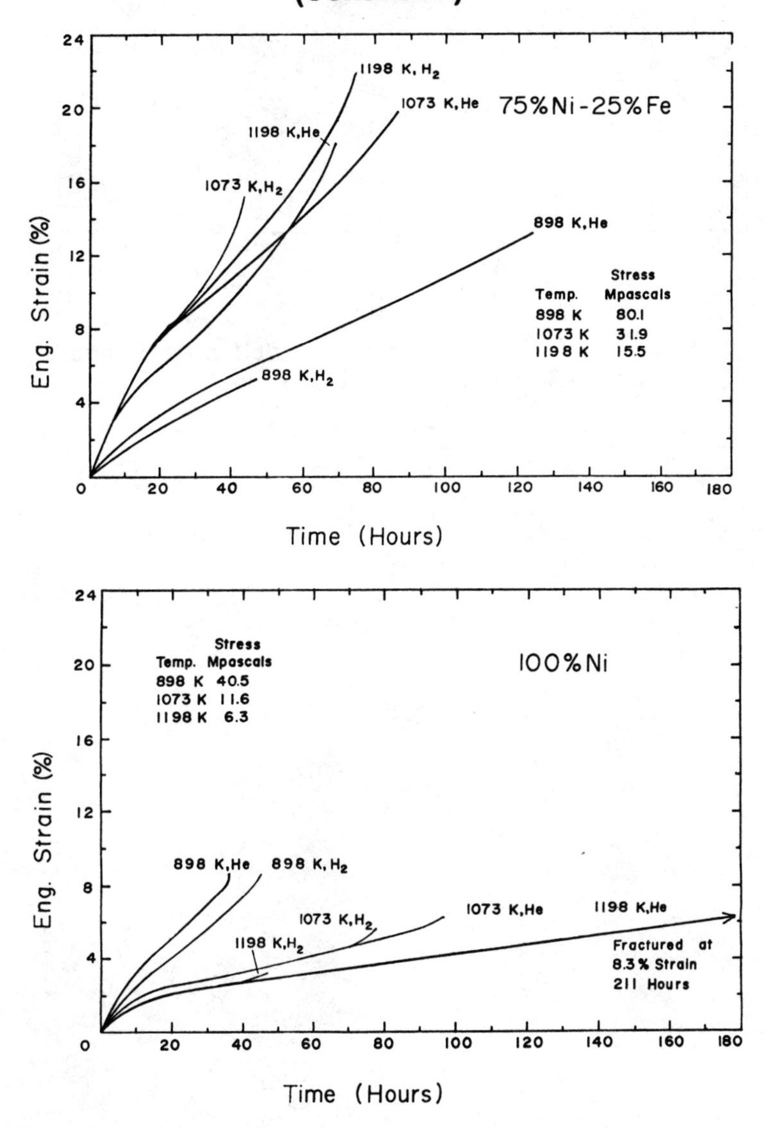

75Ni-25Fe (top) and 100Ni (bottom) tested at several temperatures and stresses in hydrogen and in helium.

Source: G.B.A. Schuster, R.A. Yeske, and C.J. Altstetter, "The Effects of Hydrogen on the Creep Rupture Properties of Fe-Ni Alloys," Met. Trans. A, Vol 11A, Oct. 1980, p 1659-1660

GMR-235: Air- and Vacuum-Melted

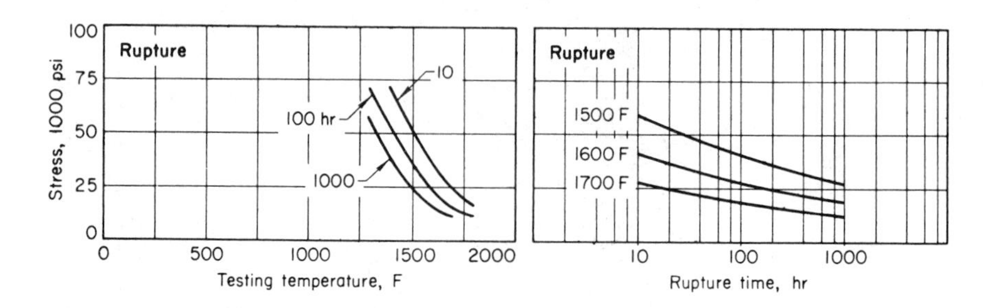

Stress-rupture properties of air-melted (left) and vacuum-melted (right) GMR-235.

Source: Source Book on Materials for Elevated-Temperature Applications, E.F. Bradley, Ed., American Society for Metals, Metals Park, OH, 1979, p 236

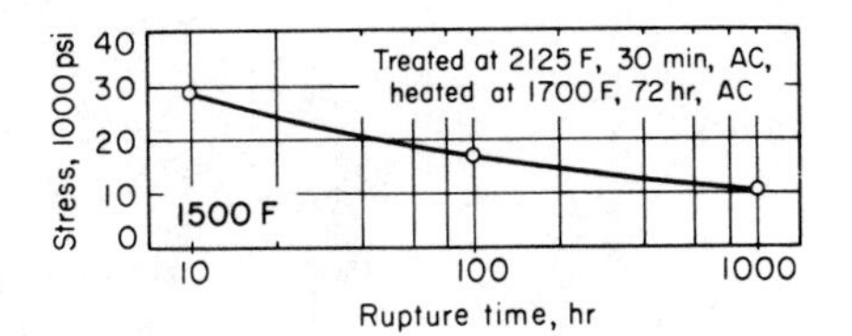

Stress-rupture strength of Hastelloy B investment-cast specimens.

Source: Source Book on Materials for Elevated-Temperature Applications, E.F. Bradley, Ed., American Society for Metals, Metals Park, OH, 1979, p 234

Hastelloy C: As-Investment-Cast

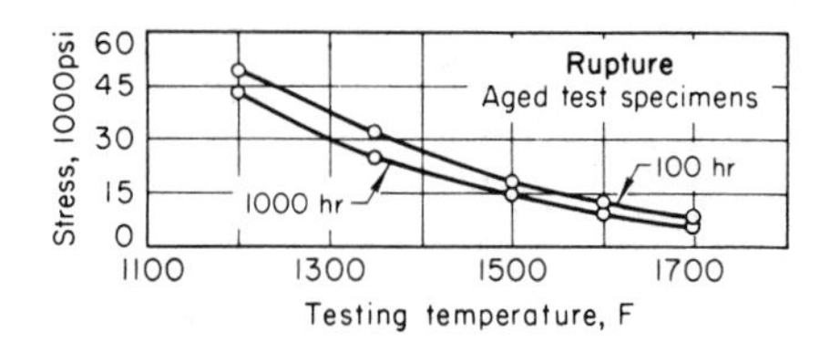

100- and 1000-h rupture strength of Hastelloy C as-investment-cast specimens.

Source: Source Book on Materials for Elevated-Temperature Applications, E.F. Bradley, Ed., American Society for Metals, Metals Park, OH, 1979, p 234

Hastelloy X: As-Investment-Cast

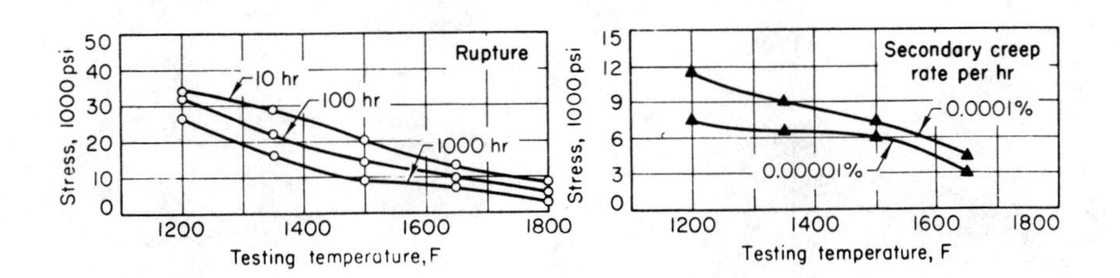

As-investment-cast.

Source: "Properties of Heat-Resistant Alloy Castings," in Source Book on Materials for Elevated-Temperature Applications, E.F. Bradley, Ed., American Society for Metals, Metals Park, OH, 1979, p 234

Hastelloy X: Effect of HTR Core Outlet Temperature

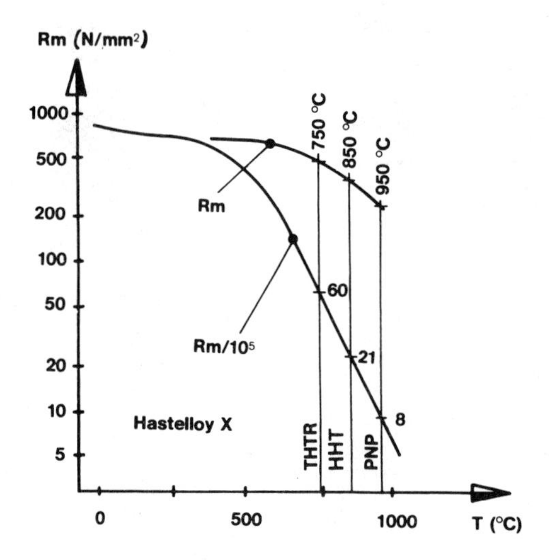

Rupture and ultimate tensile strengths of Hastelloy X at different HTR core outlet temperatures.

Source: H.W. Grünling, "High Temperature Alloys in Power Plant Systems: Current Status and Future Demands," In Behavior of High Temperature Alloys in Aggressive Environments (in Proceedings of the Petten International Conference, Petten, The Netherlands), I. Kirman, J.B. Marriott, M. Merz, P.R. Sahm, and D.P. Whittle, Eds., The Metals Society, London, 1980, p 164

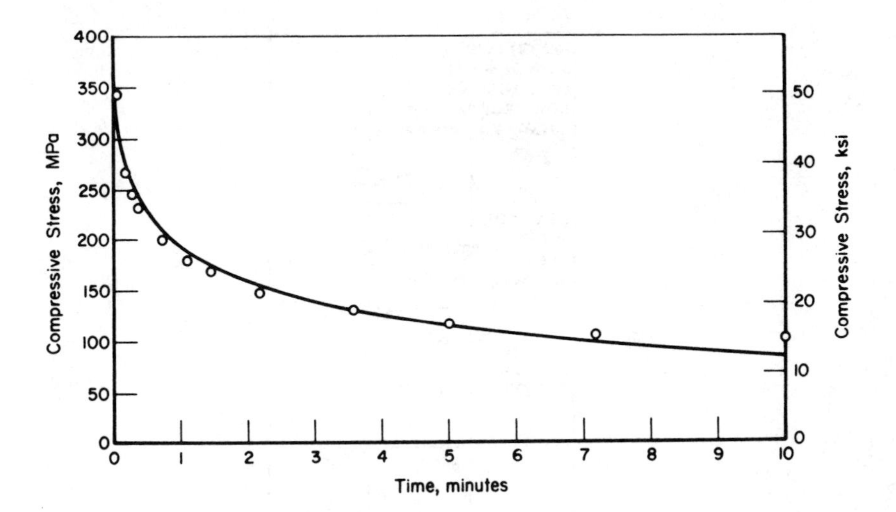

Relaxation-stress response of Hastelloy X tested in hydrogen gas at 760 °C (1400 °F) and at a total axial strain range of 3%.

Source: C.E. Jaske and R.C. Rice, "Low-Cycle Fatigue of Two Austenitic Alloys in Hydrogen Gas and Air at Elevated Temperatures," in 1976 ASME-MPC Symposium on Creep-Fatigue Interaction, American Society of Mechanical Engineers, New York, 1976, p 113

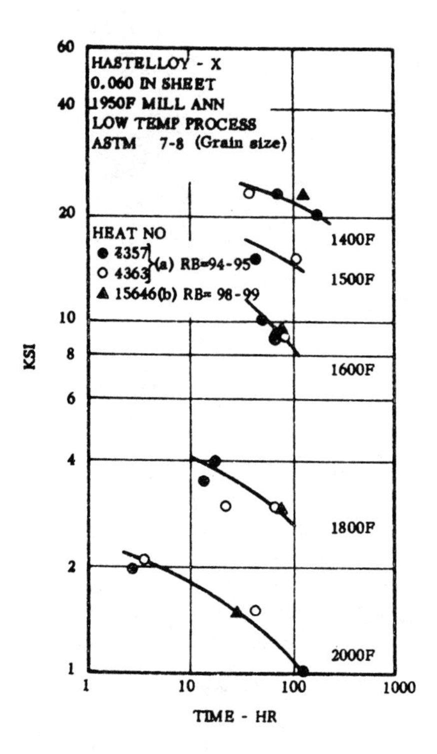

Creep-rupture curves at 1400 to 2000 °F for sheet. (a) Haynes Stellite. (b) Universal Cyclops.

Source: Structural Alloys Handbook, Volume 4, Mechanical Properties Data Center, Battelle Columbus Laboratories, revised 1980, p 4112-10

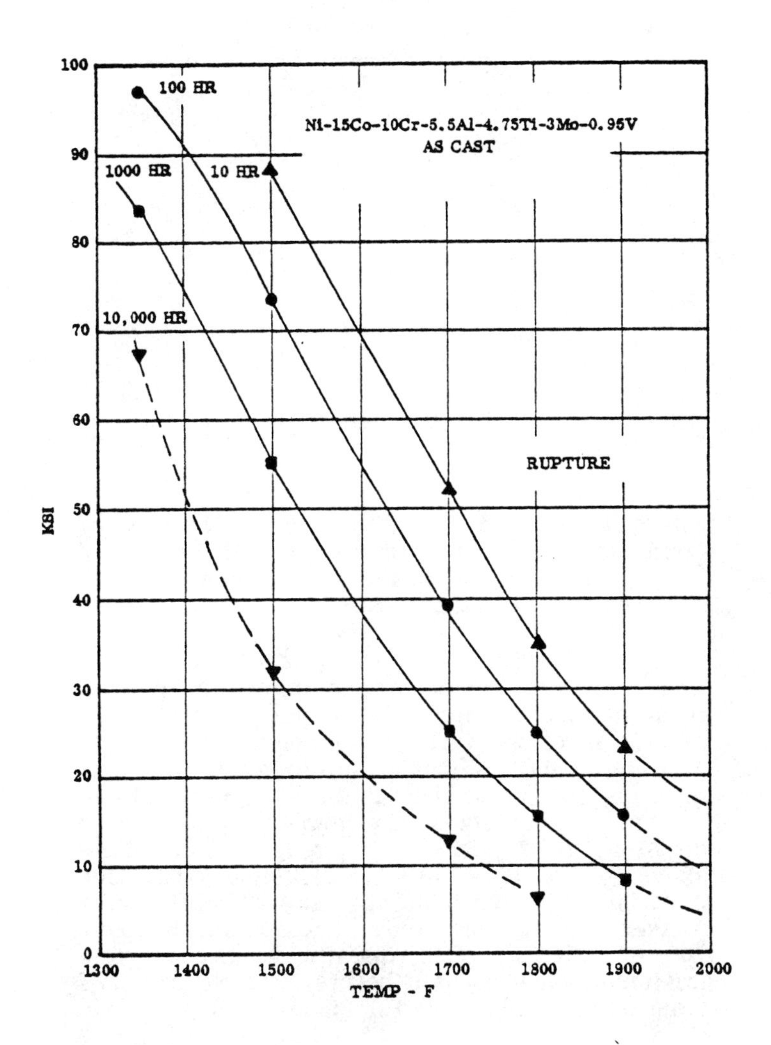

Creep-rupture properties for 10 to 10,000 h at 1300 to 2000 °F.

Source: Structural Alloys Handbook, Volume 5, Mechanical Properties Data Center, Battelle Columbus Laboratories, revised 1978, p 4212-37

IN 100: P/M vs Cast Material

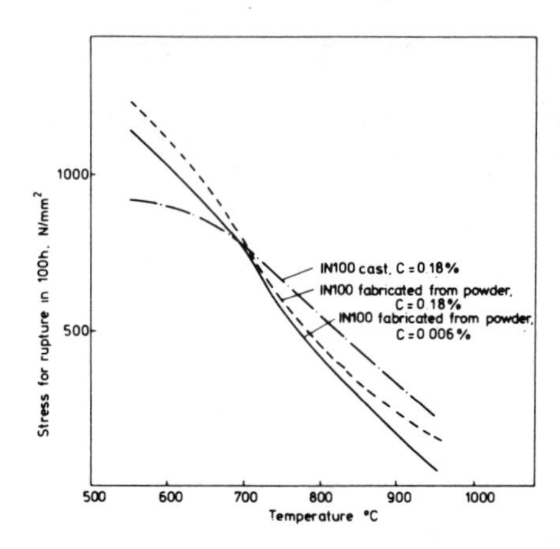

Stress to rupture in 100 h as a function of temperature for two forms of IN 100 fabricated from powder and differing in carbon content vs similar data for the cast alloy.

The improved purity of superalloys in powder form has led to significant developments in the processing of gas turbine components from alloy powders. One difficulty, however, is that the resulting fine grain size, an advantage in disks in which good resistance to high-cycle fatigue is essential, gives poor creep strength. The figure above illustrates this effect. Note that the material fabricated from powder was substantially weaker than the cast alloy at temperatures above 700 °C (1292 °F). Consequently, grain coarsening treatments must be used before material produced from powder is suitable for blade applications.

Source: V. Lupinc and T.B. Gibbons, "Factors Influencing the Creep Behavior of Ni-Cr-Base Alloys," in High Temperature Alloys for Gas Turbines, D. Coutsouradis, P. Felix, H. Fischmeister, L. Habraken, Y. Lindblom, and M.O. Speidel, Eds., Applied Science, London, 1978, p 339, 353

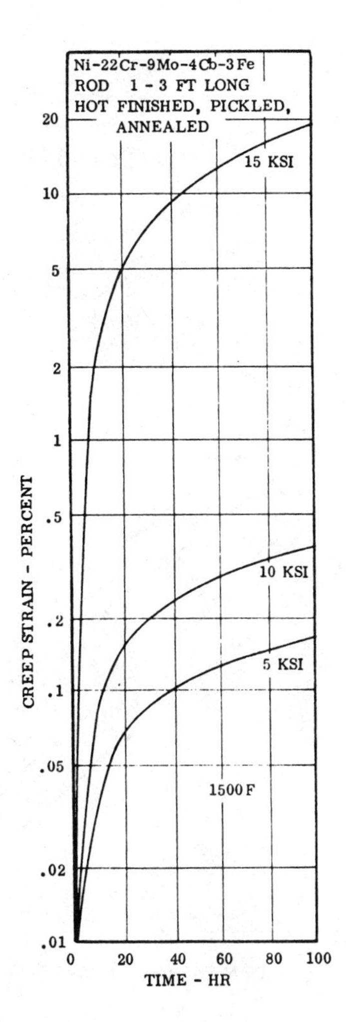

Creep curves at 1500 °F for stresses from 5 to 15 ksi for annealed alloy bar.

Source: Structural Alloys Handbook, Volume 4, Mechanical Properties Data Center, Battelle Columbus Laboratories, revised 1980, p 4117-17

IN 738: HIPed vs Cast-to-Size Bar

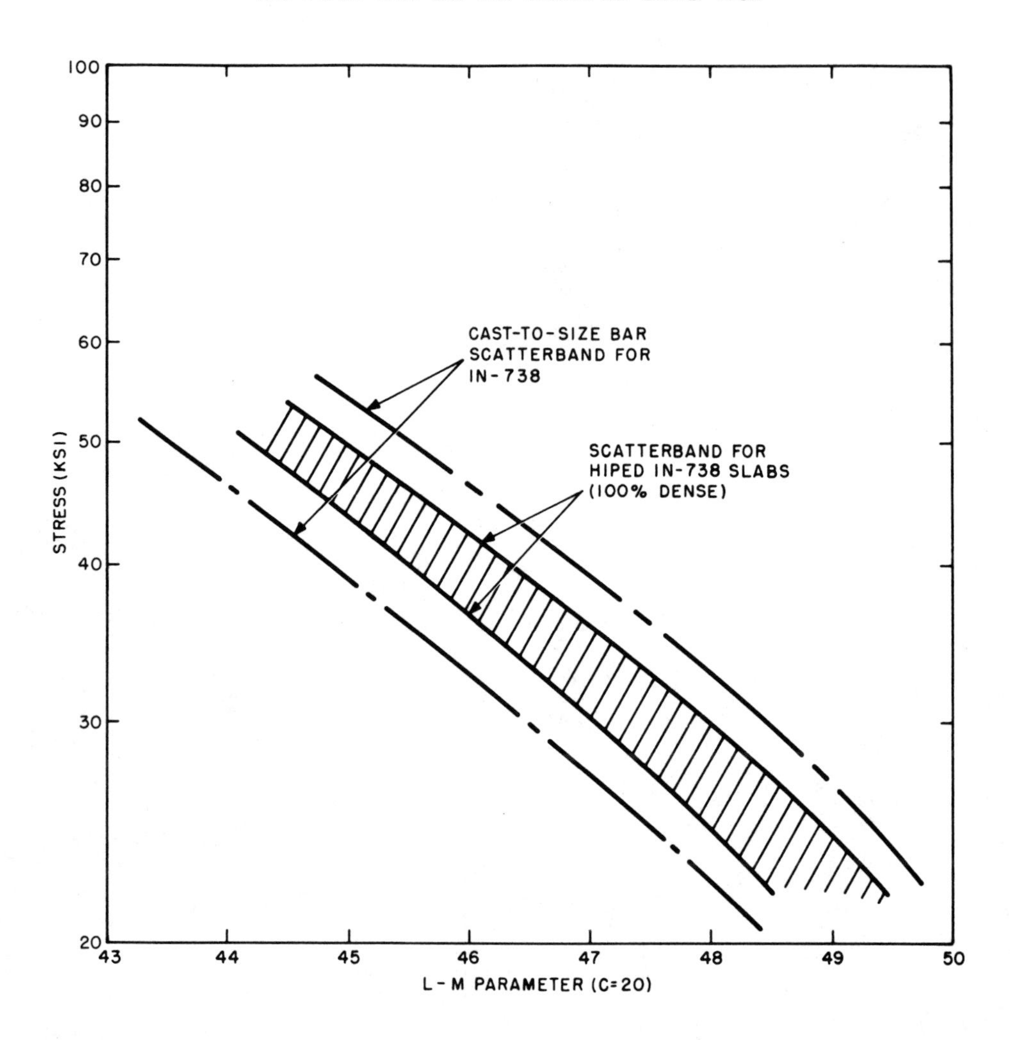

Comparison of rupture scatterband for HIPed and cast-to-size bars of IN 738. Heat treated at 1177 °C (2150 °F) for 2 h + 1121 °C (2050 °F) for 2 h + 927 °C (1700 °F) for 24 h + 843 °C (1550 °F) for 24 h.

There is significant improvement in rupture life and ductility as a result of HIP. Magnitude of the improvement is strictly a function of extent of porosity in the original casting. The above figure compares rupture properties of HIPed material with those of cast-to-size bar. Producers have developed techniques for manufacturing sound cast-to-size test bars, so these bars should represent the optimum as-cast properties obtainable. A comparison of HIPed IN 738 with these cast-to-size bar data indicates that the HIPed material has equivalent average rupture properties, but much less data scatter.

Source: G.E. Wasielewski and N.R. Lindblad, "Elimination of Casting Defects Using HIP," in Superalloys – Processing (Proceedings of the Second International Conference in Champion, PA), American Institute of Mining, Metallurgical, and Petroleum Engineers, Warrendale, PA, 1972, p D-9, D-20

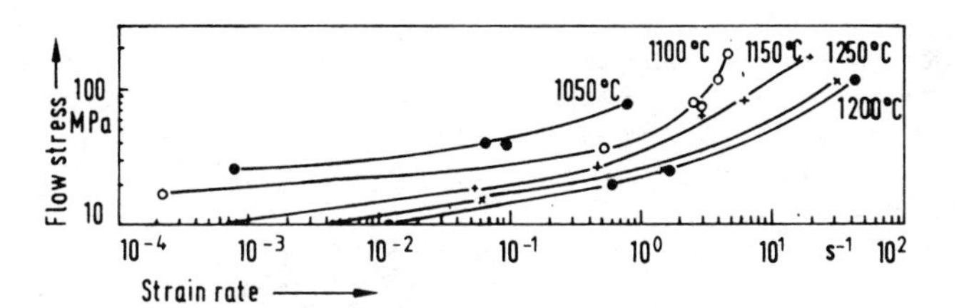

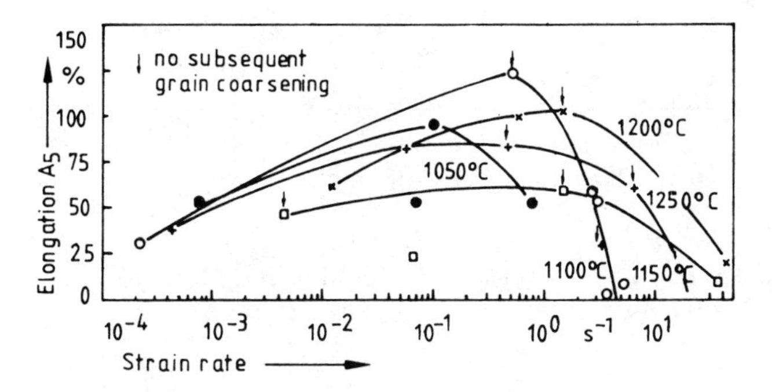

Flow stress vs strain rate (top) and elongation to rupture as a function of strain rate (bottom) in ODS IN 738.

After hot consolidation of the powders, ODS alloys have an extremely fine grain size. As a consequence, flow stress is strain-rate sensitive, and the alloy shows extended ductility and even superplasticity. At elevated temperatures, grain size has a dramatic effect on tensile properties. The flow and stress-strain rate relationships (top figure) accentuate this behavior. These hot workability tests in tension also have shown that substantial elongation can be obtained, in strong contrast with the known low ductility of coarse-grained ODS alloys. A subsequent heat treatment at 1270 °C (2318 °F) for 3 h will also result in grain growth (bottom figure).

Source: G.H. Gessinger, "Recent Developments in Powder Metallurgy of Superalloys," in Superalloys: Source Book, M.J. Donachie, Jr., Ed., American Society for Metals, Metals Park, OH, 1984, p 283-284

IN 738 + Y_2O_3: Effect of Processing and Oxygen Content

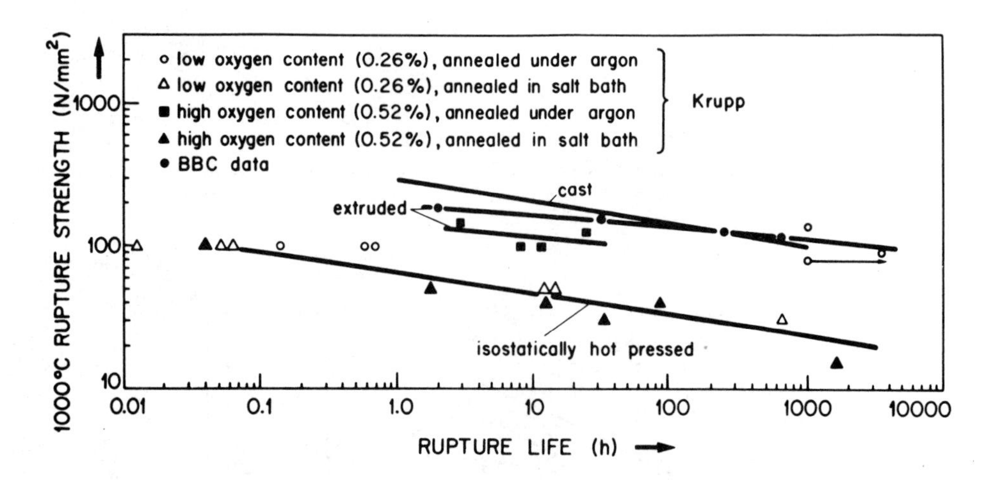

1000 °C creep-rupture strength of mechanically alloyed IN 738 + Y_2O_3 as a function of processing and oxygen content.

For comparison, the creep-rupture curve of cast IN 738 is included above. Extruded specimens demonstrated a higher creep-rupture strength and flatter slope than did cast IN 738. The Krupp data seem to indicate that a higher oxygen content has a detrimental effect on creep strength. Creep-rupture strength of the HIPed specimens lies distinctly below that of extruded specimens, in spite of optimized grain-coarsening heat treatment.

Source: G.H. Gessinger, "Oxide Dispersion Strengthened Alloys," in High Temperature Alloys for Gas Turbines, D. Coutsouradis, P. Felix, H. Fischmeister, L. Habraken, Y. Lindblom, and M.O. Speidel, Eds., Applied Science, London, 1978, p 822, 832

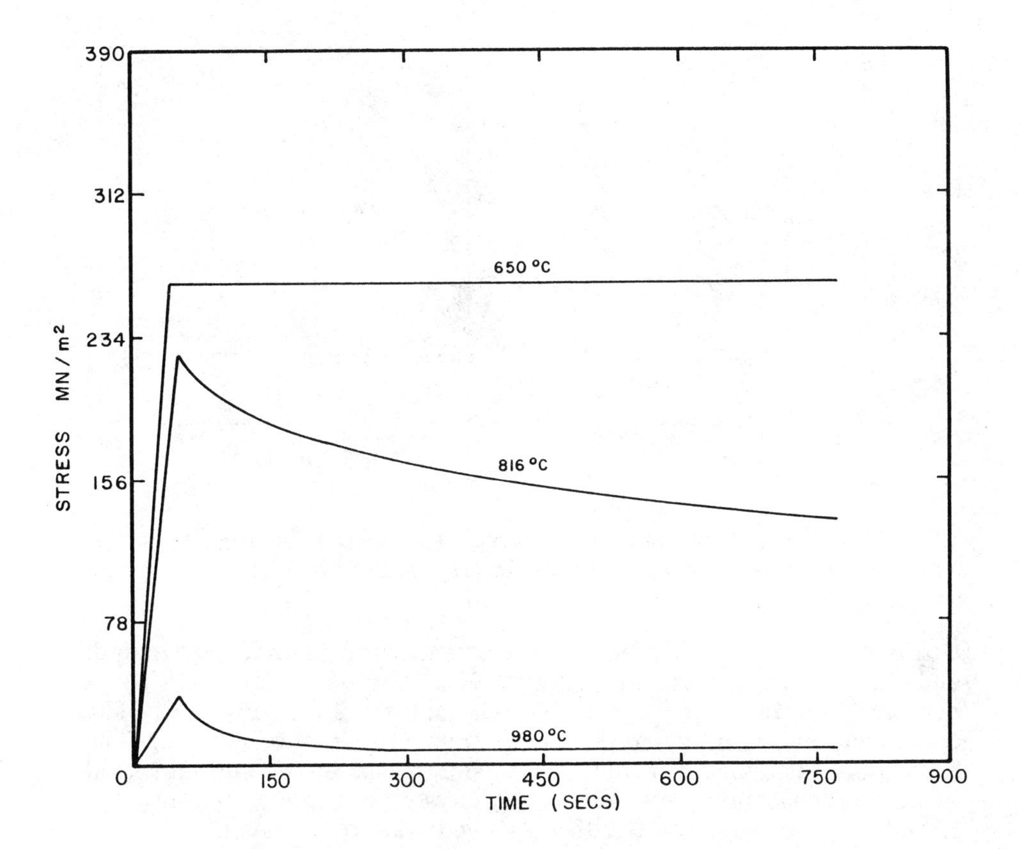

Stress-relaxation behavior of P/M IN 792.

Stress-relaxation results indicate that, at 650 °C (1202 °F), IN 792 has little capacity to relax stresses by plastic flow. However, appreciable yielding and stress relaxation are observed at 816 and 980 °C (1501 and 1796 °F). In a practical welding situation, therefore, at temperatures in the vicinity of 980 °C (1796 °F), the base metal would relax welding stresses and prevent strain concentrations in the heat-affected zone (HAZ). Therefore, if the base metal were preheated to such temperatures, a major share of the plastic strains due to welding could be accommodated in the base metal, sparing the susceptible HAZ from excessive straining.

Source: R. Thamburaj, W. Wallace, and J.A. Goldak, "Electron Beam Welding of Powder Metallurgy Nickel-Base Superalloys," in Powder Metallurgy Superalloys, Aerospace Materials for the 1980's (Proceedings of a Metal Powder Report Conference, Zürich), MPR Publishing, Shrewsbury, England, 1980, p 25-9, 25-25

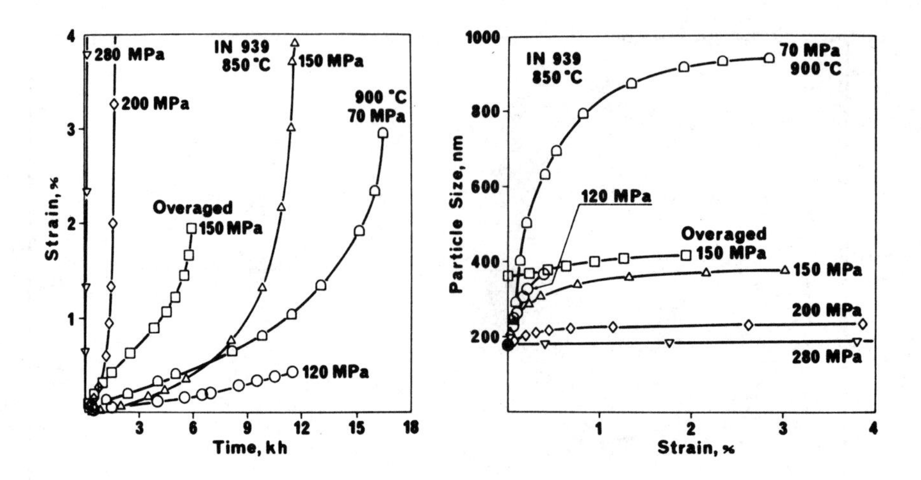

Strain vs time creep curves (left) and average particle size vs creep strain (right) for IN 939.

Creep tests were carried out on specimens with 50-mm gage length and 7.6-mm gage diam at 850 and 900 °C (1562 and 1652 °F) at constant load in air with temperature control ±2 °C (±3.6 °F). The creep curves relating strain with time are plotted above. The correction calculated for the creep rates measured at constant load instead of constant stress was nearly always negligible, because the strains and the stress sensitivities involved were quite small.

Source: S. Basso and V. Lupinc, "Particle Coarsening and Long Duration Tertiary Creep in Ni-Base Superalloy IN 939," in Strength of Metals and Alloys (Proceedings of the 7th International Conference, Montreal), Pergamon, Oxford, 1985, p 720

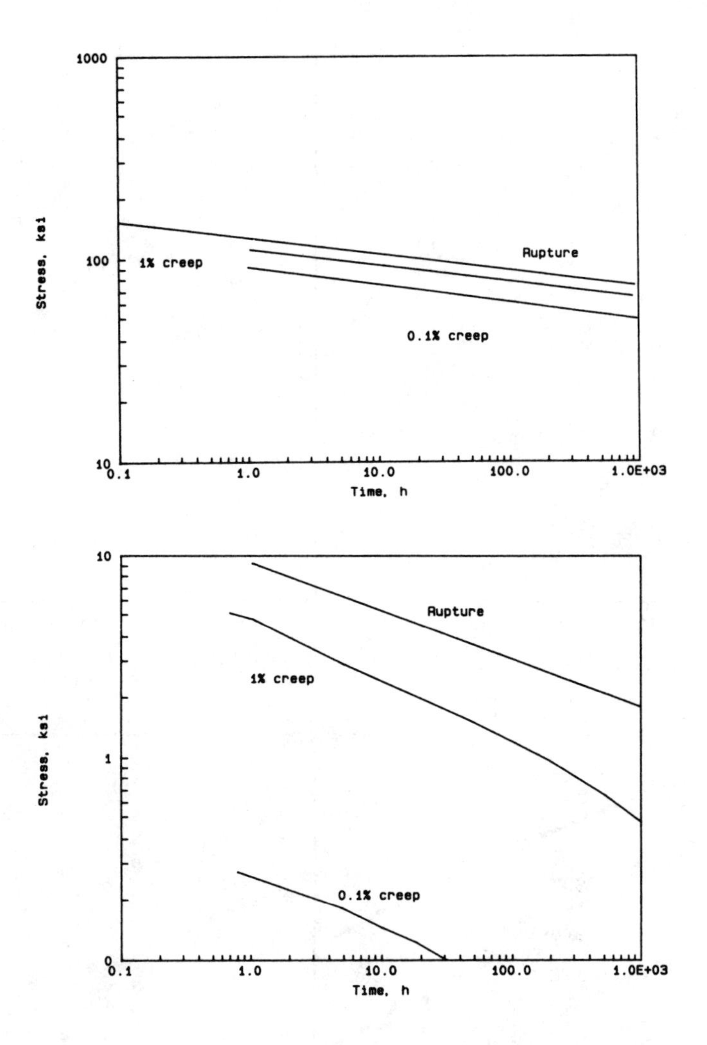

Creep and rupture lives at 1200 °F (top) and at 1800 °F (bottom) for 0.5-in. Incoloy 901 bar.

Specimens were heat treated at 2000 °F for 2 h, air cooled, aged at 1400 °F for 2 h, air cooled, aged at 1325 °F for 24 h, and air cooled.

Source: High-Temperature Property Data: Ferrous Alloys, M.F. Rothman, Ed., ASM International, Metals Park, OH, 1988, p 11.68-11.69

Inconel 617

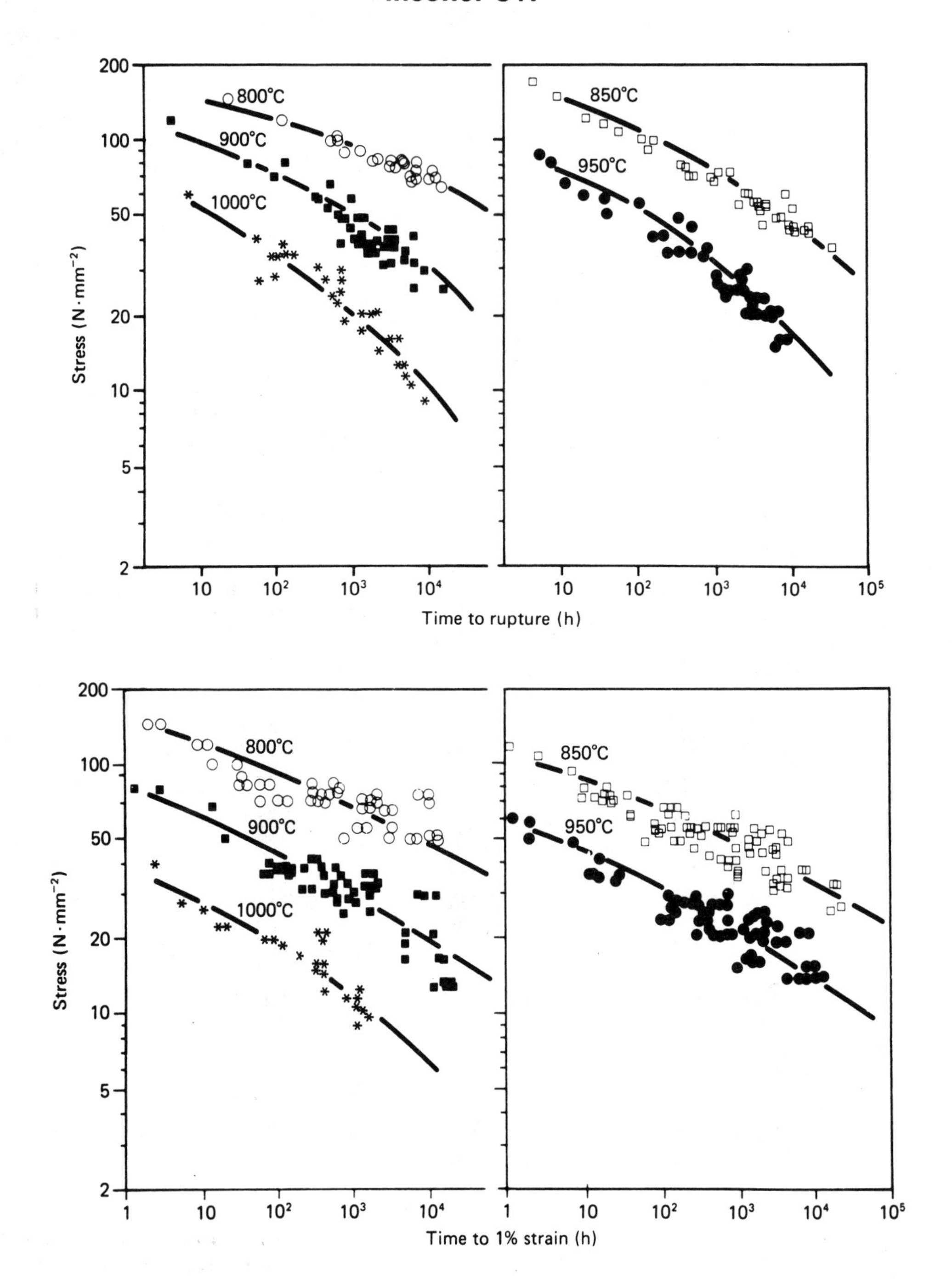

Creep-rupture properties of Inconel 617.

Source: F. Schubert, U. Bruch, R. Cook, et. al., "Creep Rupture Behavior of Candidate Materials for Nuclear Process Heat Applications," Nucl. Technol., Vol. 66, Aug. 1984, p 230

Inconel 617: Effect of Environment (Air, Helium, and Vacuum)

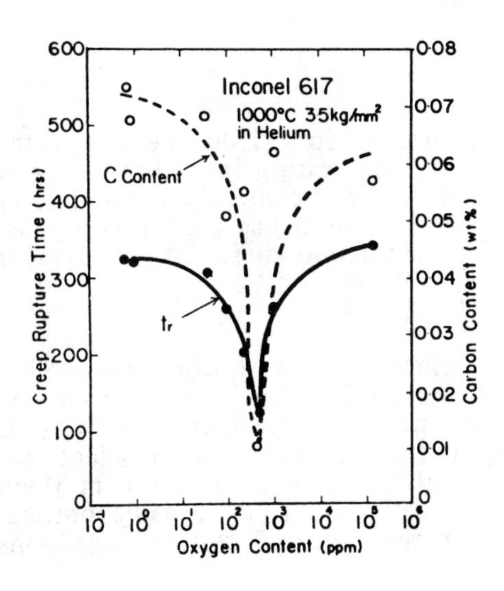

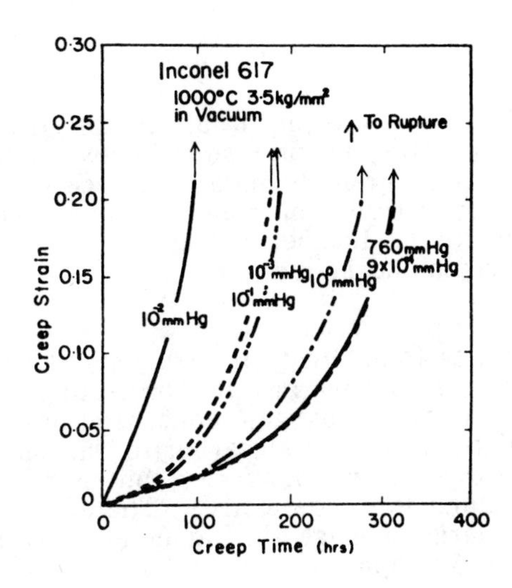

Effect of oxygen content on creep-rupture time and decarburization in helium (left) and effect of pressure on creep strain-time curves in vacuum (right) for Inconel 617.

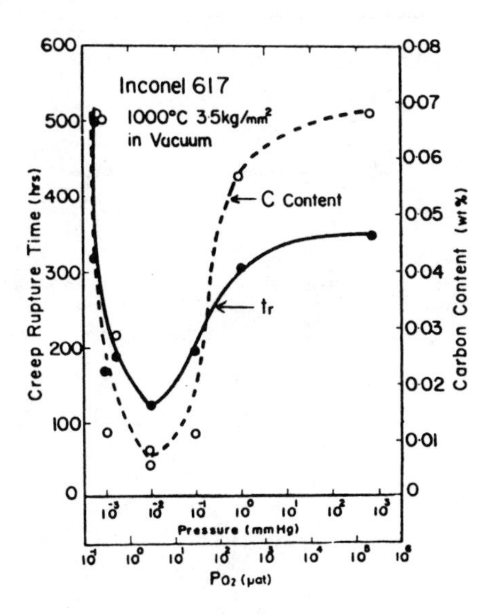

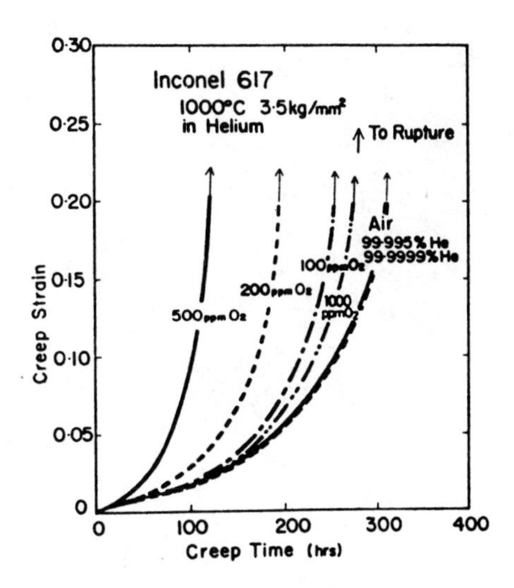

Effect of pressure (oxygen content) on creep-rupture time and decarburization in vacuum (left) and effect of oxygen content on creep strain-time curves in helium (right) for Inconel 617.

Inconel 617: Effect of Environment (Air, Helium, and Vacuum) (Continued)

When creep testing in air compared to vacuum or helium, it has been reported that testing in air results in lower creep rates than testing in helium, even if oxidized grain-boundary surface cracks appear. The seemingly contradictory creep behavior of materials in air, helium, and vacuum has been investigated by testing Inconel 617 at 1000 °C (1832 °F) in air and different mixtures of oxygen and helium.

It was found that, at high oxygen concentrations, a thick uniform oxide film is formed covering and protecting the surface against further oxidation. At intermediate oxygen concentrations in the environment, oxygen activity is high enough to decarburize the surface of the material and to selectively oxidize grain boundaries, but not high enough to form an oxide film thick enough to be protective. Therefore, there is oxygen activity causing optimal material damage during creep, which under the reported testing conditions was 500 ppm.

Source: Y. Lindblom, "Creep and Structural Stability of High Temperature Materials," in High Temperature Alloys for Gas Turbines, D. Coutsouradis, P. Felix, H. Fischmeister, L. Habraken, Y. Lindblom, and M.O. Speidel, Eds., Applied Science, London, 1978, p 311-313

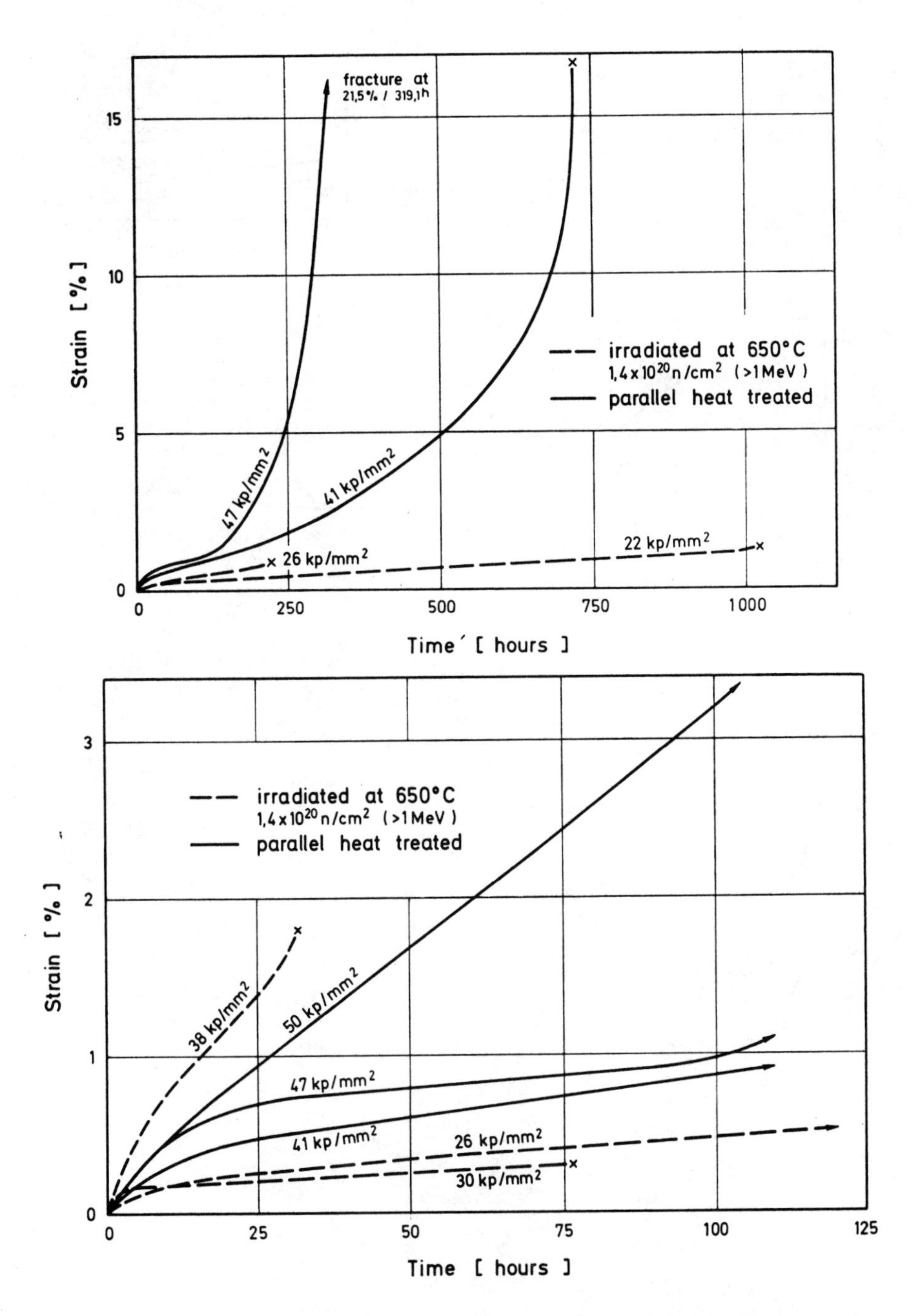

Influence of irradiation on creep characteristics (top) and primary and steady-state creep ranges (bottom) of Inconel 625 at 650 °C (1202 °F).

Irradiation causes two effects: (1) a drastic reduction of tertiary creep range with an associated decrease of stress-rupture elongation from 17-20% to 0.3-2.0% and (2) an increase in minimum creep rate by a factor of ten.

Source: F. Garzarolli, K.P. Francke, and J. Fischer, "Influence of Neutron Irradiation on Tensile and Stress Rupture Properties of Inconel 625," J. Nuclear Mat., Vol. 30, April 1969, p 245

Inconel 713C: As-Cast

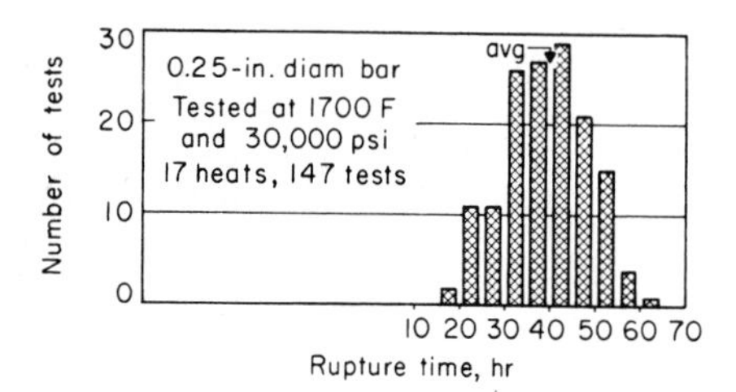

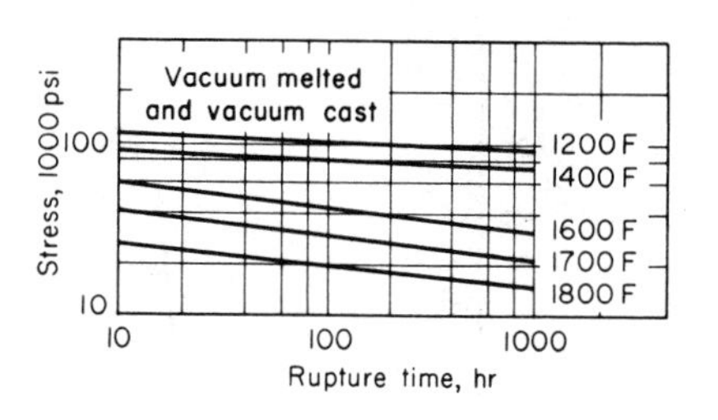

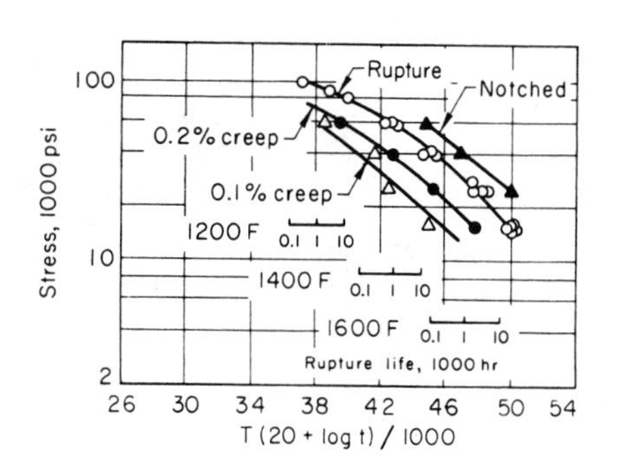

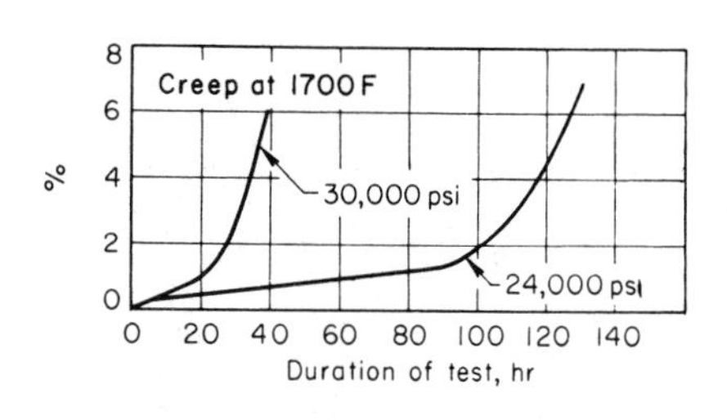

As-cast.

Source: "Heat-Resistant Alloy Castings," in Source Book on Materials for Elevated-Temperature Applications, E.F. Bradley, Ed., American Society for Metals, Metals Park, OH, 1979, p 235

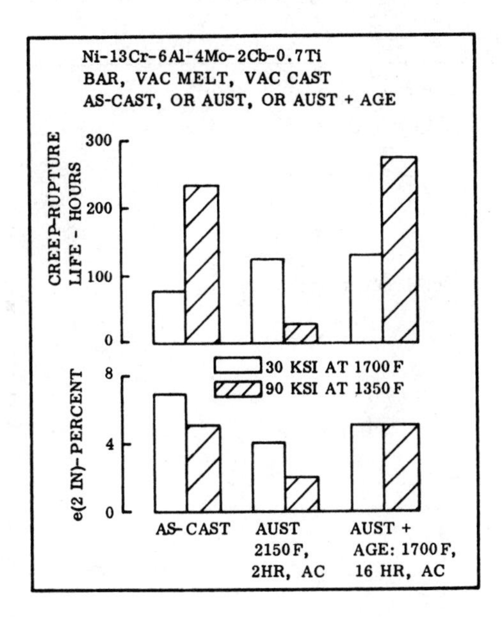

Effect of heat treatment on creep-rupture life and ductility.

Source: Structural Alloys Handbook, Volume 4, Mechanical Properties Data Center, Battelle Columbus Laboratories, revised 1980, p 4119-19

Inconel 713LC: Effect of Solidification Rate on Directionally Solidified Material

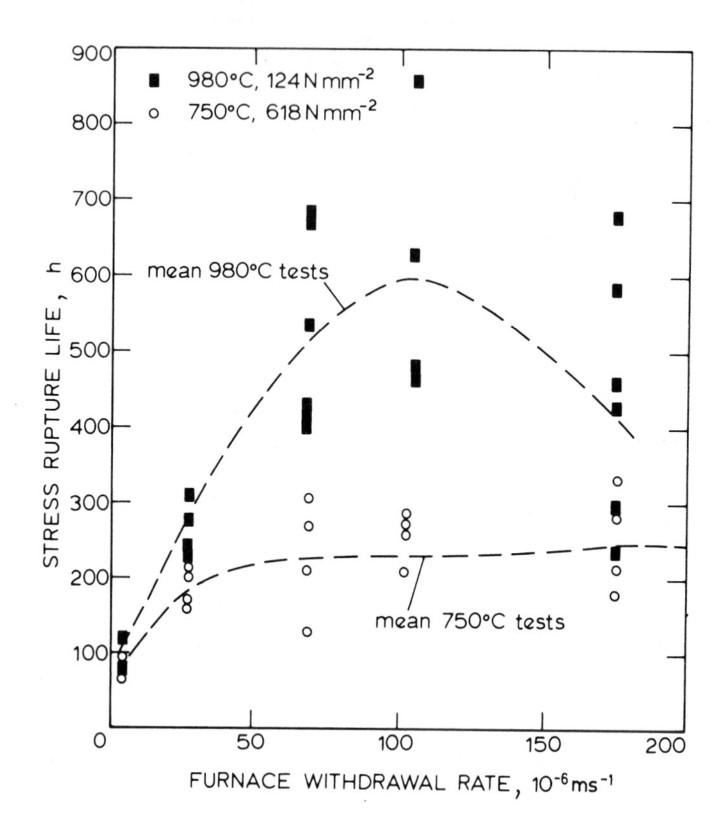

Variations in stress rupture at 750 and 980 °C of directionally solidified Inconel 713LC as a function of solidification rate. The directional solidification was carried out in a low temperature gradient typical of commercial rigs.

Source: M. McLean, Directionally Solidified Materials for High-Temperature Service, The Metals Society, London, 1983, p 170

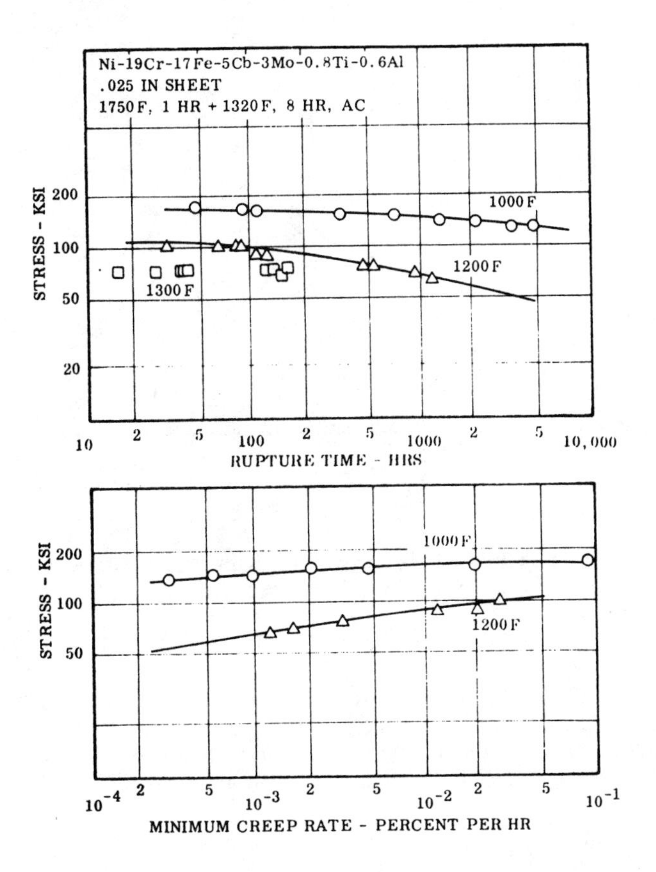

Minimum creep rate and rupture time at 1000 and 1200 °F for sheet annealed at 1750 °F and aged.

Source: Structural Alloys Handbook, Volume 4, Mechanical Properties Data Center, Battelle Columbus Laboratories, revised 1980, p 4103-41

Inconel 718: Effect of Delta-Phase Precipitation

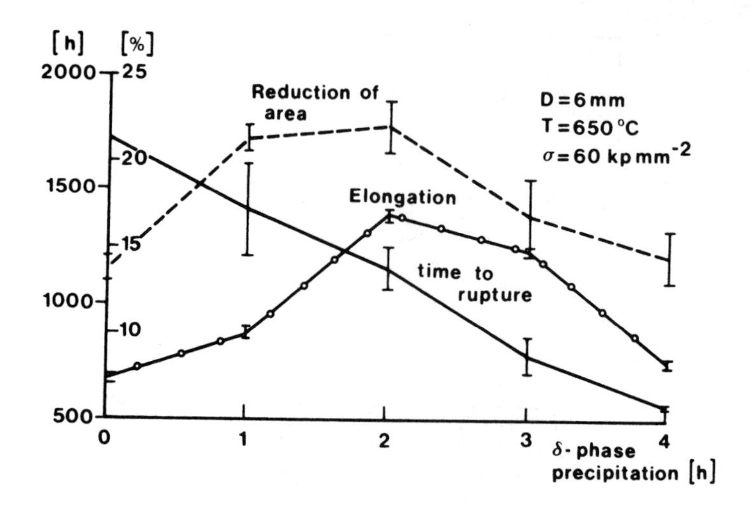

Stress-rupture of Inconel 718 unwelded at 650 °C (1202 °F) with delta-phase precipitation; F = 60 kpmm^{-2}.

The figure above shows the change in ductility and creep that occurs with δ-phase precipitation. It must be concluded that elimination of microcracking by δ-phase precipitation is offset by a decrease in creep strength.

Source: P. Adam, "Welding of High-Strength Gas Turbine Alloys," in High Temperature Alloys for Gas Turbines, D. Coutsouradis, P. Felix, H. Fischmeister, L. Habraken, Y. Lindblom, and O. Speidel, Eds., Applied Science, London, 1978, p 742, 761

Inconel 718: Effect of Trace Elements

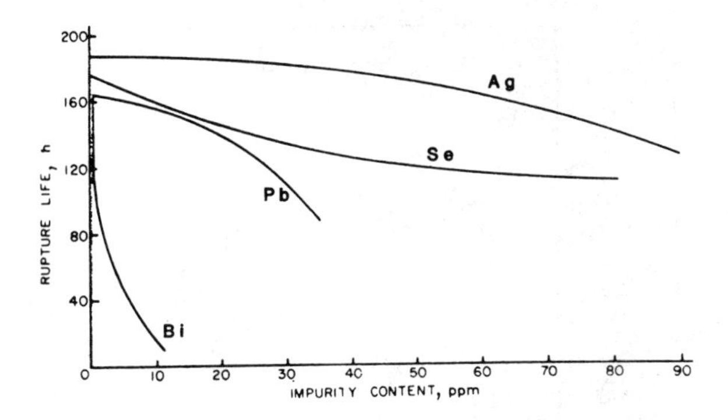

Effect of various trace elements on the stress-rupture life of Inconel 718 at 649 °C (1200 °F) and 690 MPa (100 ksi).

With the advent of very precise analytical instruments, it has been possible to accurately and quantitatively ascertain the effects of trace elements on the properties of Ni-base alloys. The removal of these elements has assumed a major importance. For example, the figure above shows how trace quantities of Ag, Se, Pb, and Bi can adversely affect the stress-rupture life of Inconel 718. Fortunately, most of the harmful trace elements have relatively high vapor pressure and are distilled under vacuum.

Source: J.W. Pridgeon, F.N. Darmara, J.S. Huntington, and W.H. Sutton, "Principles and Practices of Vacuum Induction Melting and Vacuum Arc Remelting," in Superalloys: Source Book, M.J. Donachie, Jr., Ed., American Society for Metals, Metals Park, OH, 1984, p 205

Inconel 718: Hot Rolled Bar

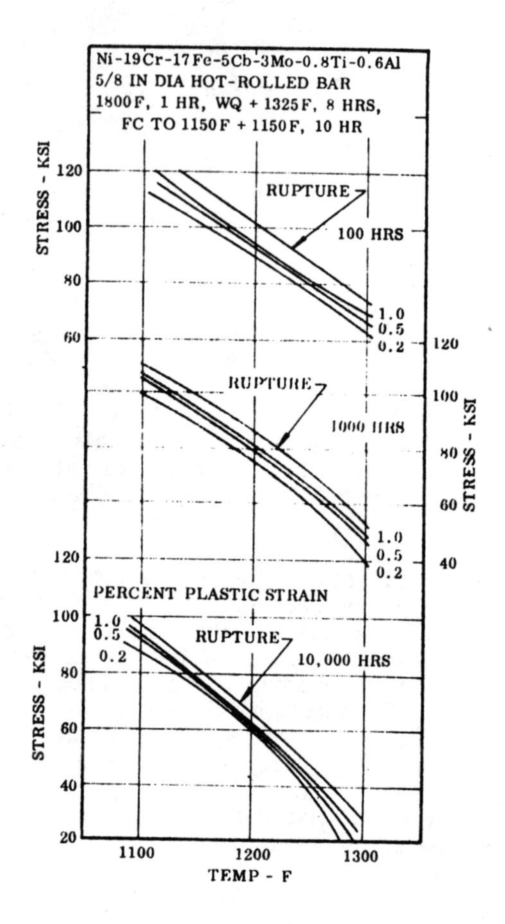

Creep and creep-rupture properties of hot rolled bar at 100, 1000, and 10,000 h.

Source: Structural Alloys Handbook, Volume 4, Mechanical Properties Data Center, Battelle Columbus Laboratories, revised 1980, p 4103-45

Inconel MA 753: Positive and Negative Creep

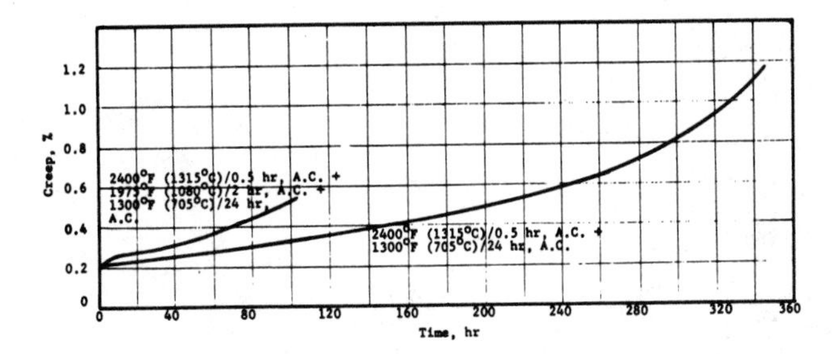

Positive creep properties of Inconel MA 753 at 1040 °C (1900 °F) and 117.2 MPa (17.0 ksi).

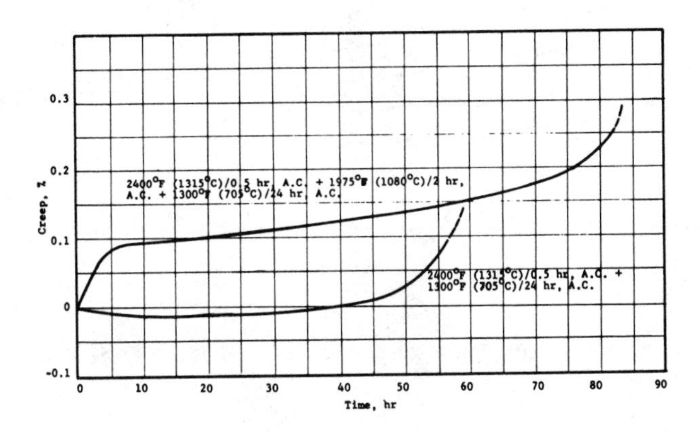

Positive and negative creep properties of Inconel MA 753 at 1150 °C (2100 °F) and 89.6 MPa (13.0 ksi).

Inconel MA 753 exhibits negative creep contraction when exposed to conditions of uniaxial tensile creep. (Positive creep is the reverse - specimen extension during exposure to uniaxial tensile-creep conditions.) Negative creep may be of concern in applications where part assemblies are involved, because contraction can increase stress on a part.

When tested at 1040 °C (1900 °F), MA 753 exhibited positive creep (top figure). At 1150 °C (2100 °F), however, the alloy exhibited either positive or negative creep, depending on the heat treatment given prior to testing (bottom figure).

Source: G.R. Strobel, "A Nickel-Base Alloy With Unusual Creep Behavior," in Source Book on Materials Selection, Vol. II, R.B. Gunia, Ed., American Society for Metals, Metals Park, OH, 1977, P 97-98

Inconel MA 6000

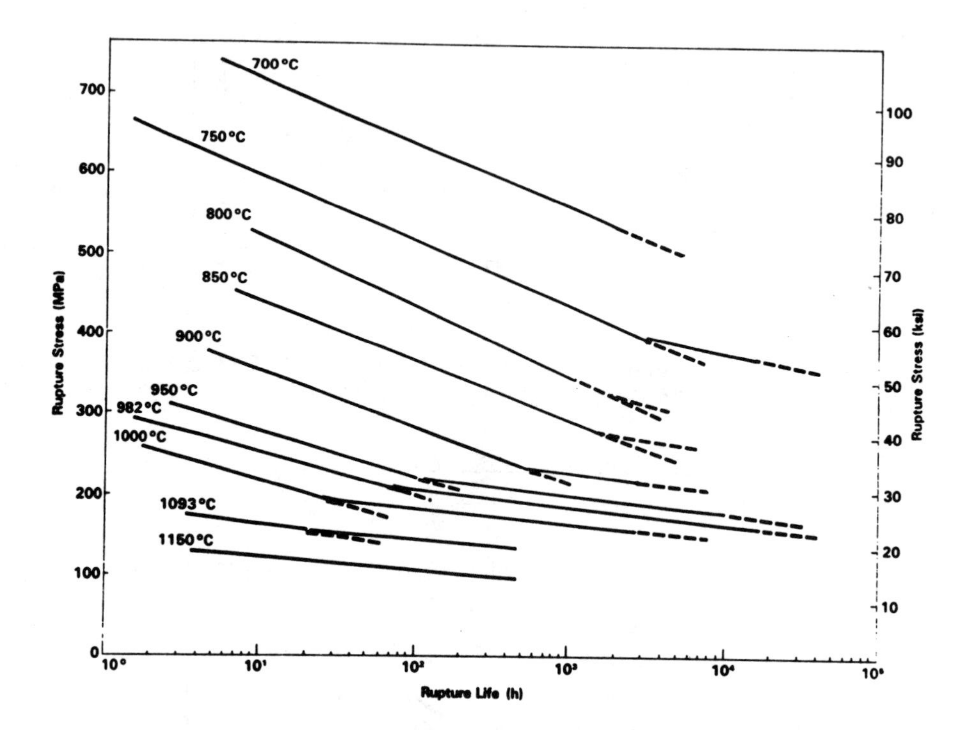

Stress vs log time rupture data for Inconel MA 6000.

Recent stress- and creep-rupture tests on Inconel MA 6000 have yielded insights into the mechanisms that give rise to its outstanding long-term high-temperature strengths and, in particular, have identified an upward break in the strength characteristics of the alloy, as shown above.

Source: R.C. Benn, J.S. Benjamin, and C.M. Austin, "Design of Oxide Dispersion Strengthened Superalloys," in High Temperature Alloys: Theory and Design, J.O. Stiegler, Ed., American Institute of Mining, Metallurgical, and Petroleum Engineers, New York, 1984, p 441-442

Inconel X-550: Plain vs Notched Specimens

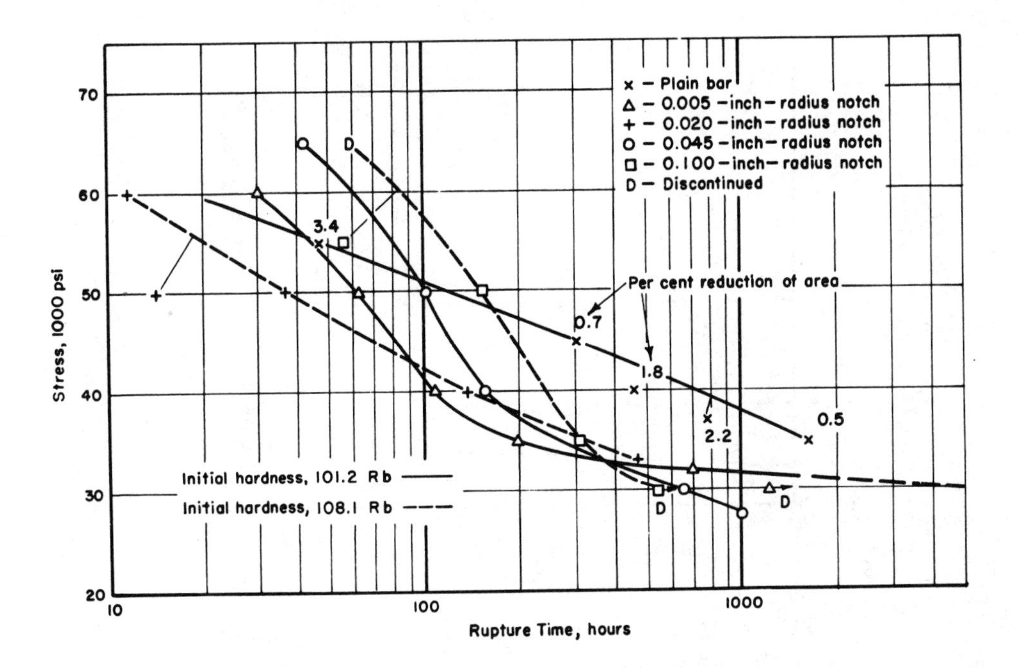

Stress-rupture data for Inconel X-550 at 732 °C (1350 °F).

The plain-bar data and 0.005- and 0.045-in.-radius notched-bar data were obtained on specimens with an initial hardness of 101 HRB. Both of these notches produced strengthening at the shorter times at this hardness level, where the plain-bar ductility was highest. Data for the 0.005-in.-radius notches indicated notch sensitivity for rupture times in excess of about 40 h and for the 0.045-in.-radius notches for rupture times in excess of about 100 h. The slopes of the curves for both notches began to decrease at about 40,000 psi; the greater change in slope was shown by the sharper notch, causing the curves to cross at about 370 h.

Source: Creep and Fracture of Metals at High Temperatures, Philosophical Library, New York, 1957, p 406, 408

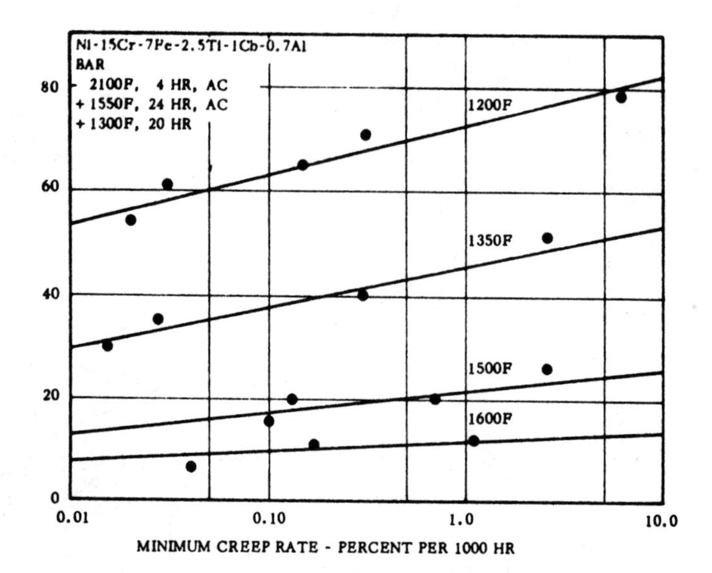

Effect of stress and temperature on minimum creep rate of bar.

Source: Structural Alloys Handbook, Volume 4, Mechanical Properties Data Center, Battelle Columbus Laboratories, revised 1980, p 4105-18

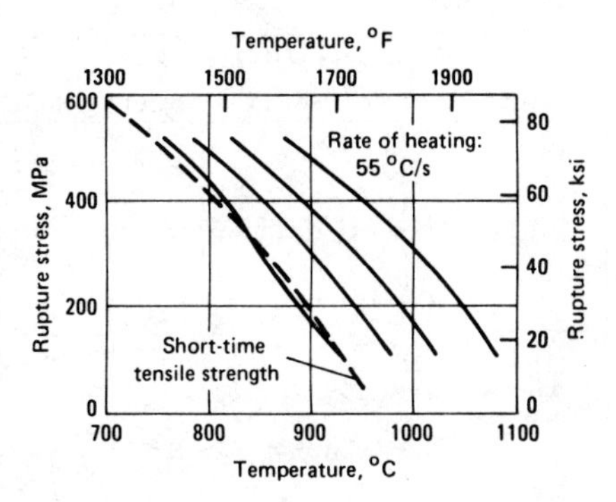

Rupture strength of Inconel X-750 sheet under conditions of rapid heating at a constant rate.

Specimens were machined from annealed 1.3-mm (0.050-in.) thick sheet, with specimen axes parallel to rolling direction.

Source: Metals Handbook, Ninth Edition, Volume 3, Properties and Selection: Stainless Steels, Tool Materials and Special-Purpose Metals, American Society for Metals, Metals Park, OH, 1980, p 230

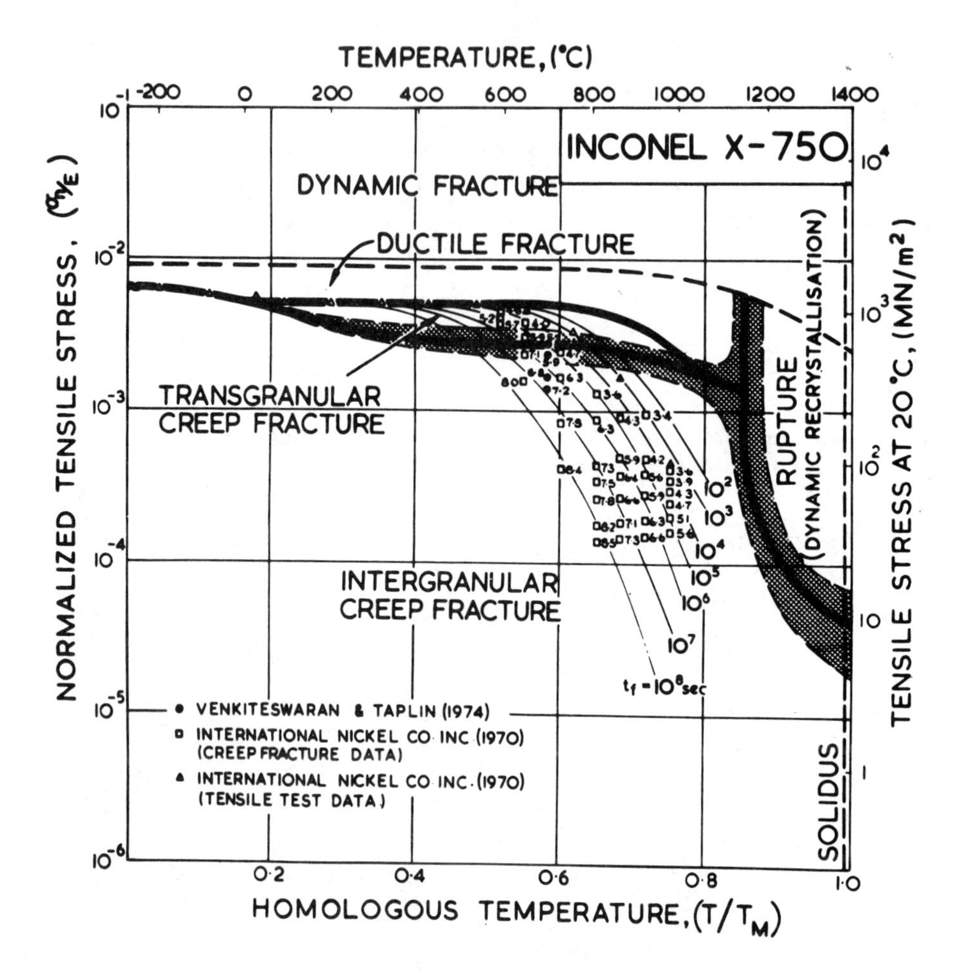

Fracture-mechanism map for Inconel X-750.

The effect of solid-solution strengthening, along with precipitation strengthening, is seen in the shrinkage of transgranular creep-fracture and rupture fields. The higher stresses observed in this alloy also may suggest that ductility may be the major concern in the intergranular creep-fracture field.

Source; C. Gandhi, "Fracture Mechanism Map for Metals and Alloys," in Flow and Fracture at Elevated Temperatures, R. Raj, Ed., American Society for Metals, Metals Park, OH, 1985, p 100-103

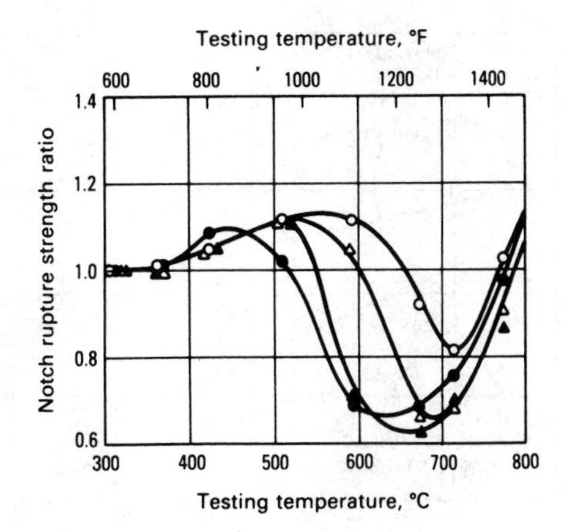

Notch-rupture strength ratio vs temperature at four different rupture times for Inconel X-750.

Source: Metals Handbook, Ninth Edition, Volume 8, Mechanical Testing, American Society for Metals, Metals Park, OH, 1985, p 316

Inconel X-750: Sheet

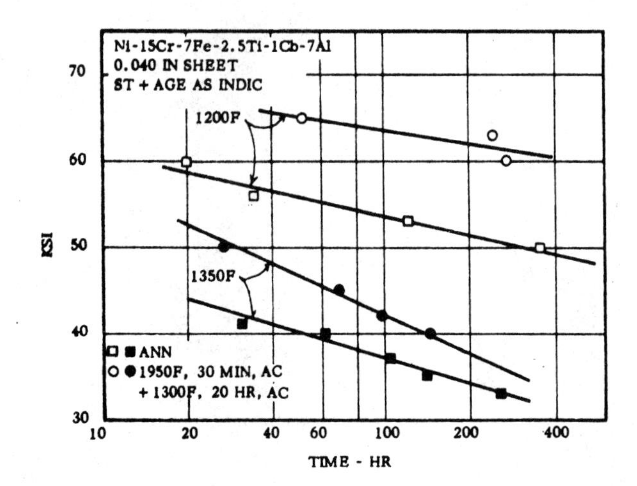

Stress-rupture curves at 1200 and 1350 °F for sheet in the annealed and heat treated condition.

Source: Structural Alloys Handbook, Volume 4, Mechanical Properties Data Center, Battelle Columbus Laboratories, revised 1980, p 4105-14

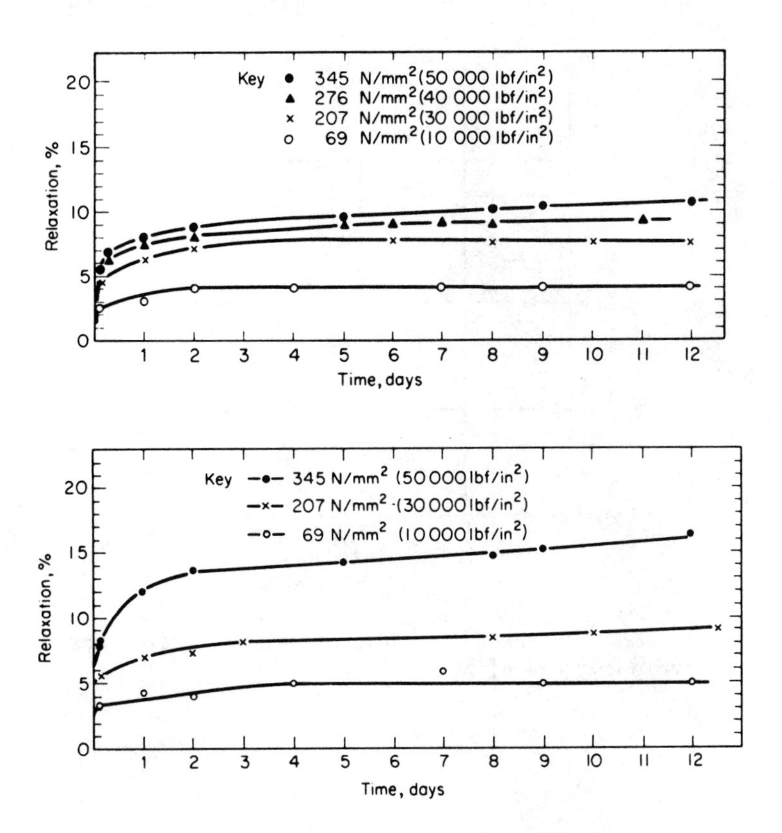

Stress relaxation values for Inconel X-750 springs at 550 °C (1022 °F) (top) and 600 °C (1112 °F) (bottom).

Springs were made from 3.6-mm (0.144-in.) diam wire with a final reduction of 65%. Heat treated 2 h at 1150 °C (2102 °F), air cooled followed by aging for 24 h at 840 °C (1544 °F), air cooled and held 20 h at 700 °C (1292 °F), air cooled.

Source: The Nimonic Alloys, W. Betteridge and J. Heslop, Eds., Crane, Russak & Co., New York, 1974, p 272

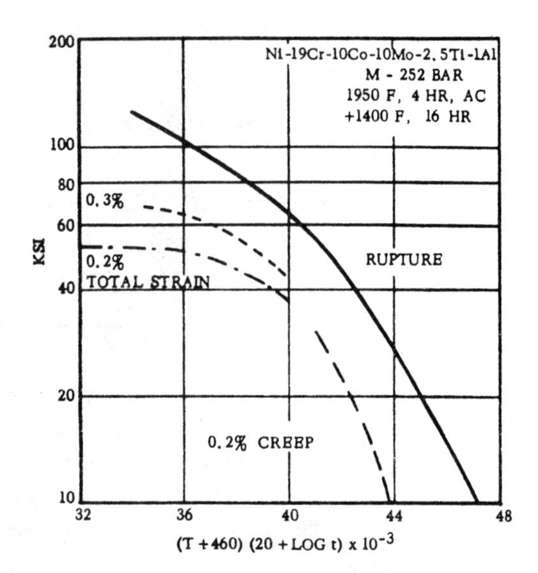

Master curves for total strain creep and creep-rupture of M-252 bar.

Source: Structural Alloys Handbook, Volume 5, Mechanical Properties Data Center, Battelle Columbus Laboratories, revised 1978, p 4202-3

MA 755E: Conventional vs Zone Annealing

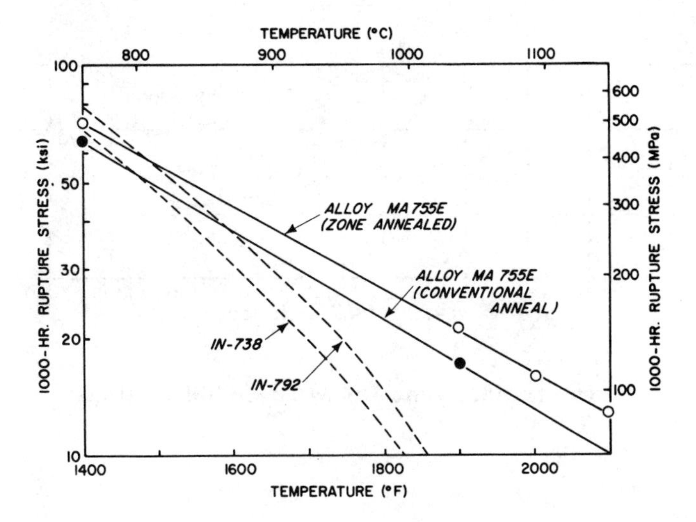

1000-h rupture stress/temperature capability for MA 755E.

As shown above, the 1000-h temperature capability of MA 755E is markedly higher than that of either IN 738 or IN 792 at temperatures above 870 °C (1600 °F). Its strength at intermediate temperatures of 760 to 870 °C (1400 to 1600 °F) compares favorably with that of the two cast alloys.

Source: J.W. Schultz and W.R. Hulsizer, "Corrosion-Resistant Nickel-Base Alloys for Gas Turbines," in Source Book on Materials Selection, Volume II, R.B. Gunia, Ed., American Society for Metals, Metals Park, OH, 1977, p 86-87

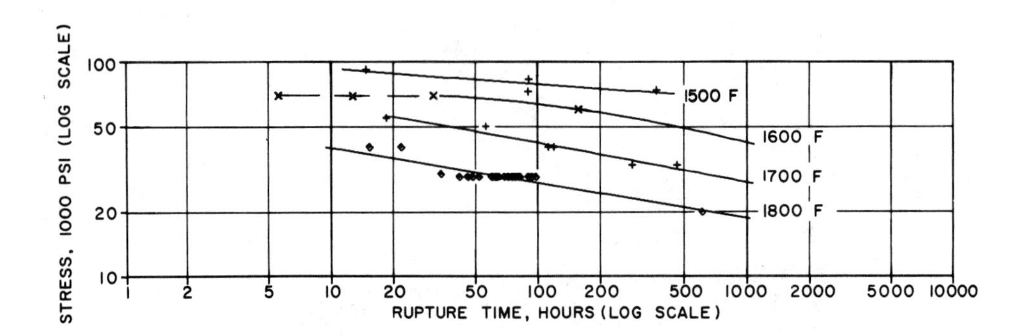

Stress-rupture time for MAR-M200 castings.

Source: The Elevated-Temperature Properties of Selected Superalloys, ASTM Data Series DS 7-S1, American Society for Testing and Materials, Philadelphia, 1968, p 89

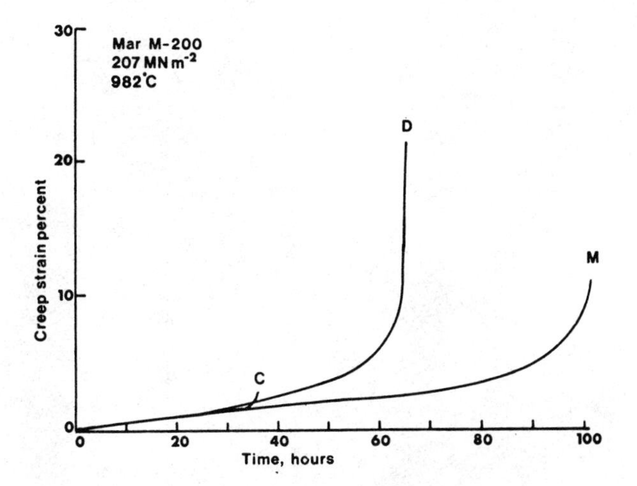

Comparison of the creep properties at 982 °C (1800 °F) of conventional (C), directional (D), and monocrystal (M) MAR-M200.

Source: C.D. Desforges, "Metals and Alloys for High Temperature Applications – Current Status and Future Prospects," in Source Book on Materials for Elevated-Temperature Applications, E.F. Bradley, Ed., American Society for Metals, Metals Park, OH, 1979, p 11

MAR-M200: Effect of Stress on Primary Creep of Directionally Solidified Material at 760 °C (1400 °F)

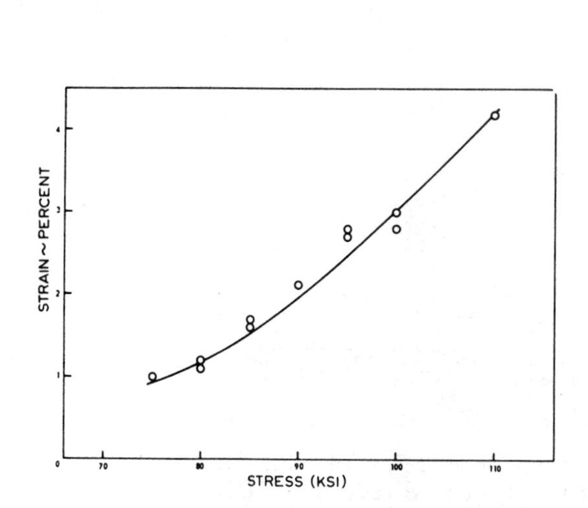

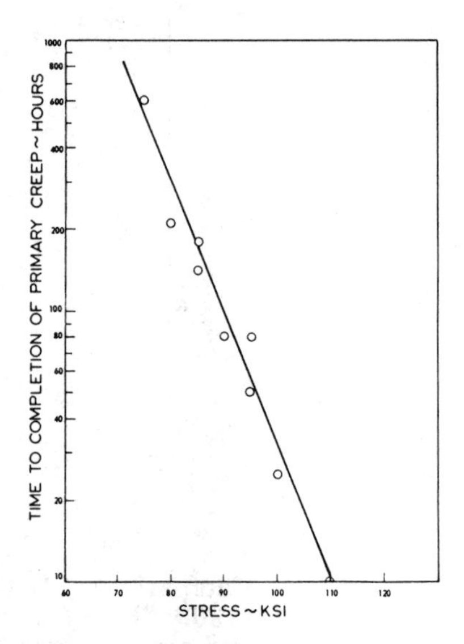

Amount of primary creep (left) and time to completion of primary creep (right) as a function of stress at 760 °C (1400 °F) for solutionized and aged directionally solidified MAR-M200.

The amount of primary creep at 760 °C (1400 °F) is highly stress dependent (left figure). At stress levels found in turbine parts, the amount of primary creep should be less than 0.5%. At the lower stress levels, it is no longer proper to consider primary creep as a rapidly occurring amount of plastic deformation. Rapid primary creep only occurs at the higher stresses. The time to completion of primary creep increases exponentially as the stress is lowered (right figure). At a stress of 345 MPa (50 ksi), it should take 10,000 h to reach the completion of primary creep in directionally solidified material at 760 °C (1400 °F).

Source: F.L. Versnyder and M.E. Shank, "The Development of Columnar Grain and Single Crystal High Temperature Materials Through Directional Solidification," in Source Book on Materials for Elevated-Temperature Applications, E.F. Bradley, Ed., American Society for Metals, Metals Park, OH, 1979, p 356, 357

MAR-M200: Transverse and Longitudinal Rupture Properties of Directionally Solidified and Conventionally Cast Material

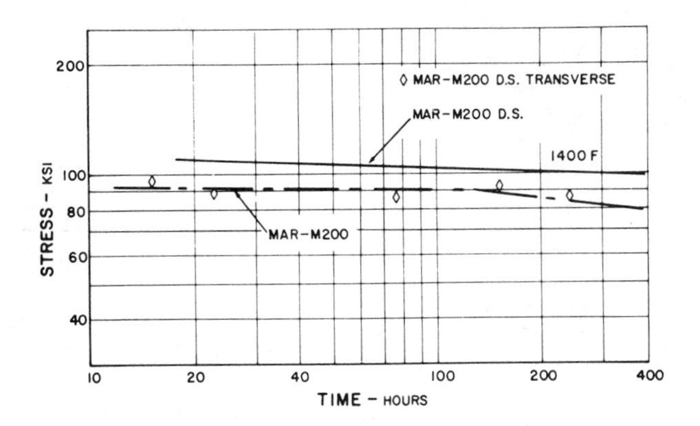

Transverse and longitudinal stress-rupture properties of directionally solidified and conventionally cast MAR-M200 at 1400 °F.

To document more fully the properties of columnar-grained MAR-M200, several tests were carried out on specimens cut from transverse sections of unidirectionally solidified ingots. In these specimens, the columnar grains were oriented normal to the axis of the specimens. As might be expected, the transverse properties of columnar-grained MAR-M200 were found to be equivalent to those of the conventionally cast material.

Source: F.L. Versnyder and M.E. Shank, "The Development of Columnar Grain and Single Crystal High Temperature Materials Through Directional Solidification," in Source Book on Materials for Elevated-Temperature Applications, E.F. Bradley, Ed., American Society for Metals, Metals Park, OH, 1979, p 356

MAR-M246: Effect of Rate of Directional Solidification and Different Temperature Gradients

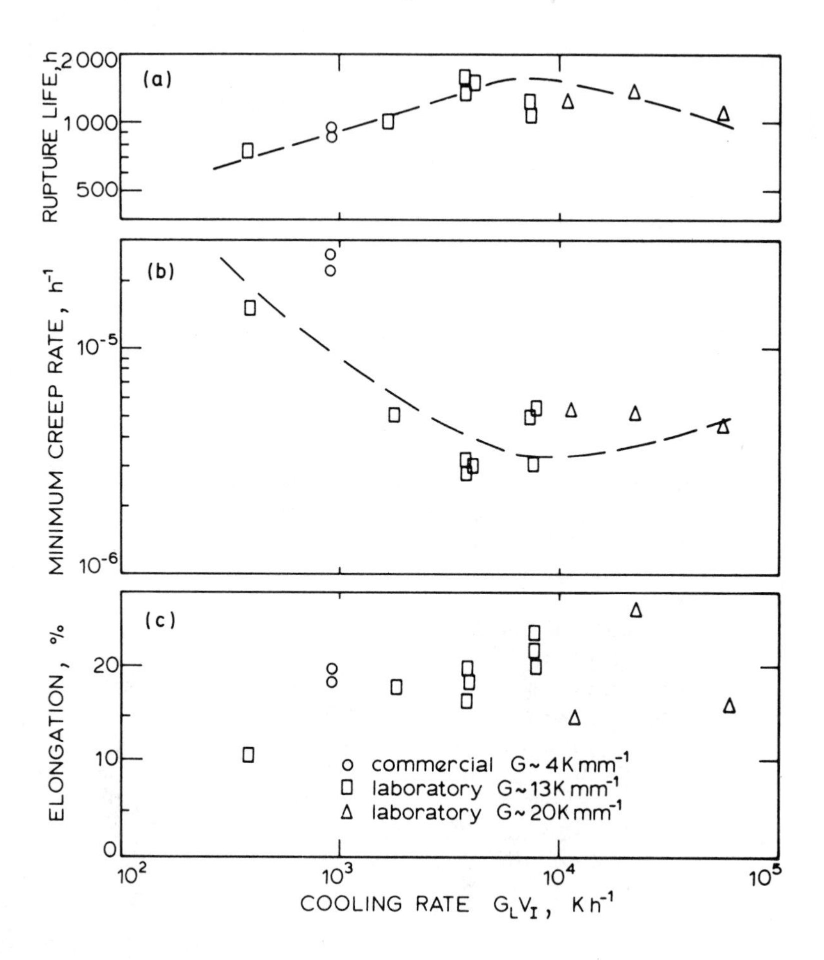

Summary of creep data for MAR-M246 directionally solidified at different rates and in different temperature gradients plotted as functions of the cooling rate $G_L V_I$: (a) rupture life, (b) minimum creep rate, and (c) extension at fracture (350 MPa, 1123 K).

Source: M. McLean, Directionally Solidified Materials for High Temperature Service, The Metals Society, London, 1983, p 172

MAR-M246: Heat Treated, Single-Crystal, UDS and Conventionally Cast

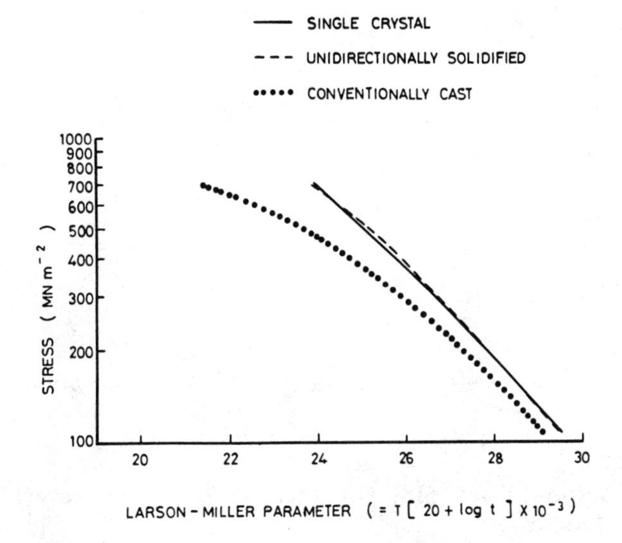

Larson-Miller curves for heat treated MAR-M246.

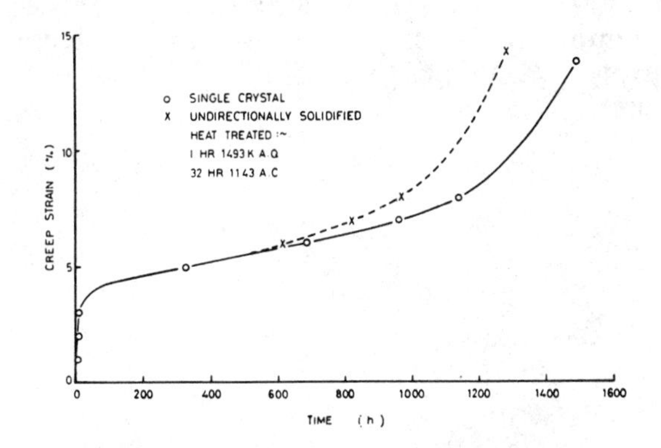

Creep curves for heat treated MAR-M246 ($10^6\ \mu m^{-1}$) at 760 °C (1400 °F).

Although the creep-rupture life of UDS material was superior to that of conventionally cast material, only at 760 °C (1033 K) in some tests in the heat treated condition was any marked improvement observed in single-crystal material over the UDS material. Except for these conditions, where increased life resulted from delayed onset of tertiary creep, the absence of grain boundaries did not result in any significant increase in creep-rupture life.

Source: J.E. Northwood, "Improving Turbine Blade Performance by Solidification Control," in Superalloys: Source Book, M.J. Donachie, Jr., Ed., American Society for Metals, Metals Park, OH, 1984, p 295

MAR-M247: Effect of Cobalt

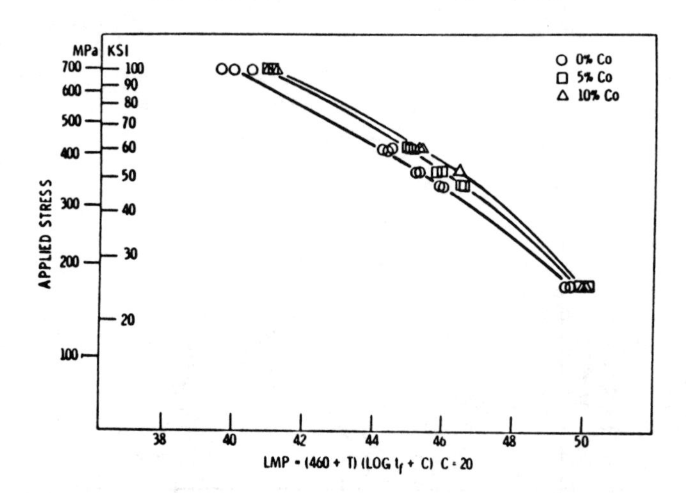

Larson-Miller plot for MAR-M247 with several cobalt levels.

This figure summarizes the stress-rupture tests performed at 760, 871, and 982 °C (1400, 1600, and 1800 °F). Note that rupture life is reduced at all temperatures as Co is removed. (Rupture ductility, however, was not affected by Co content.)

Source: M.V. Nathal, R.D. Maier, and L.J. Ebert, "The Influence of Cobalt on the Tensile and Stress-Rupture Properties of the Nickel-Base Superalloy MAR-M247," Met. Trans. A, Vol. 13A, Oct. 1982, p 1769

MAR-M247: Effect of Orientation on Creep of Single Crystals

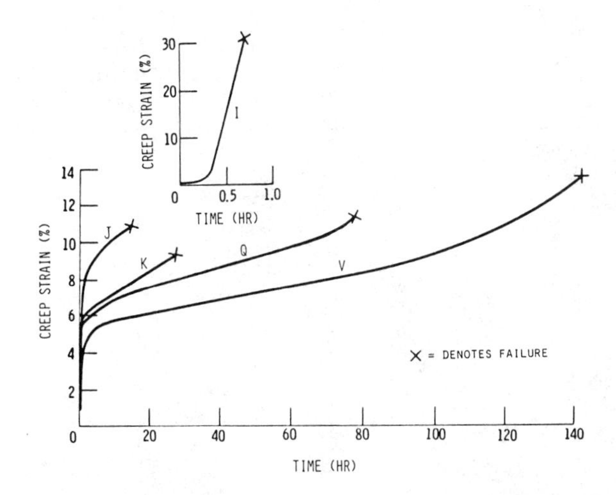

Creep curves of selected MAR-M247 single crystals tested at 774 °C (1425 °F) and 724 MPa (105 ksi). The letter associated with each curve is the specimen designation.

MAR-M247: Effect of Orientation on Creep of Single Crystals (Continued)

Compositions of MAR-M247 single crystals (wt%)

Elements	Single Crystals MAR-M247[a]	MAR-M247[b]
C	0.13	0.14
Cr	8.3	7.9
Co	9.9	9.5
Ti	0.93	0.96
Al	5.6	5.8
W	10.2	9.6
Zr	0.06	0.03
B	0.018	0.013
Fe	0.10 max	0.10 max
Nb	0.05 max	0.05 max
Mo	0.72	0.76
Ta	2.96	3.06
Hf	1.23	1.50
Ni	bal.	bal.

[a]Grown by exothermic process
[b]Grown by withdrawal process

Rupture-life data of MAR-M247 single crystals tested at 774 °C (1425 °F) and 724 MPa (105 ksi)

Specimen	Life, Hours	Elongation Pct	R. A., Pct	Orientation, Degrees from: [001]	[011]
A	179	16	15	3	42
B	26	9	12	7	38.5
C	21	11	13	10	36
D	37	12	15	8	37.5
E	20	8	12	28	17
F	138	8	15	45	28
G	1242	11	12	52.5	34.5
H	1	24	28	39	7
I	0.5	35	31	43	6.5
J	14	11	11	8	39.5
K	26	13	13	9.5	36
L	41	12	10	5.5	40.5
M	53	13	13	10.5	35
N	53	13	12	5.5	41
O	59	10	9	2	43.5
P	74	12	10	4	42
Q	80	12	10	5.5	41
R	82	12	11	7.5	39
S	90	11	14	4	41.5
T	98	12	10	5	40
U	109	12	10	1.5	43.5
V	141	14	13	2	43

The creep curve of Specimen I shows that the crystal actually failed during first-stage creep. Thus, no transition to steady-state creep occurred prior to failure, even though the specimen had a primary creep strain and total elongation of over 30%.

Source: R.A. MacKay and R.D. Maier, "The Influence of Orientation on the Stress Rupture Properties of Nickel-Base Superalloy Single Crystals," Met. Trans. A, Vol. 13A, Oct. 1982, p 1748-1750

MAR-M247: Relationship Between Time to Failure and Time to Onset of Tertiary Creep

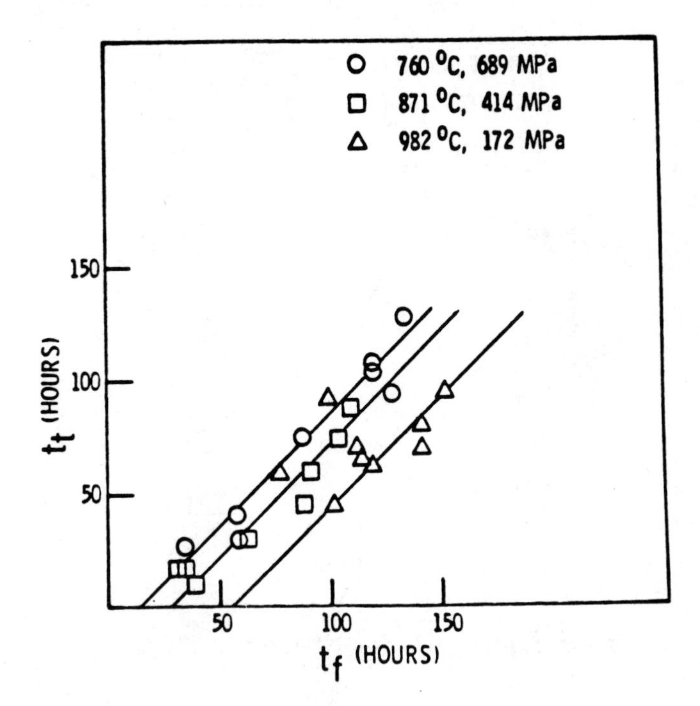

Relationship between time to failure (t_f) and time to onset of tertiary creep (t_t) for MAR-M247-type alloys.

Source: M.V. Nathal, R.D. Maier, and L.J. Ebert, "The Influence of Cobalt on the Tensile and Stress-Rupture Properties of Nickel-Base Superalloy MAR-M247," Met. Trans. A, Vol. 13A, Oct. 1982, p 1773

Ni-Cr-Nb-TiC Alloys

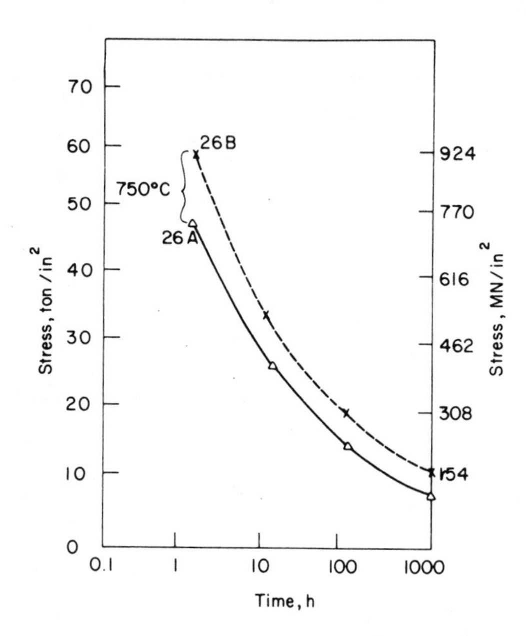

Stress-rupture life for Ni-Cr-Nb-TiC alloys.

Source: R.P.H. Fleming, "High Temperature Brazing and Infiltration of Some Nickel-Base Alloy Composites," Weld. Met. Fabr., Vol. 48, Sept. 1980, p 481

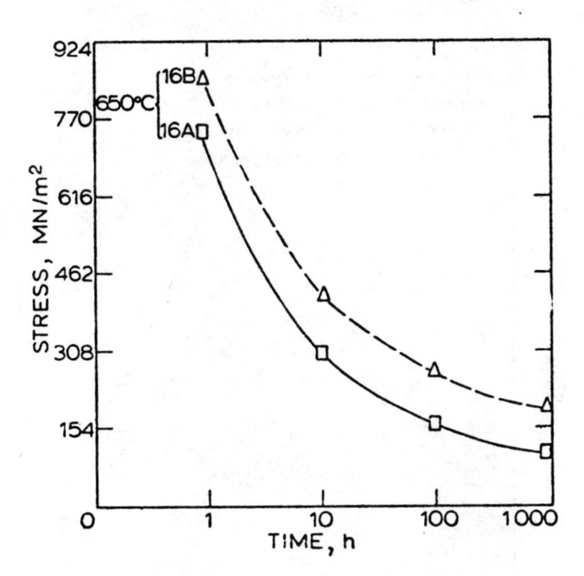

Stress-rupture life of Ni-Cr-Ta alloys 16A and 16B at 650 °C (1202 °F).

Source: R.P.H. Fleming, "Vacuum Sintering and Heat Treatment of Some Nickel Base Alloys Produced by Powder Metallurgy Techniques," Pulvermetallurgie, Vol. 28, Mar-June 1980, p 63

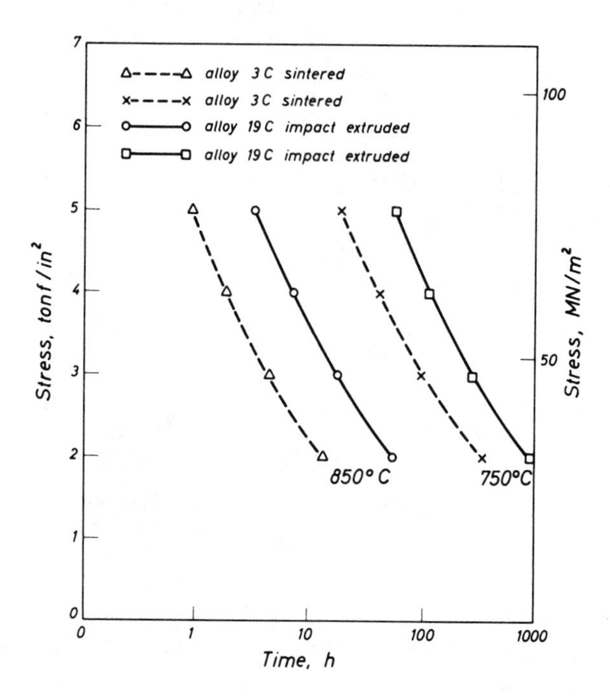

Stress-rupture life of Ni-TiC and Ni-Cr-TiC alloys 3C and 19C at 750 and 850 °C (1382 and 1562 °F) in the sintered and impact extruded conditions.

Source: R.P.H. Fleming, "Hot Impact Extrustion and Subsequent Processing of Some High Temperature Nickel Base Alloys," Pulvermetallurgie, Vol. 28, Mar-June 1980, p 80

Nimocast 738LC: Comparison of Normal, Fine-Grained, and HIPed Material

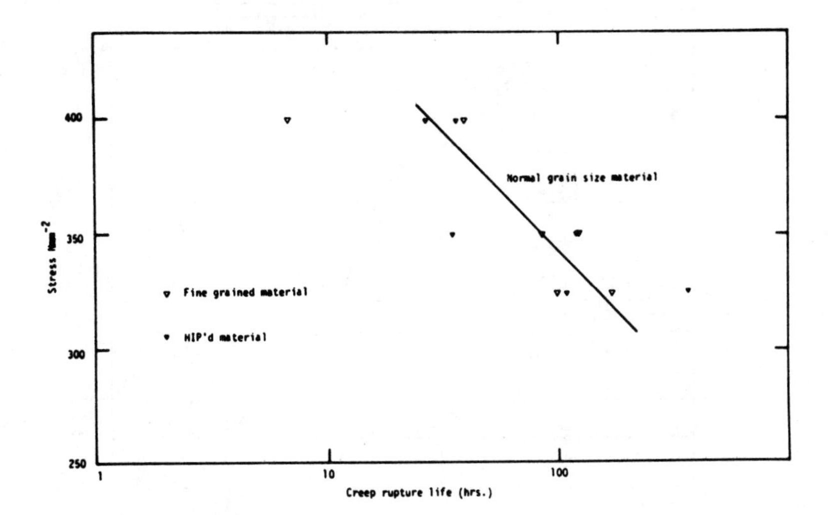

Comparison of the creep-rupture properties of normal, fine-grained, and HIPed Nimocast 738LC.

Creep-rupture properties at 850 °C (1562 °F) are compared above. There does not appear to be much difference between the fine-grained and HIPed material, although the scatter is marginally greater for the fine-grained material.

Source: G.M. McColvin, "High Cycle Fatigue of Nickel-Base Alloys," in High Temperature Alloys for Gas Turbines, D. Coutsouradis, P. Felix, H. Fischmeister, L. Habraken, Y. Lindblom, and M.O. Speidel, Eds., Applied Science, London, 1978, p 605, 625

Nimonic 80: Manson-Haferd Rupture and Creep Curves

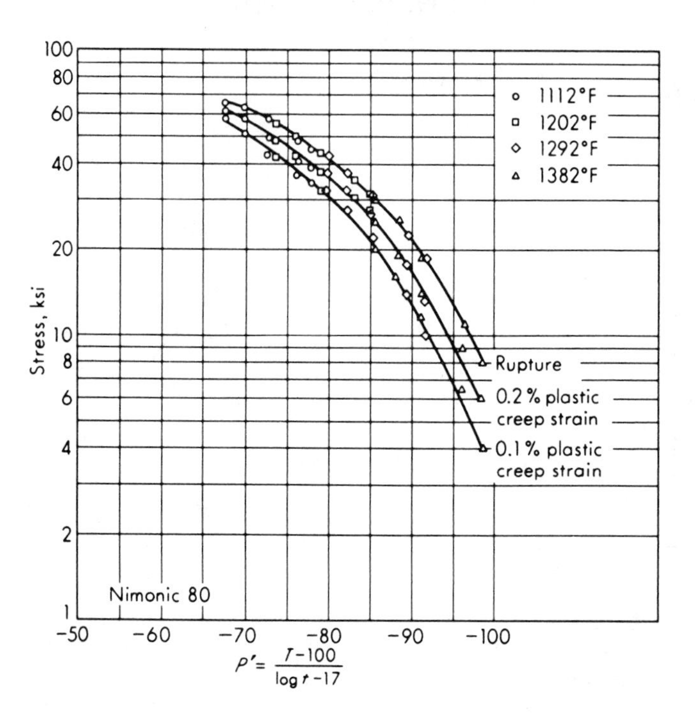

Manson-Haferd rupture and creep curves for Nimonic 80. *T* = temperature, °F; *t* = time to rupture for given creep strain, h.

Source: R.C. Juvinall, Stress, Strain, and Strength, McGraw-Hill, New York, 1967, p 423

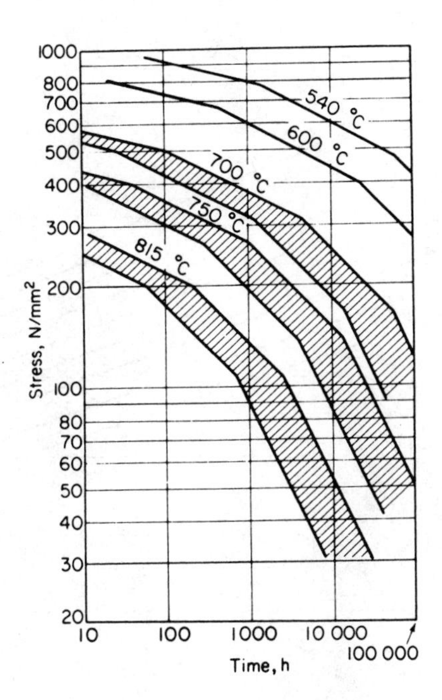

Stress-rupture properties of Nimonic 80A forged bar showing upper and lower 98% confidence limits. Heat treated 8 h at 1080 °C (1976 °F), air cooled + 16 h at 700 °C (1292 °F), air cooled.

Source: The Nimonic Alloys, W. Betteridge and J. Heslop, Eds., Crane, Russak & Co., New York, 1974, p 248

Nimonic 80A: Relaxation Characteristics of Hot Rolled Bar

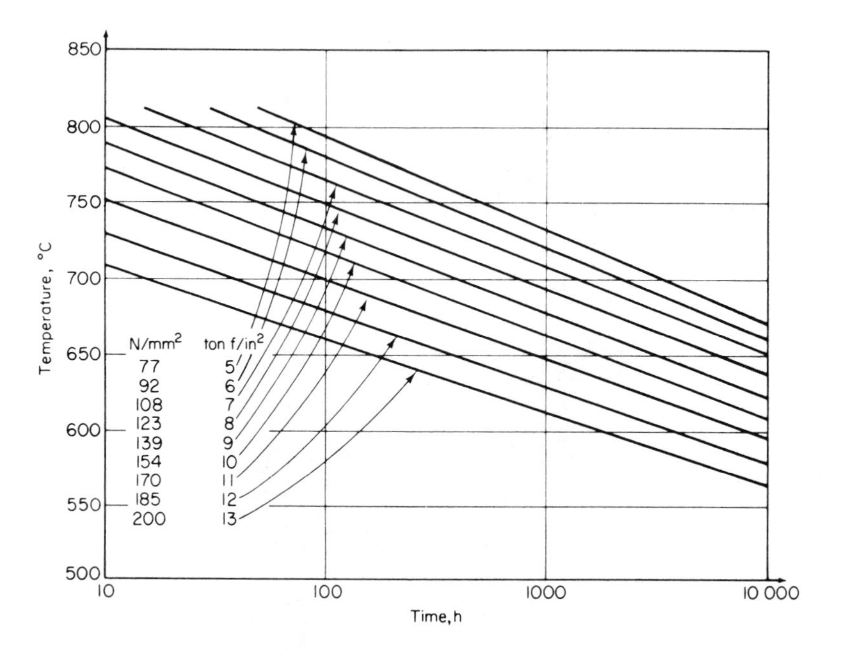

Mean relaxation characteristics of Nimonic 80A hot rolled bar. Heat treated 8 h at 1080 °C (1976 °F), air cooled + 16 h at 800 °C (1472 °F), air cooled.

Relaxation characteristics of Nimonic 80A for bolt applications were determined using an initial strain of 0.15%. Results are shown in the figure above, which details the relationship between temperature, time, and residual stress.

Source: The Nimonic Alloys, W. Betteridge and J. Heslop, Eds., Crane, Russak & Co., New York, 1974, p 266

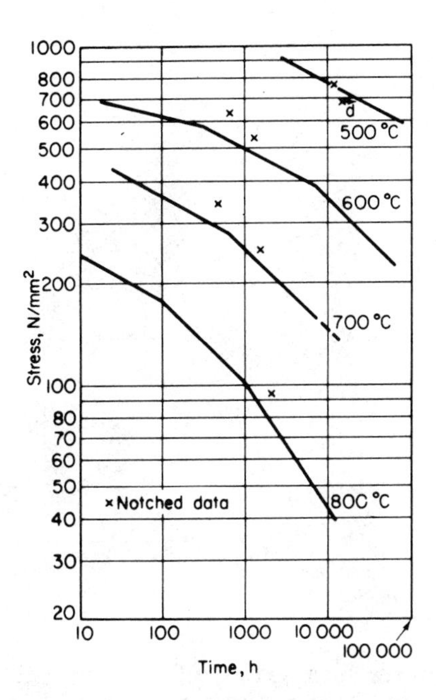

Representative stress-rupture properties of Nimonic 81 cold stretched bar, including notched data. Heat treated 4 h at 1100 °C (2012 °F), air cooled + 16 h at 700 °C (1292 °F), air cooled.

Source: The Nimonic Alloys, W. Betteridge and J. Heslop, Eds., Crane, Russak & Co., New York, 1974, p 250

Nimonic 90: Forged Bar

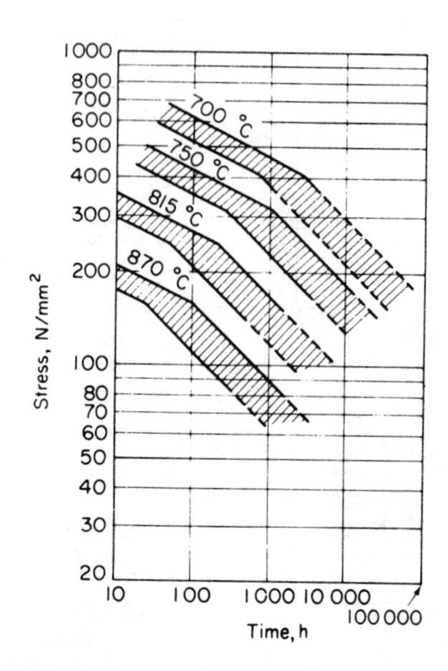

Stress-rupture properties of Nimonic 90 forged bar showing upper and lower 98% confidence limits. Heat treated 8 h at 1080 °C (1976 °F), air cooled + 16 h at 700 °C (1292 °F), air cooled.

Source: The Nimonic Alloys, W. Betteridge and J. Heslop, Eds., Crane, Russak & Co., New York, 1974, p 249

Nimonic 105: Effect of One Reheat Treatment

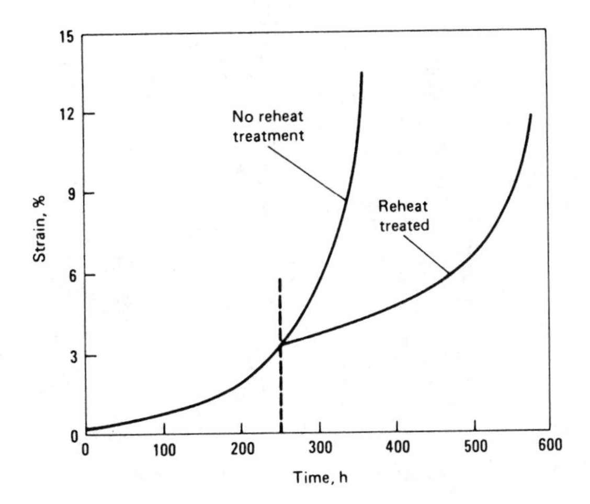

Effect of one reheat treatment on creep of Nimonic 105 at 870 °C (1600 °F). Curve at right is for specimen reheat treated after 250-h creep exposure.

Source: M.J. Donachie, Jr., "Overheating, Creep, and Alloy Stability (Appendix 1)," in Superalloys: Source Book, M.J. Donachie, Jr., Ed., American Society for Metals, Metals Park, OH, 1984, p 39

Nimonic 105: Forged Bar

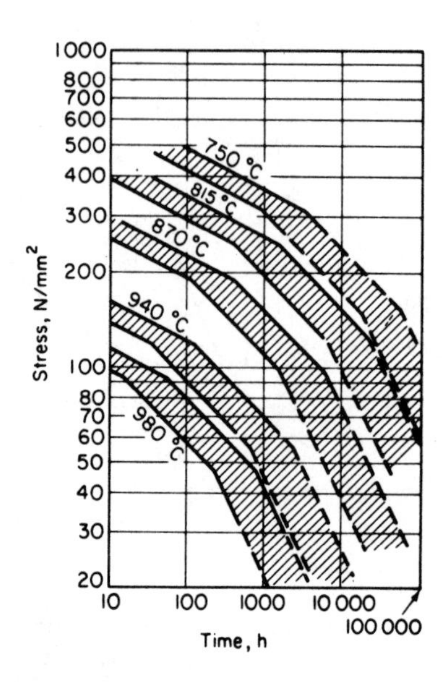

Stress-rupture properties of Nimonic 105 forged bar showing upper and lower 98% confidence limits. Heat treated 4 h at 1150 °C (2102 °F), air cooled + 16 h at 1050 °C (1922 °F), air cooled + 16 h at 850 °C (1562 °F), air cooled.

Source: The Nimonic Alloys, W. Betteridge and J. Heslop, Eds., Crane, Russak & Co., New York, 1974, p 249

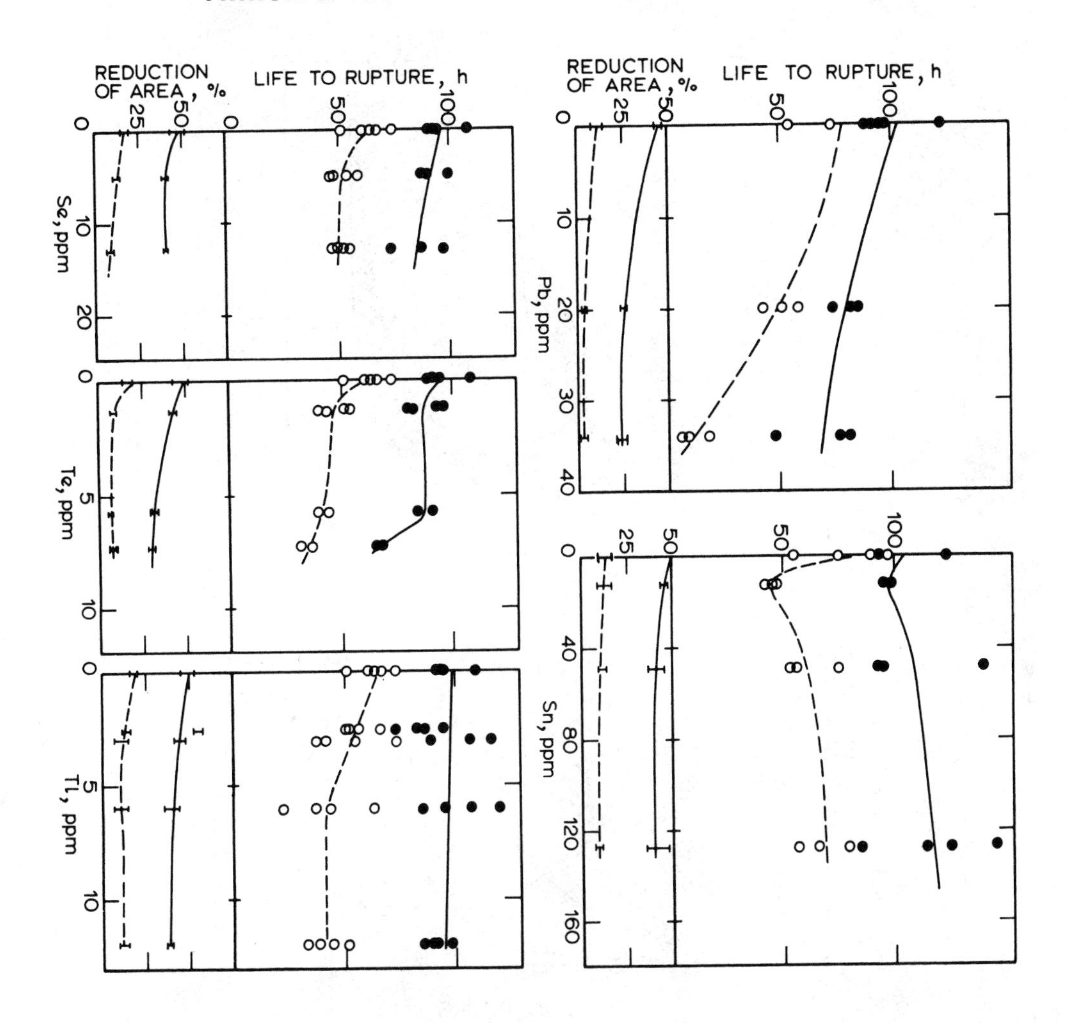

Stress-rupture properties of Nimonic 105 as a function of impurity content. Solid line represents plain specimens; broken line represents notched specimens.

Results of stress-rupture tests on plain and notched specimens at 815 °C (1499 °F) are shown above. In each case, there was some reduction in life to rupture as a consequence of the presence of the impurity element, although for Sn the effect was very slight and occurred only at 12 ppm, the lowest concentration of impurity examined. At the higher concentrations, Sn increased the life to rupture, but there was no increase in reduction of area, suggesting that the higher concentrations of Sn did have a strengthening effect under these conditions.

The two most detrimental elements were Pb and Te, which both reduced the life to rupture of plain specimens by about 30% at the maximum concentrations, 34 ppm Pb and 7.2 ppm Te. The effect of Te was marked only at the maximum concentration, and up to 5.4 ppm of this element had little influence on life to rupture. However, ductility was reduced progressively as the content of Te increased. Influence of Se and Tl on the properties of plain specimens was slight.

Source: G.B. Thomas and T.B. Gibbons, "Influence of Trace Elements on Creep and Stress-Rupture Properties of Nimonic 105," Met. Technol., Vol. 6, Mar. 1979, p 96-97

Nimonic 108: Effect Of Environment

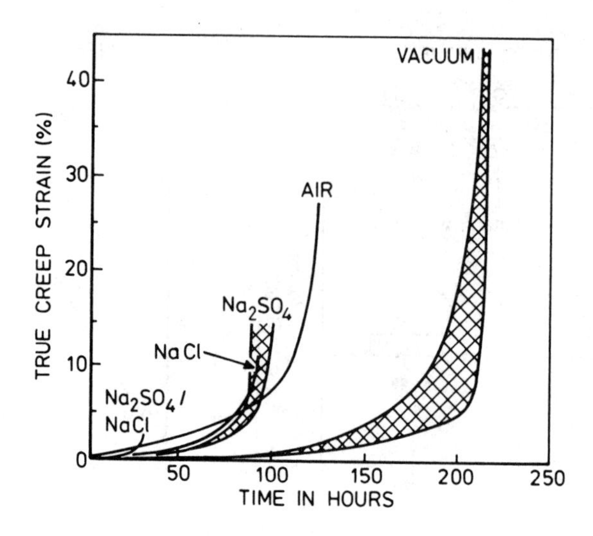

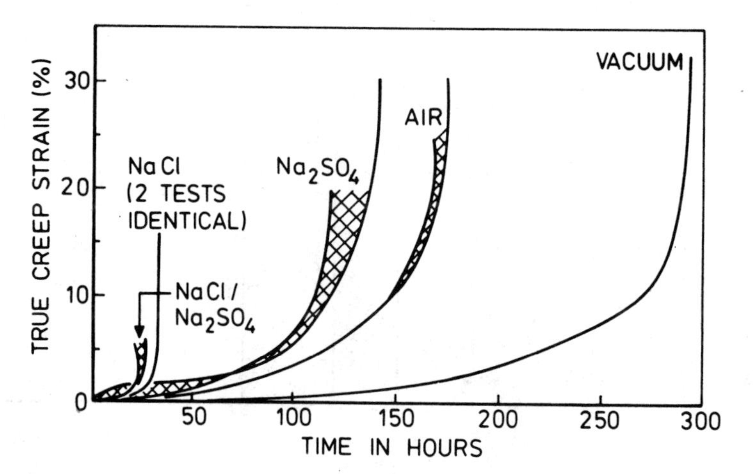

Creep curves for Nimonic 108 in various environments at 850 °C (1562 °F), 252 MN/m² (top) and 950 °C (1742 °F), 84 MN/m² (bottom).

The observed increase in creep rate with environment is also related to grain-boundary strength. At temperatures above 800 °C (1472 °F), both grain-boundary sliding and recrystallization play an increasing role in determining creep behavior of Nimonic 108. Therefore, weakening of the grain boundaries by corrosive environments will, in addition to modifying rupture properties, lead to enhanced creep rates through a greater component due to grain-boundary sliding.

Source: J.R. Nicholls, J. Samuel, R.C. Hurst, and P. Hancock, "The Influence of Hot Corrosion and Oxidation on the High Temperature Creep Behavior of 316 Stainless Steel and Nimonic 108," in Behavior of High Temperature Alloys in Aggressive Environments (Proceedings of the Petten International Conference, Petten (NH), The Netherlands), The Metals Society, London, 1980, p 916 and 923

Nimonic 115: Forged Bar

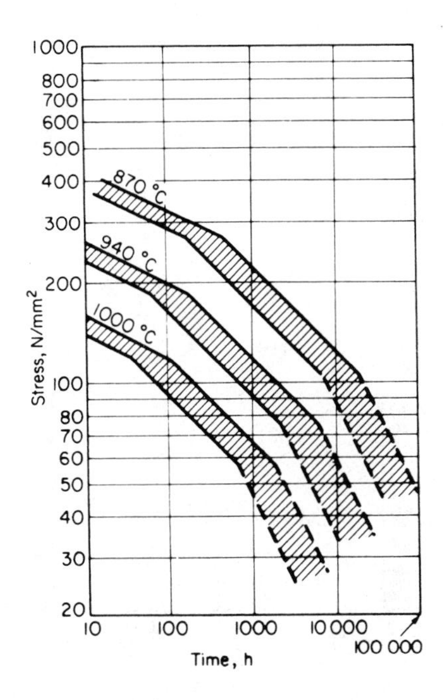

Stress-rupture properties of Nimonic 115 forged bar showing upper and lower 98% confidence limits. Heat treated 1½ h at 1190 °C (2174 °F), air cooled + 6 h at 1100 °C (2012 °F), air cooled.

Source: The Nimonic Alloys, W. Betteridge and J. Heslop, Eds., Crane, Russak & Co., New York, 1974, p 250

Nimonic 263: Extruded Section

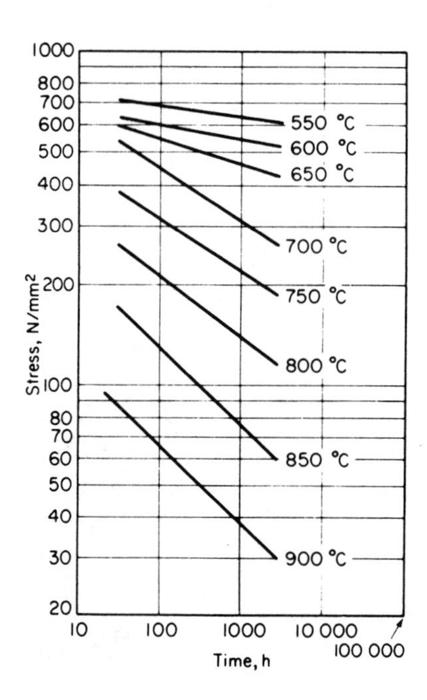

Representative stress-rupture properties of Nimonic 263 extruded section. Heat treated 2 h at 1150 °C (2102 °F), water quenched + 8 h at 800 °C (1472 °F), air cooled.

Source: The Nimonic Alloys, W. Betteridge and J. Heslop, Eds., Crane, Russak & Co., New York, 1974, p 252

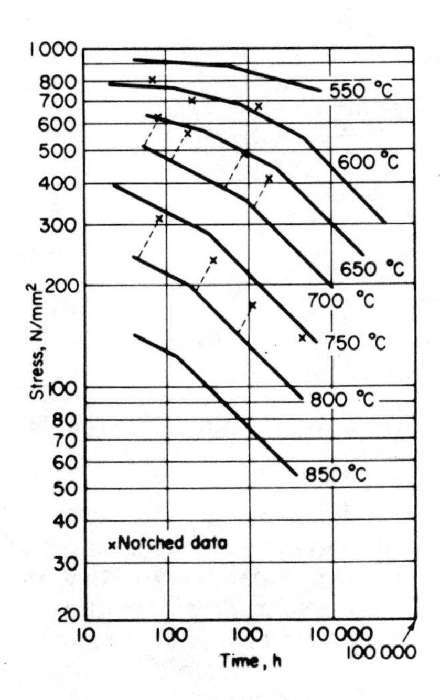

Representative stress-rupture properties of forged Nimonic 901, including notched data. Heat treated 2 h at 1090 °C (1994 °F), water quenched + 2 h at 775 °C (1427 °F), air cooled + 24 h at 720 °C (1328 °F), air cooled.

Source: The Nimonic Alloys, W. Betteridge and J. Heslop, Eds., Crane, Russak & Co., New York, 1974, p 252

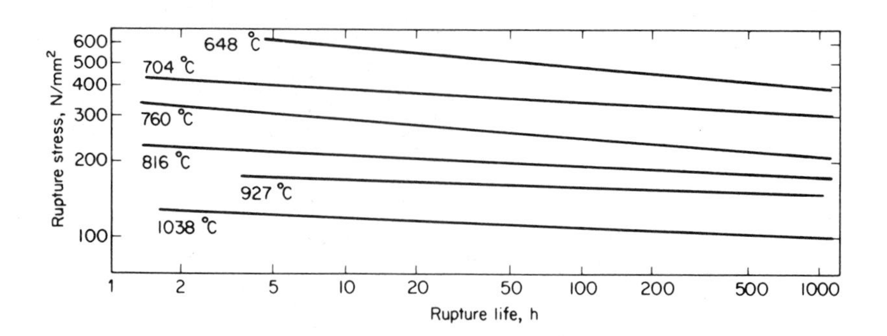

Rupture stress vs rupture life for Nimonic MA 753.

A very important characteristic of dispersion-strengthened superalloys is that the slopes of logarithmic plots of rupture stress against rupture life become smaller at higher temperatures (above about 815 °C for Nimonic MA 753) than at intermediate temperatures.

Source: The Nimonic Alloys, W. Betteridge and J. Heslop, Eds., Crane, Russak & Co., New York, 1974, p 338

Nimonic PE 16: Cold Stretched Bar

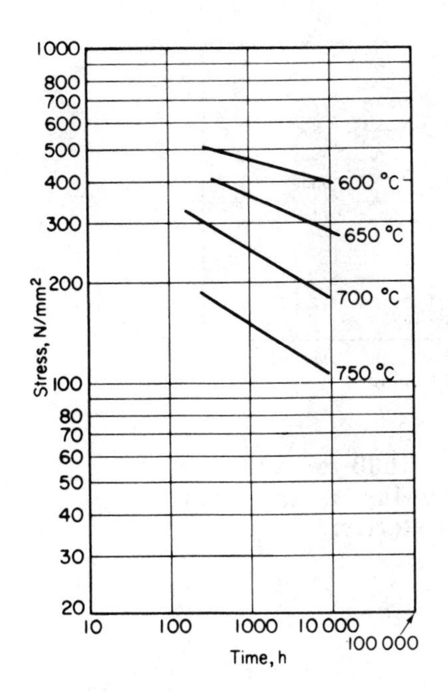

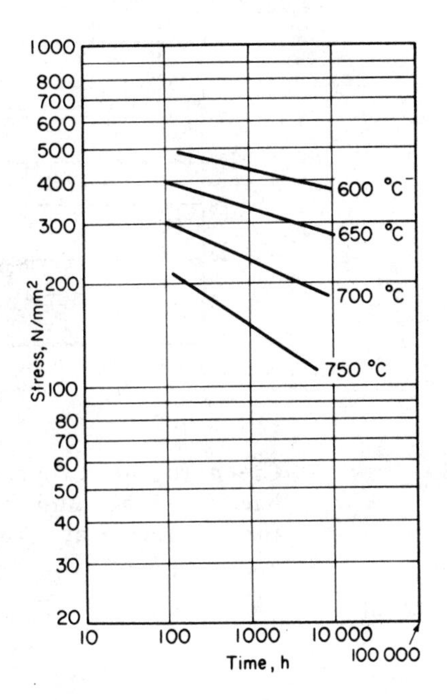

Representative stress-rupture properties of Nimonic PE 16 cold stretched bar. (Left) Heat treated 8 h at 1040 °C, air cooled + 1 h at about 900 °C, air cooled + 8 h at 750 °C, air cooled. (Right) Heat treated 2 h at 1040 °C, air cooled + 2 h at 800 °C, air cooled + 16 h at 700 °C, air cooled.

Source: The Nimonic Alloys, W. Betteridge and J. Heslop, Eds., Crane, Russak & Co., New York, 1974, p 250

Rene′ 41: Bar, Forgings, and Billet

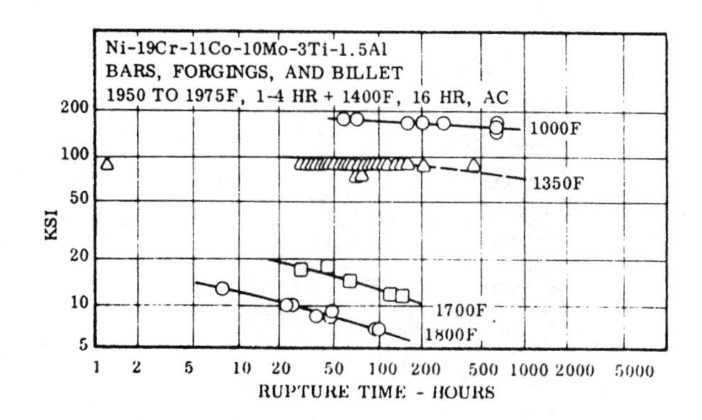

Creep-rupture properties from 1000 to 1800 °F of bar, forgings, and billet, showing a scatterband for lots tested at several laboratories.

Source: Structural Alloys Handbook, Volume 5, Mechanical Properties Data Center, Battelle Columbus Laboratories, revised 1978, p 4205-49

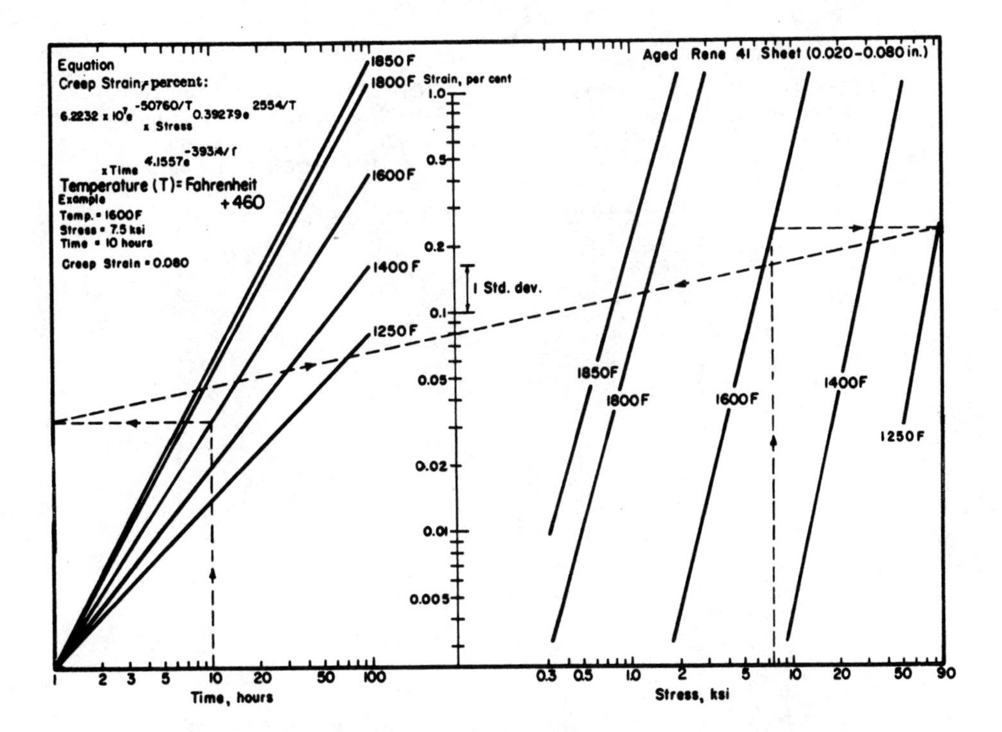

Creep properties of Rene′ 41 alloy sheet.

Source: Military Standardization Handbook MIL-HDBK-5D, Volume 1, Metallic Materials and Elements for Aerospace Vehicle Structures, U.S. Department of Defense, Washington, D.C., 1983, p 6-87

Rene′ 77: Effect of Heat Treatment

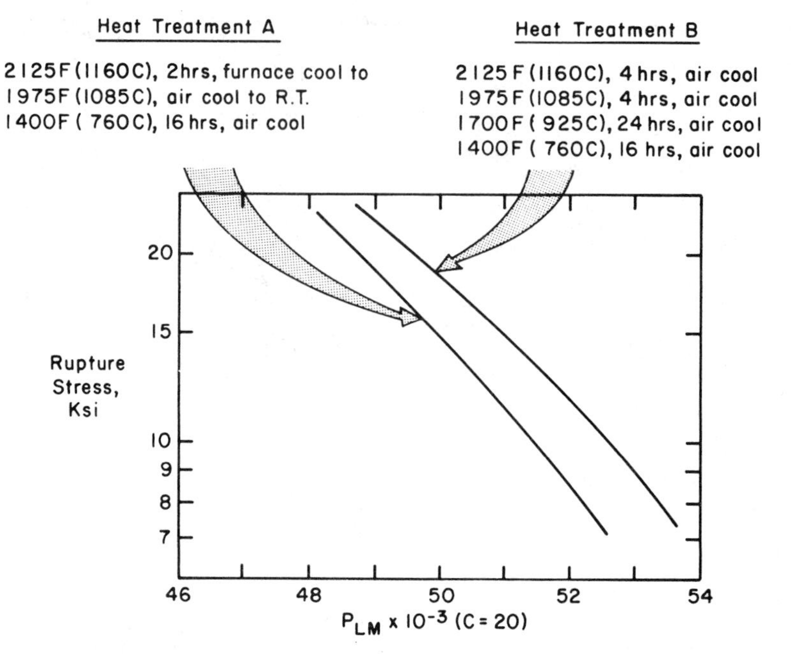

Effect of heat treatment on rupture properties of Rene′ 77.

Treatment B results in a significant improvement in rupture properties; γ′ was taken into solution at 1100 °C (2125 °F) by both treatments but began to precipitate at about 1140 °F (2075 °F). The slow cool of treatment A generates a few large γ′ nuclei at temperatures above 1085 °C (1975 °F). Considerable fine γ′ is then generated in the 16-h treatment at 760 °C (1400 °F). While treatment A results in good tensile properties, creep rupture is fair.

Treatment B, rapidly cooled from solution, gives an opportunity for nucleation of large γ′ particles but little growth. However, in 4 h at 1085 °C (1975 °F), γ′ does grow and a large number of medium-to-large particles nucleate homogeneously, using up much of the precipitation potential.

Source: R.F. Decker and C.T. Sims, "The Metallurgy of Nickel-Base Alloys," in The Superalloys, C.T. Sims and W. C. Hagel, Eds., John Wiley & Sons, New York, 1972, p 71-72

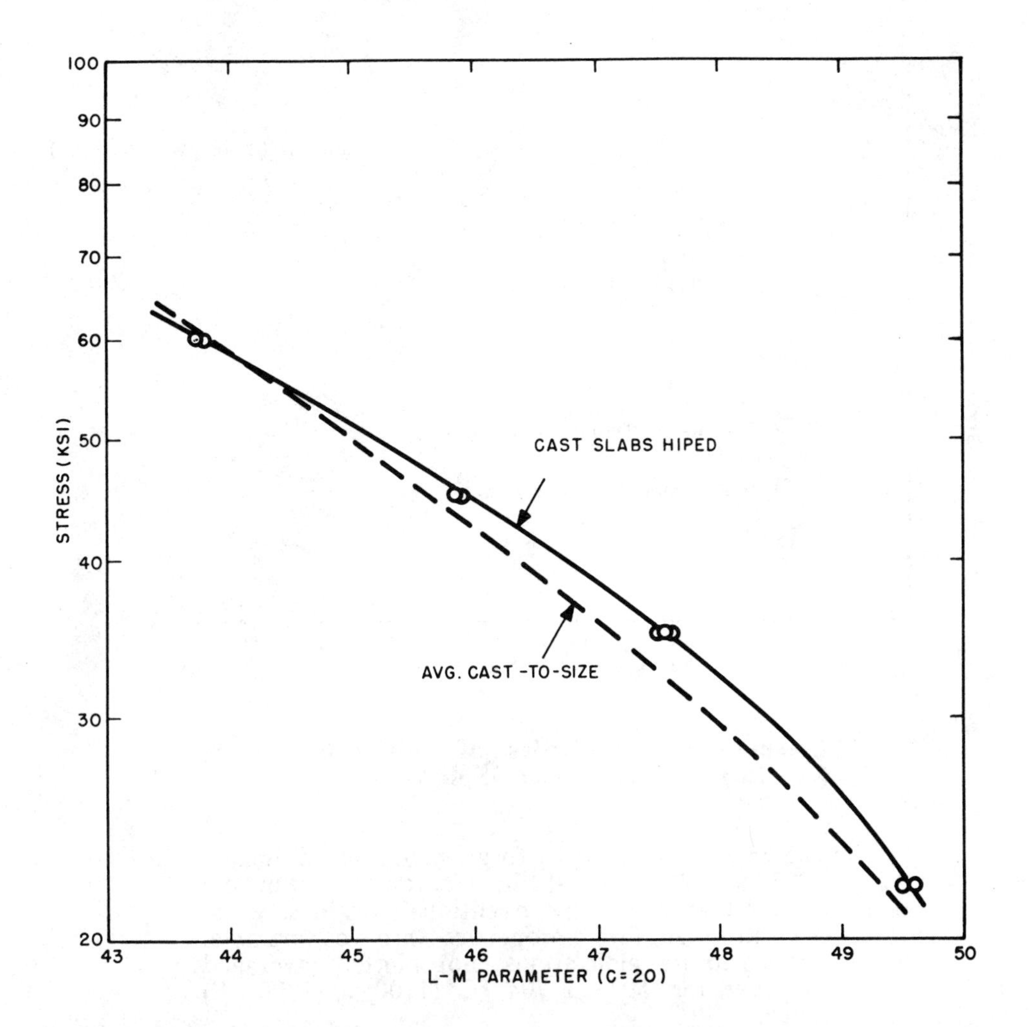

Comparison of HIPed Rene' 77 slabs and average cast-to-size-bar properties. Heat treated at about 1191 °C (2175 °F) for 4 h + 1079 °C (1975 °F) for 4 h + 927 °C (1700 °F) for 24 h + 760 °C (1400 °F) for 16 h.

There is a significant improvement in rupture life and ductility as a result of HIPing. Magnitude of the improvement is strictly a function of extent of porosity in the original casting. This figure compares rupture properties of HIPed material with those of cast-to-size bars. Producers have developed techniques for manufacturing sound cast-to-size test bars, so these bars should represent the optimum as-cast properties obtainable. A comparison of HIPed Rene' 77 slabs with these cast-to-size bars indicates some improvement in rupture properties as a result of HIP.

Source: G.E. Wasielewski and N.R. Lindblad, "Elimination of Casting Defects Using HIP," in Superalloys – Processing (Proceedings of the Second International Conference in Champion, PA), American Institute of Mining, Metallurgical, and Petroleum Engineers, 1972, Warrendale, PA, p D-9, D-21

Rene′ 95: P/M vs Conventionally Forged Disks

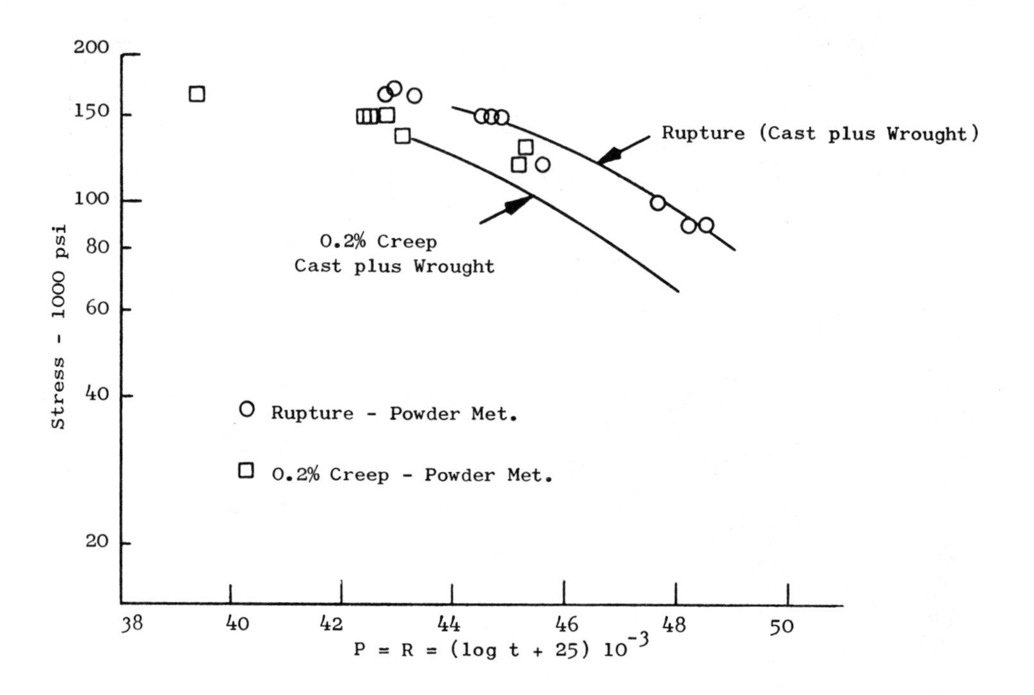

Creep-rupture properties of full-scale 6.4-cm (2.5-in.) thick rim Rene′ 95 disks.

The creep lives of P/M forgings are approximately 1 to 2 Larson-Miller parameter numbers greater than those of conventional forgings, while the stress-rupture properties are comparable. Stress-rupture elongations, not plotted, averaged 7% over the 593 to 704 °C (1100 to 1300 °F) range.

Source: J.F. Barker and E.H. VanDer Molen, "Effect of Processing Variables on Powder-Metallurgy Rene′ 95," in Superalloys – Processing (Proceedings of the Second International Conference, Champion, PA), American Institute of Mining, Metallurgical, and Petroleum Engineers, Warrendale, PA, 1972, p AA-6, AA-23

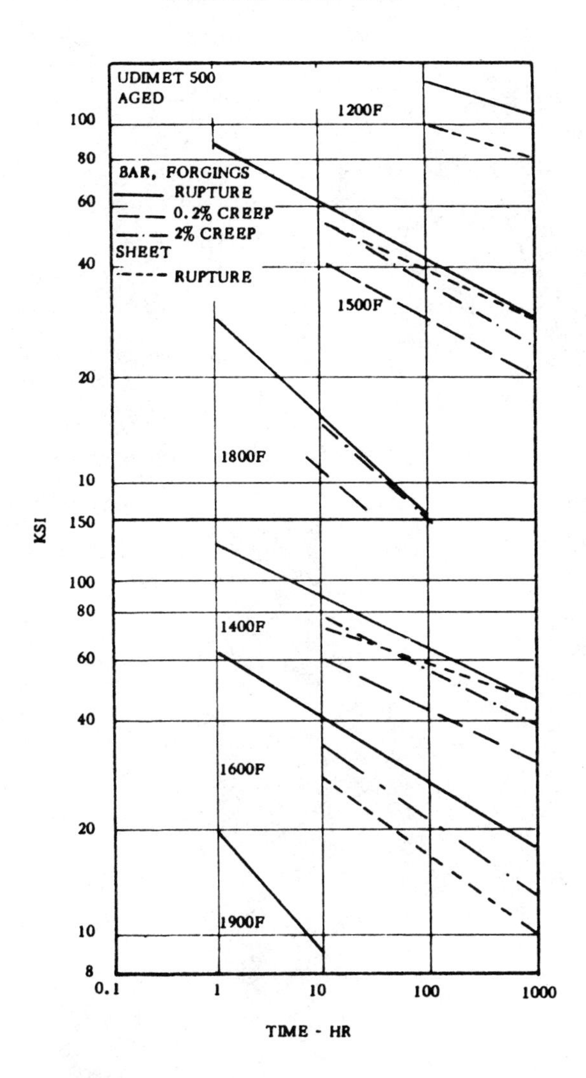

Creep and creep-rupture curves for bar and sheet at 1200 to 1900 °F.

Source: Structural Alloys Handbook, Volume 5, Mechanical Properties Data Center, Battelle Columbus Laboratories, revised 1978, p 4206-6

Udimet 500: Cast and Heat Treated

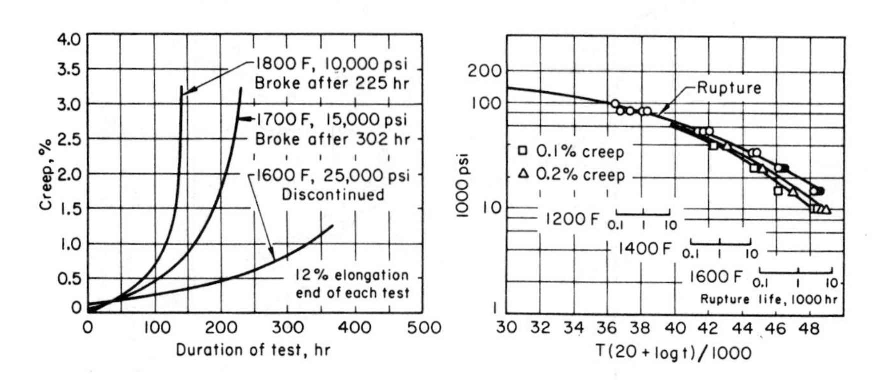

Cast and heat treated conditions.

Source: "Properties of Heat-Resistant Alloy Castings," in Source Book on Materials for Elevated-Temperature Use, E.F. Bradley, Ed., American Society for Metals, Metals Park, OH, 1979, p 235

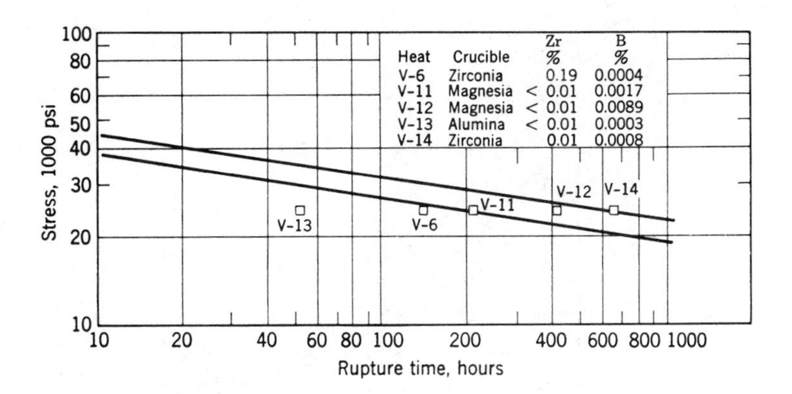

Average rupture life of experimental heats plotted on rupture band at 871 °C (1600 °F) of two heats of Udimet 500.

Reported by Utica Drop Forge & Tool Corporation. Treatment of experimental heats was 2 h at 1177 °C (2150 °F), followed by air cooling. Treatment of Utica heats was 2 h at 1177 °C (2150 °F), followed by air cooling plus 4 h at 1079 °C (1975 °F), and air cooling, plus 24 h at 843 °C (1550 °F), plus 16 h at 760 °C (1400 °F).

Source: R.F. Decker, J.P. Rowe, and J.W. Freeman, "Relations of High-Temperature Properties of a Nickel-Base Precipitation-Hardening Alloy to Contamination by Crucibles," in High Temperature Materials (Proceedings of a Conference in Cleveland, OH), R.F. Hehemann and G.M. Ault, Eds., John Wiley & Sons, New York, 1959, p 398

Udimet 500: Effect of Reduced Oxygen Content

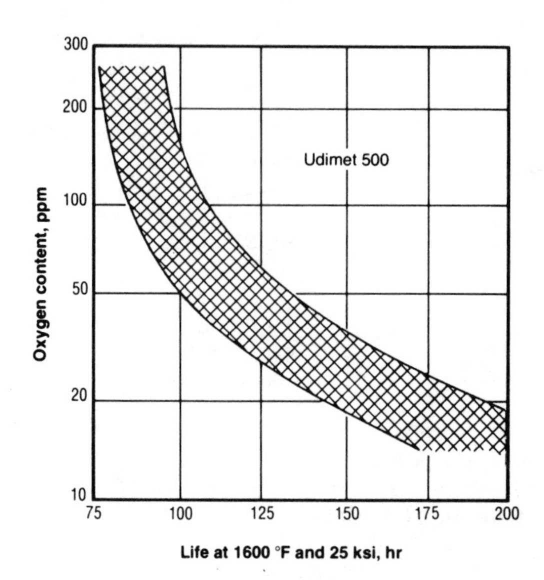

Improvement of rupture life at 1145 K (1600 °F) and 172 MPa (25 ksi) resulting from reduced oxygen content produced by vacuum melting.

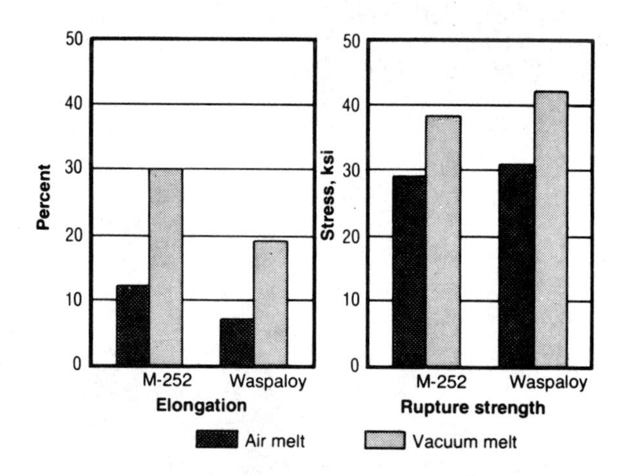

Effect of vacuum melting, incorporating beneficial modifications in chemical composition, on the properties of two nickel-base superalloys.

A number of superalloys, particularly those with a Co and Fe-Ni base, are air melted by various methods applicable to stainless steels. However, for most Ni- or Fe-Ni-base superalloys, vacuum induction melting (VIM) is required as the primary melting process. The use of VIM reduces interstitial gases (O_2, N_2) to low levels; enables higher and more controllable levels of Al and Ti (along with other relatively reactive elements) to be achieved; and results in less contamination from slag or dross formation than does air melting. The benefits of reduced gas content and ability to control Al plus Ti are shown above. Vacuum arc remelting and electroslag remelting are the commonly employed secondary melting techniques.

Source: M.J. Donachie, Jr., "Introduction to Superalloys," in Superalloys: Source Book, M.J. Donachie, Jr., Ed., American Society for Metals, Metals Park, OH, 1984, p 11

Udimet 700: Effect of Environment on Polycrystalline Material

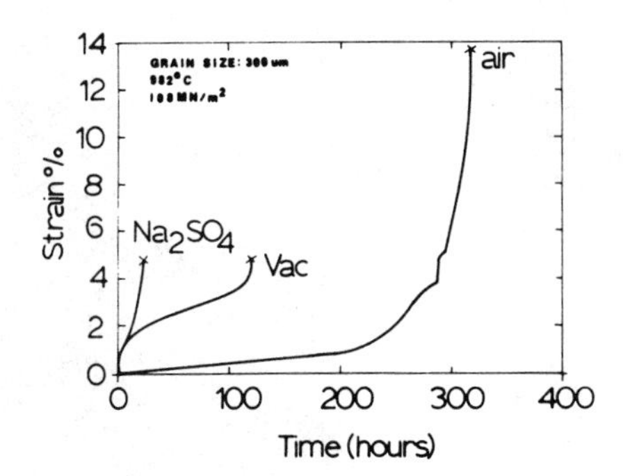

Comparison of creep curves in air, vacuum (10^{-5} torr), and air with sodium-sulfate deposition of polycrystalline Udimet 700, 300-μm grain size, at 760 °C (1400 °F) (left) and 982 °C (1800 °F) (right).

Source: J.M. Davidson, K. Aning, and J.K. Tien, "Hot Environment Effects on Alloy Mechanical Properties," in Properties of High Temperature Alloys, Z.A. Foroulis and F.S. Pettit, Eds., Electrochemical Society, Princeton, NJ, 1976, p 194-195

Udimet 700: Effect of Sigma Formation

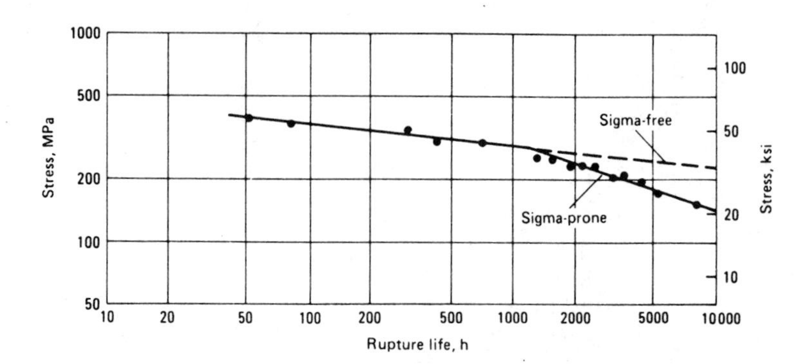

Log stress vs log rupture life at 815 °C (1500 °F) for Udimet 700.

This figure compares the strength of Udimet 700 with and without sigma (σ) formation. As σ forms, creep-rupture strength of the alloy is reduced. Furthermore, room-temperature ductility is greatly reduced and ductility at temperature may be reduced.

Source: F.R. Morral, "Wrought Superalloys," in Superalloys: Source Book, M.J. Donachie, Jr., Ed., American Society for Metals, Metals Park, OH, 1984, p 35, 38

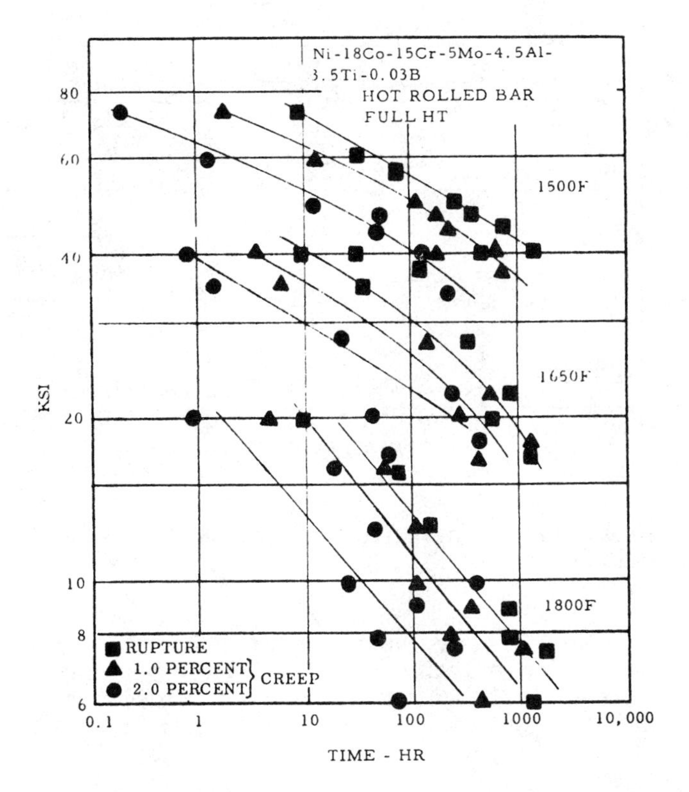

Creep and creep-rupture curves for hot rolled bar.

Source: Structural Alloys Handbook, Volume 5, Mechanical Properties Data Center, Battelle Columbus Laboratories, revised 1978, p 4207-13

Udimet 700: Wrought vs Cast

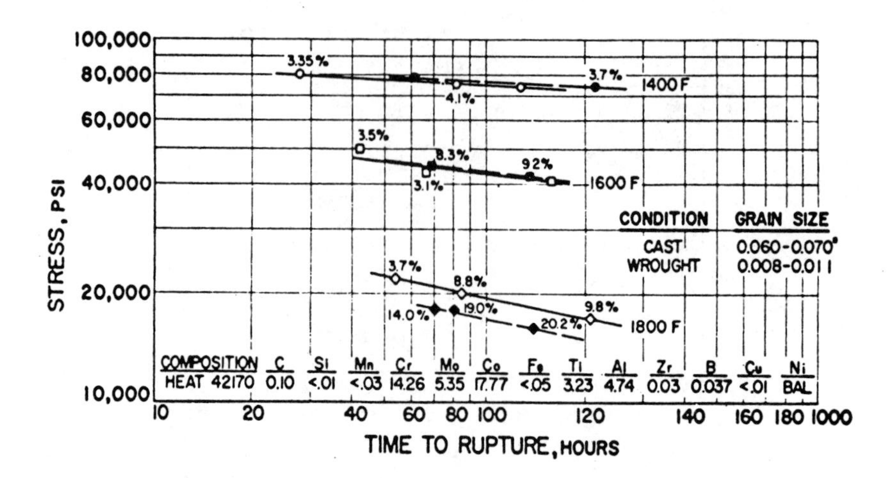

Stress-rupture properties of wrought and cast Udimet 700.

Note that the cast material has a coarser grain size, which probably enhances its creep strength at 1800 °F; hence, the comparison between cast and wrought conditions is not exactly on the same terms.

Source: C.P. Sullivan and M.J. Donachie, Jr., "Some Effects of Microstructure on the Mechanical Properties of Nickel-Base Superalloys," in Source Book on Materials for Elevated-Temperature Applications, E.F. Bradley, Ed., American Society for Metals, Metals Park, OH, 1979, p 254

Udimet 720: Effect of Environment on Creep and Fatigue Strength

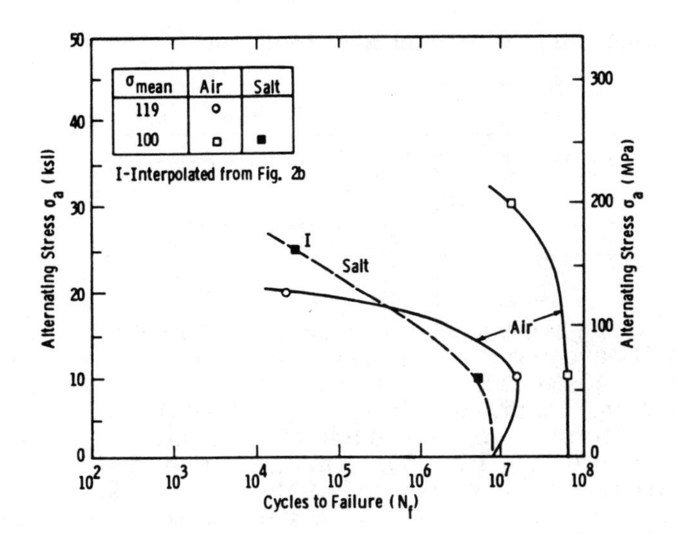

Creep and fatigue strength of Udimet 720 in air and salt under constant mean stress at 704 °C (1300 °F).

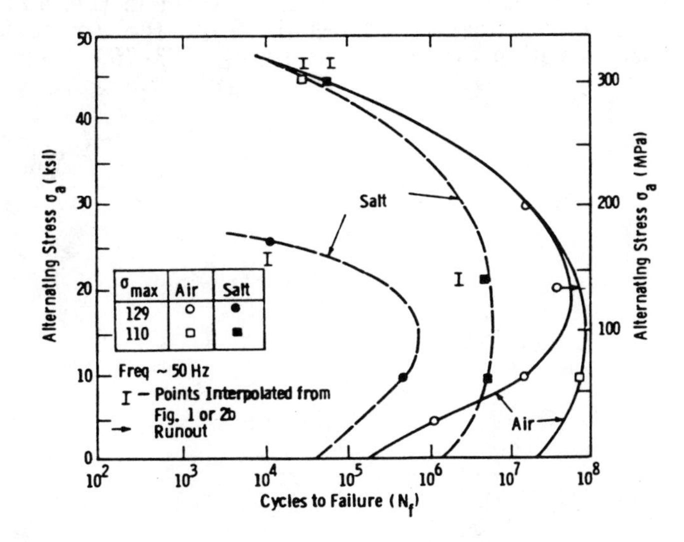

Effect of environment on creep and fatigue strength of Udimet 720 at 704 °C (1300 °F) and constant maximum stress.

Udimet 720: Effect of Environment on Creep and Fatigue Strength (Continued)

Chemistry (wt%) and heat treatment of Udimet 720

Ni	Cr	Co	Ti	Mo	Al	W	Zr	C	B
55	18	15	5	3	2.5	1.4	.04	.035	.035

- Solution 4h at 2135°F (1168°C), air cool (AC).
- Stabilize 4h at 1975°F (1079°C), AC.
- Age 24h at 1550°F (843°C), AC.
- Final age 16h at 1400°F (760°C), AC.

High-frequency, high-mean-load fatigue tests were conducted on wrought Udimet 720. Conventional, load-controlled, axial fatigue tests, in the frequency range 40-65 Hz, were carried out at 704 °C (1300 °F) using a 6.35-mm (0.25-in.) diam by 19.1-mm (0.75-in.) gage-length specimen. Testing was performed in two environments: air and molten salt. Salt environment was provided by a 60/40 mole fraction mixture of Na_2SO_4/$MgSO_4$, with a melting point of approximately 660 °C (1220 °F), containing 1 wt% NaCl. The salt was contained in a stainless steel cup integral with the bottom grip of the fatigue unit. The cup was filled to a level coincident with the midpoint of the specimen gage length, and at temperature, a thin film of liquid salt covered a small portion of the upper half of the gage length by capillary action. This simulates the deposition of such films on turbine air foils that occurs when these contaminants are present in fuel and air. At 704 °C (1300 °F), this environment produces the Type II form of hot corrosion attack, which occurs within the temperature range 593-760 °C (1100-1400 °F).

Source: J.M. Allen and G.A. Whitlow, "Observations on the Interaction of High Mean Stress and Type II Hot Corrosion on the Fatigue Behavior of a Nickel Base Superalloy," J. Eng. Gas Turbines Power (Trans. ASME), Vol. 107, Jan. 1985, p 220-222

Unitemp C-300: Effect of Extrusion Temperature on As-Extruded Material

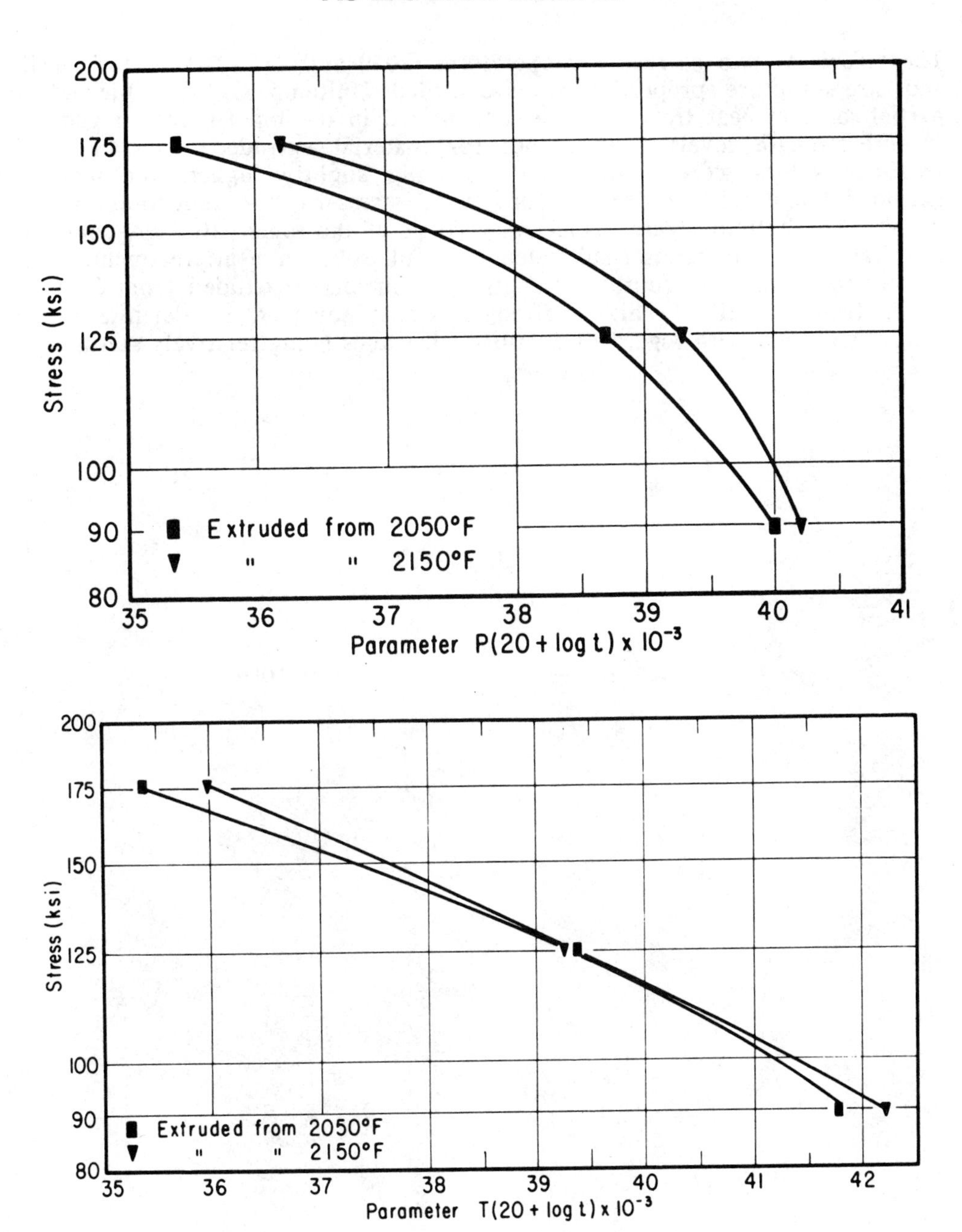

Effect of extrusion temperature on the stress-rupture properties of as-extruded Unitemp C-300 in the partial (top) and full (bottom) solution-heat-treated conditions.

Unitemp C-300: Effect of Extrusion Temperature on As-Extruded Material (Continued)

The effects of two extrusion temperatures (2050 and 2150 °F) on the tensile and stress-rupture properties of as-extruded Unitemp C-300 subjected to partial solution heat treatment are summarized in the top figure. In general, rupture-strength levels were higher for material extruded from 2150 °F. Ductility values were similar, but generally slightly higher, for material extruded from 2150 °F, except for several stress-rupture reduction in area values. The bottom figure depicts the effects of the same extrusion-temperature variations on material subjected to a full solution heat treatment. With respect to tensile and rupture strengths, the material extruded from 2150 °F again displayed slightly higher strength levels; however, no definite trends were established with regard to ductility, all values being relatively similar.

Source: W.B. Kent, "The Effect of Thermomechanical Processing Variations on the Mechanical Properties and Structural Characteristics of Unitemp C-300," in Superalloys – Processing (Proceedings of the Second International Conference, Champion, PA), Warrendale, PA, 1972, p M-4, M-13, M-15

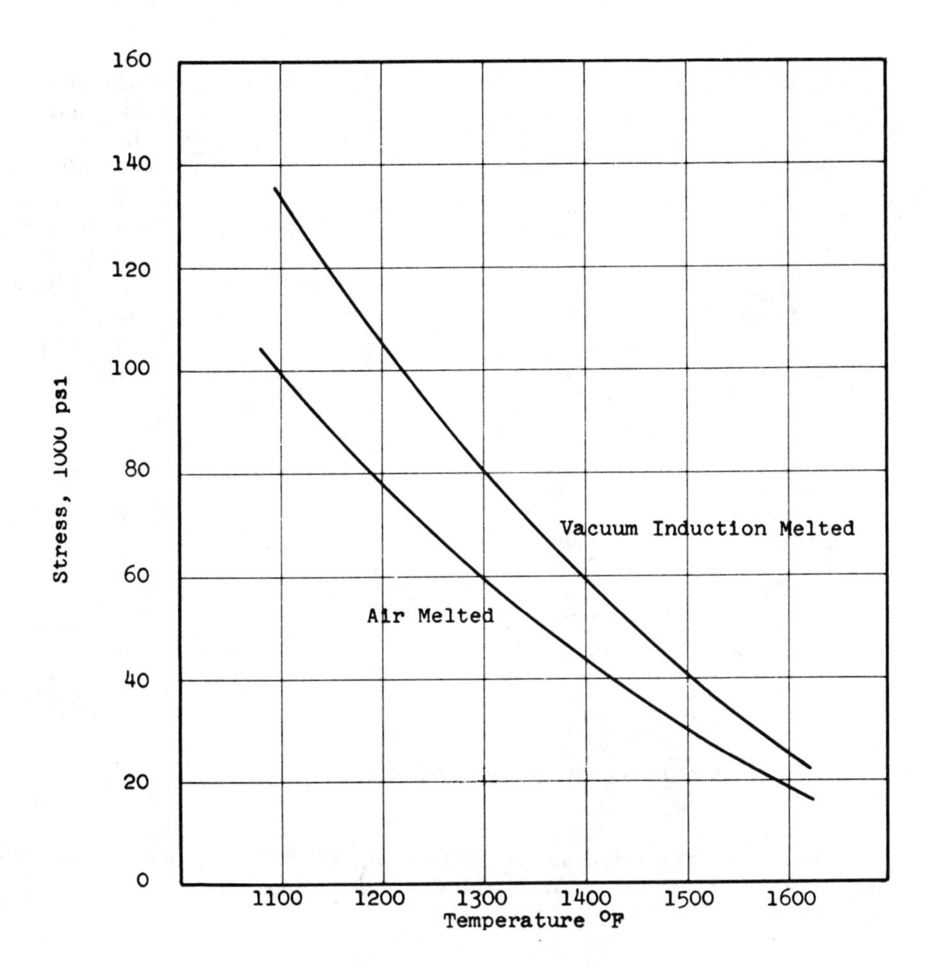

One-hundred-hour rupture strength of air-melted and vacuum-induction-melted Waspaloy tested at various temperatures.

At 816 °C (1500 °F), vacuum-induction-melted material can be used at a stress level approximately 10,000 psi higher than the corresponding level for air-melted material. In this case, the improvement is due not only to the vacuum melting process itself, but also to the higher Ti and Al contents that it permits.

Source: A.M. Aksoy, "Quality Aspects and Properties of Vacuum-Melted Super-Duty Steels," in Quality Requirements of Super-Duty Steels, R.W. Lindsay, Ed., Interscience, New York, 1959, p 282

Waspaloy: Effect of Notches

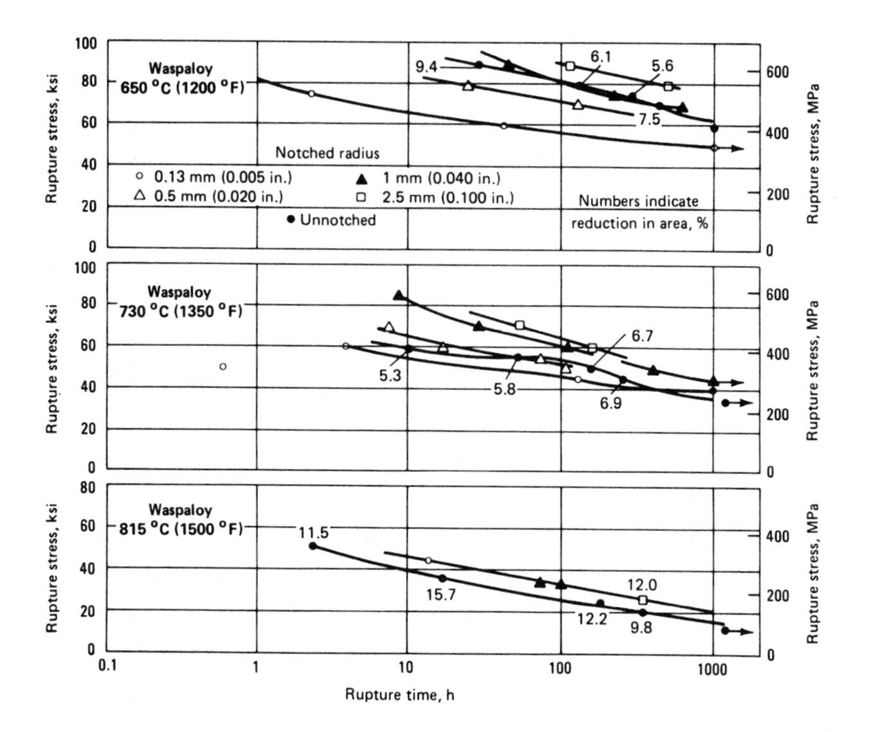

Effect of notches on the rupture life of Waspaloy.

Waspaloy exhibited notch weakening at 650 and 730 °C (1200 and 1350 °F) and notch strengthening at 815 °C (1500 °F).

Source: Metals Handbook, Ninth Edition, Volume 3, Properties and Selection: Stainless Steels, Tool Materials and Special-Purpose Metals, American Society for Metals, Metals Park, OH, 1980, p 233

Waspaloy: Forged Bar

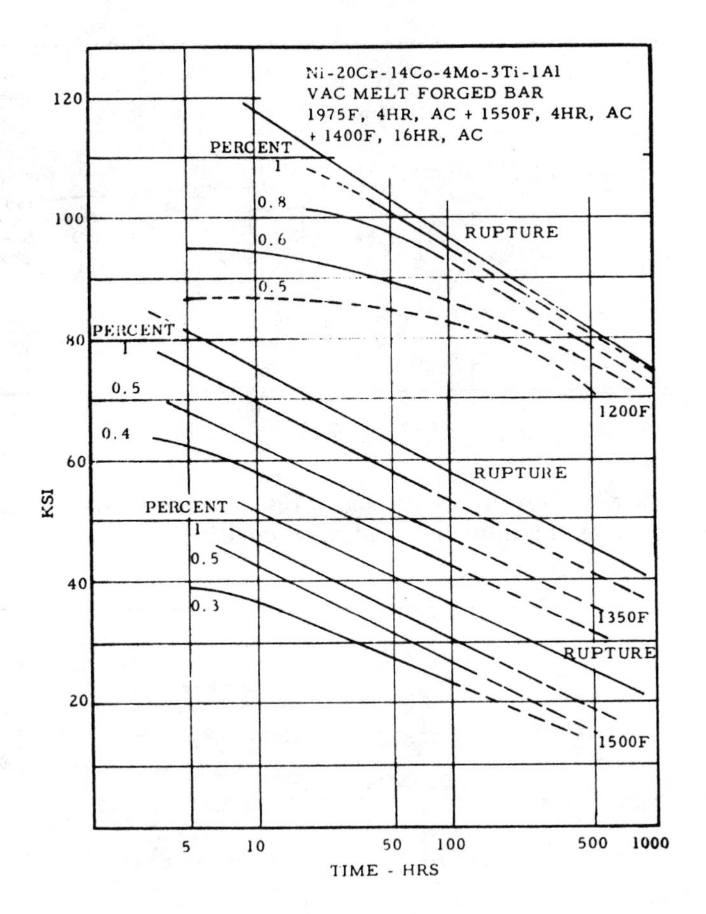

Design curves based on minimum values for creep and rupture of forged bar from 1200 to 1500 °F.

Source: Structural Alloys Handbook, Volume 5, Mechanical Properties Data Center, Battelle Columbus Laboratories, revised 1978, p 4208-14

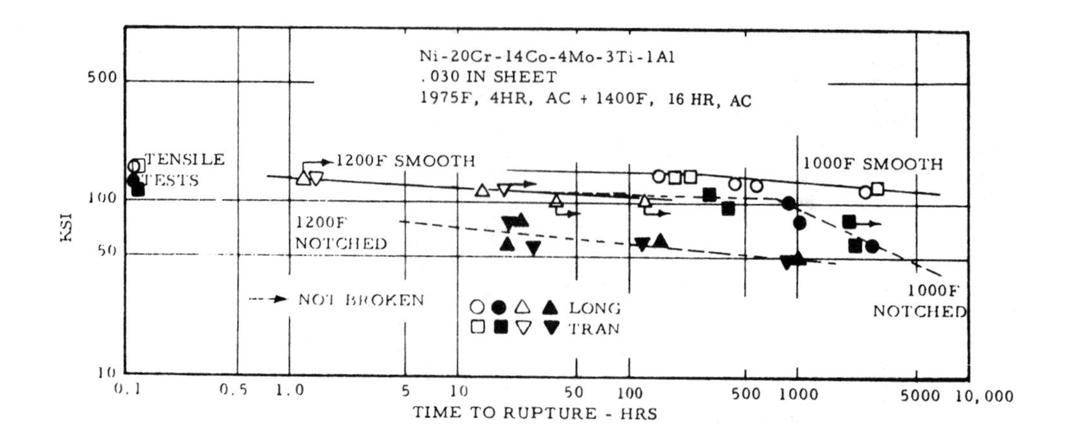

Creep-rupture curves at 1000 and 1200 °F for smooth and sharp notched sheet in the annealed and aged condition.

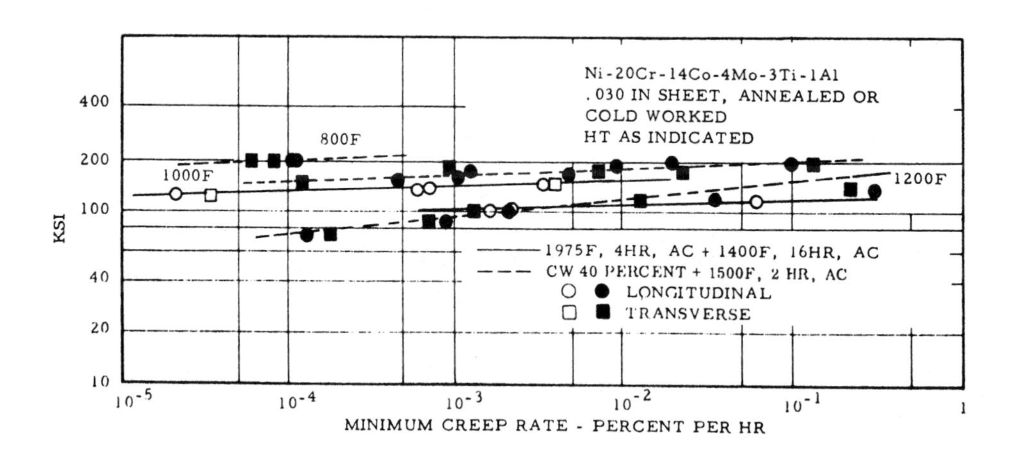

Stress vs minimum creep rate for sheet in the annealed plus aged and in the cold worked and aged condition at 800 to 1200 °F.

Source: Structural Alloys Handbook, Volume 5, Mechanical Properties Data Center, Battelle Columbus Laboratories, revised 1978, p 4208-12, 13

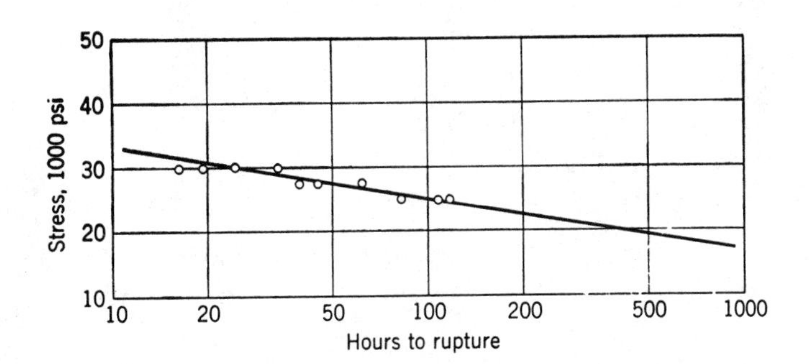

Stress-rupture properties at 871 °C (1600 °F) of vacuum-investment-cast Waspaloy.

Waspaloy begins to lose strength rapidly at 871 °C (1600 °F). The stress for 100-h life at 816 °C (1500 °F) is well above 30,000 psi.

Source: P.W. Beamer and J.J. Eisenhauer, "Evaluation of Vacuum-Melted Vacuum-Investment-Cast Nickel-Base Alloys," in High-Temperature Materials (Proceedings of the Conference in Cleveland, OH), R.F. Hehemann and G.M. Ault, Eds., Wiley & Sons, New York, 1959, p 371

WAZ-20: Comparison With ODS Alloy WAZ-D

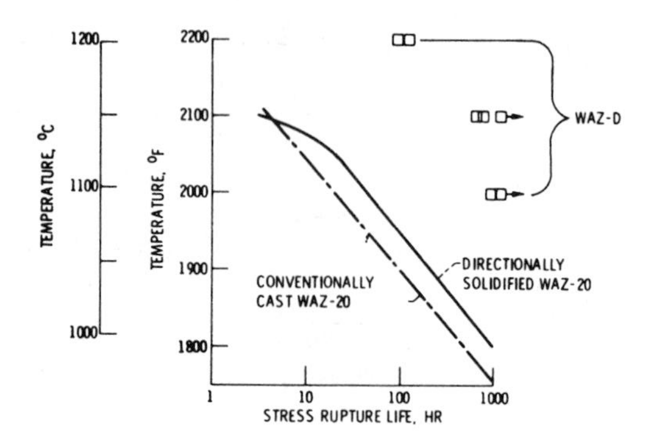

Stress-rupture lives of WAZ-D and WAZ-20.

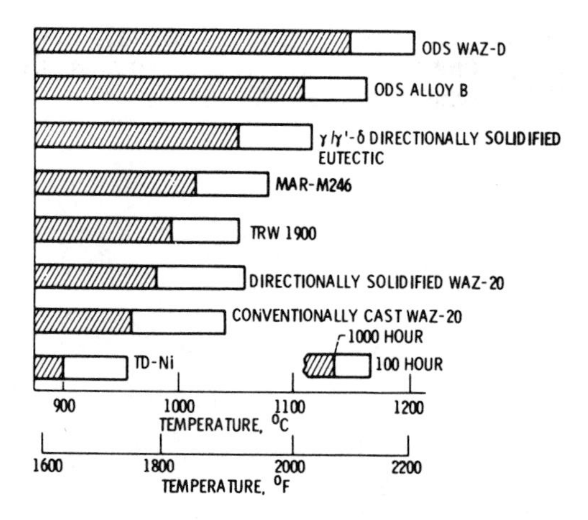

Temperature to produce rupture at 103 MPa (15 ksi) in 100 and 1000 h for advanced and conventional alloys.

The large, elongated-grain, dispersion-strengthened alloy WAZ-D provides a considerable stress-rupture-life advantage over the comparable cast alloy. Since comparison was made to the cast alloy of even larger and more elongated grains prepared by directional solidification, most of the advantage may be ascribed to the hard-phase dispersion. A comparison at 103 MPa (15 ksi) and 1150 °C (2100 °F) indicates a rupture-life improvement of two orders of magnitude, from 5 h for the cast material to approximately 1000 h for WAZ-D. Alternatively, this may be considered as a use-temperature advantage of 130 °C (234 °F) over the strongest conventional cast alloys, or 90 °C (194 °F) over the γ-γ'-δ directionally solidified eutectic, itself a candidate material for advanced gas turbine blades.

Source: T.K. Glasgow, "An Oxide Dispersion Strengthened Ni-W-Al Alloy with Superior High Temperature Strength," in Superalloys: Metallurgy and Manufacture (Proceedings of the Third International Symposium, Seven Springs, PA), B.H. Kear, D.R. Muzyka, J.K. Tien, and S.T. Wlodek, Eds., Claitor's, Baton Rouge, 1976, p 388, 394

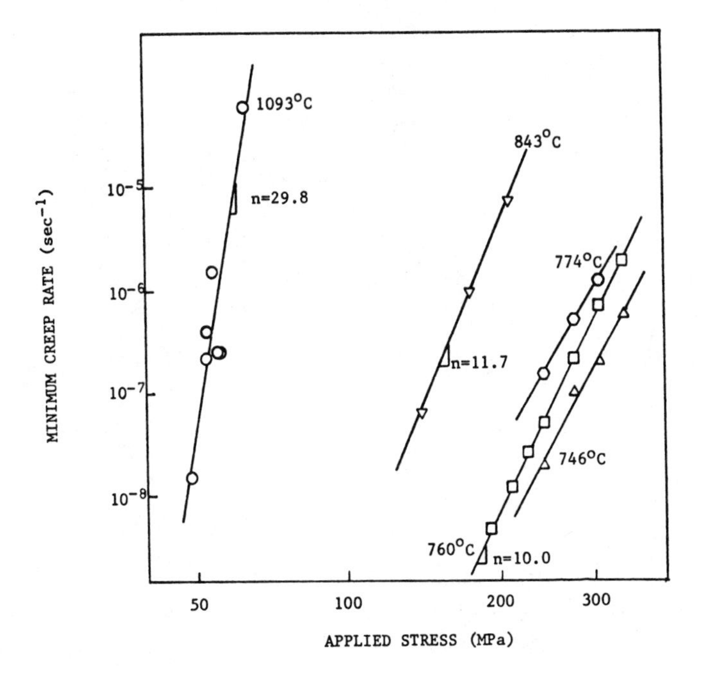

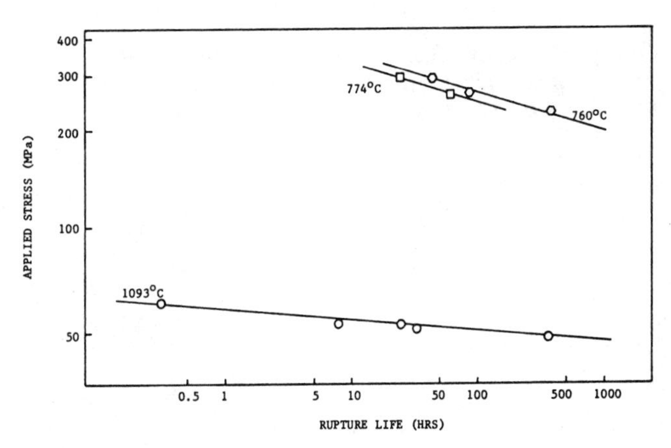

Applied stress vs minimum creep rate (top) and rupture life (bottom) for YDNiCrAl.

Although testing was limited, it appears that the 100-h rupture life is ≈275.8 MPa (≈40 ksi) at 760 °C (1400 °F) and ≈51 MPa (≈7.4 ksi) at 1093 °C (1999 °F). Extrapolation to 1000 h suggests that the rupture stress would be 214.8 MPa (31.2 ksi) at 760 °C (1400 °F) and 48 MPa (7.0 ksi) at 1093 °C (1999 °F).

Source: M.E. McAlarney, R.M. Arons, T.E. Howson, J.K. Tien, and S. Baranow, "Creep and Rupture of an ODS Alloy with High Stress Rupture Ductility," Met. Trans. A, Vol. 13A, Aug. 1982, p 1458

5.100 Comparison of Effect of Directional Solidification on the Rupture Ductility of Nickel-Base Superalloys

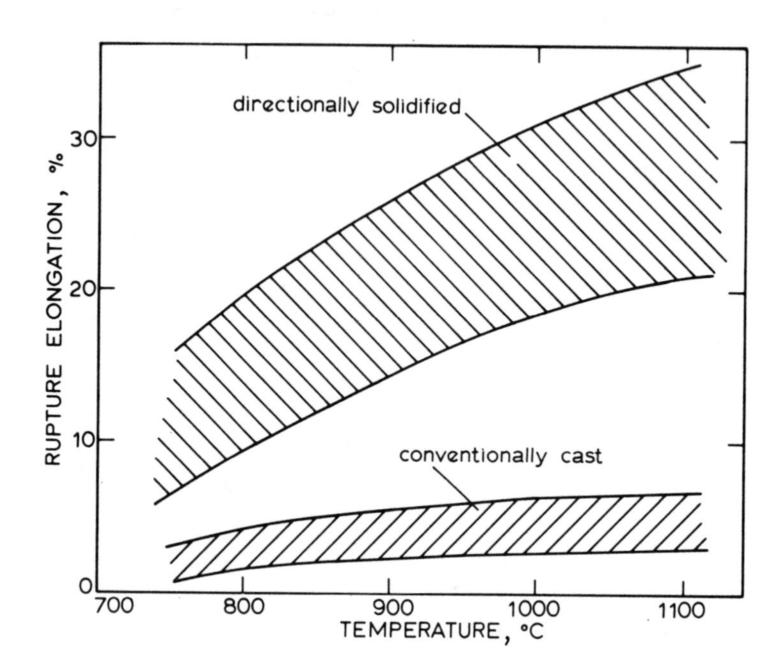

Comparison of the rupture ductilities of several nickel-base superalloys in the conventionally cast and directionally solidified conditions.

Source: M. McLean, Directionally Solidified Materials for High Temperature Service, The Metals Society, London, 1983, p 155

Comparison of Effects of Rapid Solidification Processing on the Creep Strength of Selected Nickel-Base Superalloys

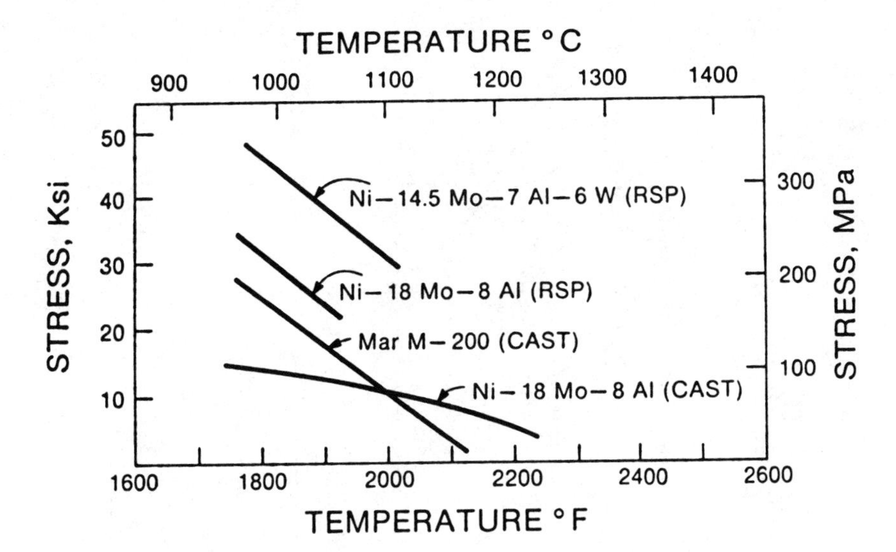

Creep strength comparisons of several nickel-base superalloys (1% creep in 100 h), showing some effects of rapid solidification processing.

Source: M. Cohen, "Rapid Solidification Processing and the Control of Structure/Property Relationships," in Specialty Steels and Hard Materials (Proceedings of the International Conference on Recent Developments in Specialty Steels and Hard Materials, Pretoria), N.R. Comins and J.B. Clark, Eds., Pergamon, Oxford, 1983, p 19

5.102 Comparison of Hot Corrosion Degradation on Udimet 700, Alloy X-750, Udimet 500, and IN 738

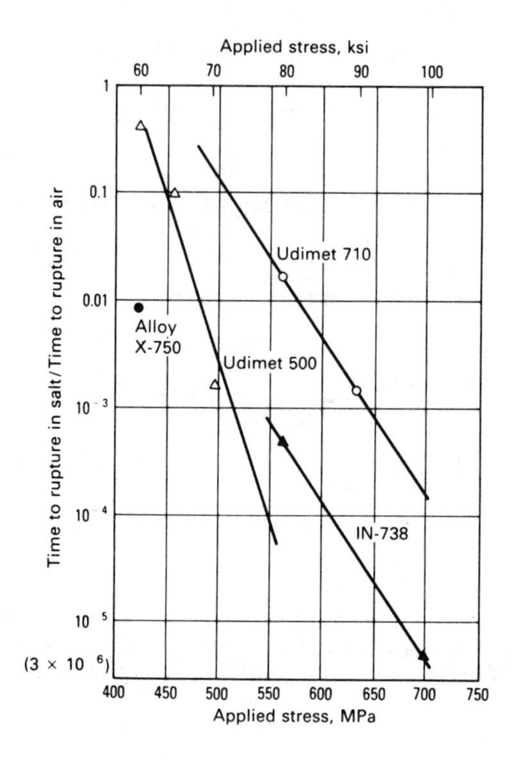

Degradation in rupture life occurring as a result of hot corrosion for four nickel-base superalloys.

The figure above, based on laboratory tests, depicts for several alloys the variation of the ratio of time to rupture in a salt environment to the time to rupture in air with applied stress at a temperature of 705 °C (1300 °F). The pronounced degradation of stress-rupture life (sometimes by a factor of as much as 10^5) due to the corrosive action of the salt mixture is evident.

Source: Metals Handbook, Ninth Edition, Volume 13, Corrosion, American Society for Metals, Metals Park, OH, 1987, p 1001

Comparison of Nimonic 80A, M-252, and Udimet 500: Effect of Increasing Intermediate Aging

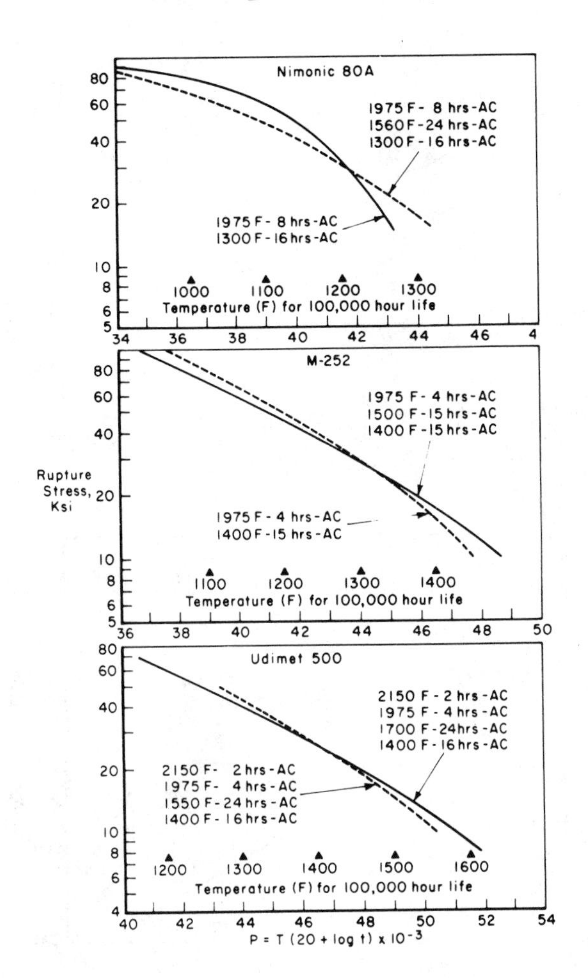

Effect of increasing intermediate aging on rupture properties of wrought superalloys.

Early heat treatments for wrought alloys such as M-252 and Nimonic 80A consisted principally of only a high-temperature solution followed by a low-temperature age. This generated good tensile and short-time rupture properties, but did not stabilize the structure sufficiently to produce optimized long-time rupture properties. Heat treatments for Nimonic 80A and M-252 were studied to correct this failing. An additional intermediate-temperature age was added to each respective heat treatment to drive the MC degeneration reaction forward; grain boundaries of coarse particulate $M_{23}C_6$ engloved in a layer of γ' developed. The results are shown above. In each case, rupture strength at low stresses and high P_{LM} (time-temperature parameter) was increased. In effect, the alloy structure was "stabilized." This philosophy was then applied successfully to Udimet 500. Because this alloy already underwent a four-part heat treatment, the effect was obtained by increasing the temperature of the 845 °C (1500 °F) age to 925 °C (1700 °F) to optimize the carbide degeneration reaction.

Source: R.F. Decker and C.T. Sims, "The Metallurgy of Nickel-Base Alloys," in The Superalloys, C.T. Sims and W.C. Hagel, Eds., John Wiley & Sons, New York, 1972, p 69-70

Comparison of Udimet 500, 710, and 720: Air and Salt Environments

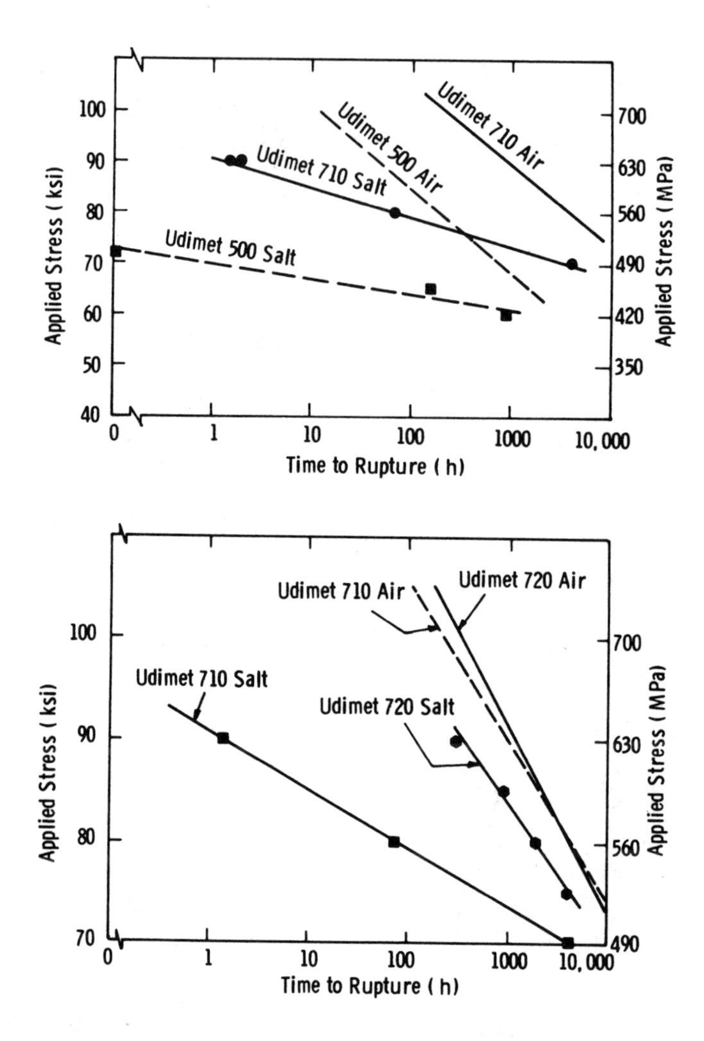

Comparison of the rupture strength at 704 °C (1300 °F) for Udimet 710 and Udimet 500 (top) and Udimet 710 and Udimet 720 (bottom) in air and salt.

All alloys were evaluated in the fully heat-treated condition as specified by the manufacturer. Conventional stress-rupture tests were performed in both air and in the corrosive salt environment at 704 °C (1300 °F). In the latter, a stainless steel cup welded to the bottom grip of the creep unit was filled with the powdered salt mixture prior to the test so that, at 704 °C (1300 °F), the melting of the mixture resulted in a liquid level at the midpoint of the specimen 25.4 by 6.35 mm (1.0 by 0.25 in.) gage length. A eutectic salt mixture consisting of 63 wt% Na_2SO_4 and 36 wt% $MgSO_4$ with an addition of 1 wt% NaCl having a melting point of 660 °C (1200 °F) was used with an air environment above the melt surface.

Measured stress-rupture data were supplemented by optical and scanning metallography and electron microprobe analyses of the failed specimens.

Source: G.A. Whitlow, C.G. Beck, R. Viswanathan, and E.A. Crombie, "The Effects of a Liquid Sulfate/Chloride Environment on Superalloy Stress Rupture Properties at 1300 °F (704 °C)," Metallurgical Transactions A, Vol. 15A, Jan 1984, p 24

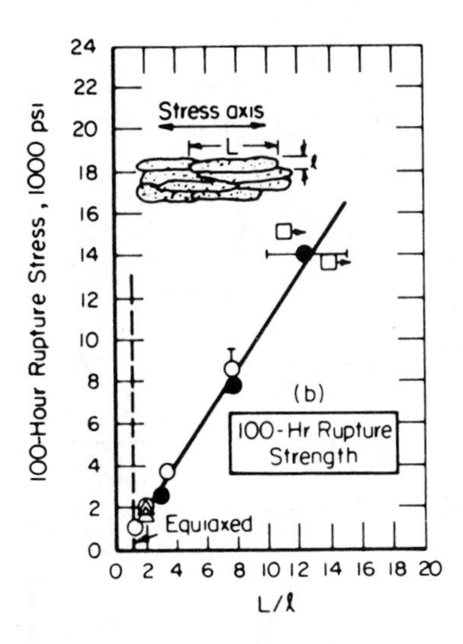

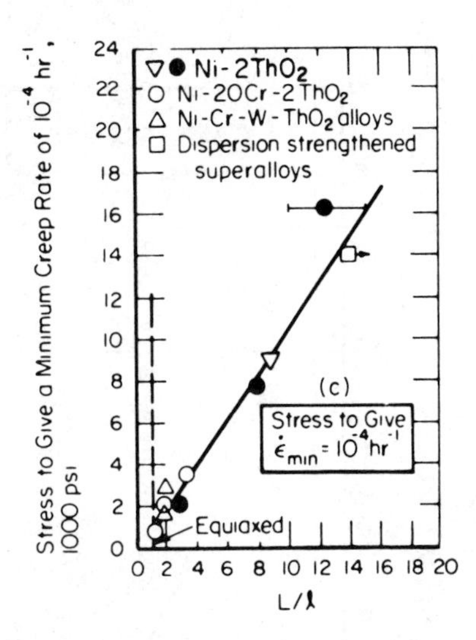

Effect of GAR L/l in dispersion-strengthened nickel alloys on 100-h rupture strength (left) and stress to a minimum creep rate of 10^{-4} h^{-1} (right) at 1100 °C (2000 °F).

Values of the GAR coefficient, K, and equiaxed grain strength, σ_e, for thoriated Ni alloys deformed at 1100 °C (2000 °F) in tension and creep

Material	0.2% Yield Strength		Stress for 100-hr rupture life		Stress to give a minimum creep rate of 10^{-4} hr^{-1}	
	K (psi)	σ_e (psi)	K (psi)	σ_e (psi)	K (psi)	σ_e (psi)
Ni-2ThO$_2$	1200	3000	1100	1000	1100	1000
Dispersion-strengthened Ni-20Cr-base alloys	~2800	2000-4000	1100	1000	1100	1000

The influence of GAR on high-temperature strength can be formulated as follows: The yield strength or creep strength σ is given by:

$$\sigma = \sigma_e + K\,(L/l - 1)$$

where σ_e is the strength of $L/l = 1$ (equiaxed grains) and K is defined as the GAR coefficient. Values of σ_e and K are listed in the table above for results plotted in the figures. For creep conditions, the matrix composition does not influence the GAR effect because points for Ni-Cr-ThO$_2$, Ni-Cr-W-ThO$_2$, and dispersion-strengthened superalloys fall on the same plot as Ni-2ThO$_2$.

Source: B.A. Wilcox and A.H. Clauer, "Dispersion Strengthening," in The Superalloys, C.T. Sims and W.C. Hagel, Eds., John Wiley & Sons, New York, p 212, 214

Influence of Composition on the Creep Strength of Nickel-Base Superalloys

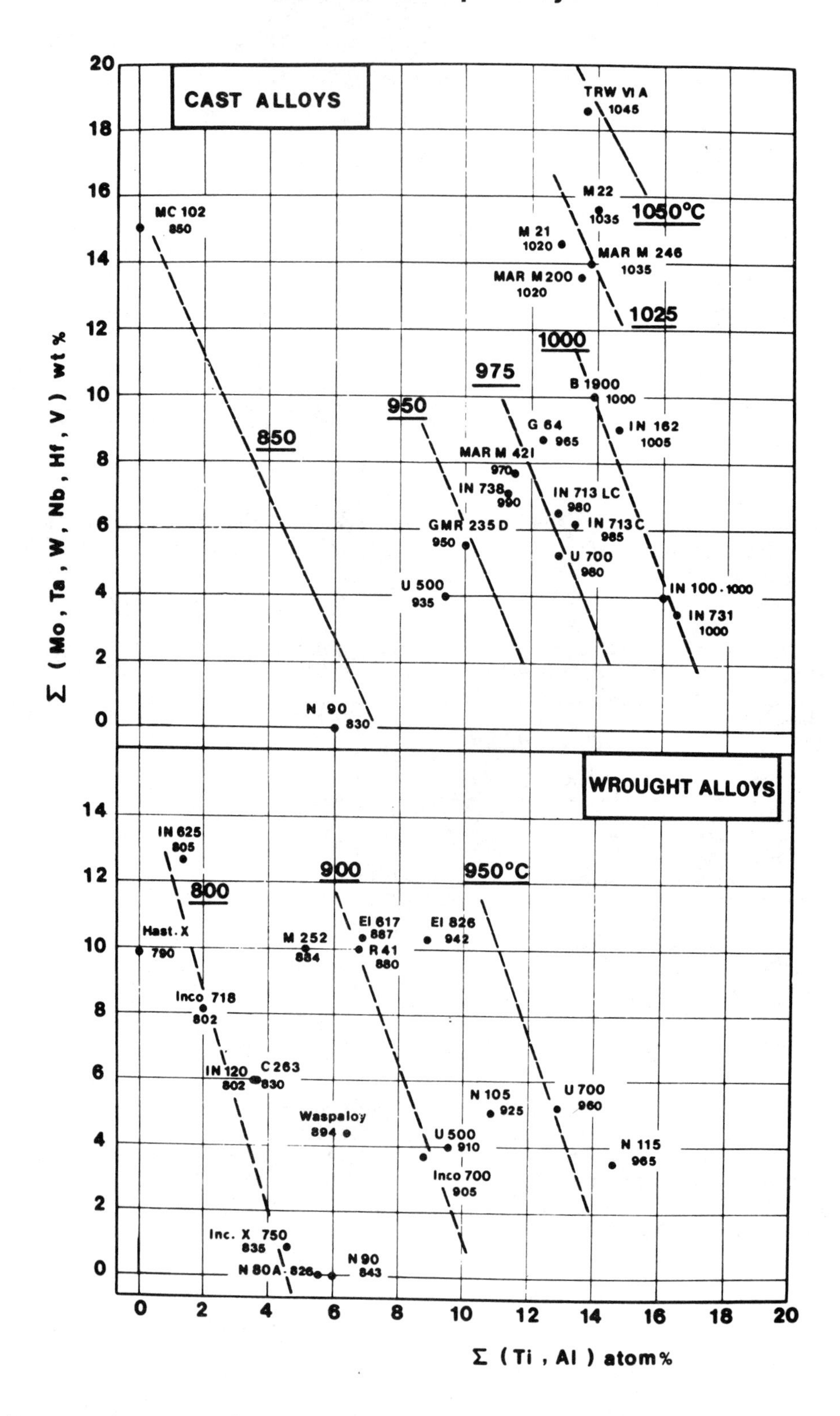

Influence of composition on the creep strength of nickel-base superalloys. Temperatures are for 100-h rupture life at 14 kg/mm^{-2}.

Influence of Composition on the Creep Strength of Nickel-Base Superalloys (Continued)

The problem in hot working superalloys, obviously, is that they have been designed to resist high-temperature deformation. This resistance is derived from a combination of solid-solution strengthening and precipitation hardening by intermetallics. The preceding figure shows that temperature capability increases with the total hardener content.

As regards solution hardening, the most important aspect in creep is the reduction in diffusion rate. This is greater the higher the melting point of the element added, and for the elements considered, the melting point increases with atomic weight. For this reason, the solid-solution hardeners are plotted on a weight percent basis, whereas atom percentages are used for elements contributing to γ' precipitation hardening, since volume fraction is of major importance in this case.

Source: J.H. Davidson, "High Temperature Materials Requirements of the Metallurgical Industries," in Source Book on Materials for Elevated-Temperature Applications, E.F. Bradley, Ed., American Society for Metals, Metals Park, OH, 1979, p 47

Nickel-Base Forging and Casting Superalloys: Effect of Al Plus Ti Content

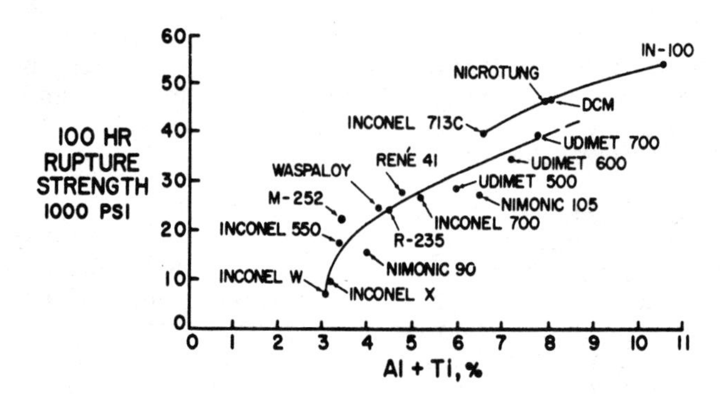

Stress-rupture strength vs aluminum plus titanium content (100-h life at 871 °C, or 1600 °F) for commercial nickel-base forging and casting alloys.

High-strength Ni alloys, for the most part, owe their existence to the remarkable strengthening effect of $Ni_3Al(Ti)$ compound, the significance of which becomes obvious from examination of the figure above.

Source: W.H. Sharp, "Status and Future of Cobalt- and Nickel-Base Alloys," in High Temperature Materials II (Proceedings of a Technical Conference in Cleveland), G.M. Ault, W.F. Barclay, and H.P. Munger, Eds., Interscience, New York, 1963, p 191

Nimonic 80A, 90, 105, 115, and 120: Stress-Rupture Comparisons

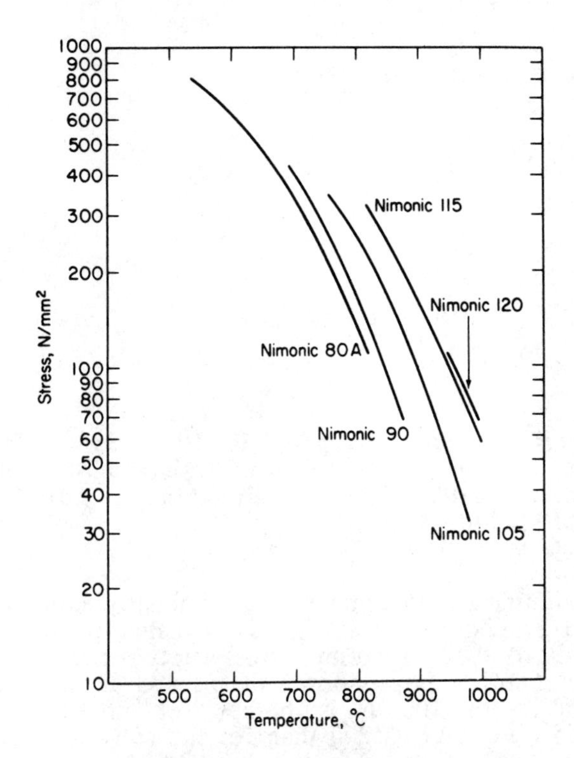

Stress to produce rupture in 100 h for Nimonic 80A, 90, 105, 115, and 120.

Source: The Nimonic Alloys, W. Betteridge and J. Heslop, Eds., Crane, Russak & Co., New York, 1974, p 245

Wrought Udimet 700 vs Conventionally Cast and Directionally Solidified MAR-M200

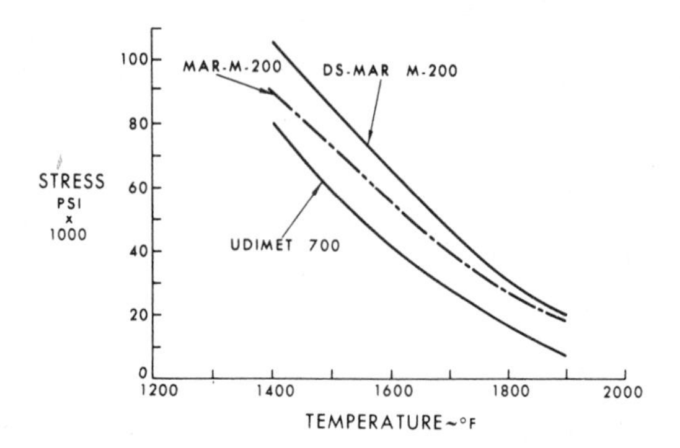

Stress to produce rupture in 100 h at various temperatures for wrought Udimet 700 and conventionally cast and directionally solidified MAR-M200.

A significant advantage of directionally solidified and monocrystal materials is revealed through a study of their creep and stress-rupture properties. The figure above shows that stress to produce rupture in 100 h is higher for directionally solidified MAR-M200 than for the conventionally cast alloy and an advanced wrought alloy (Udimet 700). At all temperatures, minimum rupture ductilities of directionally solidified MAR-M200 were greater than average values of conventionally cast MAR-M200.

Source: F.L. Versnyder and M.E. Shank, "The Development of Columnar Grain and Single Crystal High Temperature Materials Through Directional Solidification," in Source Book on Materials for Elevated-Temperature Applications, E.F. Bradley, Ed., American Society for Metals, Metals Park, OH, 1979, p 356

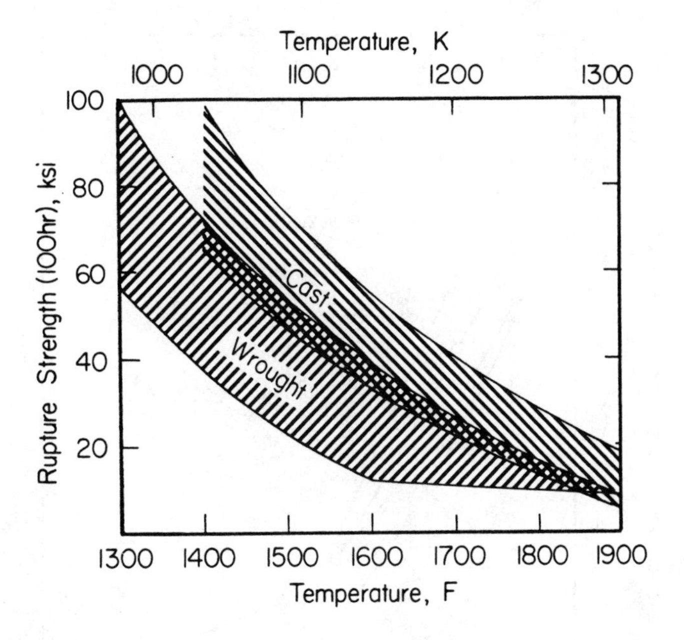

Range of 100-h rupture strengths for nickel-base superalloys.

An indication of the range of stress-rupture properties that can be developed in Ni-base superalloys is given above. The cast alloys tend to have higher rupture strengths than the wrought alloys, but there is some overlapping.

Source: J.L. Everhart, Engineering Properties of Nickel and Nickel Alloys, Plenum, New York, 1971, p 83

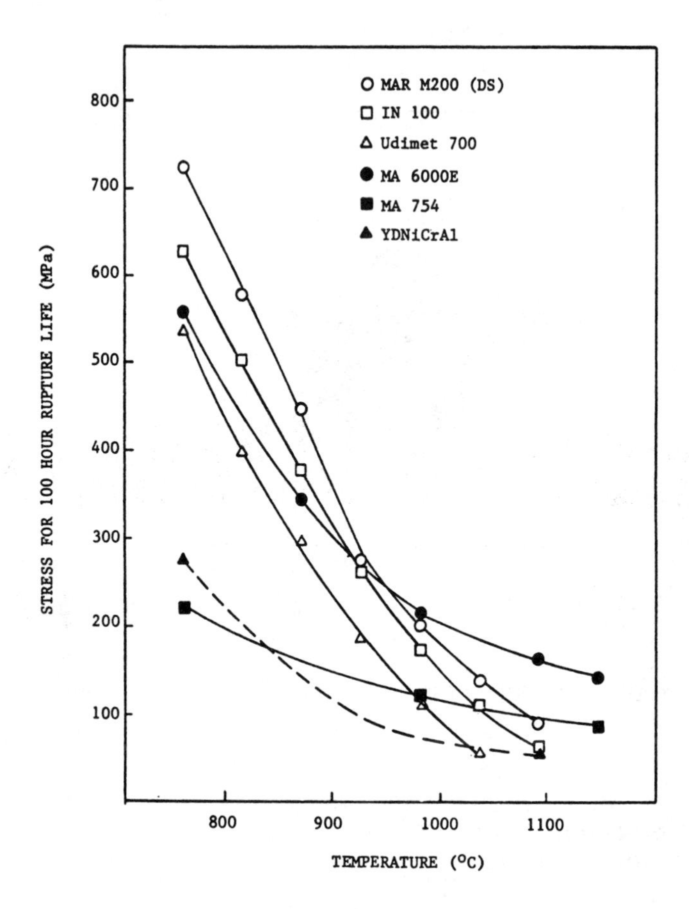

Stress for 100-h rupture life vs temperature.

Although YDNiCrAl is strengthened by high volume fractions of both inert dispersoids and γ' precipitates, it does not attain either the intermediate temperature strength characteristic of Ni-based superalloys or the elevated temperature strength characteristic of other Ni-base ODS alloys.

Source: M.E. McAlarney, R.M. Arons, T.E. Howson, J.K. Tien, and S. Baranow, "Creep and Rupture of an ODS Alloy with High Stress Rupture Ductility," Metallurgical Transactions A, Vol. 13A, Aug. 1982, p 1460-1461

6

Cobalt-Base Superalloys

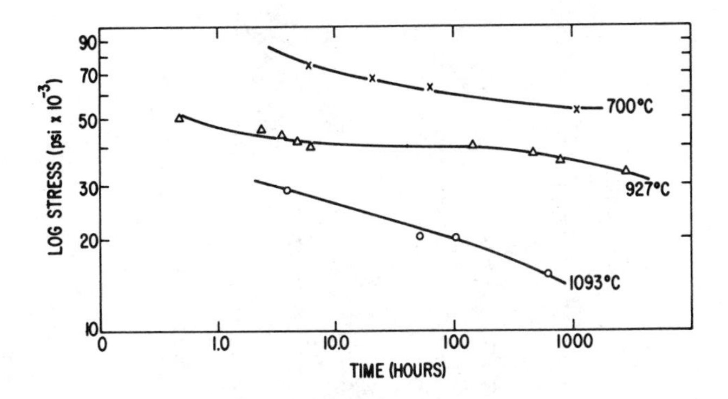

Stress-rupture strength of Co-15Cr-13TaC eutectic alloy on a log stress-log time plot.

The figure above illustrates stress versus time to failure for Co-15Cr-13TaC specimens tested at 700, 927, and 1093 °C (1292, 1700, and 2000 °F). Curves drawn through the data points exhibit different characteristics at the three test temperatures. The 700 °C (1292 °F) curve is concave upward. This characteristic also is observed for the 871 °C (1600 °F) data. The 927 °C (1700 °F) curve is characterized by a flat region in which the lives of stress-rupture specimens vary by almost two orders of magnitude in time at a single stress level of 40,000 psi (276 MN/m^2). The portions of the curves above and below this stress level are characterized by a much steeper slope. This characteristic has also been observed for the 1000 °C (1832 °F) data. The 1093 °C (2000 °F) curve is slightly concave downwards, characteristic of most superalloys.

Source: E.R. Buchanan and L.A. Tarshis, "Strengths and Failure Mechanisms of a Co-15Cr-13TaC Directionally Solidified Eutectic Alloy," Metallurgical Transactions, Vol. 5, June 1974, p 1416

Haynes 88: Notched vs Smooth Bars

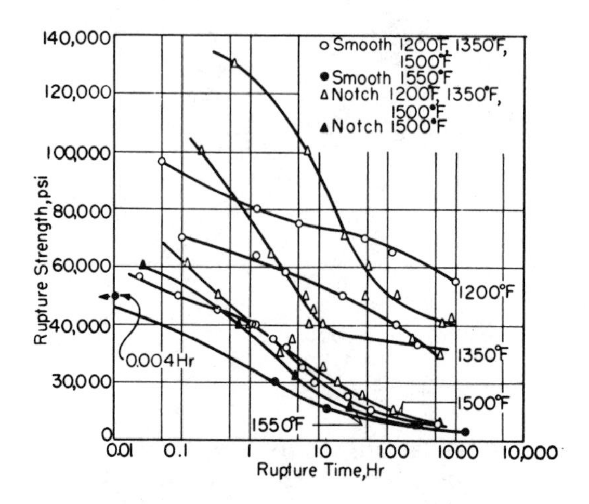

Rupture strength of Haynes 88 smooth and notched bars at several temperatures.

It was determined that a minimum of approximately 5% elongation to rupture is desirable to avoid notch sensitivity, particularly if a rapid drop in the unnotched rupture ductility with testing time is observed. Where notched-bar rupture strength is less than the standard rupture strength, a notch-sensitive condition is said to exist. Thus, the figure above indicates that a given material with a given history may be notch sensitive at one temperature and not at another, or at one period during the test and not at another.

Source: J.W. Freeman, C.L. Corey, and A.I. Rush, "Metallurgical Variables Influencing Properties of Heat Resistant Alloys," in Utilization of Heat Resistant Alloys, American Society for Metals, Metals Park, OH, 1954, p 252

Haynes 188: Air vs Coal Gasifier Atmosphere

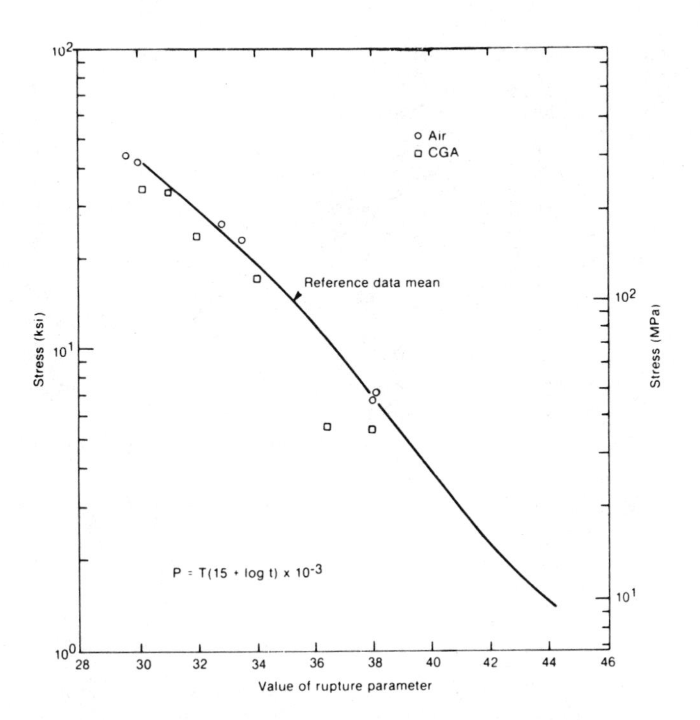

Biaxial stress-rupture values in air and coal gasifier atmosphere for Haynes 188.

Biaxial stress-rupture values in air were essentially identical to the mean of the reference uniaxial data. The biaxial rupture strength/life in a coal gasifier atmosphere is generally below that in air, especially at the highest temperature.

Source: R.M. Horton, "Effect of a Simulated Coal Gasifier Atmosphere on the Biaxial Stress Rupture Strength and Ductility of Selected Candidate Coal Gasifier Alloys," in Behavior of High Temperature Alloys in Aggressive Environments (Proceedings of the Petten International Conference, Petten, The Netherlands), I. Kirman, J.B. Marriott, M. Merz, P.R. Sahm, and D.P. Whittle, Eds., The Metals Society, London, 1980, p 901, 908

Haynes 188: Sheet

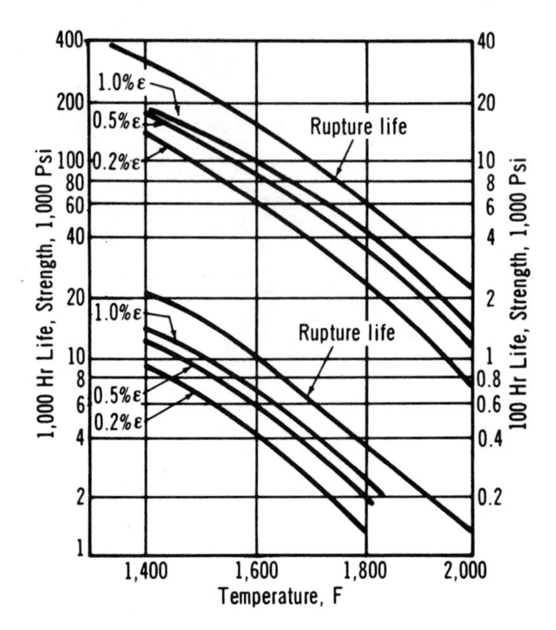

Creep and stress-rupture properties for Haynes 188. Average data for 0.076- to 0.203-cm (0.030- to 0.080-in.) sheet.

In terms of creep and rupture parameters, Haynes 188 is similar to Haynes 25, exhibiting approximately a 55 °C (100 °F) improvement over Hastelloy X. This relationship is true for comparable gages. However, load-bearing properties of 0.041-0.020 cm (0.016-0.008 in.) thin-gage sheet are lower than those of sheet in the range of 0.076 to 0.203 cm (0.030 to 0.080 in.).

Source: R.B. Herchenroeder, S.J. Matthews, J.W. Tackett, and S.T. Wlodek, "A Versatile High-Temperature Alloy," in Source Book on Materials Selection, Volume II, R.B. Gunia, Ed., American Society for Metals, Metals Park, OH, 1977, p 102-103

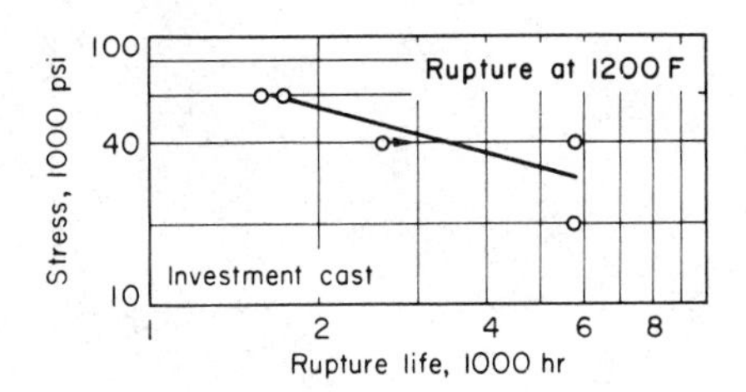

Stress-rupture life for investment-cast HS-3.

Source: Source Book on Materials for Elevated-Temperature Applications, E.F. Bradley, Ed., American Society for Metals, Metals Park, OH, 1979, p 238

HS-21

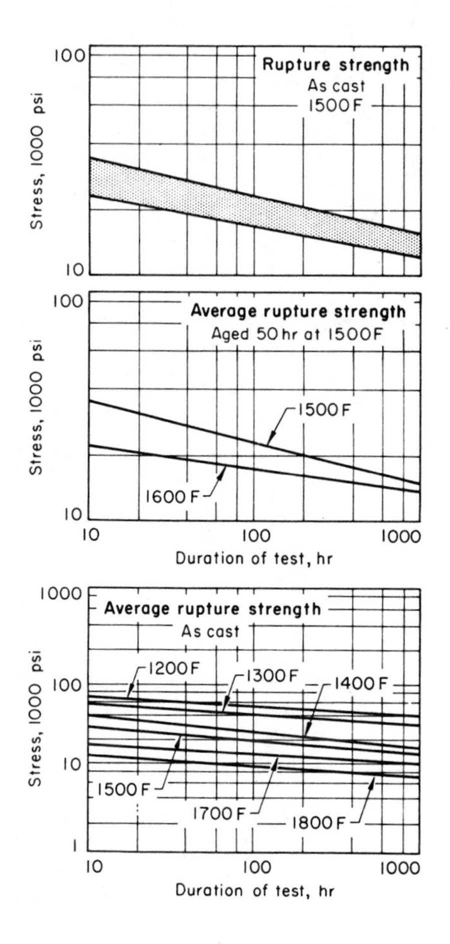

Stress-rupture properties of HS-21.

Source: Source Book on Materials for Elevated-Temperature Applications, E.F. Bradley, Ed., American Society for Metals, Metals Park, OH, 1979, p 238

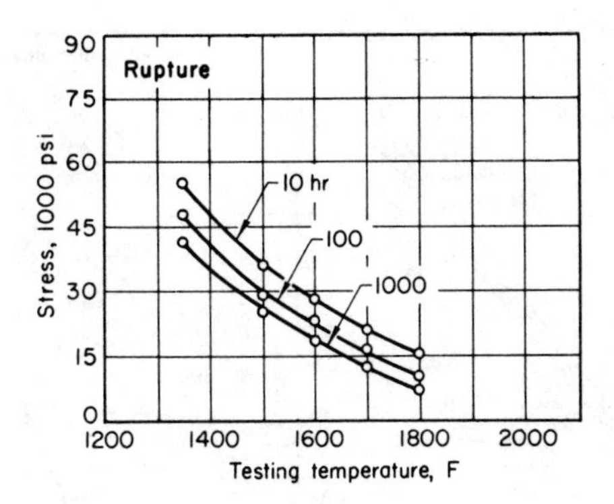

Stress-rupture life for HS-36 specimens aged 16 h at 732 °C (1350 °F).

Source: Source Book on Materials for Elevated-Temperature Applications, E.F. Bradley, Ed., American Society for Metals, Metals Park, OH, 1979, p 236

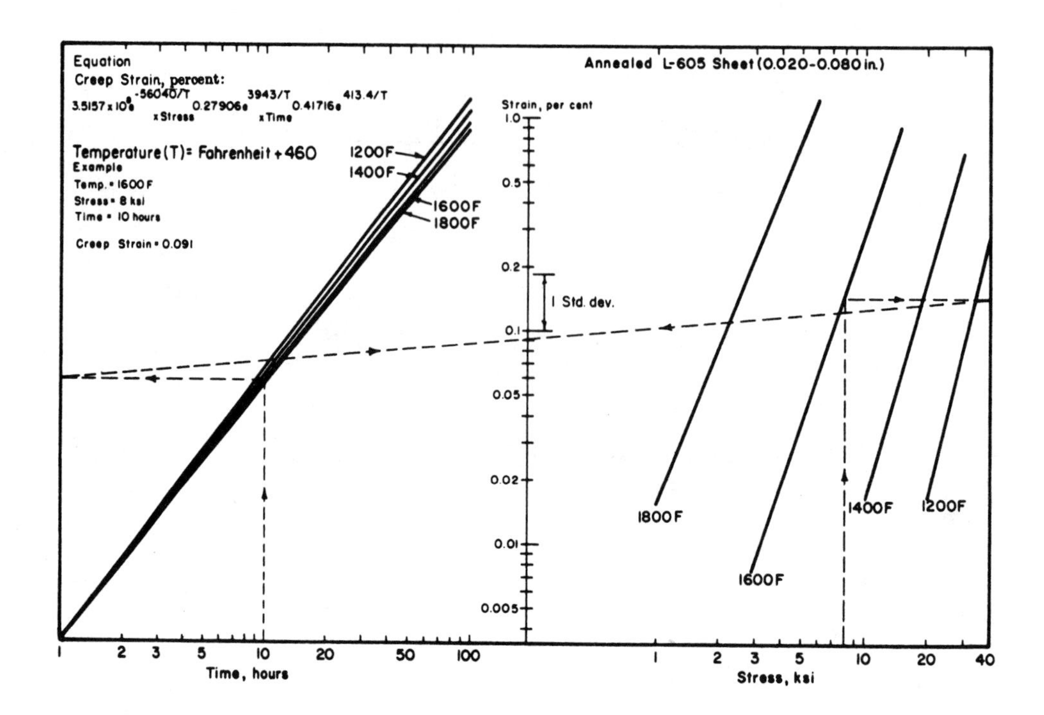

Creep properties of L-605 alloy sheet.

Source: Military Standardization Handbook MIL-HDBK-5D, Volume 1, Metallic Materials and Elements for Aerospace Vehicle Structures, U.S. Department of Defense, Washington, D.C., 1983, p 6-108

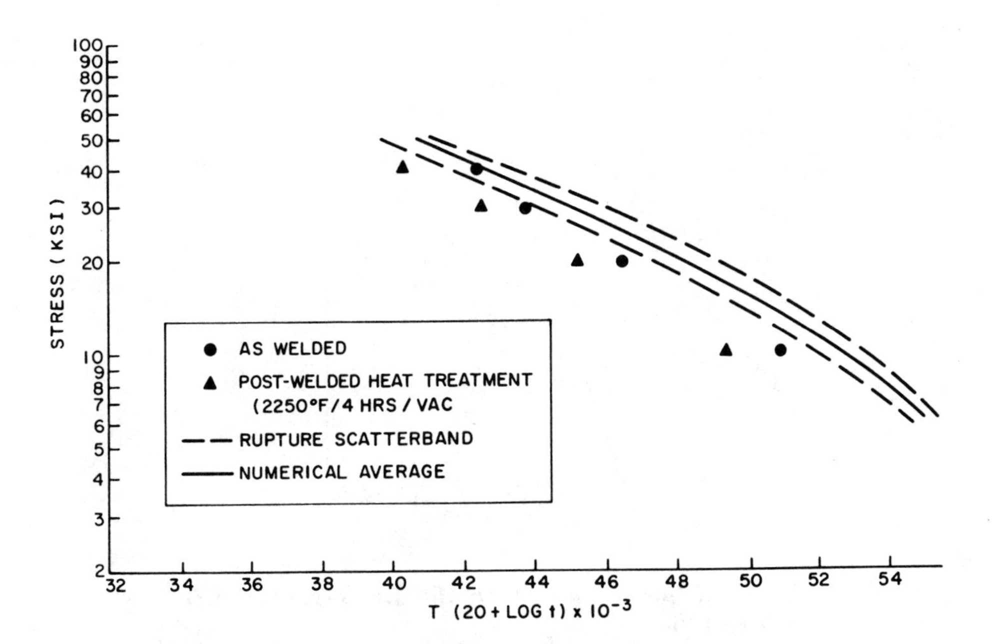

Stress-rupture properties of electron-beam-welded MAR-M509 cast material.

Stress-rupture test results show the post-weld solution-treated material to be stronger than the as-welded material. All of the as-welded bars broke in the base material, while the solution-treated bars failed in the heat-affected zone (HAZ), with lower rupture ductility. This is the reverse of gas tungsten-arc results, where lower strengths were obtained in HAZ-fractured solution-treated material.

Source: T.F. Chase and A.M. Beltran, "The High-Temperature Properties of Welded Cast Co-Base Alloys," in Superalloys – Processing (Proceedings of the Second International Conference, Champion, PA), American Institute of Mining, Metallurgical, and Petroleum Engineers, Warrendale, PA, 1972, p R-8, R-22

ML-1700: Effect of Boron Content

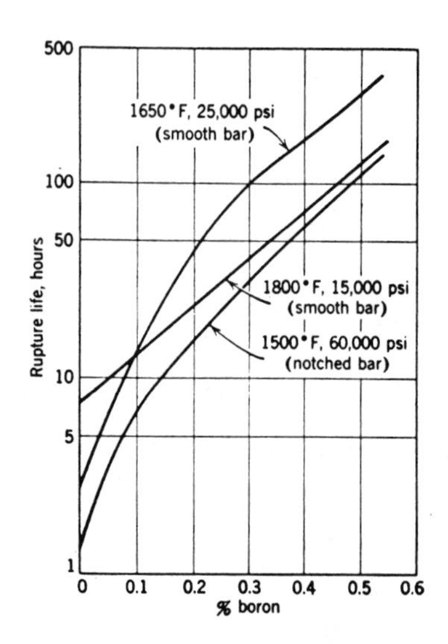

Effect of boron content on the stress-rupture life of ML-1700.

The figure above shows the beneficial effect increased B has on the rupture life of ML-1700. Excessive amounts of B promote extreme brittleness both in tension and impact and are reported to decrease thermal fatigue resistance.

Source: C.P. Sullivan, J.D. Varin, and M.J. Donachie, Jr., "Relationship of Properties to Microstructure in Cobalt-Base Superalloys," in Source Book on Materials for Elevated-Temperature Applications, E.F. Bradley, Ed., American Society for Metals, Metals Park, OH, 1979, p 307

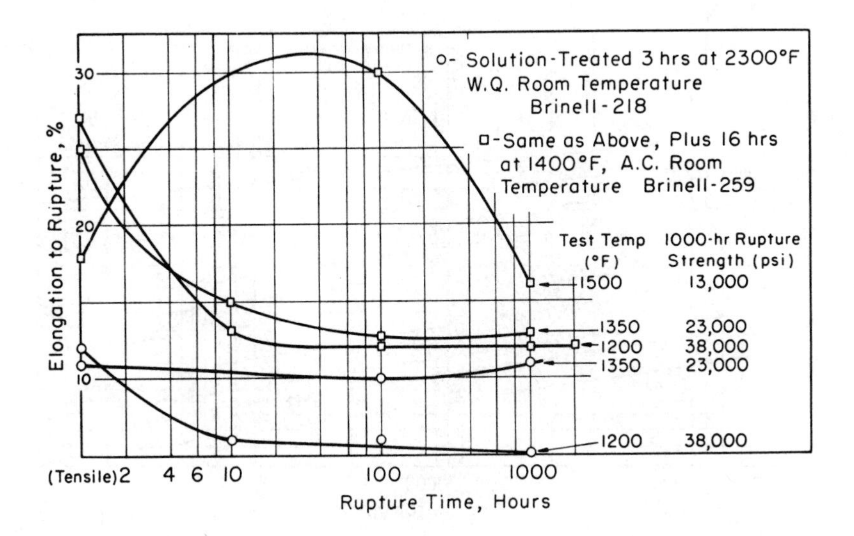

Effect of aging on the elongation to rupture of S-590.

Whether rupture ductility or strength of aged material is improved over that obtainable from solution-treated material depends generally on the test temperature, the test time, and the particular alloy. Thus, the figure above shows that rupture strength was not increased at 1200 and 1350 °F by aging at 1400 °F, but that the elongation to rupture was improved.

Source: J.W. Freeman, C.L. Corey, and A.I. Rush, "Metallurgical Variables Influencing Properties of Heat Resistant Alloys," in Utilization of Heat Resistant Alloys, American Society for Metals, Cleveland, OH, 1954, p 248, 251

S-590: Rupture Properties

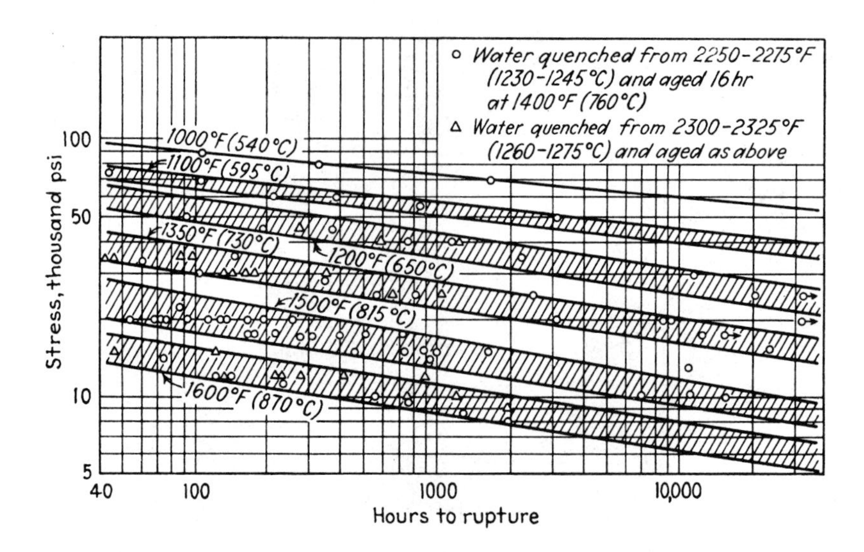

Stress-rupture properties of S-590.

Source: R.A. Grange, F.J. Shortsleeve, D.C. Hilty, W.O. Binder, et al., Boron, Calcium, Columbium, and Zirconium in Iron and Steel, John Wiley & Sons, New York, 1957, p 370

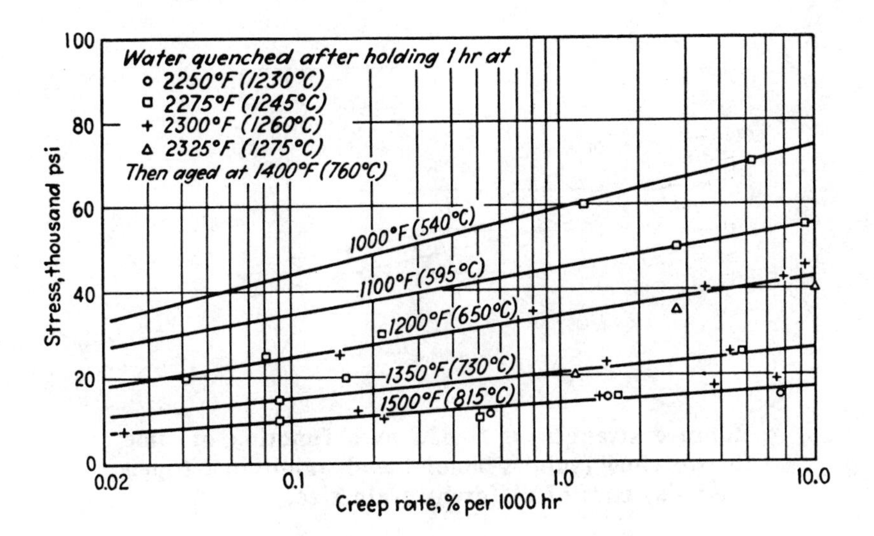

Secondary creep rates of S-590, heat treated as shown.

Source: R.A. Grange, F.J. Shortsleeve, D.C. Hilty, W.O. Binder, et al., Boron, Calcium, Columbium, and Zirconium in Iron and Steel, John Wiley & Sons, New York, 1957, p 372

S-816: Effect of Grain Size on Notched and Unnotched Rupture Strength

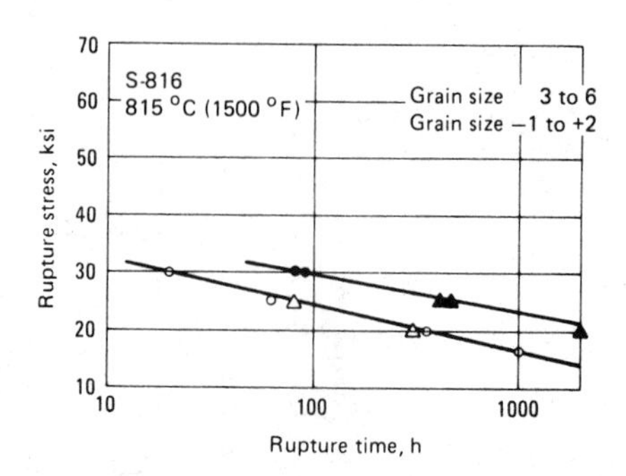

Rupture strength of S-816 as a function of time for notched (solid symbols) and unnotched (open symbols) bars of different grain size.

S-816 was heated to 1175 °C (2150 °F) and water quenched, reheated to 760 °C (1400 °F), held 12 h and air cooled. Smaller grain sizes were produced by cold reducing 1%, cold rolling at about 24 °C (75 °F), and then heat treating. Diameter of specimens was 12.7 mm (0.5 in.), diameter at base of notch was 8.9 mm (0.35 in.), root radius was 0.1 mm (0.004 in.), and notch angle was 60°. Data are a composite of results from two laboratories.

Note that notches had a strengthening effect on S-816 tested at 815 °C (1500 °F). There was no measured difference in the effect of grain size on either the notched or unnotched specimens.

Source: Metals Handbook, Ninth Edition, Volume 3, Properties and Selection: Stainless Steels, Tool Materials and Special-Purpose Metals, American Society for Metals, Metals Park, OH, 1980, p 232

S-816: Rupture Properties

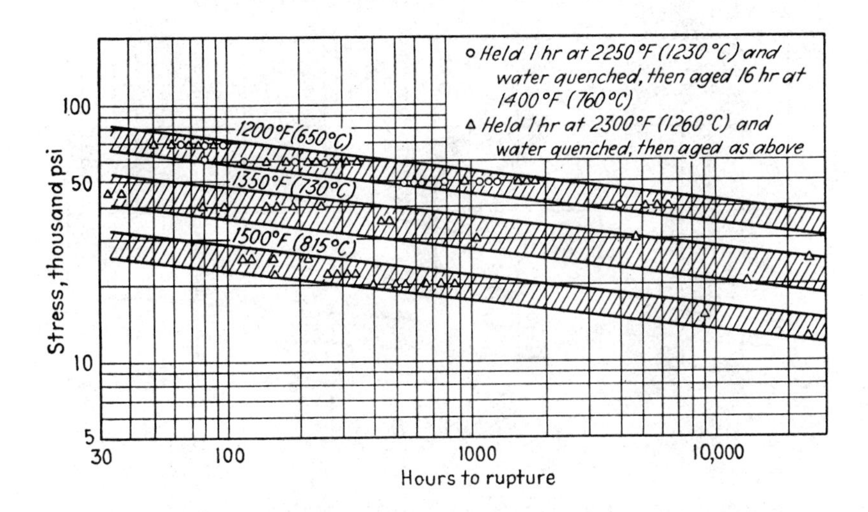

Rupture properties of S-816.

Source: R.A. Grange, J. Shortsleeve, D.C. Hilty, W.O. Binder, et. al., Boron, Calcium, Columbium, and Zirconium in Iron and Steel, John Wiley & Sons, New York, 1957, p 366

S-816: Secondary Creep Rates

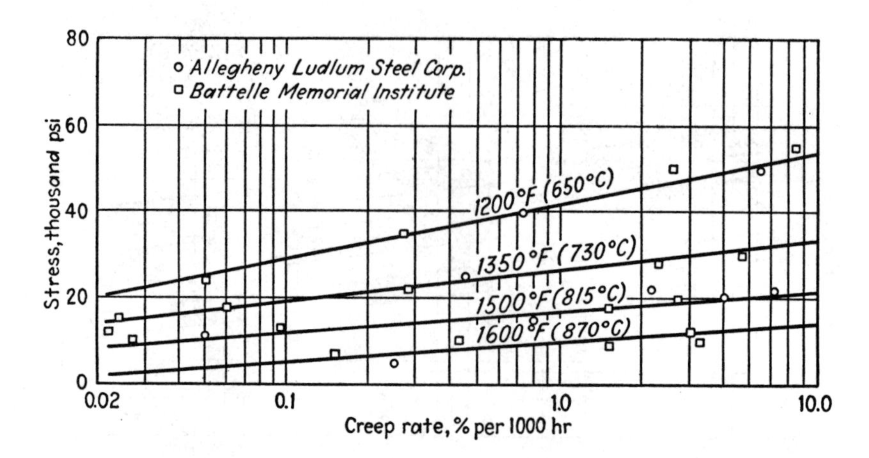

Secondary creep rates of S-816, water quenched from 1260 °C (2300 °F) and aged at 760 °C (1400 °F).

Source: R.A. Grange, J. Shortsleeve, D.C. Hilty, W.O. Binder, et. al., Boron, Calcium, Columbium, and Zirconium in Iron and Steel, John Wiley & Sons, New York, 1957, p 368

S-816: Wrought vs Cast

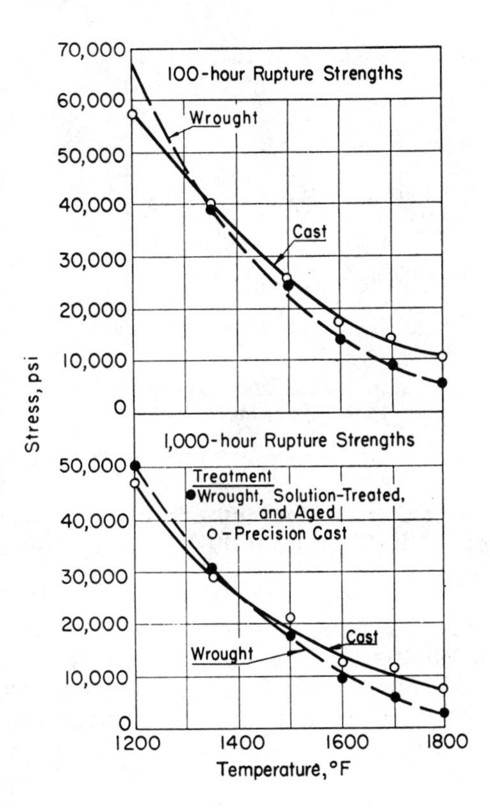

Effect of temperature on the rupture strength of wrought and cast S-816.

Source: J.W. Freeman, C.L. Corey, and A.I. Rush, "Metallurgical Variables Influencing Properties of Heat Resistant Alloys," in Utilization of Heat Resistant Alloys, American Society for Metals, Metals Park, OH, 1954, p 259

Vitallium: Effect of Aging

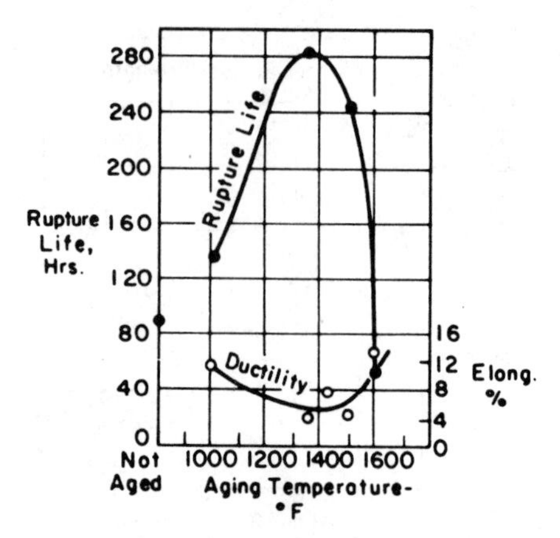

Effect of aging on the rupture life and ductility of as-cast Vitallium at 1500 °F and 20,000 psi.

The effect of aging on generation of improved rupture strength by carbide precipitation is illustrated above. $M_{23}C_6$ precipitation plays a dominant role at about 1400 °F, probably as a result of degeneration of M_7C_3 and MC.

Source: C.T. Sims, "Cobalt-Base Alloys," in The Superalloys, C.T. Sims and W.C. Hagel, Eds., John Wiley & Sons, New York, 1972, p 161-162

Vitallium: Effect of Carbon Content

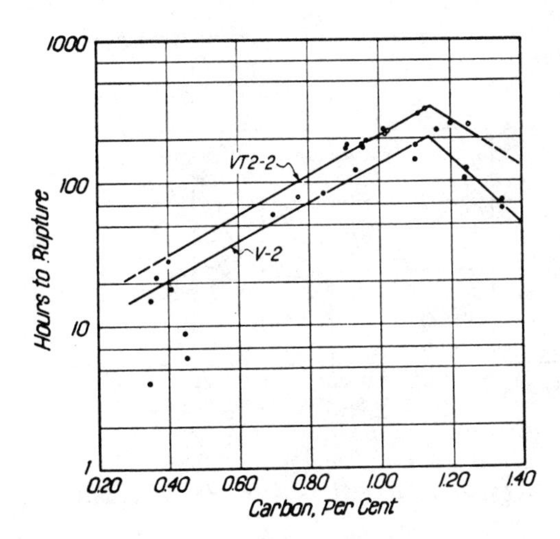

Effect of carbon content on the rupture life of Vitallium and a modified Vitallium alloy at a temperature of 815 °C (1500 °F) and stress of 207 MPa (30,000 psi).

Source: C.P. Sullivan, J.D. Varin, and M.J. Donachie, Jr., "Relationship of Properties to Microstructure in Cobalt-Base Superalloys," in Source Book on Materials for Elevated-Temperature Applications, E.F. Bradley, Ed., American Society for Metals, Metals Park, OH, 1979, p 304

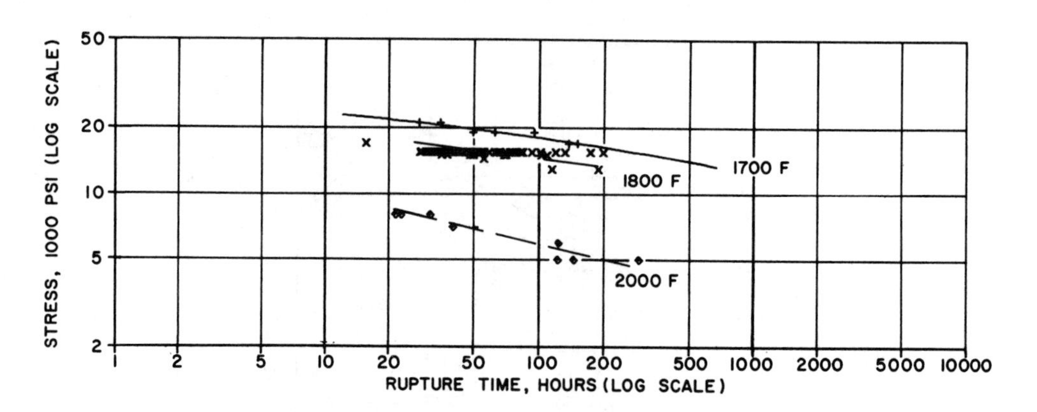

Stress-rupture time for WI-52 castings.

Source: The Elevated-Temperature Properties of Selected Superalloys, ASTM Data Series DS-7S1, American Society for Testing and Materials, Philadelphia, 1968, p 65

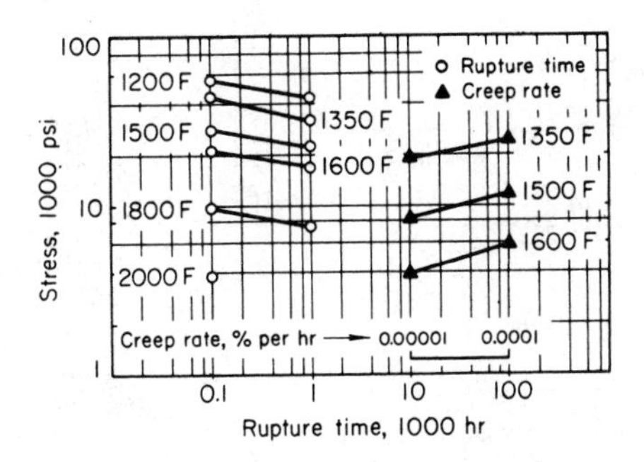

Creep and stress-rupture properties of X-40.

Source: Source Book on Materials for Elevated-Temperature Applications, E.F. Bradley, Ed., American Society for Metals, Metals Park, OH, 1979, p 236

X-40: As-Cast

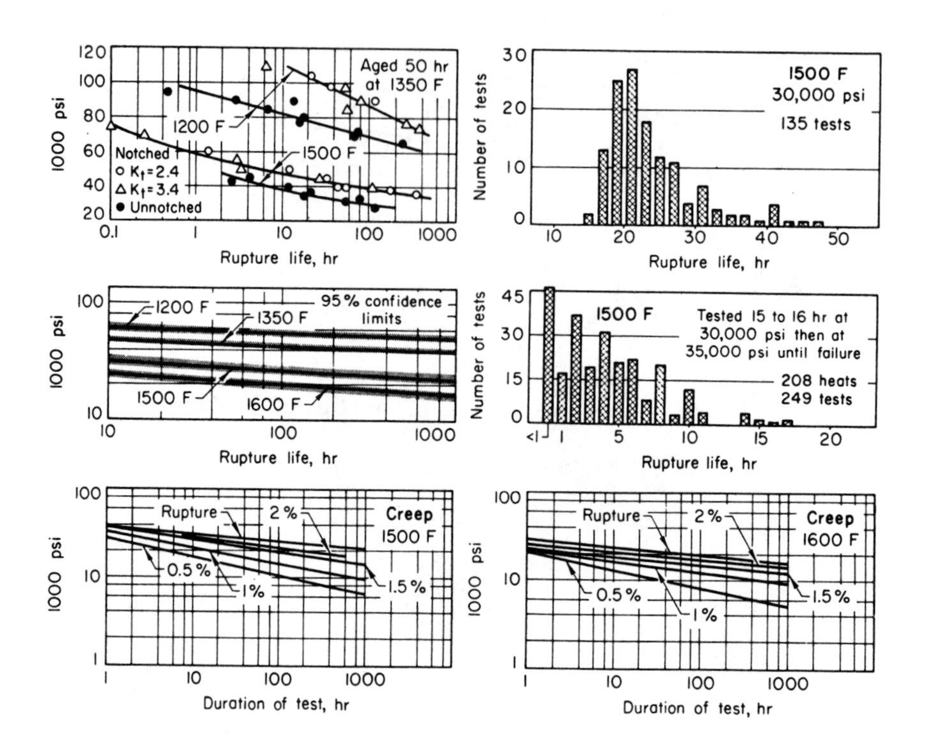

Creep and stress-rupture data for as-cast X-40.

Source: Source Book on Materials for Elevated-Temperature Applications, E.F. Bradley, Ed., American Society for Metals, Metals Park, OH, 1979, p 237

X-40: Effect of Heat Treatment

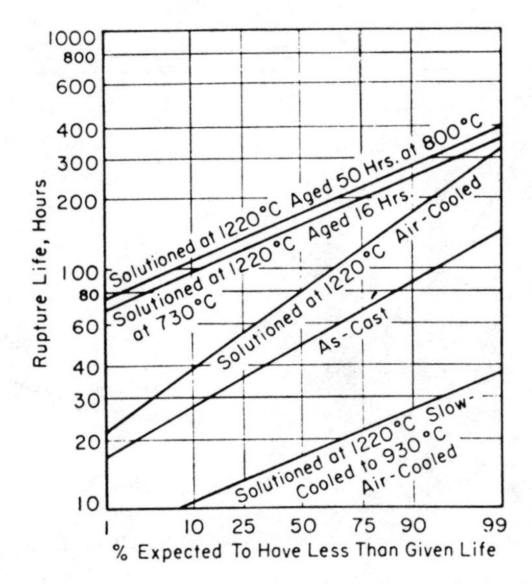

Effect of heat treatment on rupture behavior of X-40 tested at 815 °C (1500 °F) and 207 MPa (30,000 psi).

Source: C.P. Sullivan, J.D. Varin, and M.J. Donachie, Jr., "Relationship of Properties to Microstructure in Cobalt-Base Superalloys," in Source Book on Materials for Elevated-Temperature Applications, E.F. Bradley, Ed., American Society for Metals, Metals Park, OH, 1979, p 303

6.24 Comparison of Effect of 100-h Prior Exposures at Various Temperatures on MAR-M509 and WI-52

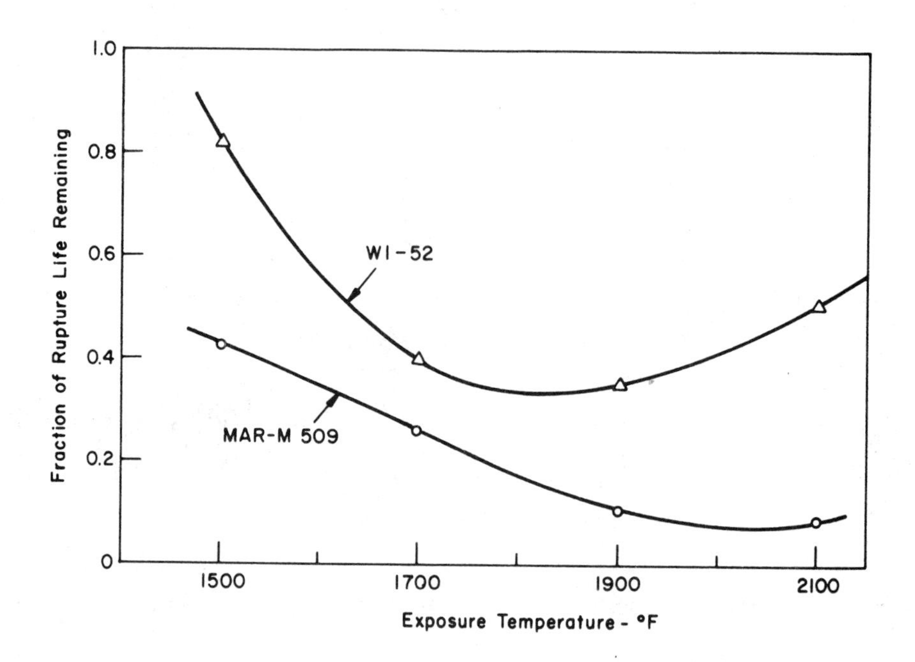

Effect of 100-h prior exposure at various temperatures on the rupture lives at 1500 °F and 35,000 psi of WI-52 and MAR-M509 cobalt-base superalloys.

Creep-rupture bars of WI-52 and MAR-M509 were exposed at no load for 100 h at temperatures ranging from 1500 to 2100 °F prior to a creep-rupture test at 1500 °F, 35,000 psi. Both alloys rapidly lost strength due to exposure to approximately 1900 °F. This loss was attributed to the combined effects of second-phase agglomeration (overaging) and partial breakdown of carbides (MC in MAR-M509, M_6C in WI-52) and reprecipitation as acicular $M_{23}C_6$, a morphology that is less efficient for strengthening. At temperatures above 1900 °F, $M_{23}C_6$ precipitation is not favored, and the principal effect of high-temperature exposure is then agglomeration and dissolution of the strengthening phases.

Source: J.D. Varin, "Microstructures and Properties of Superalloys," in The Superalloys, C.T. Sims and W.C. Hagel, Eds., John Wiley & Sons, New York, 1972, p 234-235

7

Superalloy Comparisons

Co-15Cr-13TaC Eutectic Alloy: As Compared with Rene′ 80 and MAR-M509

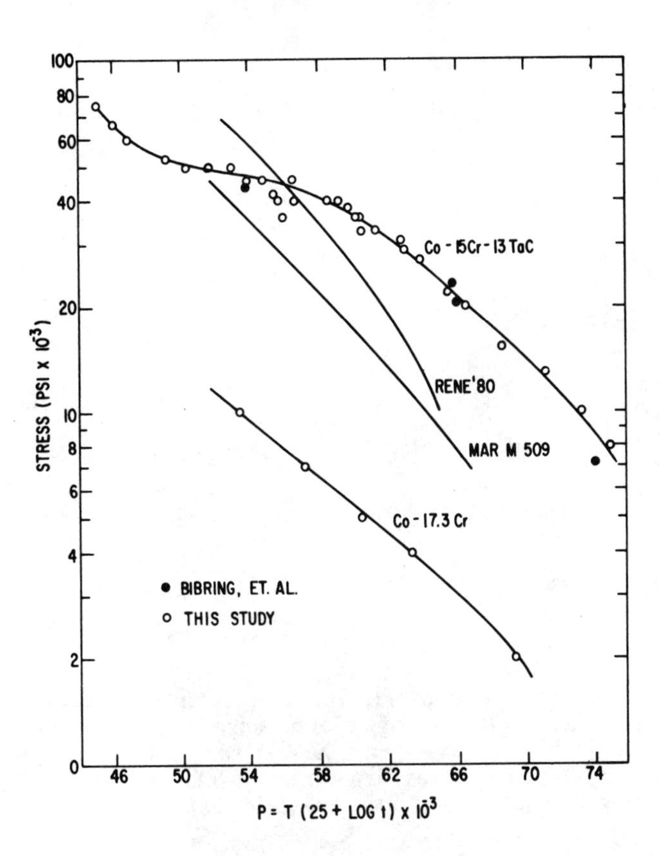

Stress-rupture strength of Co-15Cr-13TaC eutectic, Rene′ 80, and MAR-M509 on a Larson-Miller (C = 25) plot.

Thirty-four stress-rupture tests were run on specimens of the Co-15Cr-13TaC eutectic alloy at temperatures between 700 and 1260 °C (1292 and 2300 °F). The data exhibit, in general, little scatter.

Results are compared above to the stress-rupture properties of two high-strength alloys: Rene′ 80, a Ni-base superalloy, and MAR-M509, a Co-base superalloy. At 20,000 psi (138 MN/m²), the eutectic alloy stress-rupture behavior exceeds that of Rene′ 80 by about five parameter units, equivalent to about a 111 °C (200 °F) property improvement at this stress level. The eutectic stress-rupture advantage over Rene′ 80 decreases with increasing stress levels, however, becoming equal at about 40,000 psi (276 MN/m²), and inferior at higher stresses. At 50,000 psi (345 MN/m²), the stress-rupture behavior of the Co-15Cr-13TaC eutectic alloy is about three parameter units lower than that of Rene′ 80.

Source: E.R. Buchanan and L.A. Tarshis, "Strengths and Failure Mechanisms of a Co-15Cr-13TaC Directionally Solidified Eutectic Alloy," Met. Trans., Vol. 5, June 1974, p 1415-1416

7.2 Comparison of Effect of Cyclic Overtemperatures on M-252, S-816, and X-40

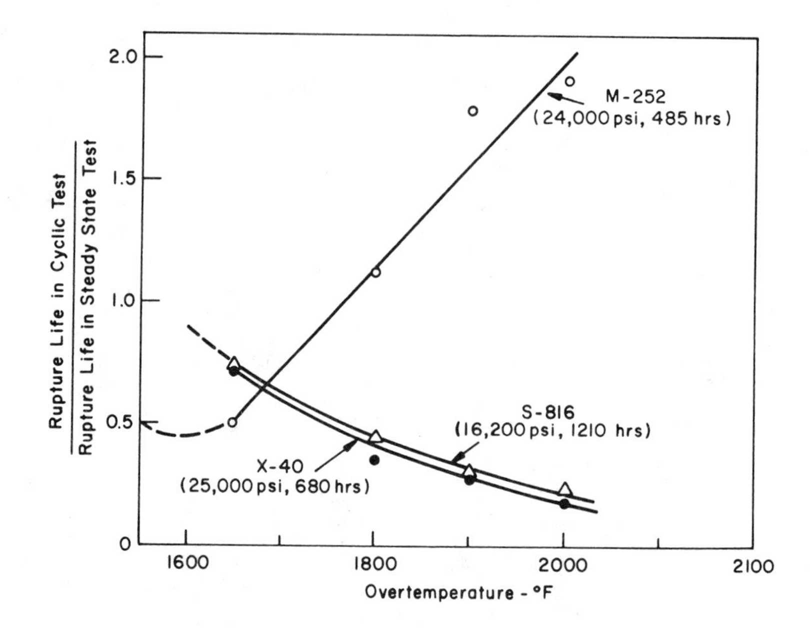

Effect of cyclic overtemperatures on the rupture lives at 1500 °F of nickel-base (M-252) and cobalt-base (S-816 and X-40) superalloys. Figures in parentheses are stress levels and rupture lives in the 1500 °F steady-state test.

The effects of brief overtemperature exposures reveal basic differences in behavior of Ni- and Co-base alloys, which can be accounted for on the basis of microstructural differences. The Co-base alloys are strengthened by a fine dispersion of $M_{23}C_6$ and/or M_6C carbide particles. The Ni-base alloy is strengthened by a fine dispersion of γ'. The differences in behavior are related directly to the response of these hardening phases to overtemperature.

Source: J.D. Varin, "Microstructures and Properties of Superalloys," in The Superalloys, C.T. Sims and W.C. Hagel, Eds., John Wiley & Sons, New York, 1972, p 232-233

Comparison of Effect of Notches on the Rupture Life of S-816 and Nimonic 80A

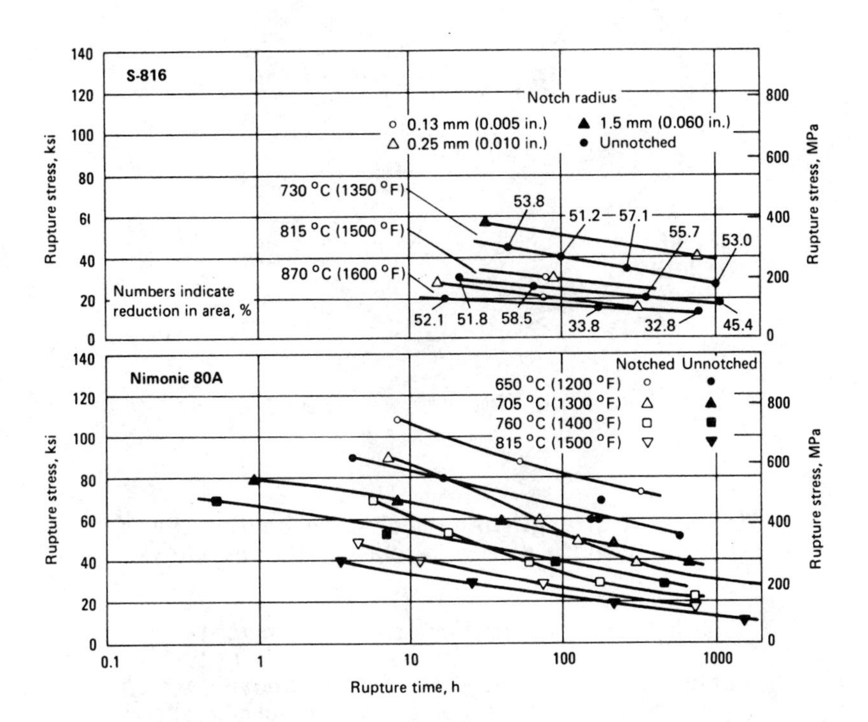

Effect of notches on the rupture life of S-816 and Nimonic 80A.

Source: Metals Handbook, Ninth Edition, Volume 3, Properties and Selection: Stainless Steels, Tool Materials and Special-Purpose Metals, F.R. Morral, Ed., American Society for Metals, Metals Park, OH, 1980, p 233

7.4

Comparison of Effects of Strengthening Method on the Rupture Strength of Nickel- and Cobalt-Base Superalloys

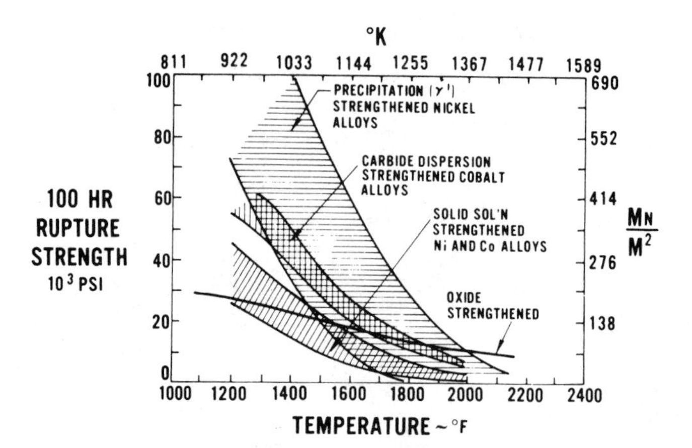

Effects of strengthening technique on the 100-h rupture strength of several classes of superalloys.

The hot section, particularly first-stage turbine blades, historically has paced aircraft turbine engine development. The figure above, which gives an overview of hot section materials, shows why γ'-strengthened Ni alloys are used for highly stressed turbine blades.

Source: E.F. Bradley and M.J. Donachie, "Changes and Evolution of Aircraft Engine Materials," in Source Book on Materials Selection, Volume I, R.B. Gunia, Ed., American Society for Metals, Metals Park, OH, 1977, p 3

Comparison of Rupture Strengths of Advanced Turbine Blade Materials

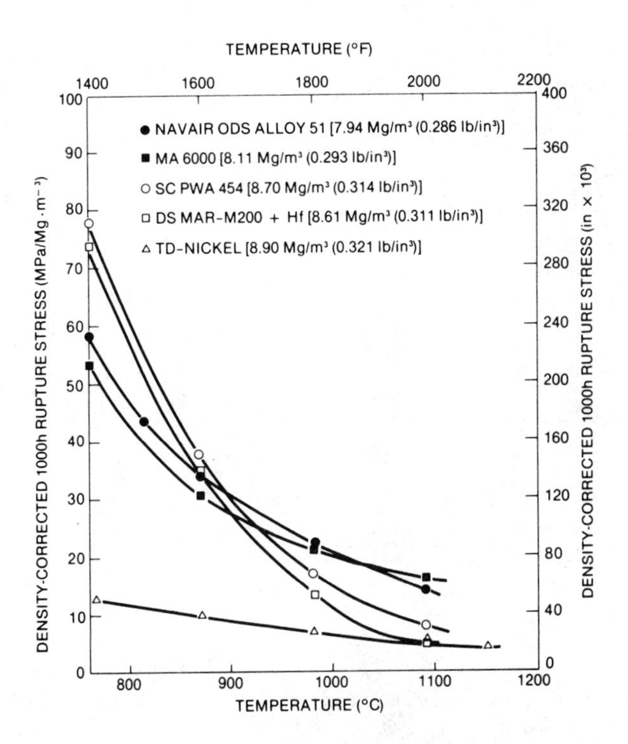

Comparison of 1000-h specific rupture strength vs temperature for advanced turbine blade materials.

Elemental Ni, Cr, W, and Mo powders were blended with preground Ni-Al, Ni-Zr, and Ni-B master alloys and Y_2O_3. Powder batches were then mechanically alloyed in attritor mills under controlled conditions. The resultant mechanically alloyed powder was characterized using chemical and screen analyses and metallographic examination. After screening to remove the coarse +12 mesh particles, powder batches were cone blended and packed into extrusion cans, which were then sealed, and extruded to round bar 20.4-mm (0.8-in.) in diameter.

Source: S.K. Kang and R.C. Benn, "Microstructural Development in High Volume Fraction Gamma Prime Ni-Base Oxide-Dispersion-Strengthened Superalloys," Met. Trans. A., Vol. 16A, July 1985, p 1286

Creep Behavior Comparison: B-1900 and SM-200 (MAR-M200) vs WI-52 and SM-302 (MAR-M302)

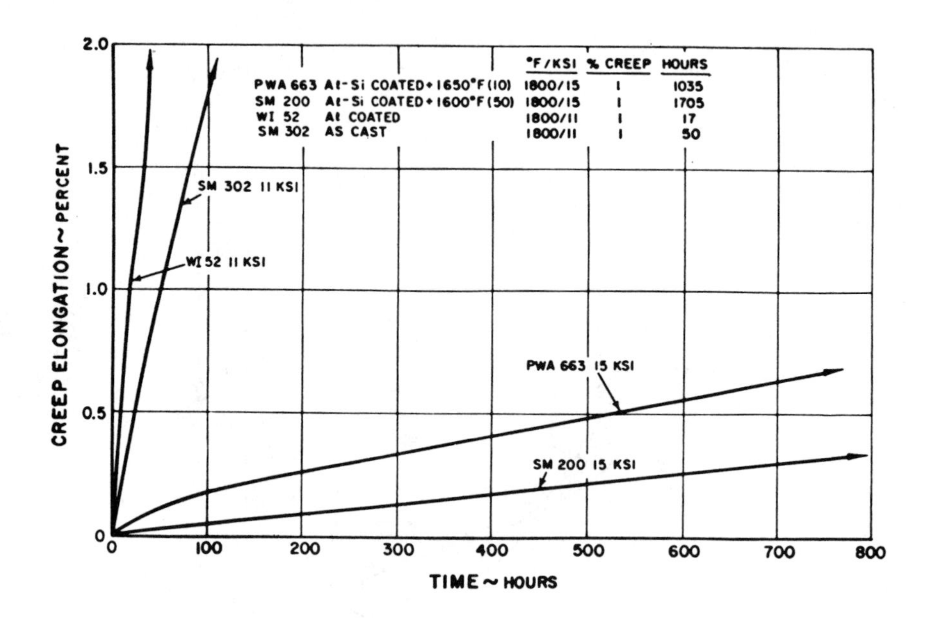

Creep behavior of two cast nickel-base alloys, B-1900 and SM-200 (MAR-M200), vs two cast cobalt-base alloys, WI-52 and SM-302 (MAR-M302).

An illustration of the extent to which creep strength of present day Ni-base alloys has outdistanced Co-base alloys is shown above. It should be noted that the cast Ni-base superalloy, B-1900, was creep tested at a stress $^1/_3$ greater than was the cast Co-base superalloy, WI-52.

Source: C.P. Sullivan, J.D. Varin, and M.J. Donachie, Jr., "Relationship of Properties to Microstructure in Cobalt-Base Superalloys," in Source Book on Materials for Elevated-Temperature Applications, E.F. Bradley, Ed., American Society for Metals, Metals Park, OH, 1979, p 300

Effect of Temperature on 100-h Rupture Strength of Fe-Cr-Ni, Ni-, and Co-Base Superalloys

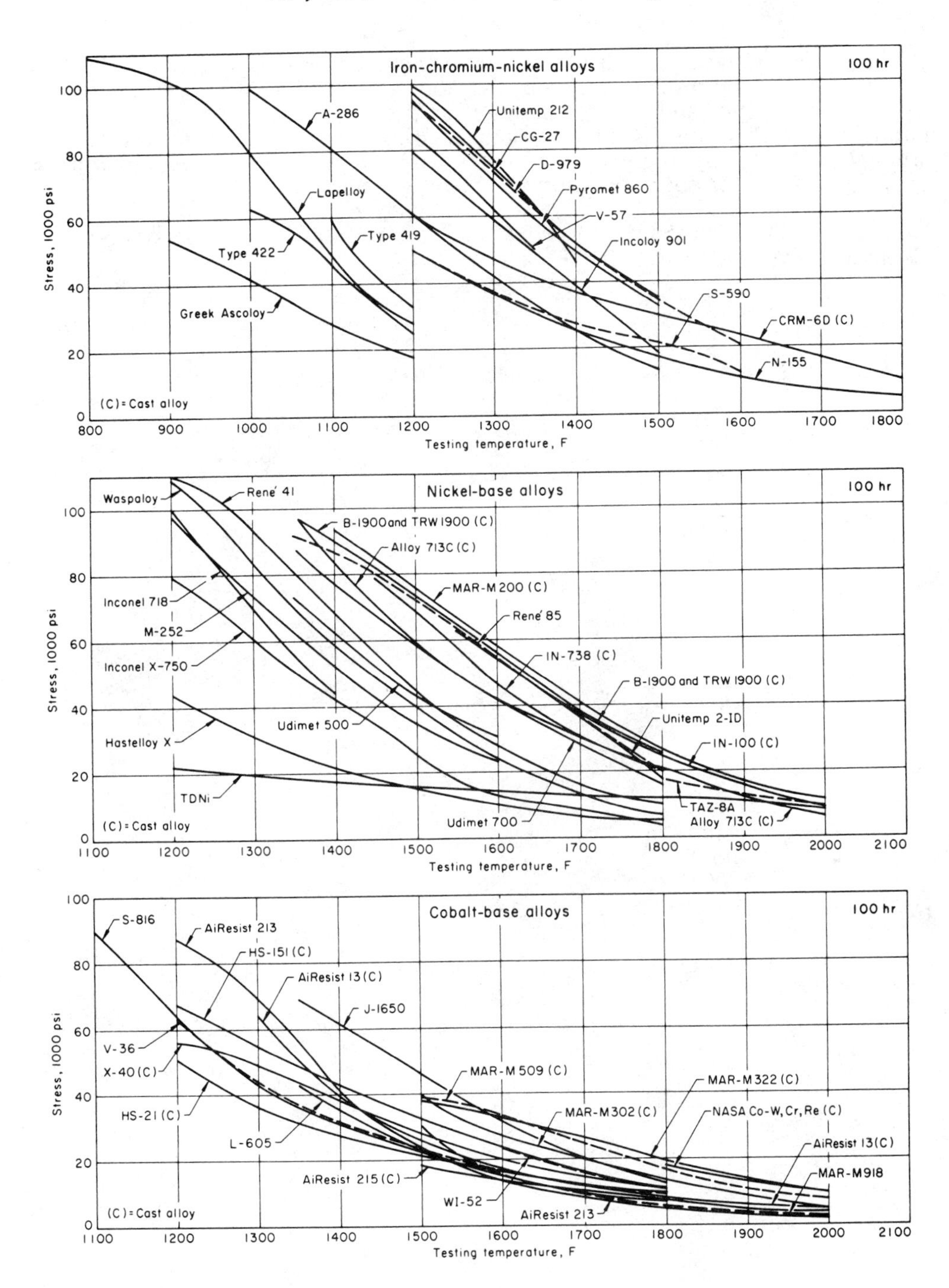

Effect of temperature on the 100-h rupture strength of iron-chromium-nickel and nickel- and cobalt-base superalloys.

7.8 Effect of Temperature on 100-h Rupture Strength of Fe-Cr-Ni, Ni-, and Co-Base Superalloys (Continued)

The rupture strengths shown above cannot be used as design criteria, because deformation at rupture far exceeds that allowable in design. Also, the curves shown are typical and reflect neither statistical distribution nor specified minimums. Variations in composition, melting, forging, and heat treatment are not reflected by these smoothed, typical curves. The curves, therefore, provide only a first approximation for material selection.

Note that curves for the Fe-Cr-Ni alloys end at 1800 °F. Those for Ni- and Co-base alloys extend to 2000 °F.

Source: "Materials for Aerospace Forgings," in Source Book on Materials Selection, Volume I, American Society for Metals, Metals Park, OH 1977, p 17

Effect of Temperature on 1000-h Rupture Strength on a Variety of Fe-Cr-Ni, Ni-, and Co-Base Superalloys

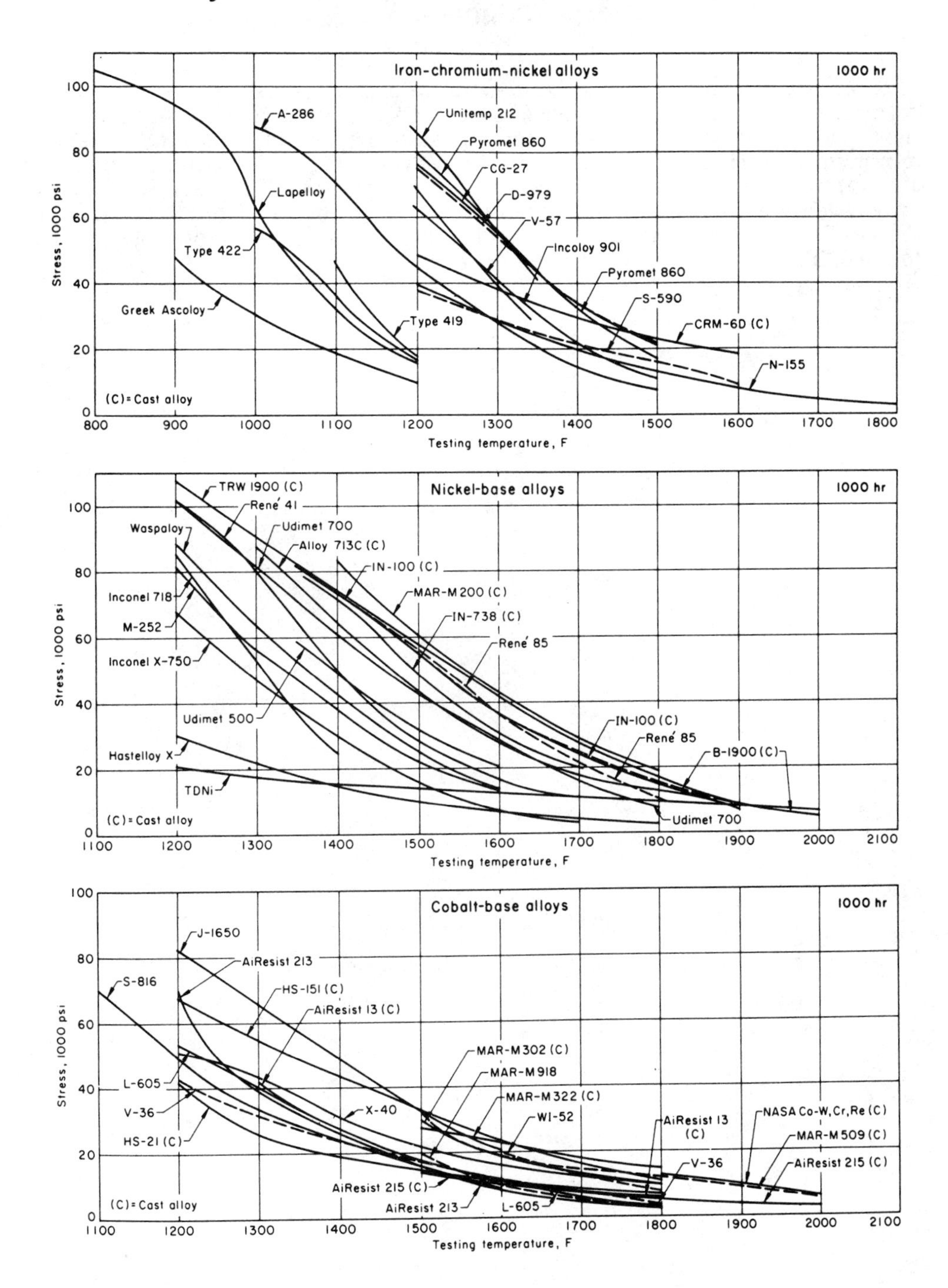

Effect of temperature on the 1000-h rupture strength of iron-chromium-nickel and nickel- and cobalt-base superalloys.

7.10 Effect of Temperature on 1000-h Rupture Strength on a Variety of Fe-Cr-Ni, Ni-, and Co-Base Superalloys (Continued)

The rupture strengths shown above cannot be used as design criteria, because deformation at rupture far exceeds that allowable in design. Also, the curves shown are typical and reflect neither statistical distribution nor specified minimums. Variations in composition, melting, forging, and heat treatment are not reflected by these smoothed, typical curves. The curves, therefore, provide only a first approximation for material selection. Note that curves for the Fe-Cr-Ni alloys end at 1800 °F. Those for Ni- and Co-base alloys extend to 2000 °F.

Source: "Materials for Aerospace Forgings," in Source Book on Materials Selection, Volume I, American Society for Metals, Metals Park, OH, 1977, p 18

Inconel 617 vs Type 316 Stainless Steel: Effect of Impure Helium

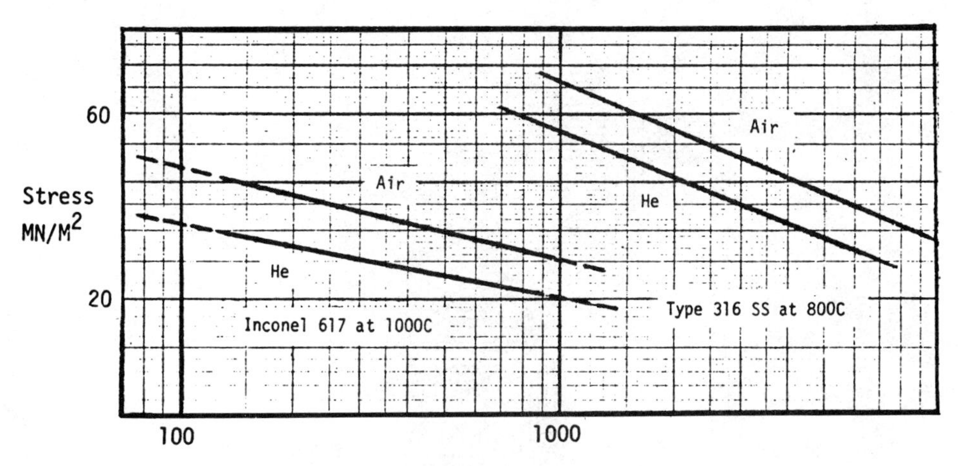

Effect of impure helium on the rupture properties of Inconel 617 and type 316 stainless steel.

The effect of gas/metal reactions on mechanical properties can be quite serious. Some alloys show a 25% reduction in stress capability, depending on severity of the atmosphere.

The significant factor in helium contamination is that P_{O2} is so low that insufficient (if any) protective oxide film forms on the metal, so it can be faster than in normal oxidative service. The formation permits internal oxidation and carburization reactions to proceed into the alloy as fast as diffusion will allow.

Source: C.T. Sims, "Superalloys: Their Use and Requirements in Advanced Energy Systems," in Superalloys: Metallurgy and Manufacture (Proceedings of the Third International Symposium, Seven Springs, PA), B.H. Kear, D.R. Muzyka, J.K. Tien, and S.T. Wlodek, Eds., Claitors, Baton Rouge, 1976, p 23

7.12 Rupture-Strength Behavior of the Three Superalloy Classes: Fe-Ni, Ni-, and Co-Base

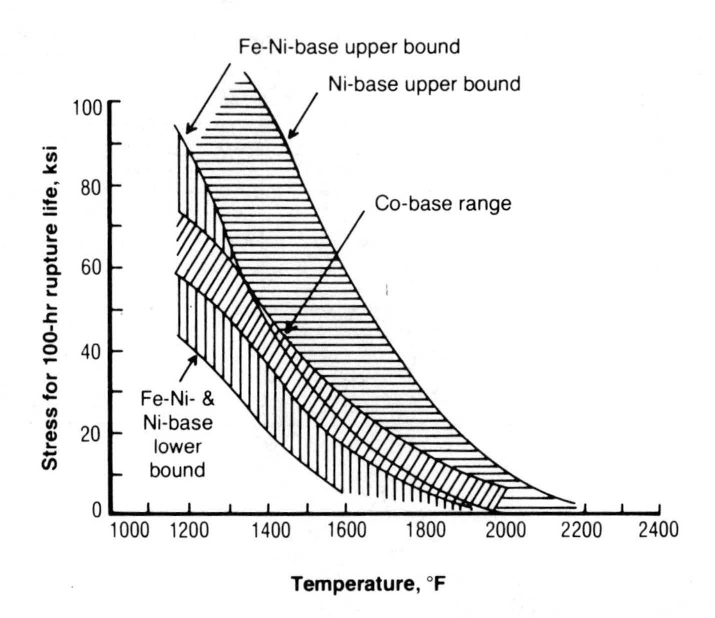

Rupture-strength behavior of superalloys.

Source: M.J. Donachie, Jr., "Introduction to Superalloys," in Superalloys: Source Book, M.J. Donachie, Jr., Ed., American Society for Metals, Metals Park, OH, 1984, p 3

Rupture Strength Comparison of γ'-Strengthened M-252 and Carbide-Strengthened X-40

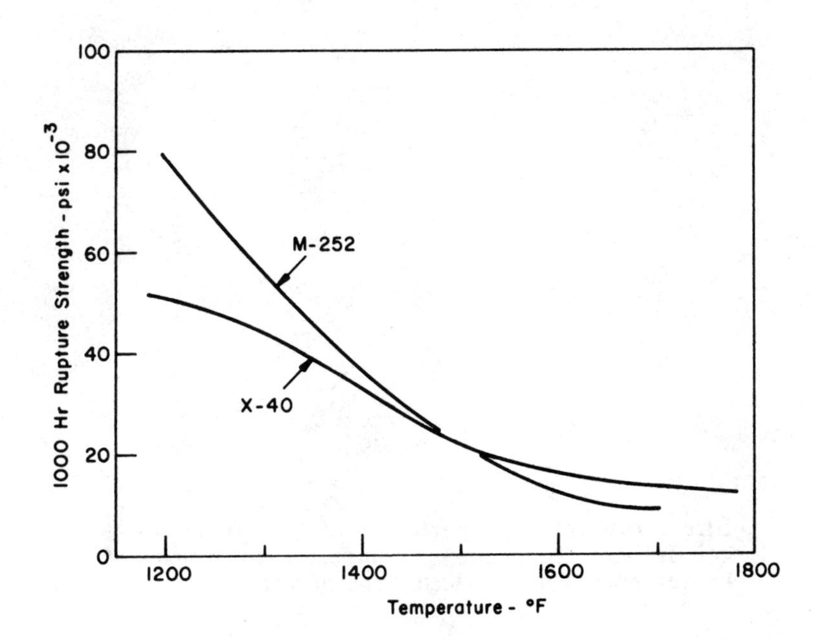

Comparison of 1000-h rupture strengths of γ'-strengthened nickel-base (M-252) and carbide-strengthened cobalt-base (X-40) superalloys.

The superior thermal stability of the carbide phases compared with γ' can be seen in a comparison of the variation of creep-rupture strengths of M-252 and X-40 with test temperature. The more rapid loss of strength of M-252 with increasing temperature is due to the more rapid agglomeration and dissolution of γ' at higher temperatures.

Source: J.D. Varin, "Microstructures and Properties of Superalloys," in The Superalloys, C.T. Sims and W.C. Hagel, Eds., John Wiley & Sons, New York, 1972, p 233-234

7.14 Rupture Strength Comparisons of MAR-M509, MAR-M302, and FSX-414/X45

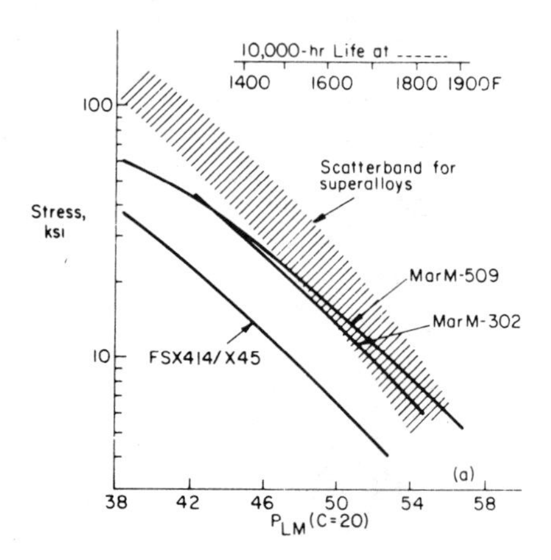

Stress-rupture properties of representative cobalt-base superalloys compared to those of contemporary nickel-base superalloys.

Rupture properties of Co alloys FSX-414, MAR-M509, and MAR-M302 are shown above. A band illustrating the rupture properties demonstrated by many leading high-strength Ni alloys also is plotted. The contrast is obvious; modern Co alloys such as MAR-M509 possess superior rupture properties where the time-temperature parameter (P_{LM}) is greater than approximately 50. However, at lower parameters and at higher strengths, rupture properties are as low as 50-75% of those of Ni alloys.

Source: C.T. Sims, "Cobalt-Base Alloys," in The Superalloys, C.T. Sims and W.C. Hagel, Eds., John Wiley & Sons, New York, 1972, p 168-169

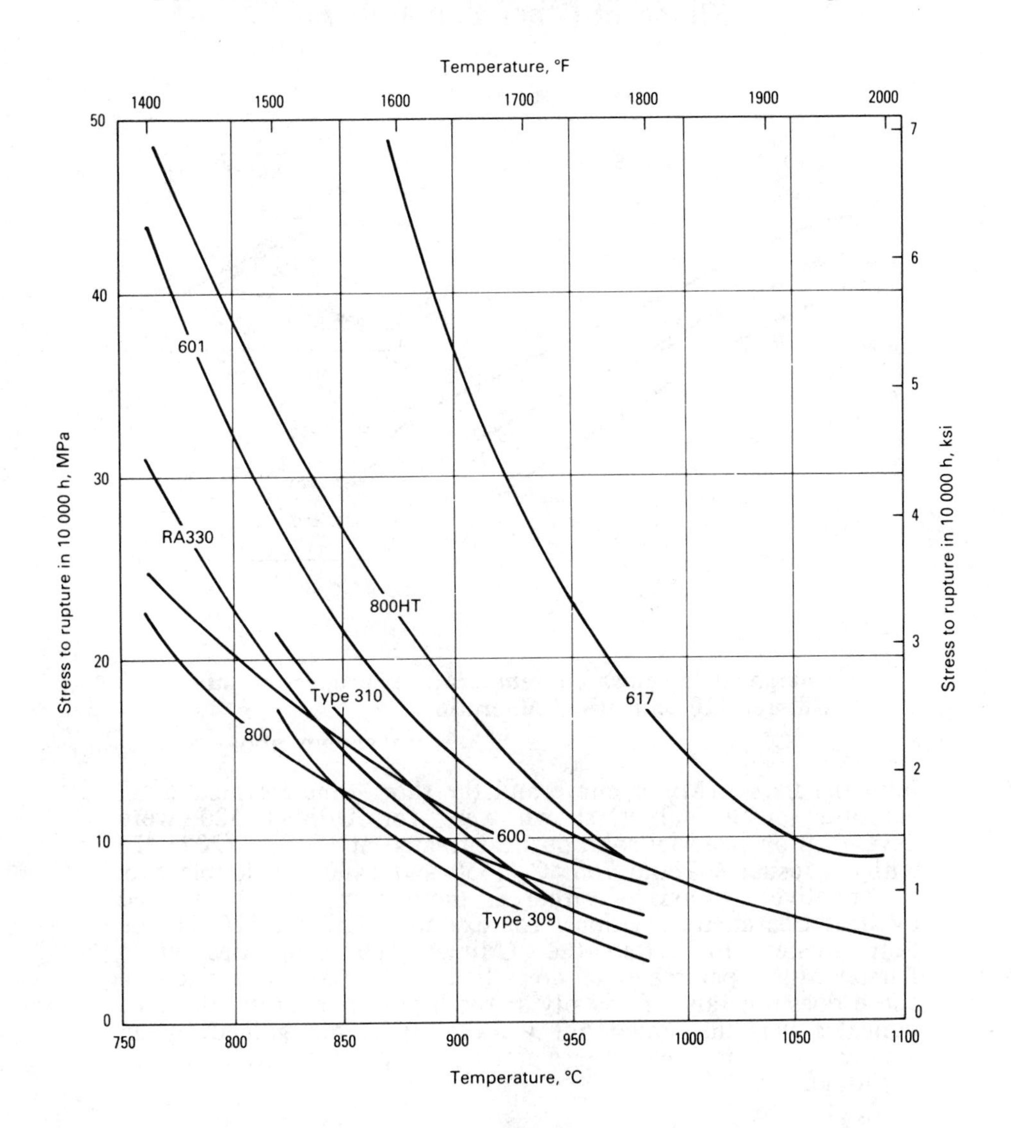

Stress-rupture data for various nickel-base and stainless steel alloys. Test duration was up to 10,000 h. (From manufacturer's published data)

Source: Metals Handbook, Ninth Edition, Volume 13, Corrosion, American Society for Metals, Metals Park, OH, 1987, p 655

Udimet 520: Elevated-Temperature Properties Compared to Those of Other Superalloys

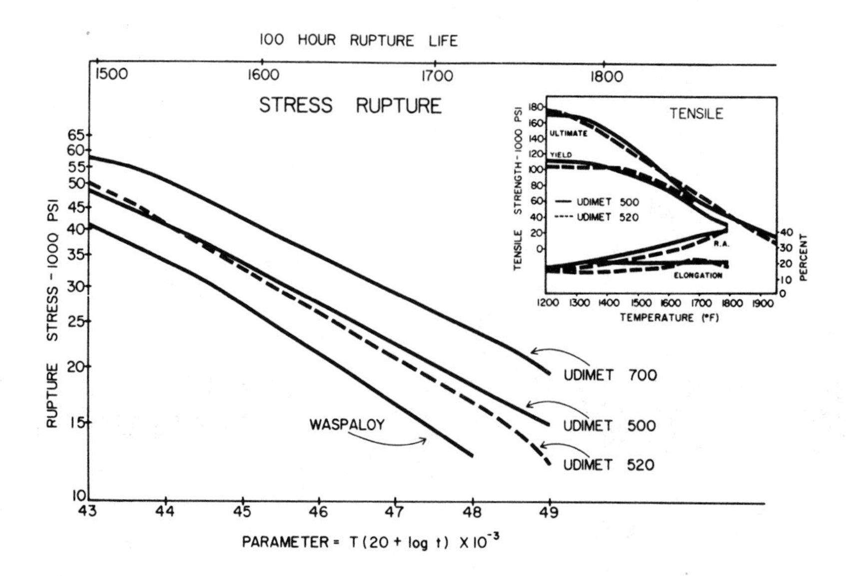

Comparative elevated-temperature properties of Udimet 520 and other superalloys.

Both the Larson-Miller curve and the short-time elevated-temperature tensile curves shown here for Udimet 520 were developed on material solution heat treated at 1107 °C (2025 °F) with the usual 843 and 760 °C (1550 and 1400 °F) double age. The relative positions of either of these curves can be shifted by heat treatment variations; for example, Udimet 520 can be heat treated to better the Udimet 500 properties at a Larson-Miller parameter of 46×10^3. Note that the hot tensile curve does not fall off sharply at the higher temperatures. This indicates that the added hot workability is not gained by an appreciable softening of the material, but probably by increased hot toughness.

Source: C.S. Freer, R.A. Woodall, and S. Abkowitz, "Nickel-Base Superalloys for the 1960's," in High Temperature Materials II (Proceedings of a Conference in Cleveland, OH), G.M. Ault, W.F. Barclay, and H.P. Munger, Eds., Interscience, New York, 1963, p 209, 212

8

Refractory Metals

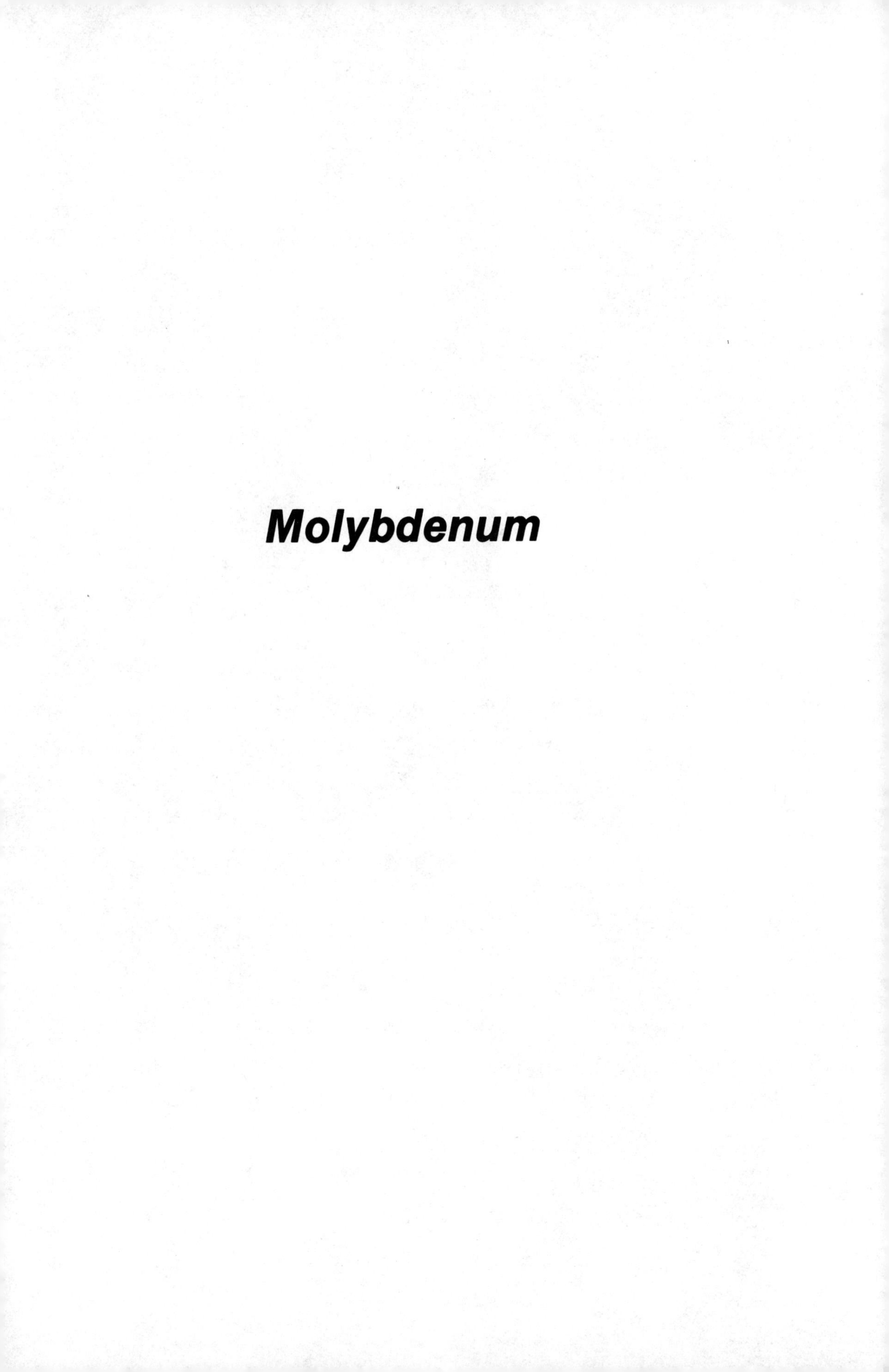

Molybdenum

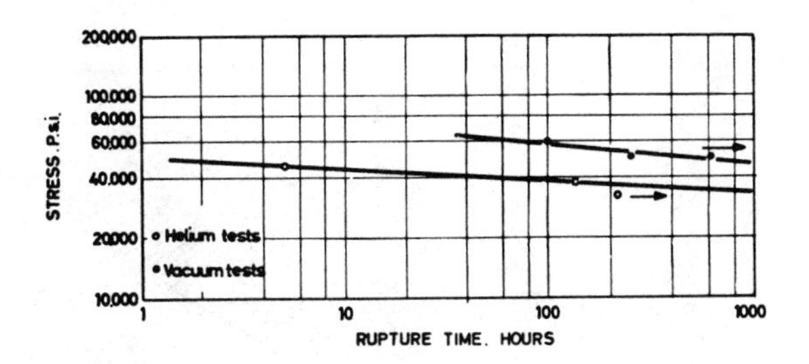

Creep-rupture properties of a P/M 1% Nb molybdenum alloy at 1800 °F in vacuum and helium atmospheres.

Source: D. Coutsouradis, "Evaluation of Test Methods for Refractory Metals," in The Science and Technology of Tungsten, Tantalum, Molybdenum, Niobium and Their Alloys, N.E. Promisel, Ed., Pergamon, New York, 1964, p 446

8.2 Arc-Cast Molybdenum: Effect of Load Application Rate and Pretest Annealing Temperature on Material Tested in Hydrogen

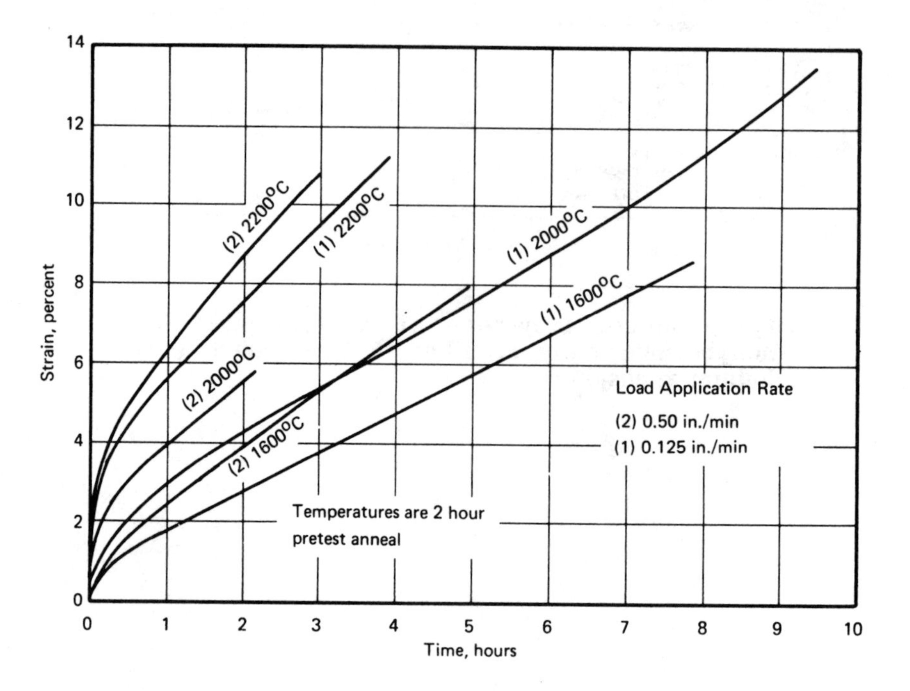

Creep curves for wrought arc-cast molybdenum sheet tested at 1600 °C and 2000 psi in hydrogen, showing the effect of load application rate and pretest annealing temperature.

Source: J.B. Conway and P.N. Flagella, Creep-Rupture Data for the Refractory Metals to High Temperatures, Gordon and Breach, New York, 1971, p 109

Arc-Cast Molybdenum: Tested in Argon

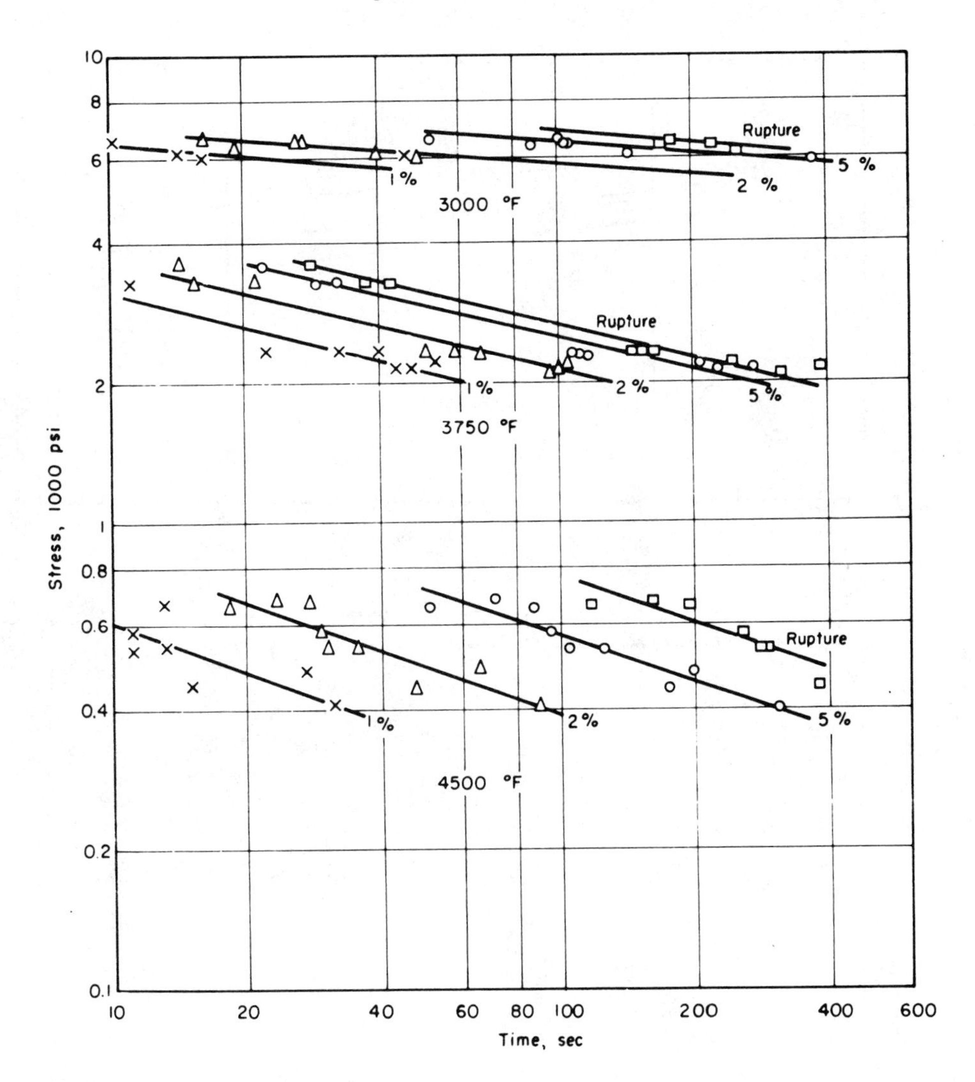

Creep stress vs time for various values of constant creep strain for arc-cast molybdenum sheet tested in argon at 3000, 3750, and 4500 °F.

The figure above summarizes short-time creep behavior at very high temperatures on sheet specimens resistance heated rapidly to test temperature after loading at room temperature. Strain values indicated above apparently refer to creep strain after test temperature is reached. Because the specimens were under load during the heating period, these values do not represent total strain.

Source: T.E. Tietz and J.W. Wilson, Behavior and Properties of Refractory Metals, Stanford University Press, Stanford, 1965, p 194

Arc-Cast (Recrystallized) Molybdenum

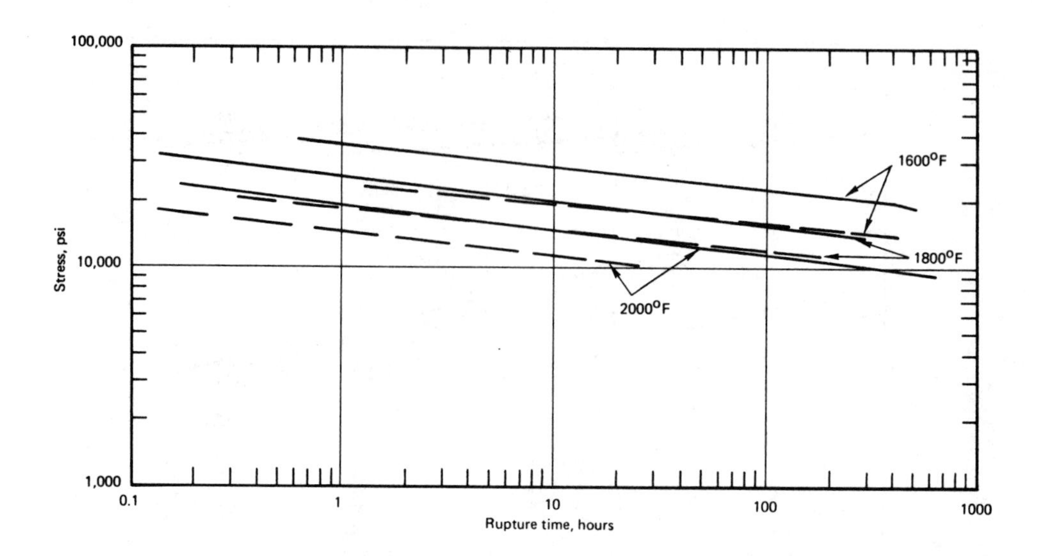

Comparison of stress-rupture data for arc-cast (recrystallized) molybdenum.

Source: J.B. Conway and P.N. Flagella, Creep-Rupture Data for the Refractory Metals to High Temperatures, Gordon and Breach, New York, 1971, p 585

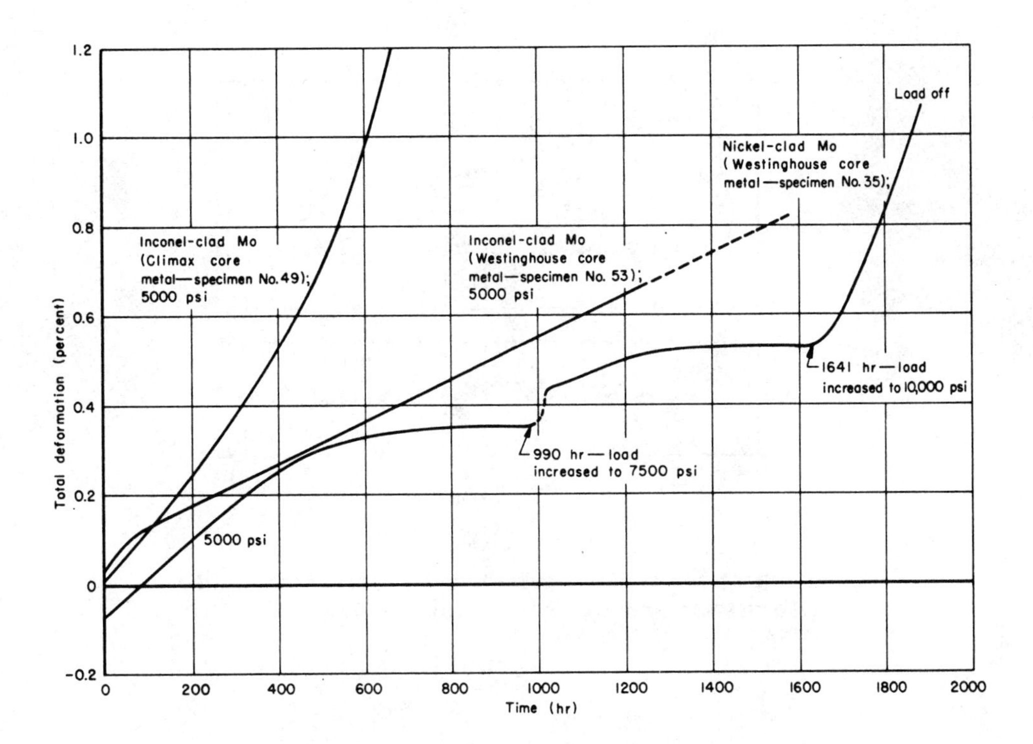

Creep curves for clad molybdenum (15% per side) at 1800 °F.

The figure above shows the creep curves for both Ni-clad and Inconel-clad Mo. The breaks in the curves of Ni-clad Westinghouse Mo at 990 and 1640 h were caused by increasing the stress at those times. A stress of 5000 psi applied to a Ni-clad Westinghouse creep specimen produced first-stage creep lasting about 450 h. The second-stage creep rate was quite low and would certainly have become nil had the initial stress been maintained. Increasing the load to 7500 psi at 990 h produced an elastic strain and a first-stage creep that shows on the curve. This lasted for nearly 200 h and the second-stage rate became nil. Further increase in the load to 10,000 psi produced third-stage creep (at a rate of 0.002% per hour); the test was discontinued after 1855 h with a final total elongation of 1.1%.

Source: W.L. Bruckart, R.I. Jaffee, S.J. Whalen, and B.W. Gonser, Molybdenum Alloys and Protection by Cladding, The Rand Corp., Santa Monica, CA, 1950, p 93, 95

8.6

Mo-0.5Ti: Creep Stress vs Time for Various Amounts of Total Deformation

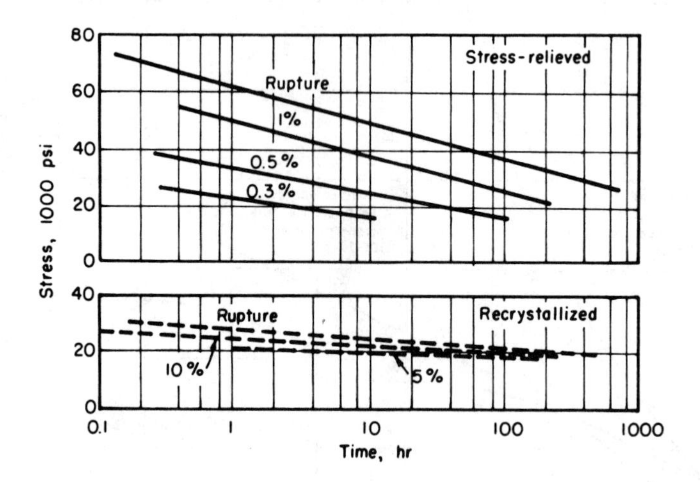

Creep stress vs time for various amounts of total deformation for Mo-0.5Ti alloy at 2000 °F.

Source: T.E. Tietz and J.W. Wilson, Behavior and Properties of Refractory Metals, Stanford University Press, Stanford, 1965, p 197

Mo-0.5Ti: Hot-Rod Buckets vs Rolled Bar Stock

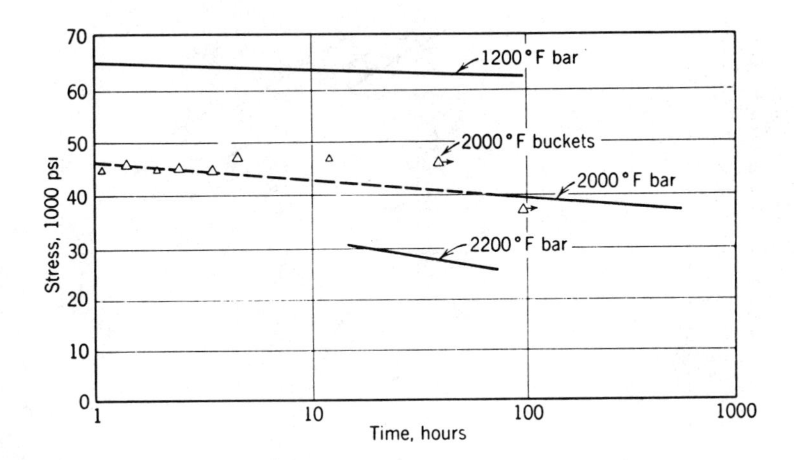

Stress-rupture properties of Mo-0.5Ti hot-rod buckets and rolled bar stock.

(1) Rolled bar-stock properties are minimum properties from 14 heats as rolled plus 2100 °F for 1 h anneal. (2) The 1200 °F 100-h strength of hot-rod buckets was >45,000 psi. (3) The 1200 °F notched stress rupture was 100,000 psi for 39 h.

Source: R.G. Frank, "Properties of Molybdenum-Alloy Turbine Buckets," in High Temperature Materials, R.E. Hehemann and G.M. Ault, Eds., American Institute of Mining, Metallurgical, and Petroleum Engineers, Warrendale, PA, 1959, p 284

Mo-0.5Ti: Short-Time Creep Properties of Arc-Cast Sheet

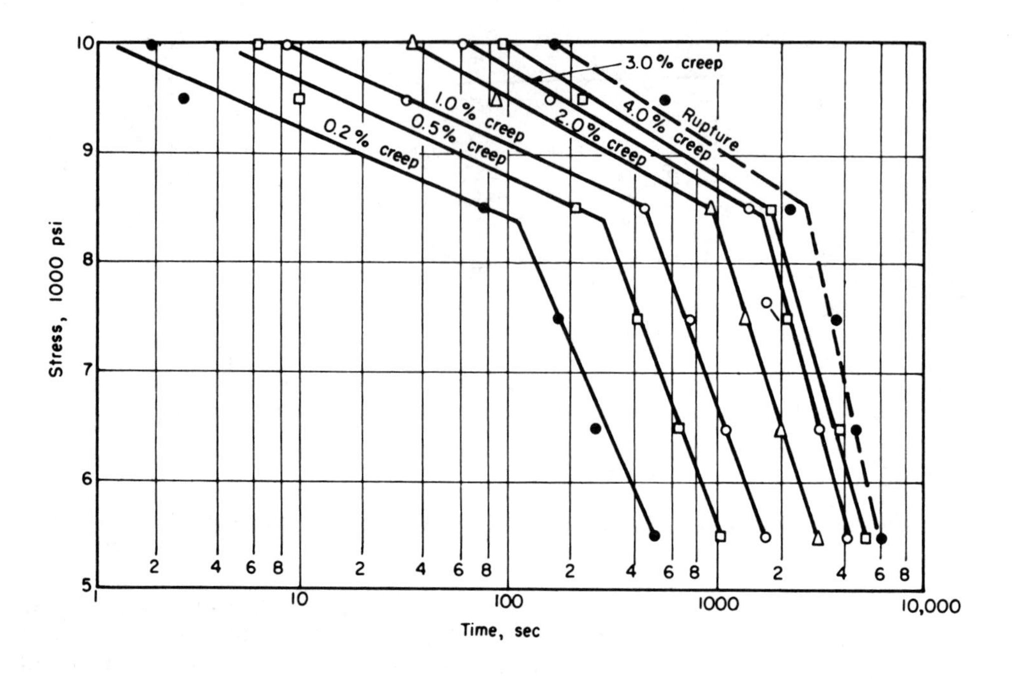

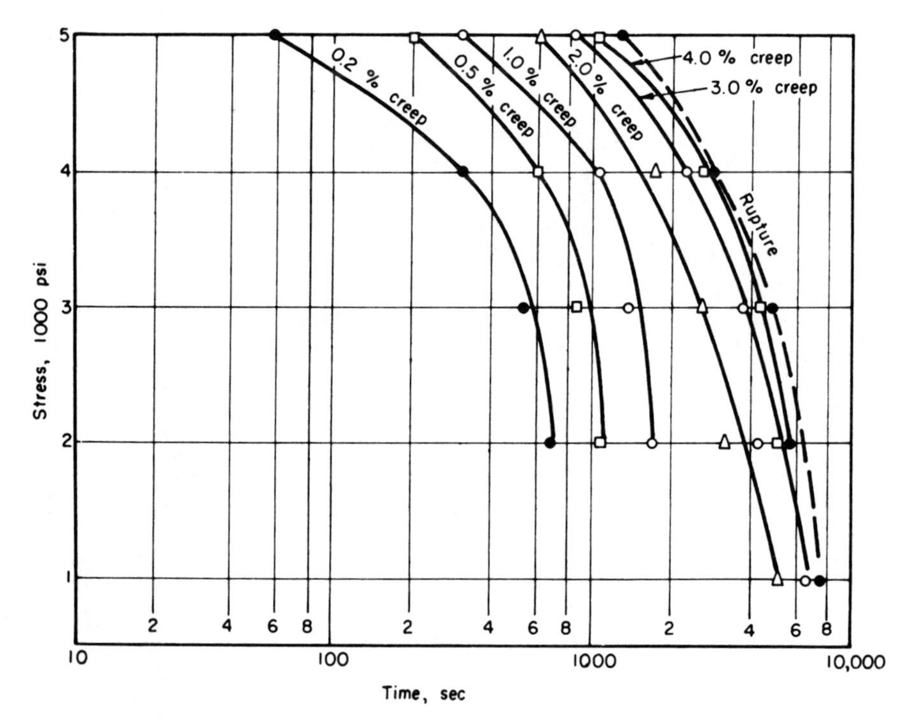

Short-time creep properties of arc-cast Mo-0.5Ti sheet at 1538 °C (2800 °F) (top) and 1705 °C (3100 °F) (bottom).

Source: T.E. Tietz and J.W. Wilson, Behavior and Properties of Refractory Metals, Stanford University Press, Stanford, 1965, p 198

Mo-1.5Nb: Arc-Cast Sheet Tested in Vacuum

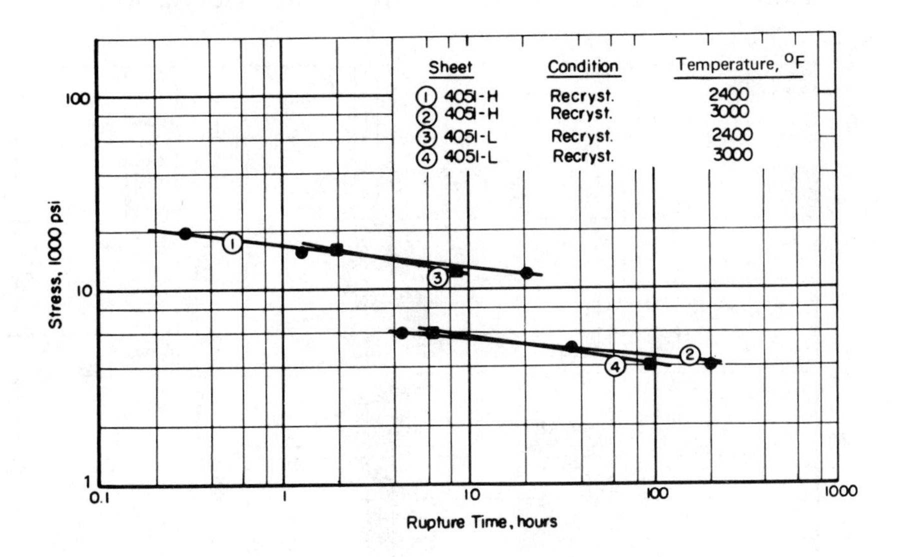

Stress-rupture data for arc-cast Mo-1.5Nb sheet tested in vacuum.

Source: J.B. Conway and P.N. Flagella, Creep-Rupture Data for the Refractory Metals to High Temperatures, Gordon and Breach, New York, 1971, p 628

8.10 Mo-50Re: Stress-Rupture and Creep Characteristics Compared to Those of Rhenium and Molybdenum

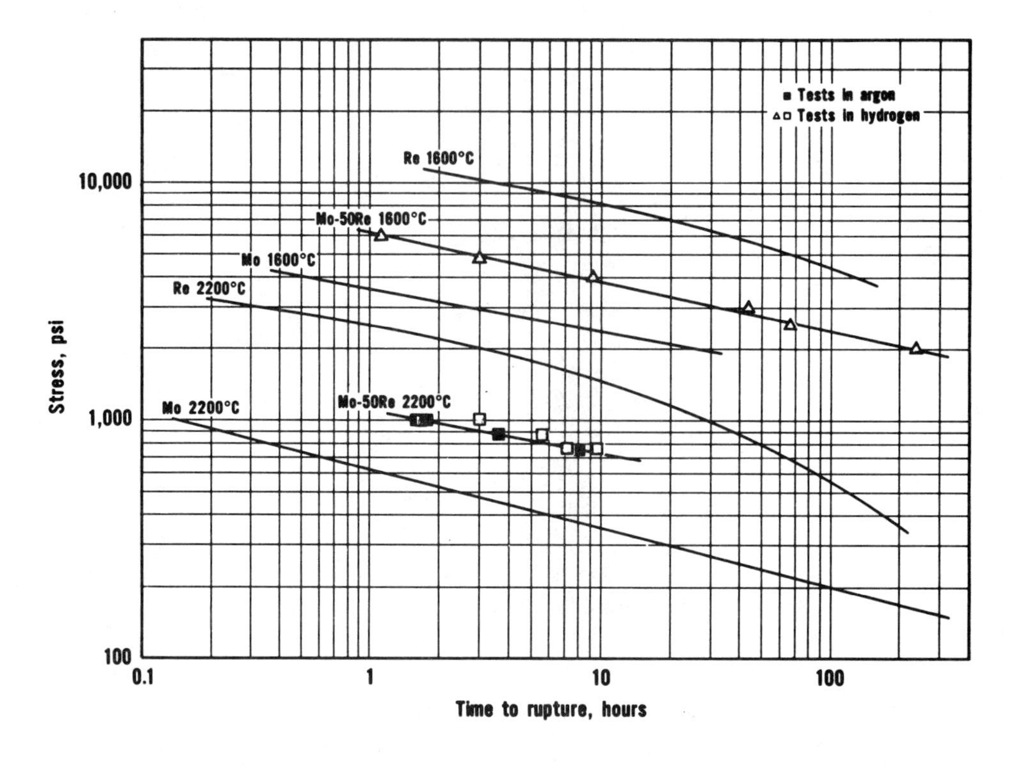

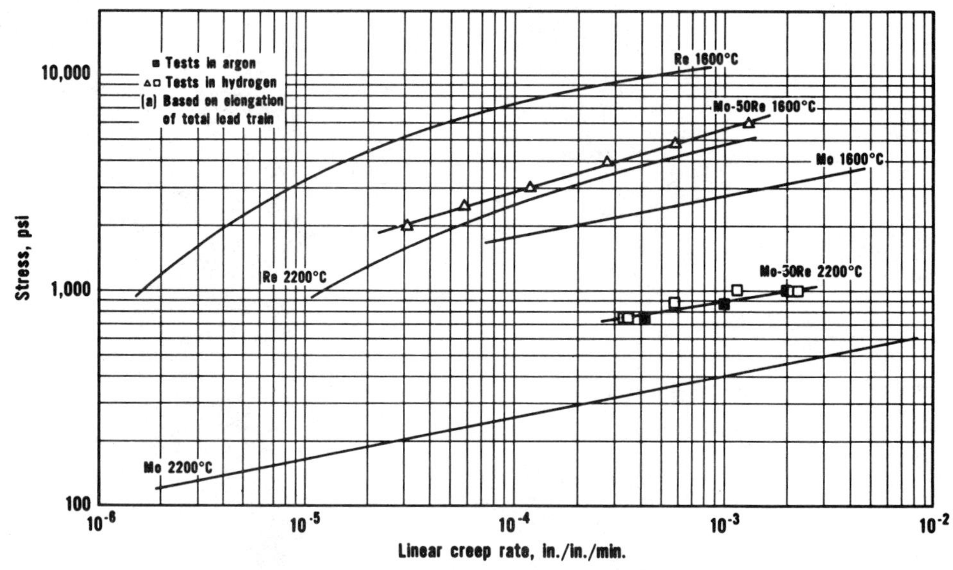

Stress-rupture characteristics (top) and creep rate characteristics (bottom) of Mo-50Re compared to rhenium and molybdenum.

Mo-50Re: Stress-Rupture and Creep Characteristics Compared to Those of Rhenium and Molybdenum (Continued)

The stress-rupture and creep data for sintered Mo-50Re obtained at 1600 and 2200 °C in hydrogen are presented in the figures above. The data are compared to the results obtained for Re and Mo at the same temperatures in hydrogen. The addition of the Re to Mo is seen to increase the rupture life and creep resistance of Mo-50Re by approximately an order of magnitude over unalloyed arc-cast Mo.

Source: P.N. Flagella and C.O. Tarr, "Creep-Rupture Properties of Rhenium and Some Alloys of Rhenium at Elevated Temperatures," in Refractory Metals and Alloys IV, Volume II (Proceedings of a Conference in French Lick, IN), R.I. Jaffe, G.M. Ault, J. Maltz, and M. Semchyshen, Eds., Gordon and Breach, New York, 1967, p 837-839

Molybdenum: Creep Characteristics

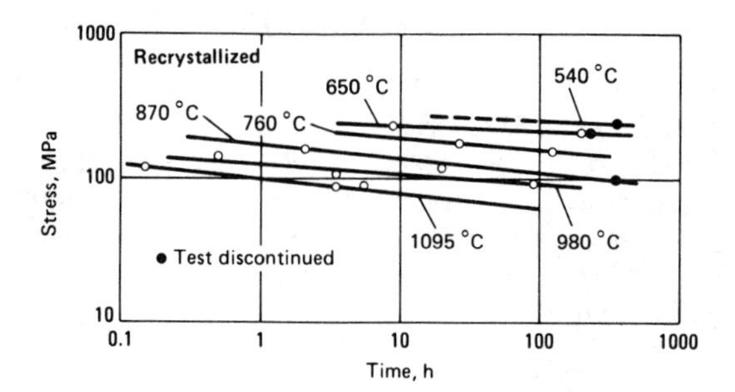

Creep characteristics of molybdenum.

Source: Metals Handbook, Ninth Edition, Volume 2, Properties and Selection: Nonferrous Alloys and Pure Metals, American Society for Metals, Metals Park, OH, 1979, p 775

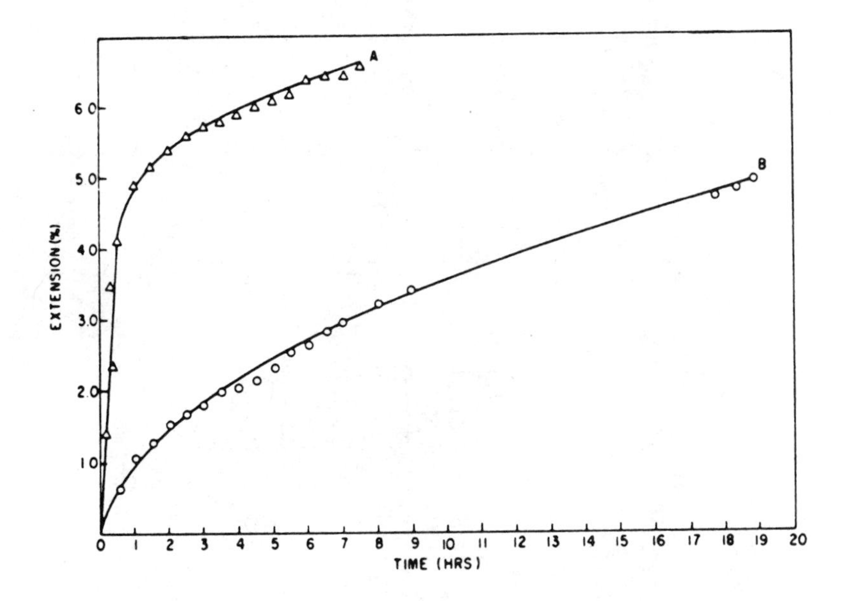

Effect of polygonization on creep properties of molybdenum at 1000 °C.

Strain hardening and stable substructure formation are very important creep-strengthening mechanisms. The effect of polygonization on the creep properties of Mo is shown above. It was found that 5% prestrain at room temperature followed by a high-temperature polygonization anneal significantly reduced primary creep at 1000 °C. In 8 h, the unpolygonized sample had twice the creep extension of the polygonized sample. The secondary (minimum) creep rates, however, appeared to be similar for the two materials.

Source: R.A. Perkins, "The Effect of Thermal-Mechanical Treatments on the Structure and Properties of Refractory Metals," in Refractory Metal Alloys (Proceedings of a Symposium in Washington, D.C.), I. Machlin, R.T. Begley, and E.D. Weisert, Eds., Plenum, New York, 1968, p 108, 110

Molybdenum: Rupture Strength

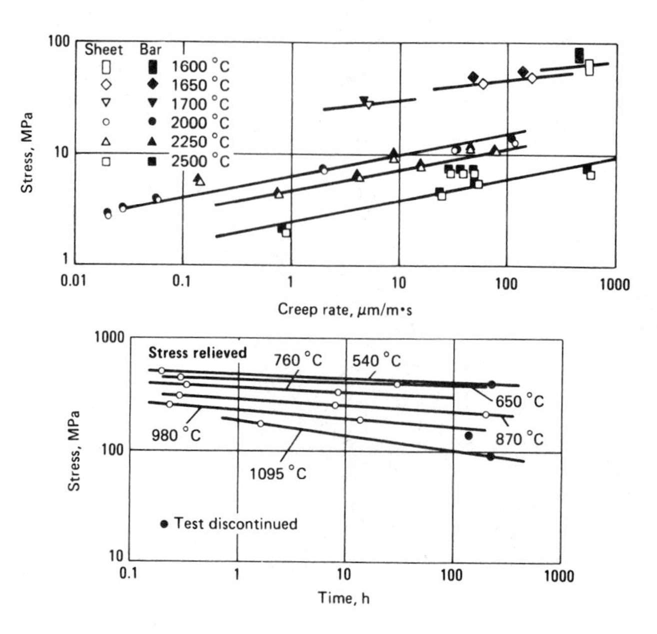

Rupture strength of molybdenum.

Source: Metals Handbook, Ninth Edition, Volume 2, Properties and Selection: Nonferrous Alloys and Pure Metals, American Society for Metals, Metals Park, OH, 1979, p 775

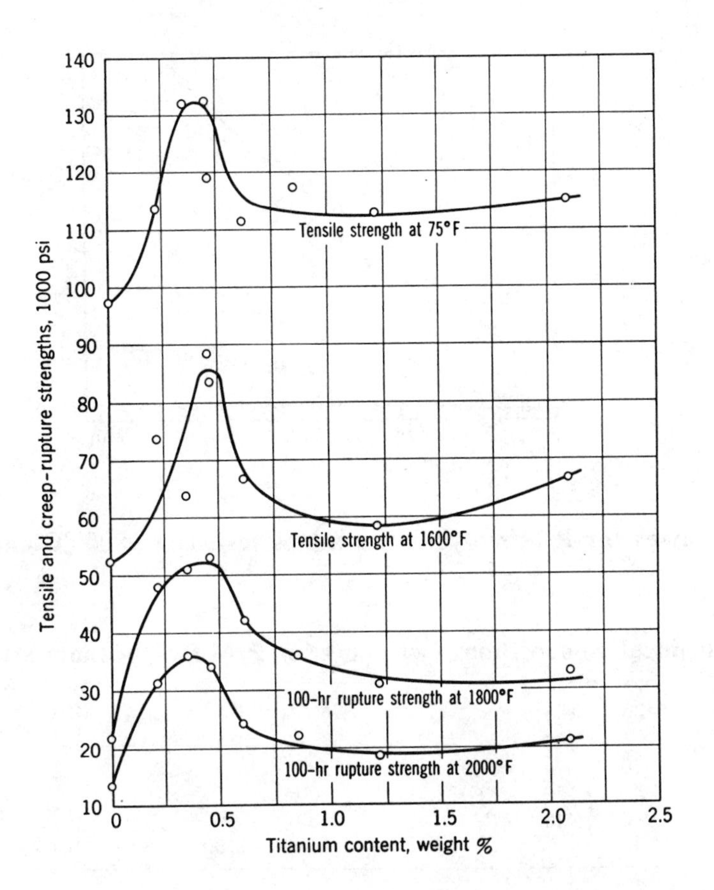

Effect of titanium content of molybdenum-titanium alloys on the tensile and creep-rupture strengths of 5/8-in. rolled stress-relieved rounds.

Source: J.J. Harwood and M. Semchyshen, "Molybdenum, Its Alloys and Its Protection," in High Temperature Materials, R.F. Hehemann and G.M. Ault, Eds., American Institute of Mining, Metallurgical, and Petroleum Engineers, Warrendale, PA, 1959, p 263

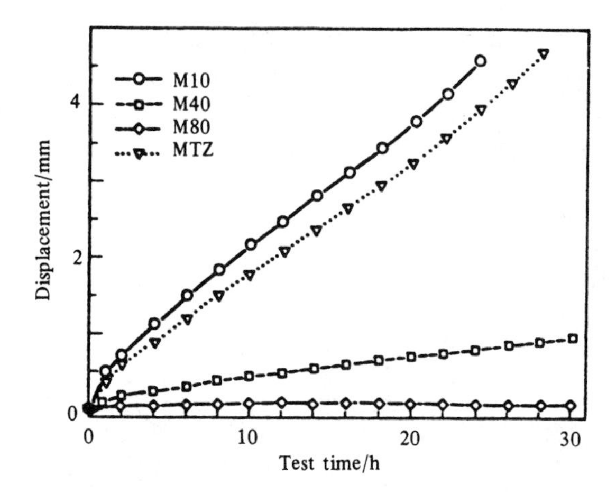

Creep curves for P/M molybdenum alloys tested at 1800 °C under 9.8 MPa.

Chemical compositions (wt ppm) for P/M molybdenum alloys

Alloy	K	Si	Al	O	Ti	Zr	C
M10	26	8	5	35			
M40	150	60	6	200			
M80	440	790	5	460			
MTZ	-	-	-	168	7700	460	101

Displacements of the M10 and MTZ samples increased very rapidly, and the samples ruptured in almost the same time. The M40 alloy was characterized by a smaller displacement and a longer time to rupture than the M10 and MTZ alloys. The M80 alloy showed a very small displacement of 0.16 mm after 250 h, thus demonstrating its excellent creep strength. The latter can be ascribed to the coarse and elongated recrystallized grain structure due to K and Si doping and rolling, and strengthened recrystallized grains through dispersion-hardening of dopant bubbles.

Source: Y. Fukasawa, S. Ogura, H. Koizumi, and T. Matsumoto, "Very High-Temperature Creep Behaviour of Powder Metallurgy Produced (P/M) Molybdenum Alloys," High Temperatures – High Pressures, Vol. 18, 1986, p 329, 331-332

Stress-Relieved vs Recrystallized Molybdenum (0.015% C): Tested in Vacuum

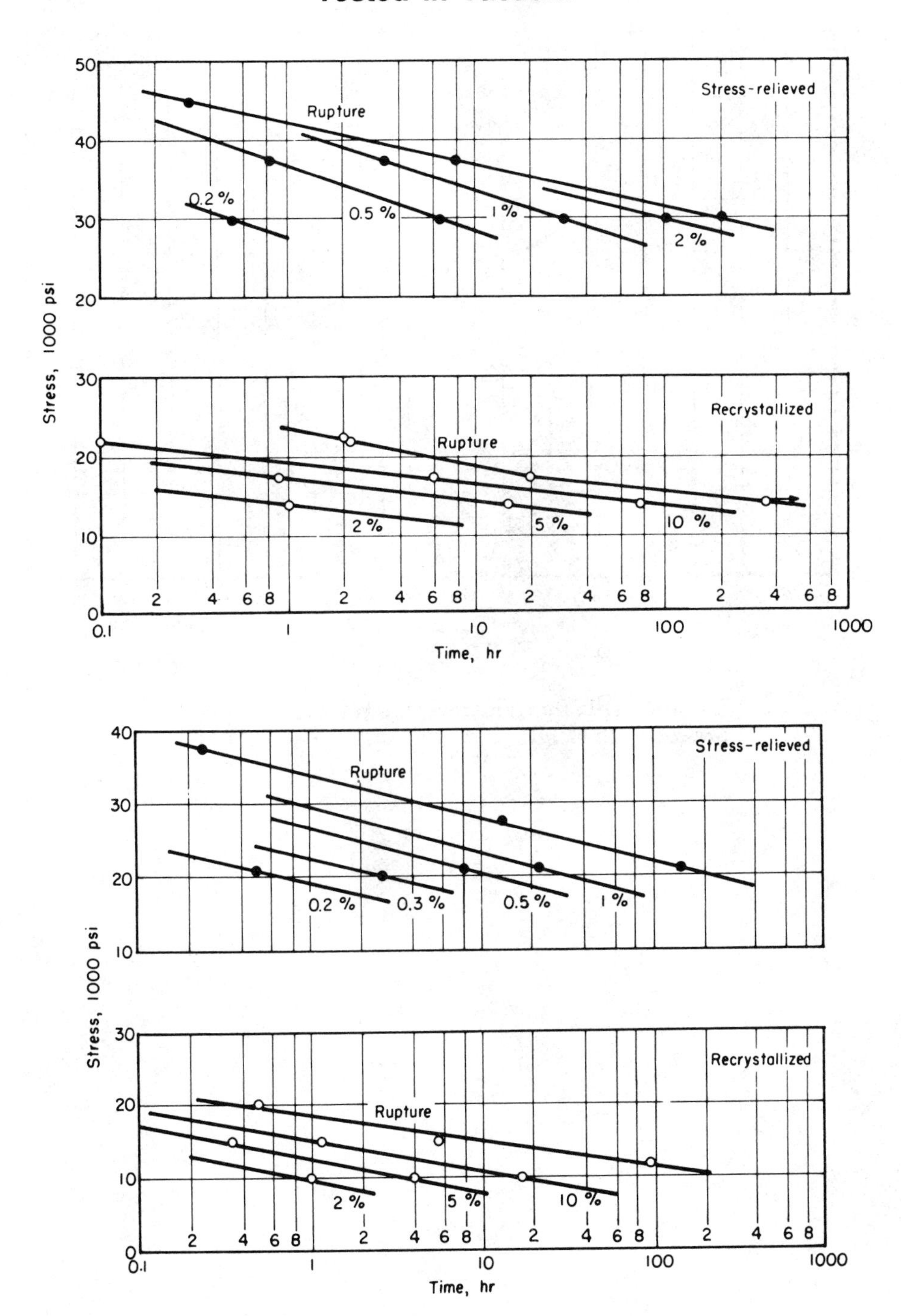

Creep stress vs time for various amounts of total deformation for molybdenum (0.015% C) tested in vacuum at 1600 °F.

Source: T.E. Tietz and J.W. Wilson, Behavior and Properties of Refractory Metals, Stanford University Press, Stanford, 1965, p 191

TZC: Arc-Cast

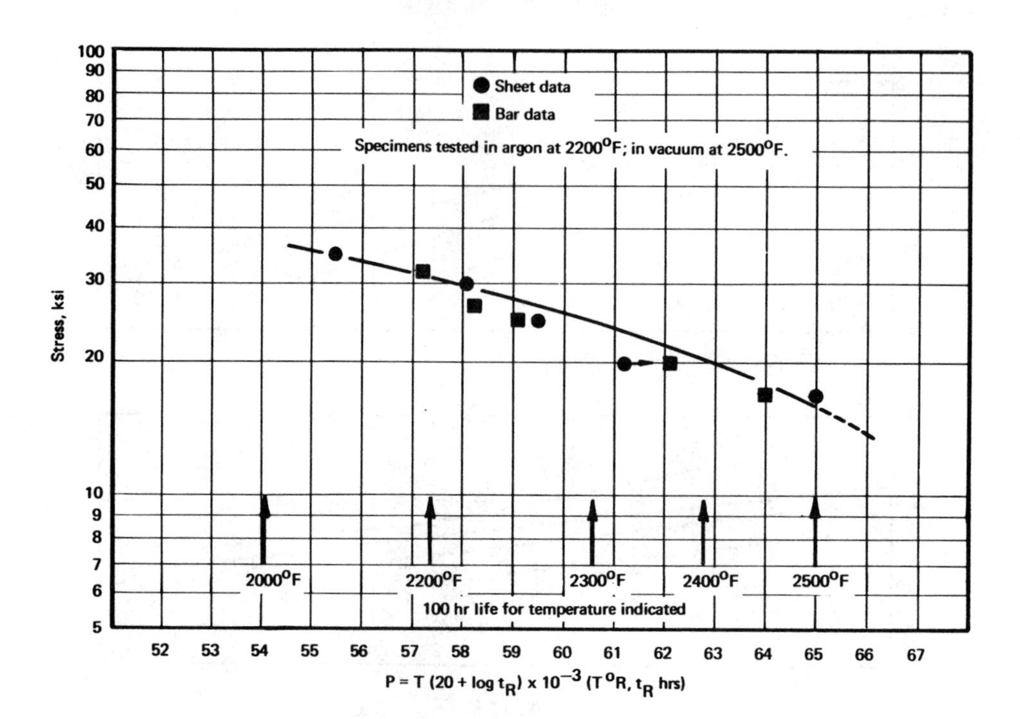

Larson-Miller parameter plot for rupture data for arc-cast TZC alloy.

Source: J.B. Conway and P.N. Flagella, Creep-Rupture Data for the Refractory Metals to High Temperatures, Gordon and Breach, New York, 1971, p 657

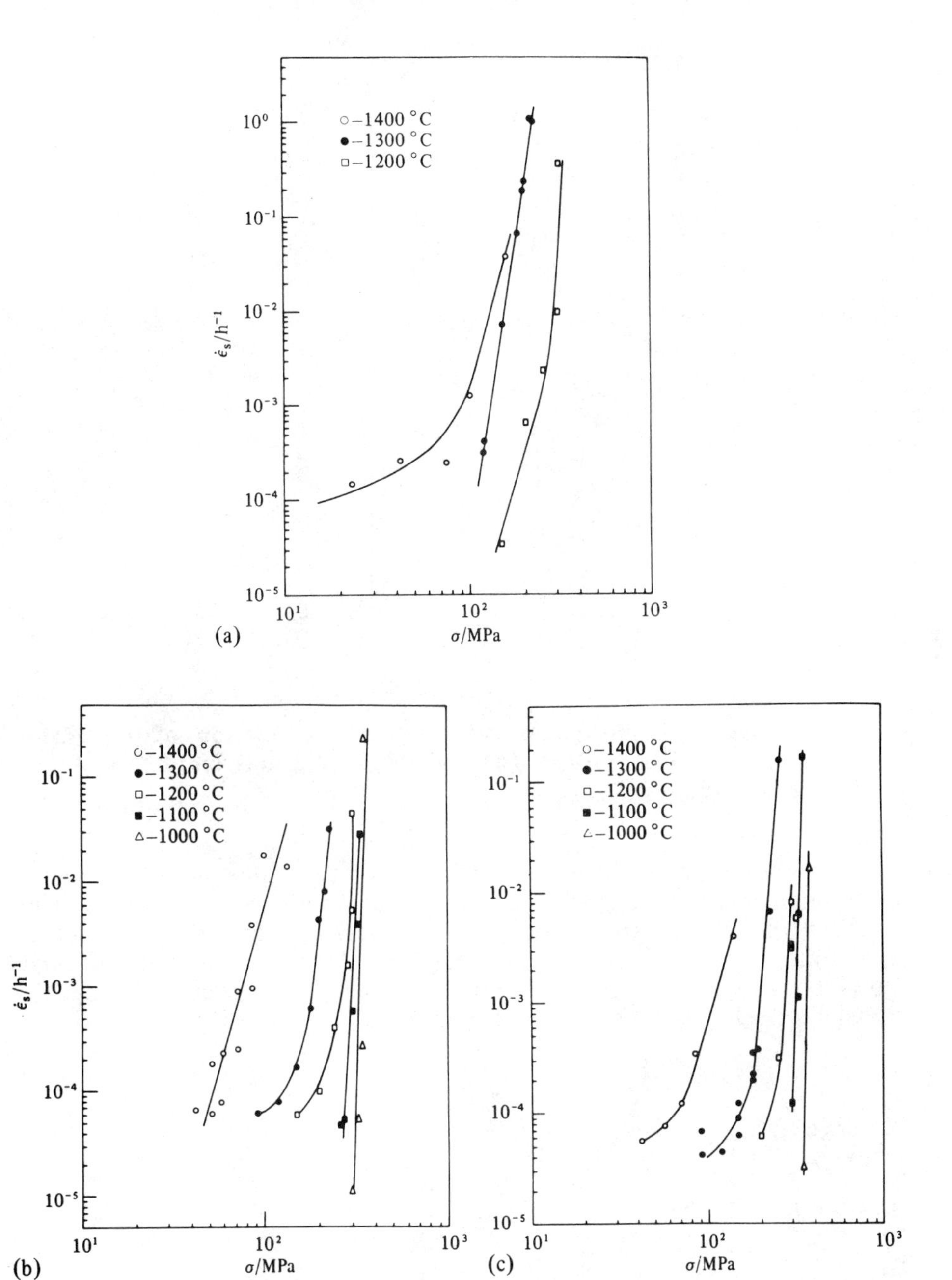

Dimensionless steady-state creep rate, $\dot{\varepsilon}_D$, as a function of dimensionless stress σ/G, for the TZM alloy: (a) recrystallized; (b) rod 222, 18.6% reduction; (c) rod 221, 36% reduction.

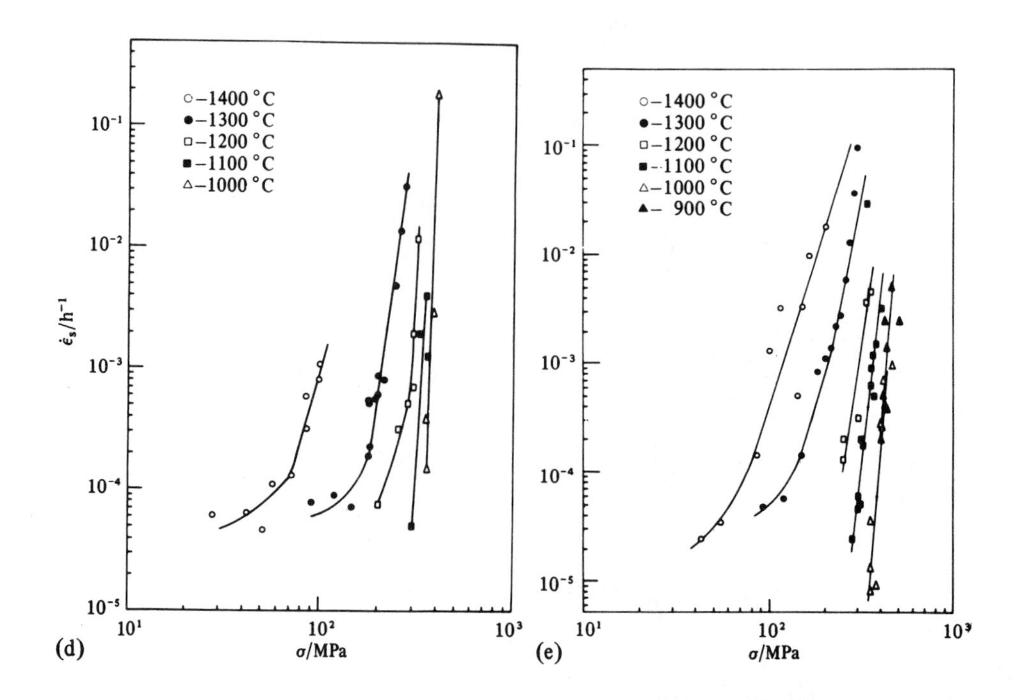

Dimensionless steady-state creep rate, $\dot{\varepsilon}_D$, as a function of dimensionless stress σ/G, for the TZM alloy: (d) rod 212, 50.2% reduction; and (e) rod 211, 75% reduction.

The figure above shows a dimensionless plot of the value $\dot{\varepsilon}_D = F\dot{\varepsilon}$ versus σ/G. Creep results obtained at 1300 and 1400 °C are not included in these figures. In spite of the rather large scatter, all the data points can be connected by straight lines for the high-stress regions. The departure from the linearity in the low-stress region can be due either to a change of creep mechanism or to some artefact, such as a very long region of primary creep.

Source: D. Agronov, E. Freund, and A. Rosen, "The Contribution of Particle Strengthening on the Creep Resistance of Molybdenum Alloys," High-Temperatures – High Pressures, Vol. 18, 1986, p 156-157

TZM: Steady-State Creep Rate

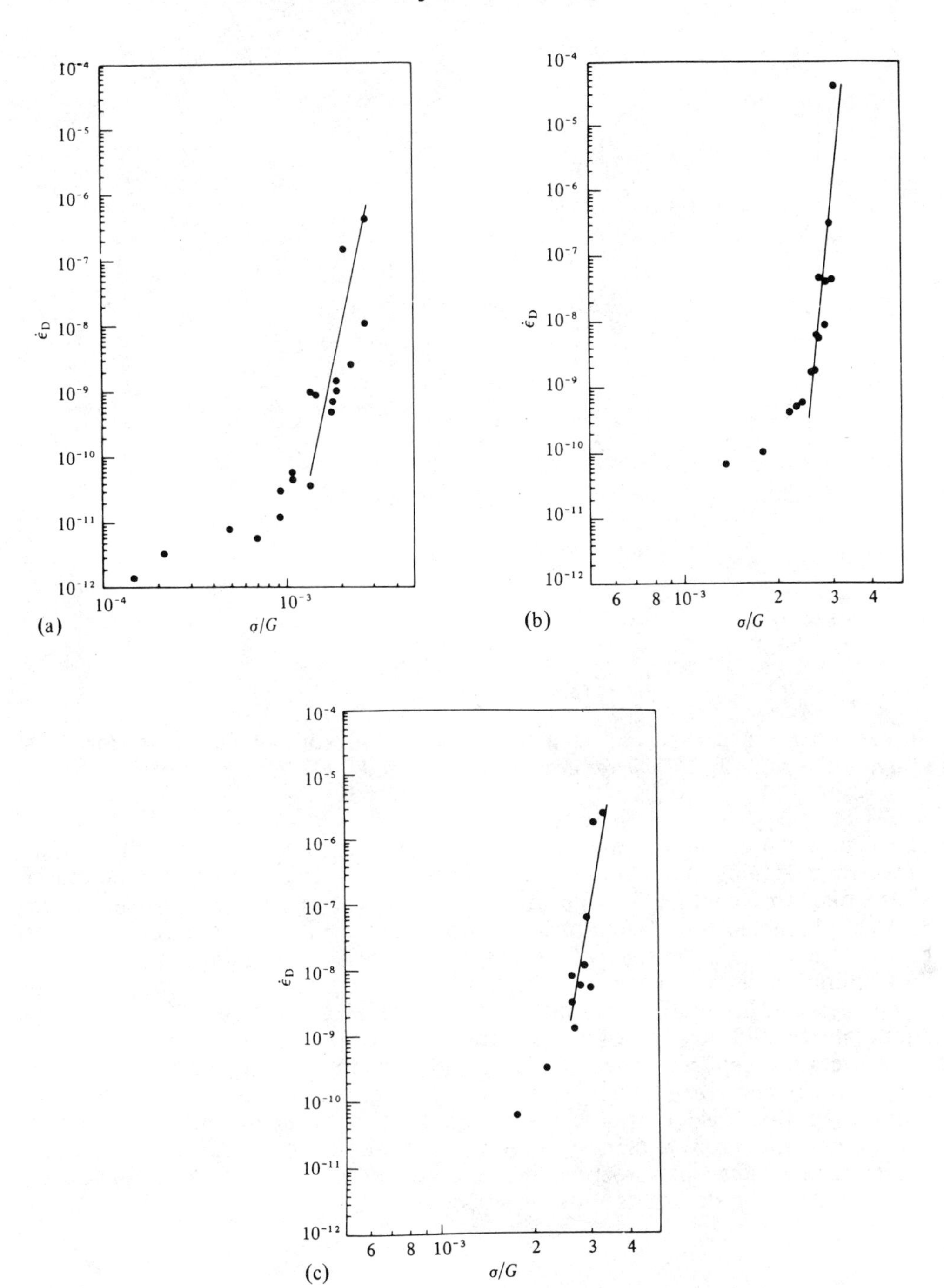

Steady-state creep rate, $\dot{\varepsilon}_s$, as a function of the applied stress, σ, for TZM alloy: (a) recrystallized; (b) rod 222, 18.6% reduction; (c) rod 221, 36% reduction.

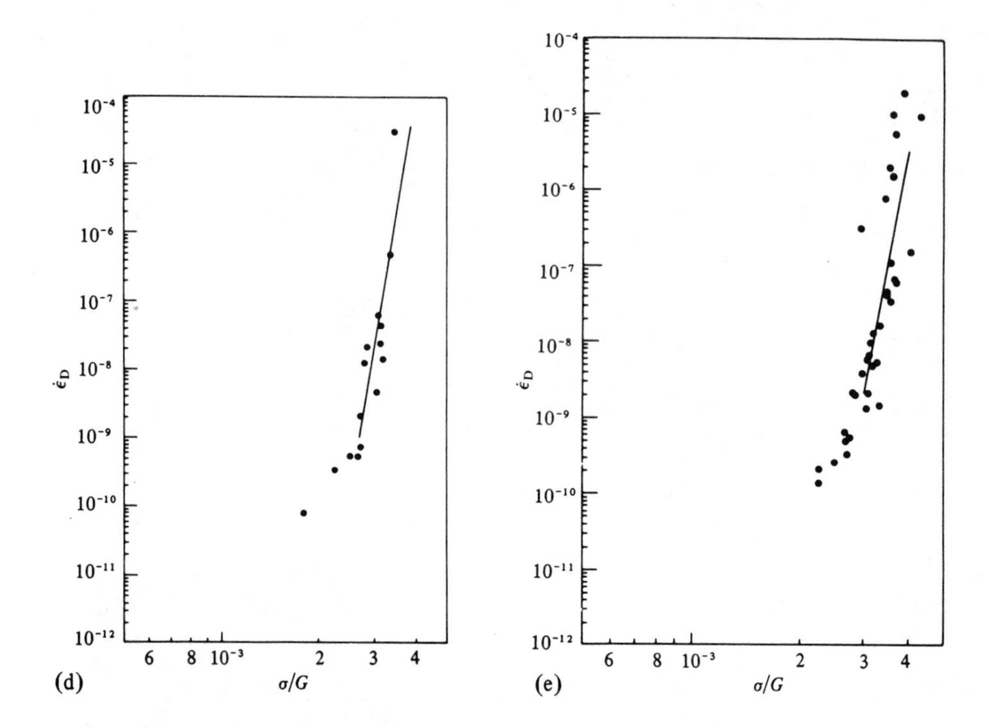

Steady-state creep rate, $\dot{\varepsilon}_s$, as a function of the applied stress, σ, for TZM alloy: (d) rod 212, 50.2% reduction; and (e) rod 211, 75% reduction.

The figure shows the variation of the steady-state creep rate with the applied stress in a log-log plot, for various degrees of reduction and temperatures. These diagrams contain a large amount of data, collected over a few years. Some of the creep tests lasted as long as 6 weeks, while others took only a few hours. The determination of steady-state creep is quite accurate for short-duration tests (high creep rate), but not so exact for the long-duration creep tests. The main reasons for this are as follows. (1) At high temperatures and low stresses, the specimen can partially recrystallize and therefore its creep resistance decreases. (2) When the applied stress is very low, the primary creep region is extremely long, and the test does not end in failure. In this case (with recrystallized specimens below 40 MPa applied stress), one measures a primary creep rate that can be orders of magnitude higher than steady-state creep. (3) The accuracy of the method for measuring steady-state creep decreases with decreasing ε.

Source: D. Agronov, E. Freund, and A. Rosen, "The Contribution of Particle Strengthening on the Creep Resistance of Molybdenum Alloys," High-Temperatures – High Pressures, Vol. 18, 1986, p 154-157

Unalloyed Arc-Cast Molybdenum: Effect of Temperature on Notch Strength

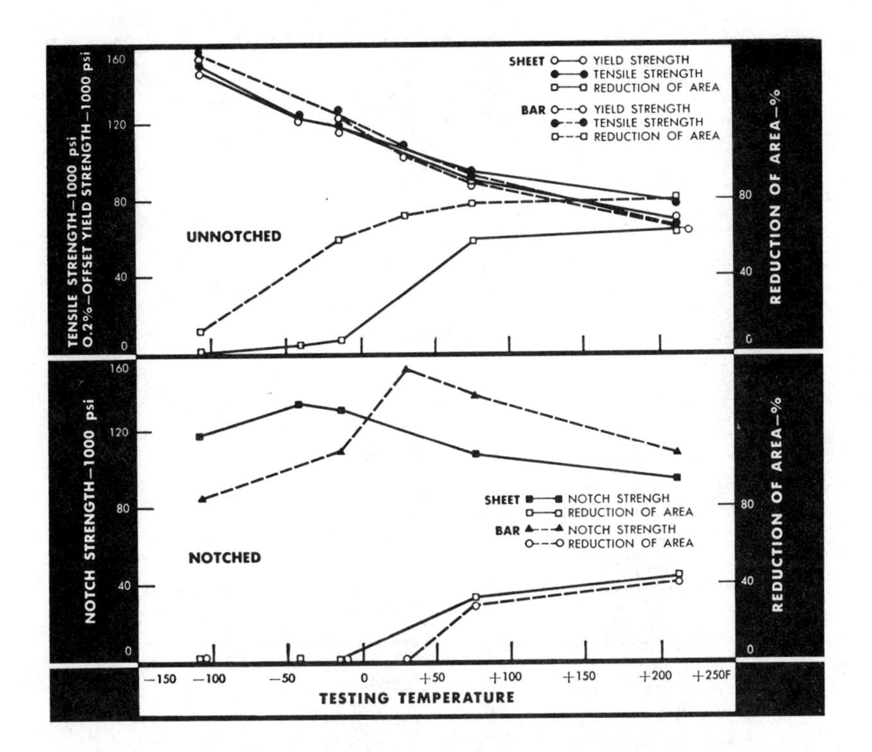

Effect of temperature on the notch strength of unalloyed arc-cast molybdenum stress relieved at 1830 °F (15 min).

7/16-in.-diam bars were machined to unnotched specimens having a 0.212-in.-diam reduced section with a 1-in. gage length and to notched specimens with a 0.300-in. diameter, and a 0.212-in. notch diameter (60° V with a 0.011-in. root radius) to give a theoretical stress concentration factor of K_T = 3. Unnotched sheet specimens had a 0.250-in.-wide reduced section with a 1-in. gage length; notched sheet specimens had a 0.500-in.-wide reduced section with a 60° V-notch (0.250-in. wide at base with a 0.021-in. root radius) to give K_T = 3. Head travel speed was 0.02 in./min for unnotched specimens and 0.005 in./min for notched specimens.

Source: Molybdenum Metal, Climax Molybdenum Co., New York, 1960, p 39

8.24 Unalloyed Arc-Cast Molybdenum: Effect of Temperature on Time Required for Various Amounts of Creep

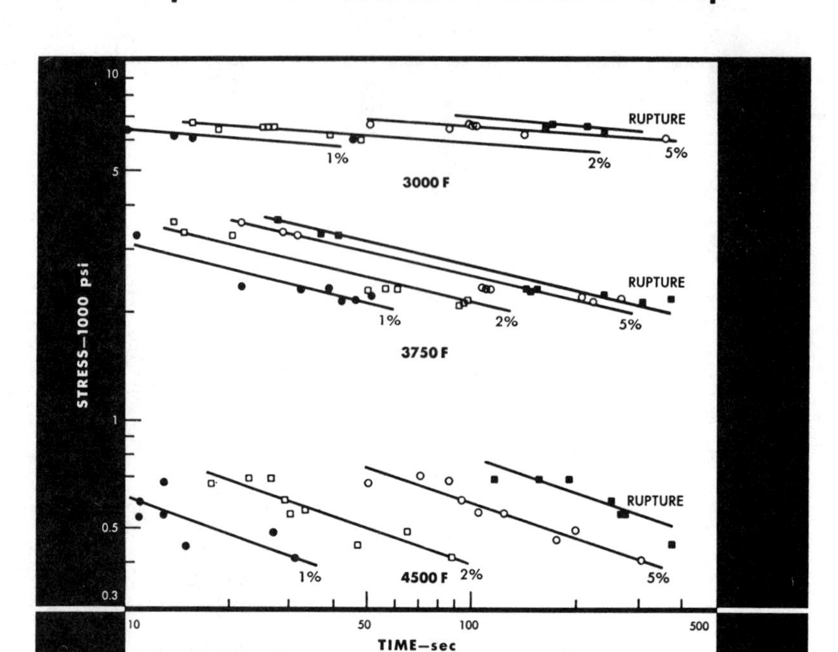

Effect of temperature on time required for various amounts of creep in unalloyed arc-cast molybdenum.

Specimens were loaded at room temperature, heated rapidly to test temperature in argon atmosphere, and creep measurements were made for a maximum time of 5 min. Longitudinal specimens with gage dimensions of $^1/_2$ in. by 2 in. were taken from 0.060-in.-thick sheet.

Source: Molybdenum Metal, Climax Molybdenum Co., New York, 1960, p 47

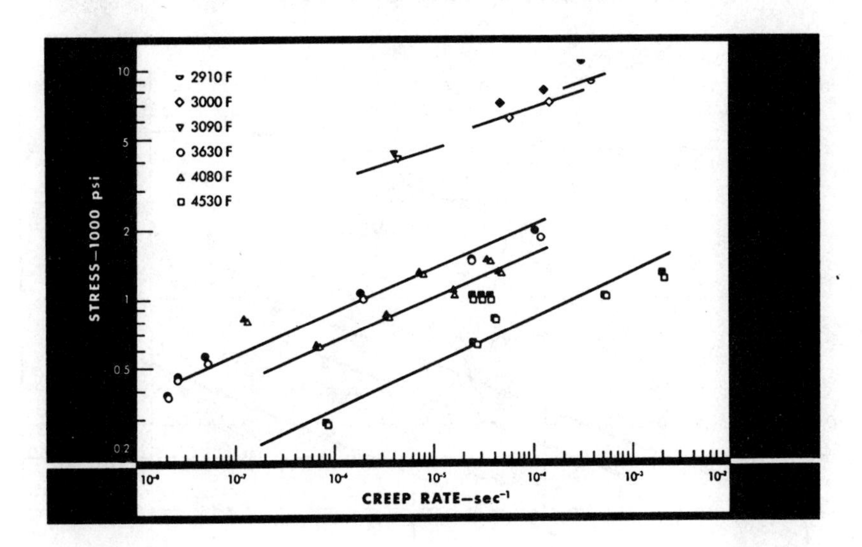

Short-time creep properties of unalloyed P/M molybdenum.

Open points represent engineering stress vs linear creep rate. Solid points represent true creep stress corresponding to minimum true creep rates. The gage dimensions of specimens 1/4-in. in diameter by 3/4-in. long were held at testing temperature for 30 min before loading. Except for one test in vacuum, all tests were run in helium.

Source: Molybdenum Metal, Climax Molybdenum Co., New York, 1960, p 46

Unalloyed P/M Molybdenum: Short-time Creep and Stress-Rupture Data

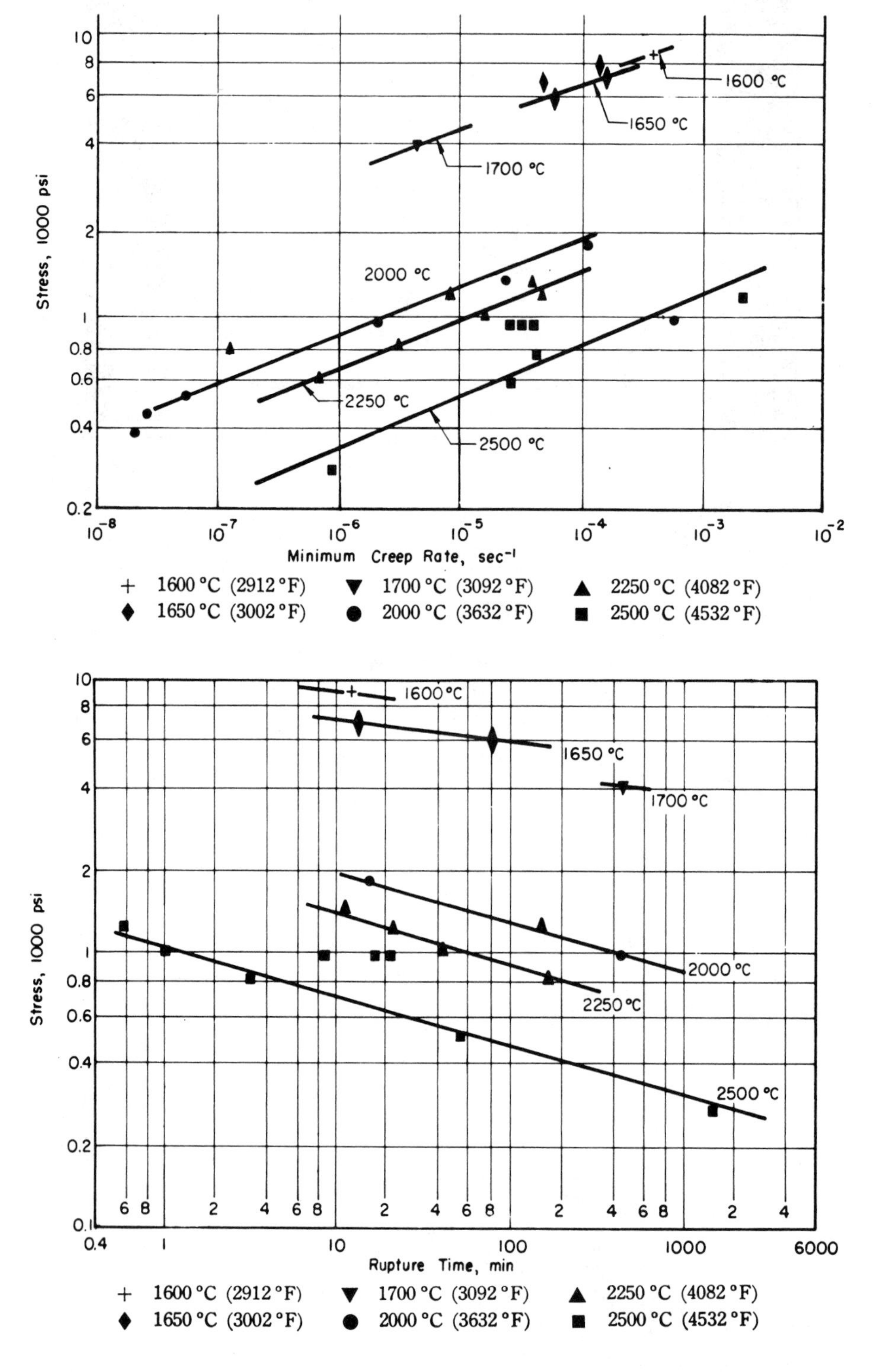

Short-time creep properties of unalloyed P/M molybdenum (top) and short-time stress-rupture data for P/M molybdenum (bottom).

Source: T.E. Tietz and J.W. Wilson, Behavior and Properties of Refractory Metals, Stanford University Press, Stanford, 1965, p 193

Mo-0.5Ti vs TZM: A Comparison

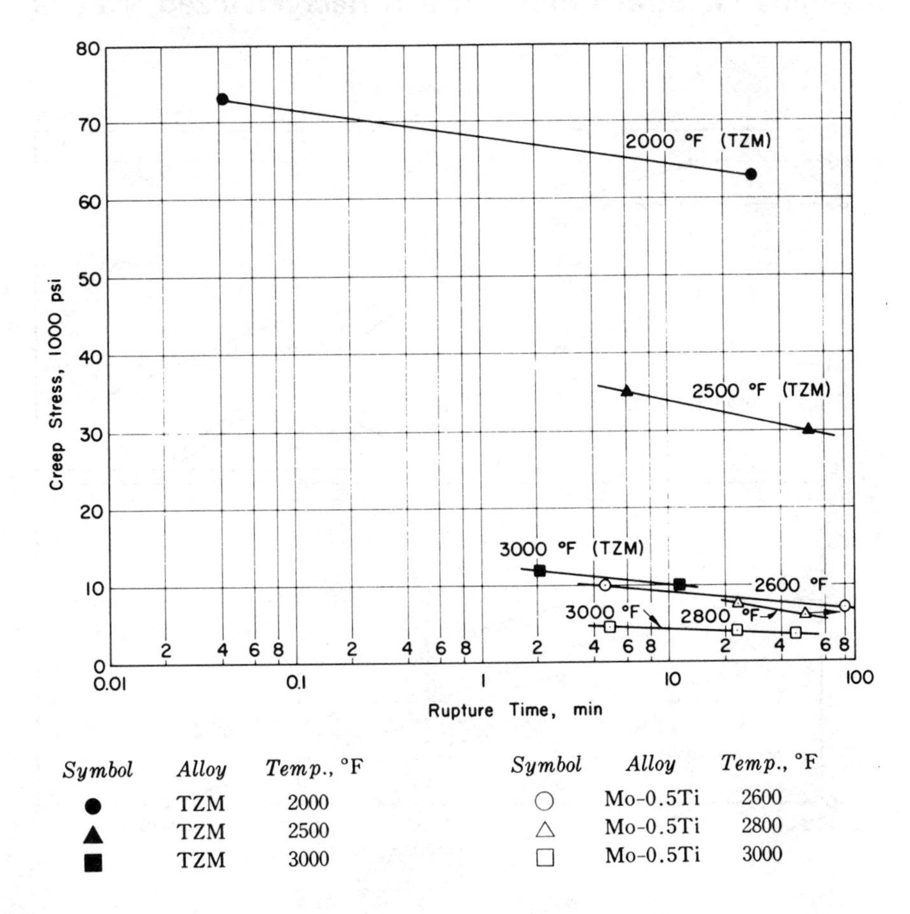

Creep stress vs rupture time for Mo-0.5Ti and TZM alloys from 2000 to 3000 °F.

The figure above presents short-time rupture data for TZM alloy based on the Marquardt tests. Data were related to a study on the effect of variations in processing procedures from ingot to sheet material. Thus, these data may not agree closely with creep properties of commercially supplied TZM.

Source: T.E. Tietz and J.W. Wilson, Behavior and Properties of Refractory Metals, Stanford University Press, Stanford, 1965, p 396

Molybdenum Alloys: Comparison of 100-h Rupture Strengths for Stress-Relieved and Recrystallized Material

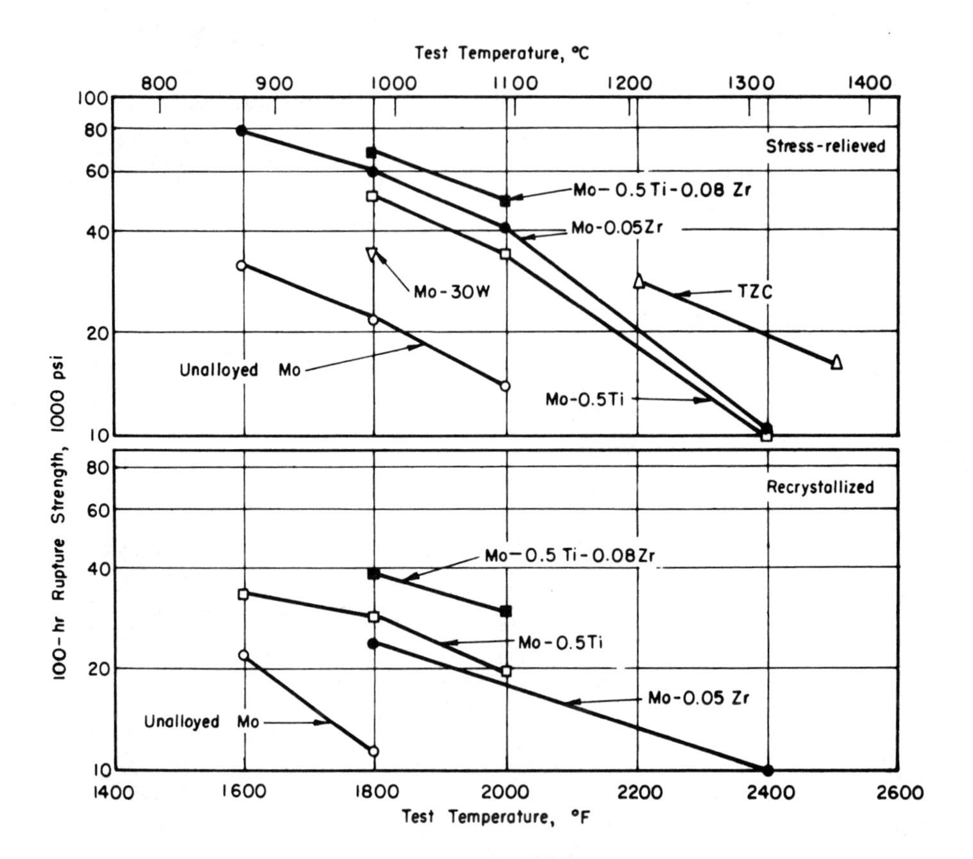

One-hundred-hour rupture strength of selected molybdenum alloys in the stress-relieved and recrystallized condition.

Source: T.E. Tietz and J.W. Wilson, Behavior and Properties of Refractory Metals, Stanford University Press, Stanford, 1965, p 197

Niobium

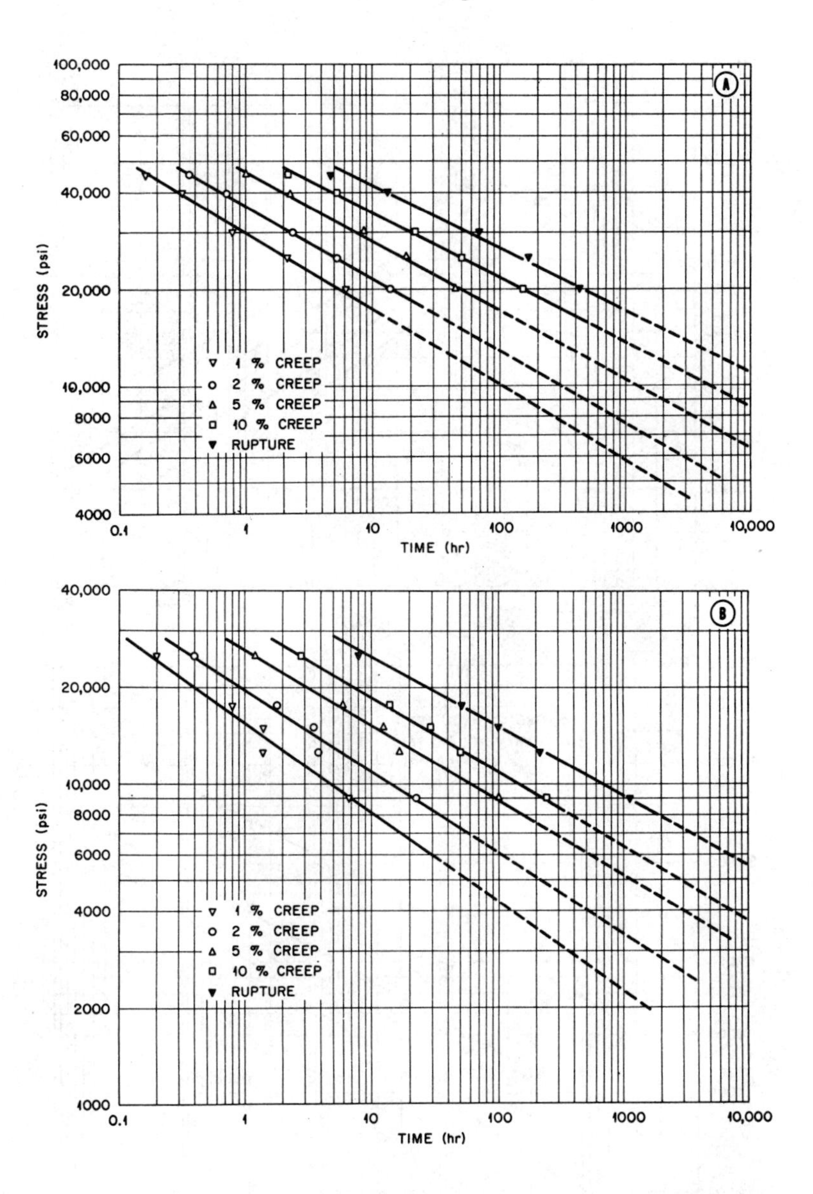

Creep-rupture properties of B-66 alloy at 982 °C (top) and at 1093 °C (bottom).

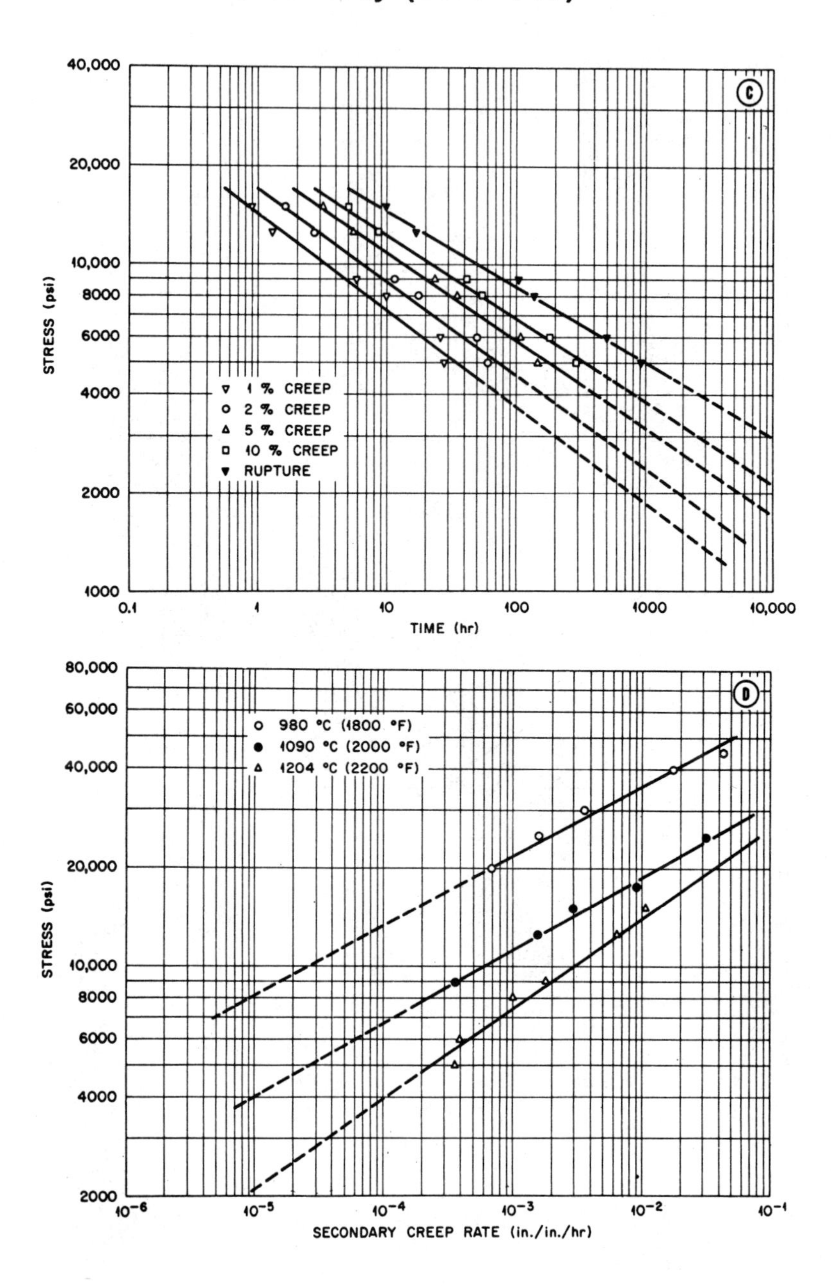

Creep-rupture data for B-66 alloy at 1204 °C (top) and secondary creep rate vs stress (bottom).

Source: R.L. Stephenson, "Comparative Creep-Rupture Properties of D-43 and B-66 Columbium Alloys," in Refractory Metals and Alloys IV, Volume II (Proceedings of a Conference in French Lick, IN), R.I. Jaffee, G.M. Ault, J. Maltz, and M. Semchyshen, Eds., Gordon and Breach, New York, 1967, p 796-798

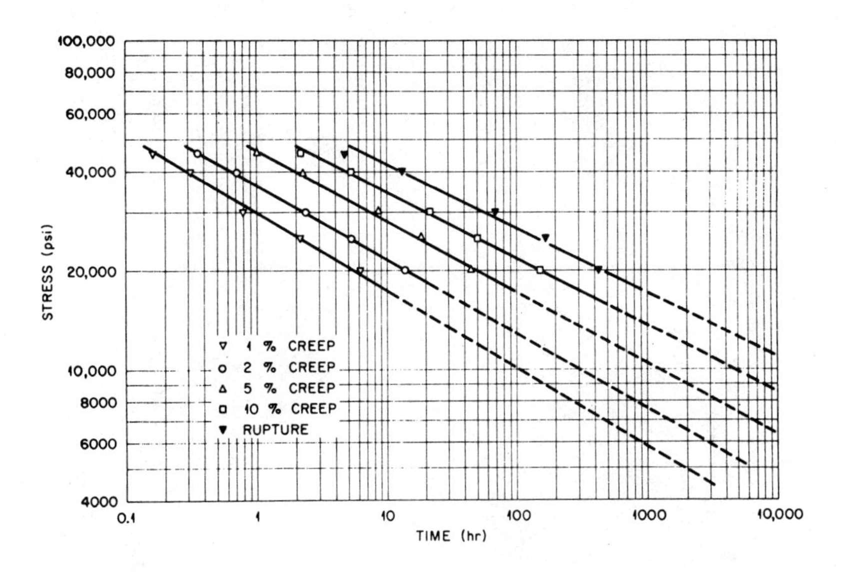

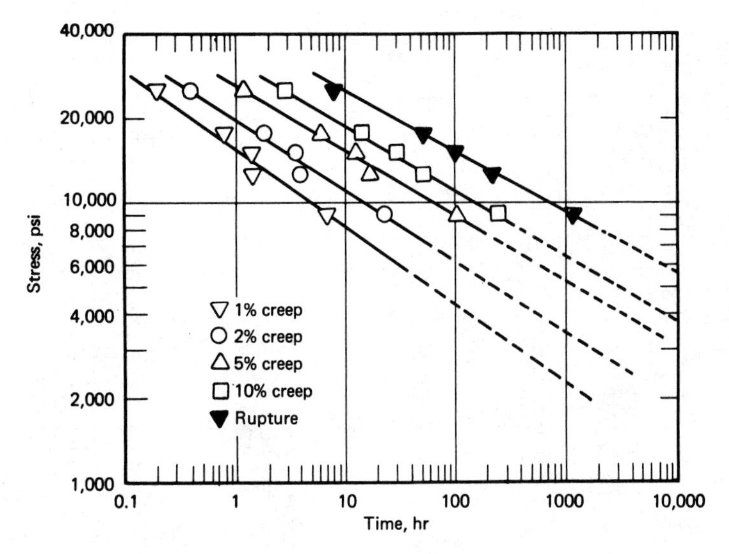

Creep-rupture data for as-rolled B-66 alloy sheet tested in vacuum at 980 °C (top) and at 1090 °C (bottom).

B-66: As-Rolled Sheet in Vacuum (Continued)

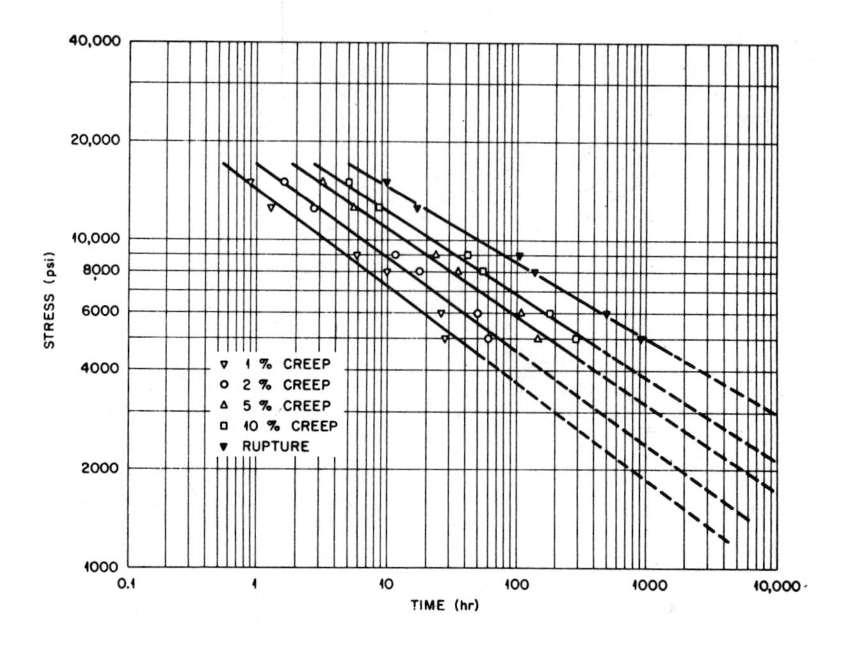

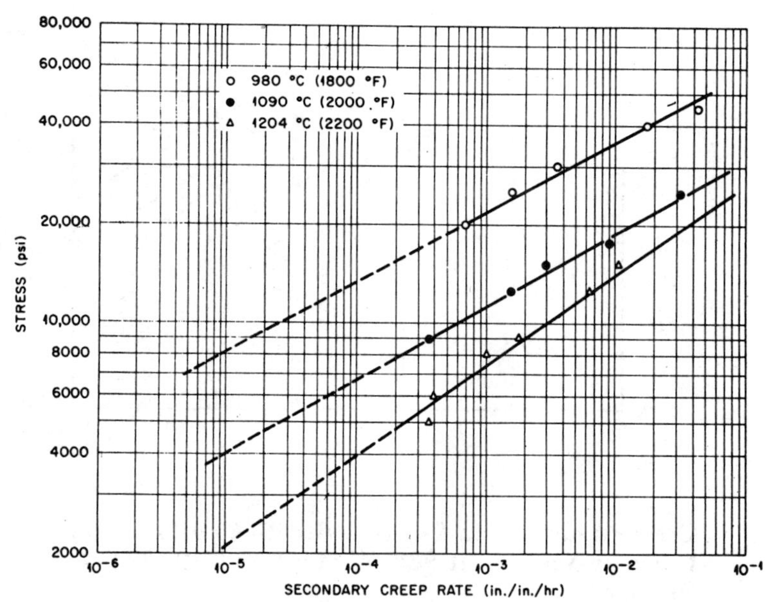

Creep-rupture data for as-rolled B-66 alloy sheet tested in vacuum at 1204 °C (top) and secondary creep rate data for as-rolled sheet tested in vacuum (bottom).

Source: J.B. Conway and P.N. Flagella, Creep-Rupture Data for the Refractory Metals to High Temperatures, Gordon and Breach, New York, 1971, p 717-718

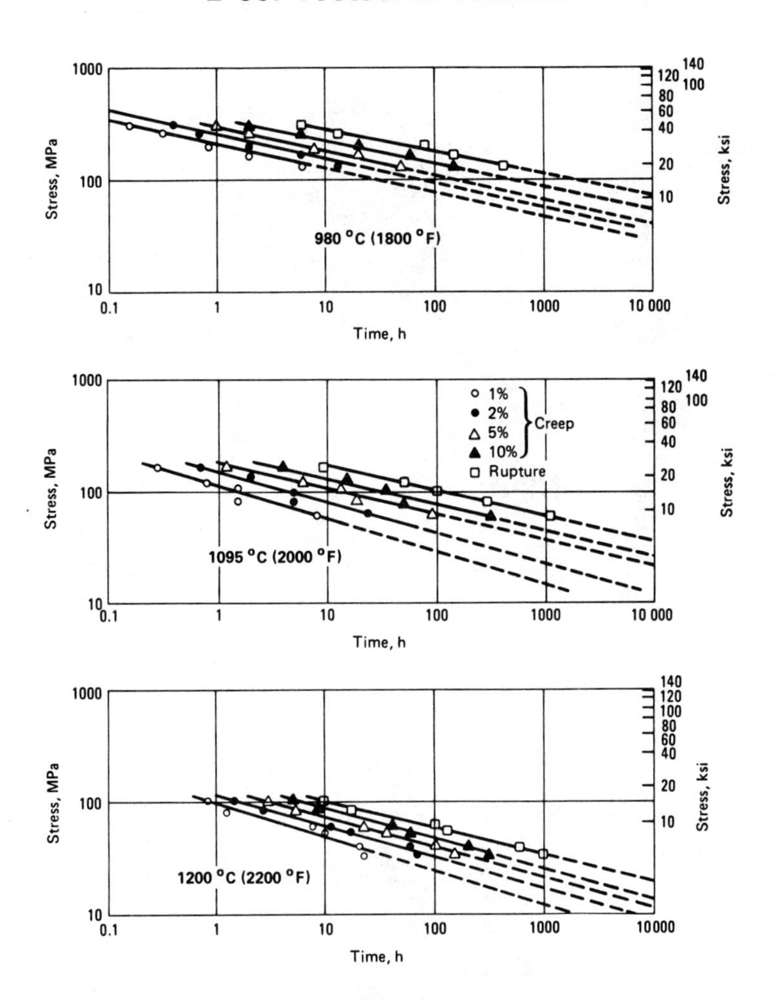

Creep and stress-rupture curves for B-66 alloy tested in vacuum.

Source: Metals Handbook, Ninth Edition, Volume 3, Properties and Selection: Stainless Steels, Tool Materials and Special-Purpose Metals, American Society for Metals, Metals Park, OH, 1980, p 340

C-103: Recrystallized

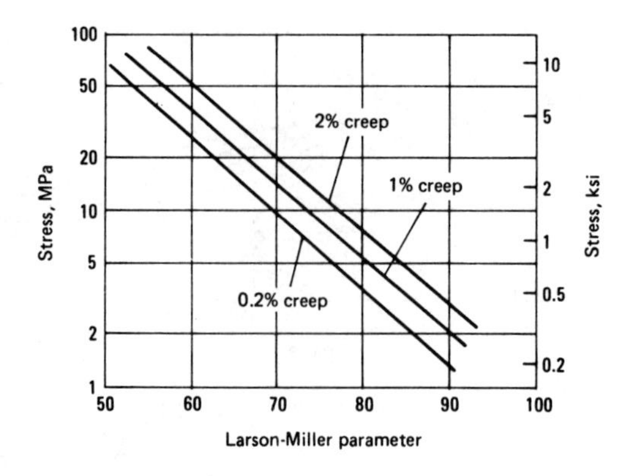

Larson-Miller plot for recrystallized C-103.

Larson-Miller parameter is $T/1000\ (20 + \log t)$.

Source: Metals Handbook, Ninth Edition, Volume 3, Properties and Selection: Stainless Steels, Tool Materials and Special-Purpose Metals, American Society for Metals, Metals Park, OH, 1980, p 335

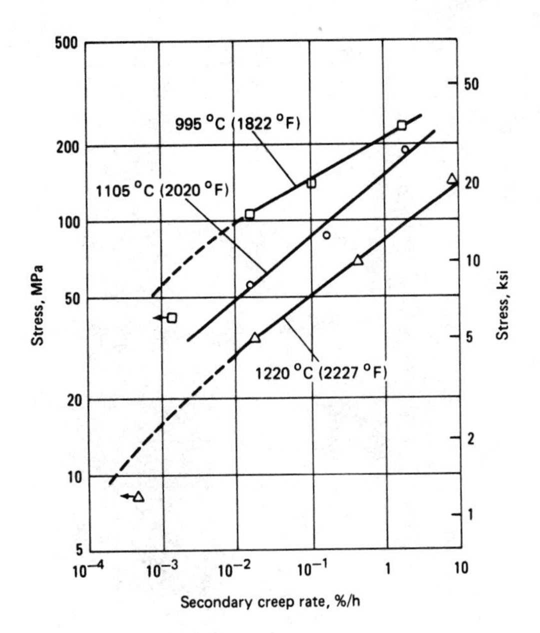

Secondary creep rate vs stress for C-129Y sheet cold worked 50%.

Source: Metals Handbook, Ninth Edition, Volume 3, Properties and Selection: Stainless Steels, Tool Materials and Special-Purpose Metals, American Society for Metals, Metals Park, OH, 1980, p 337

C-129Y: Sheet Tested in Vacuum

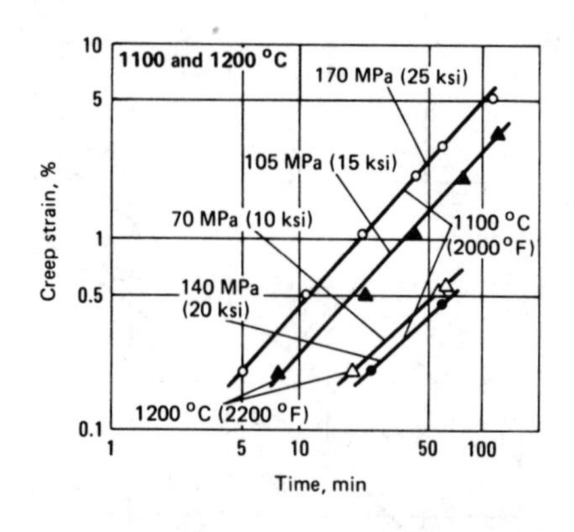

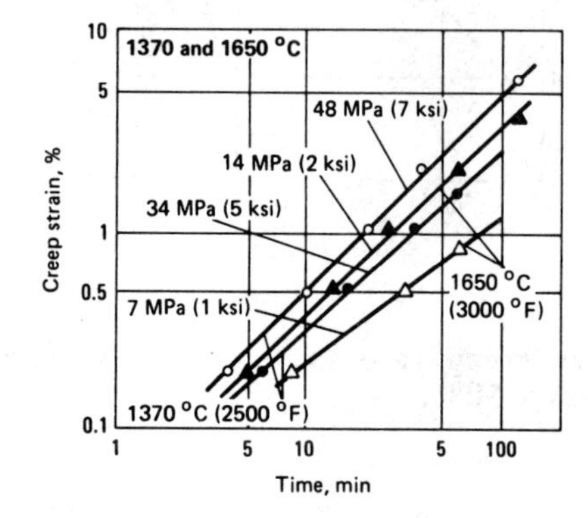

Creep curves for C-129Y sheet in vacuum.

C-129Y sheet, 1 mm (0.04 in.) thick, was annealed 1 h at 1315 °C (2400 °F) and tested in vacuum at 13 MPa (10^{-4} torr).

Source: Metals Handbook, Ninth Edition, Volume 3, Properties and Selection: Stainless Steels, Tool Materials and Special-Purpose Metals, American Society for Metals, Metals Park, OH, 1980, p 336

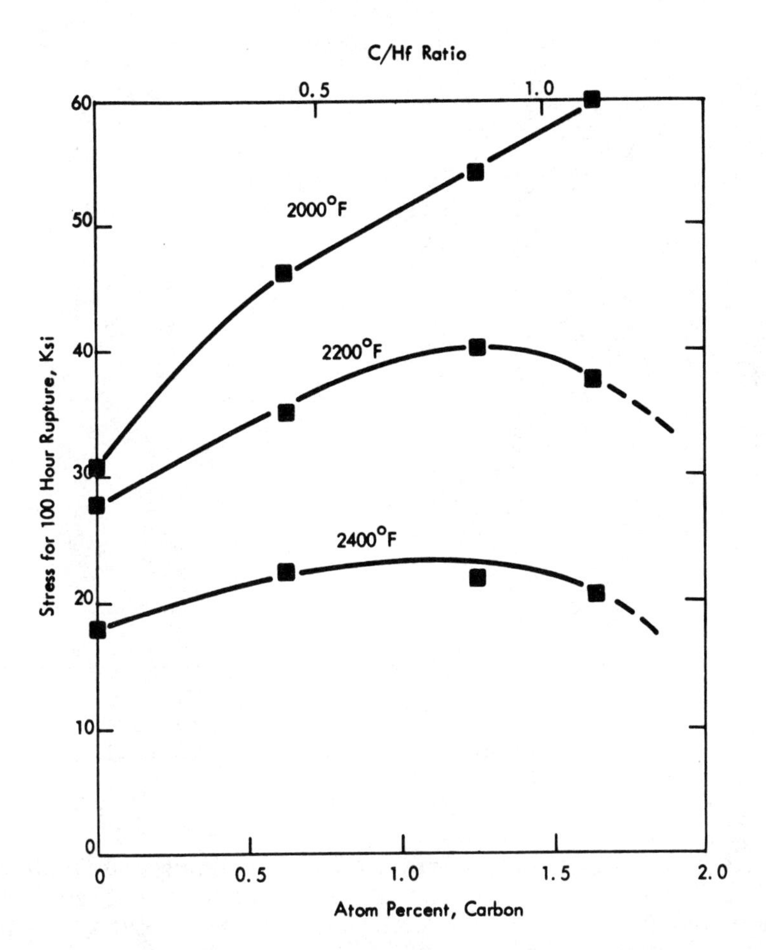

Effect of carbon and hafnium content on the stress for 100-h rupture of Cb-22W-2Hf alloy.

Phase diagrams have now been determined over major portions of the useful composition range of a number of refractory metal ternary and higher order systems. In many specific alloys, the composition and crystal structure of the precipitating phases have been determined. In some cases, the creep properties of a series of alloys in a given system have been correlated with the different phase fields of the phase diagram.

In the absence of phase diagrams, the interstitial/reactive metal atom ratio is often taken as a guide in selecting the concentrations of these elements to be used. It has frequently been observed, in carbide-strengthened alloys, that optimum strengthening is achieved with carbon/reactive metal atom ratios of approximately unity. This was observed in W-Hf-C alloys. In Nb alloys, the best results were obtained at a carbon/reactive metal atom ratio of ~0.8. Also in Nb alloys, it was found that the optimum ratio varied somewhat with temperature, as shown in the figure above. At 2000 °F, the creep strength, as indicated by 100-h rupture data, increased with the carbon/reactive metal atom ratio up to a ratio of at least 1.15. At 2400 °F, the maximum creep strength was achieved at a ratio of ~0.8.

Source: R.T. Begley, D.L. Harrod, and R.E. Gold, "High Temperature Creep and Fracture Behavior of the Refractory Metals," in Refractory Metal Alloys (Proceedings of a Symposium in Washington, D.C.), I. Machlin, R.T. Begley, and E.D. Weisert, Eds., Plenum, New York, 1968, p 71-72

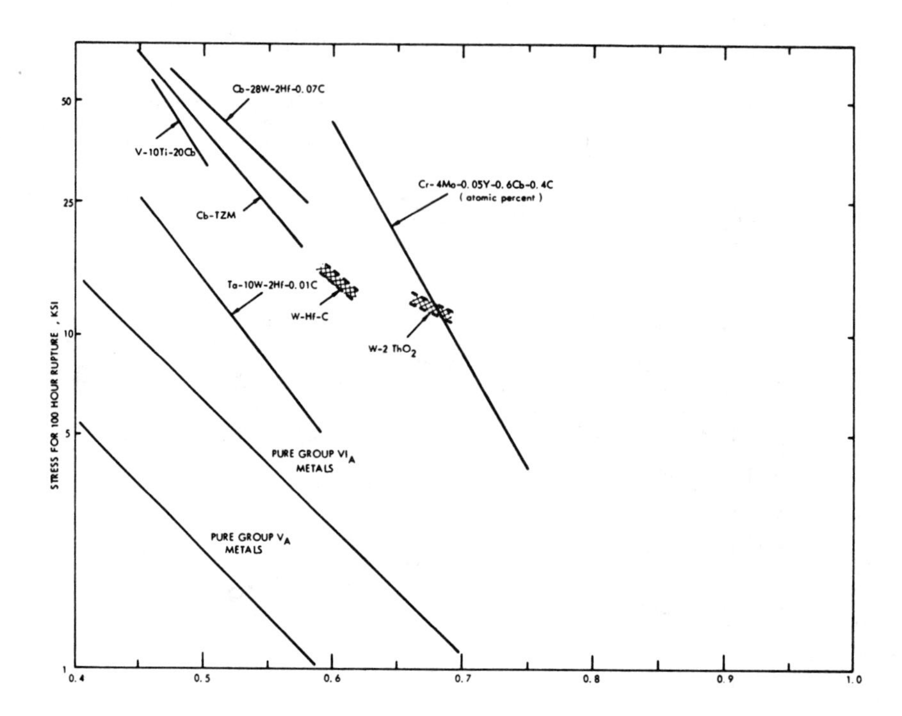

Effect of microstructure on creep-rupture characteristics of Cb-28W-2Hf-0.067C.

A study was made into the grain boundary sliding and fracture in Nb-1Zr and Nb-10W-5V-1Zr alloys. Grain boundary sliding, as measured by offsets in fiducial markings, was observed in both alloys. However, grain boundary migration and fold formation prevented intergranular fracture in the relatively weak Nb-1Zr alloy, whereas intergranular cracking was observed in the stronger alloy after only several percent creep strain. In a still more highly alloyed Nb-base alloy, it was observed that grain size and morphology had a pronounced effect on creep fracture, as illustrated in the figure above. Both material conditions gave similar creep curves through the steady-state stage, and both exhibited intergranular cracking. However, the grain boundary cracks could not propagate far in the fine grained and fibered (duplex) structure, with the result that a long third stage of creep occurred, which led to much higher ductility and more than a doubling of the rupture life.

Source: R.T. Begley, D.L. Harrod, and R.E. Gold, "High Temperature Creep and Fracture Behavior of the Refractory Metals," in Refractory Metal Alloys (Proceedings of a Symposium in Washington, D.C.), I. Machlin, R.T. Begley, and E.D. Weisert, Eds., Plenum, New York, 1968, p 75-76

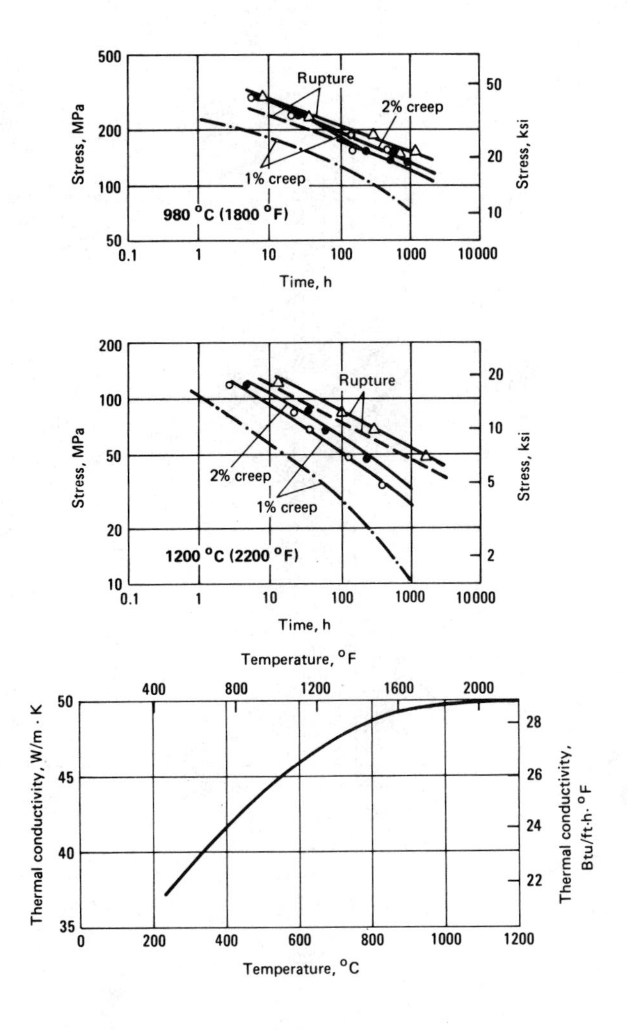

Total creep curves for Cb-752 sheet.

Points represent material duplex annealed, then aged 1 h at 1600 °C (2900 °F). Broken lines are for material duplex annealed only.

Source: Metals Handbook, Ninth Edition, Volume 3, Properties and Selection: Stainless Steels, Tool Materials and Special-Purpose Metals, American Society for Metals, Metals Park, OH, 1980, p 338

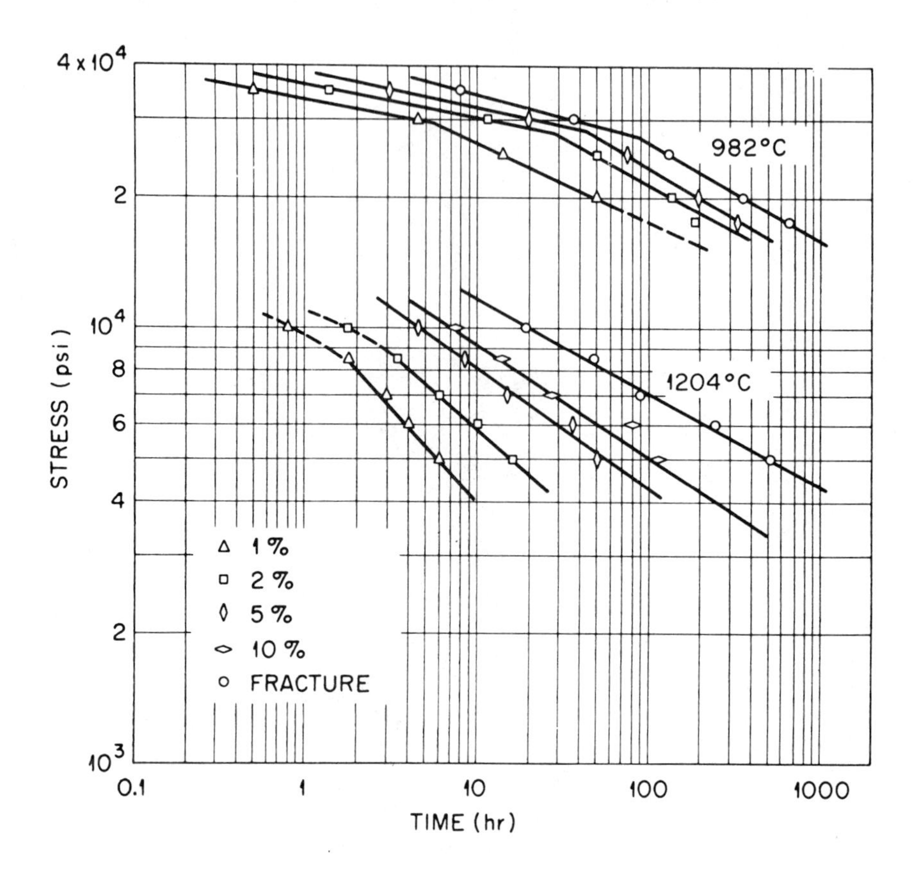

Creep-rupture properties for arc-melted Cb-753 alloy sheet tested in vacuum.

Cb-753: Arc-Melted Sheet Tested in Vacuum (Continued)

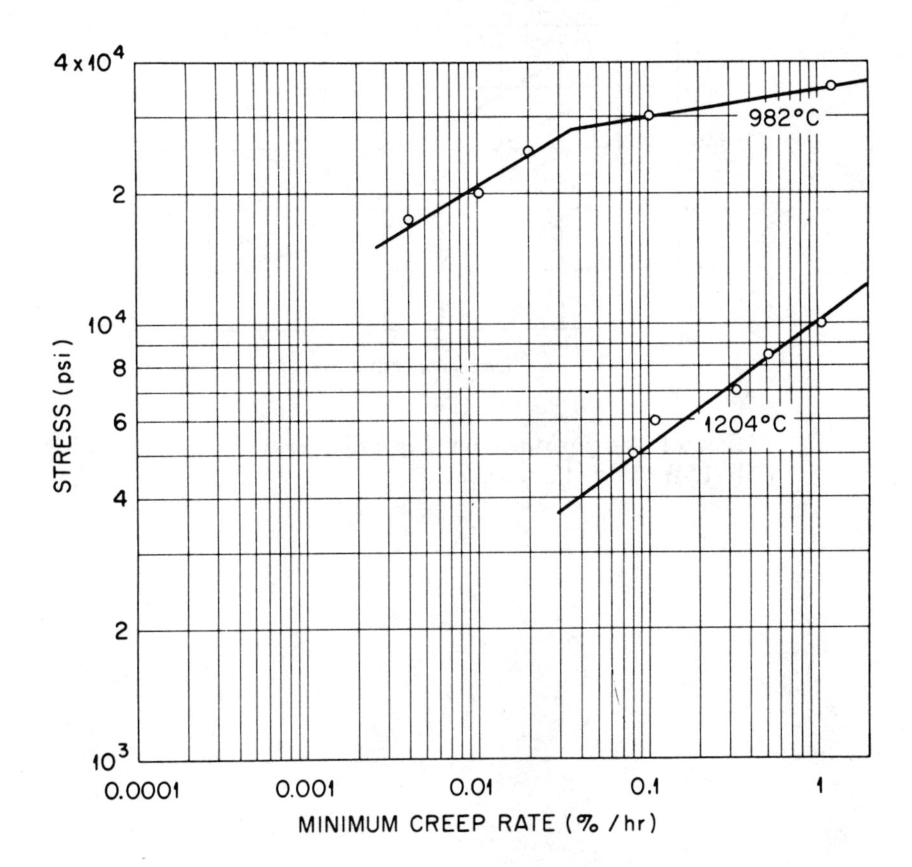

Minimum creep rate data for arc-melted Cb-753 alloy sheet tested in vacuum.

Source: J.B. Conway and P.N. Flagella, Creep-Rupture Data for the Refractory Metals to High Temperatures, Gordon and Breach, New York, 1971, p 724-725

F-48: Sheet

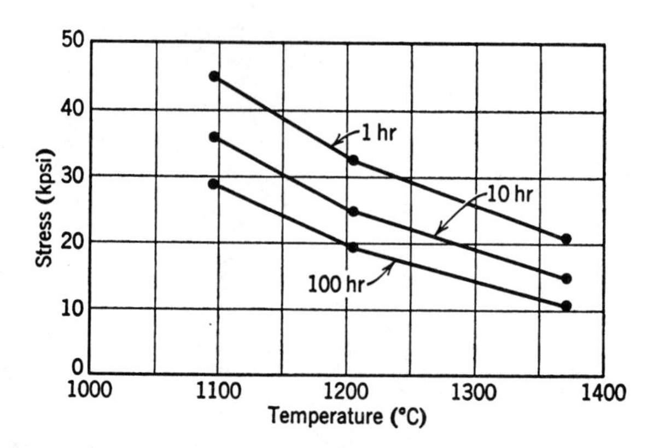

Effect of temperature on rupture stress of F-48 (Nb-15W-5Mo-1Zr) sheet.

Source: R.W. Fountain and C.R. McKinsey, "Physical and Mechanical Properties of Columbium and Tantalum and Their Alloys," in Columbium and Tantalum, F.T. Sisco and E. Epremian, Eds., John Wiley & Sons, New York, 1963, p 280

F-48: Sheet Tested in Vacuum

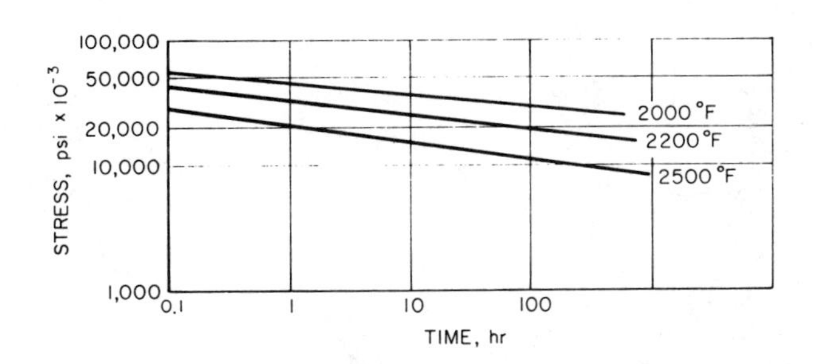

Stress-rupture properties of 0.040-in. F-48 sheet stress relieved at 1200 °C.

Stress-rupture properties of F-48 sheet

Test temp., °C	Stress, psi	Rupture life	Elong., %
1204	25,000	10.6	—
1204	30,000	5.5	29
1204	20,000	83.8	50
1204	25,000	11.4	27
1204	20,000	46.2	27
1371	12,000	55.6[a]	44
1093	25,000	136.0	Test stopped, no failure
1093	20,000	120	Test stopped, no failure

[a] One-hundred per cent recrystallized after test, ductile bend.

These stress-rupture tests were conducted in vacuum capsules at 1093 and 1204 °C. The capsules were Inconel tubes with a molybdenum liner and were inserted in a Kanthal-wound electric furnace capable of operation to 1250 °C. The vacuum system consisted of a 2-in. diffusion pump backed up by a mechanical pump. A vacuum of the order of 0.1 μm was maintained during the test, and the system had a low leak rate. Tests with Nb-base alloys have been conducted for periods up to at least 300 h in this apparatus without detectable contamination.

Source: T.K. Redden, "Processing and Properties of F-48 Columbium Alloy Sheet," in Columbium Metallurgy (Proceedings of a Symposium in Bolton Landing, NY), D.L. Douglass and F.W. Kunz, Eds., Interscience, New York, 1961, p 289-291

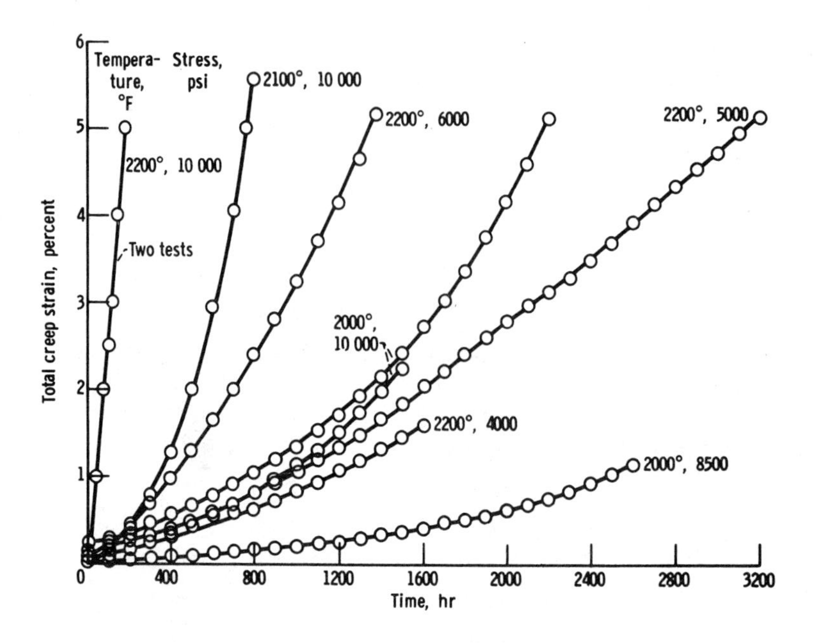

Ultrahigh-vacuum creep behavior of FS-85.

In view of the superior creep strength exhibited by FS-85 in these tests, the creep properties of this alloy were evaluated extensively. The results of the tests at 2000 to 2200 °F and at stresses ranging from 4000 to 10,000 psi are summarized above.

Source: R.H. Titran and R.W. Hall, "Ultrahigh-Vacuum Creep Behavior of Columbium and Tantalum Alloys at 2000 and 2200 °F for Times Greater than 1000 Hours," in Refractory Metals and Alloys IV, Volume II, R.I. Jaffee, G.M. Ault, J. Maltz, and M. Semchyshen, Eds., Gordon and Breach, New York, 1967, p 767

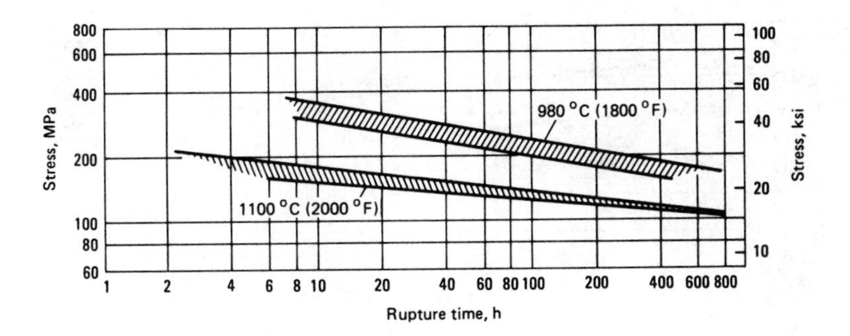

Stress-rupture properties of Nb-1Zr.

Source: Metals Handbook, Ninth Edition, Volume 3, Properties and Selection: Stainless Steels, Tool Materials and Special-Purpose Metals, American Society for Metals, Metals Park, OH, 1980, p 334

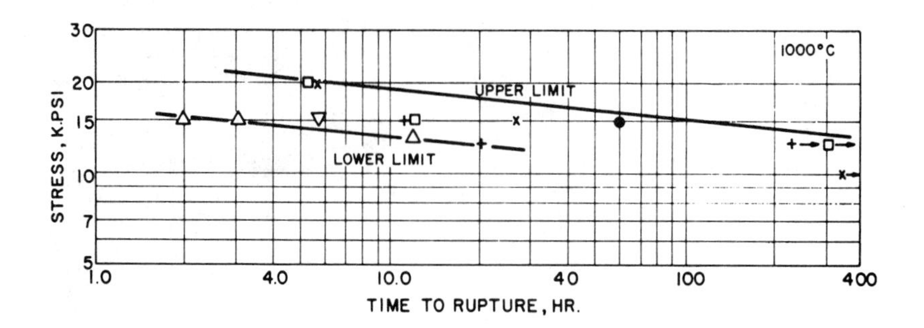

Stress-rupture properties of as-worked Nb-Al-V-Ti-Zr alloys at 1000 °C.

Mechanical properties

Nominal composition				Tensile properties[a]					1000°C stress-rupture			
Al	V	Ti	Zr	Temp., °C	0.2% YS kpsi	UTS, kpsi	Elong., %	RA, —	Stress, kpsi	Life, hr	Elong., %	RA, %
3	3	—	—	RT	126.2	132.9	7	6.2	15	50	41	86
				850	61.0	62.8	17	44.8				
				1000	31.6	32.9	88	81.9				
				1200	12.9	12.9	78	92.1				
3	3	7	—	RT	148.0	154.9	5	9.8				
				850	44.0	44.8	52	77.4				
				1000	25.0	26.6	78	92.5				
				1200	7.9	8.4	1.2	100.0				
3	3	—	1	850	76.3	86.8	19	23	10	>352.5	28	21
				1000	31.9	37.2	52	54	15	26.9	82	81
									20	5.6	48	72
3	3	5	1						15	5.7	77	91
3	3	7	1	850	65.8	73.5	24.0		12.5	20.1	37	69
				1000	25.0	29.1	52.4					
				1200	5.9	6.8	67.6		15	11.2	47	71
3	3	10	1	RT Brittle failure at 102 kpsi					12.5	63.6	30	82
				900	58.3	60.3	29.7	64.7				
				1000	25.5	29.3	57.4	98+	15	2.0	50	78
									15	3.0	125	8
3	3	10	5						12.5	>305	72	82
									15.0	12.0	70	79
									20.0	5.2	72	82

[a] Here, YS indicates yield strength; UTS, ultimate tensile strength; RA, reduction in area; and RT, room temperature.

Stress-rupture tests were done in a dynamic vacuum of about 0.01 μm on as-worked samples. Results of these tests are shown in the table and plotted in the figure.

Source: W.F. Sheely and J.L. Wilson, "The Properties of Columbium-Aluminum-Vanadium Alloys. Part II. Mechanical Properties," in Columbium Metallurgy (Proceedings of a Symposium in Bolton Landing, NY), D.L. Douglass and F.W. Kunz, Eds., Interscience, New York, 1961, p 587-589

Niobium: Effect of Environment

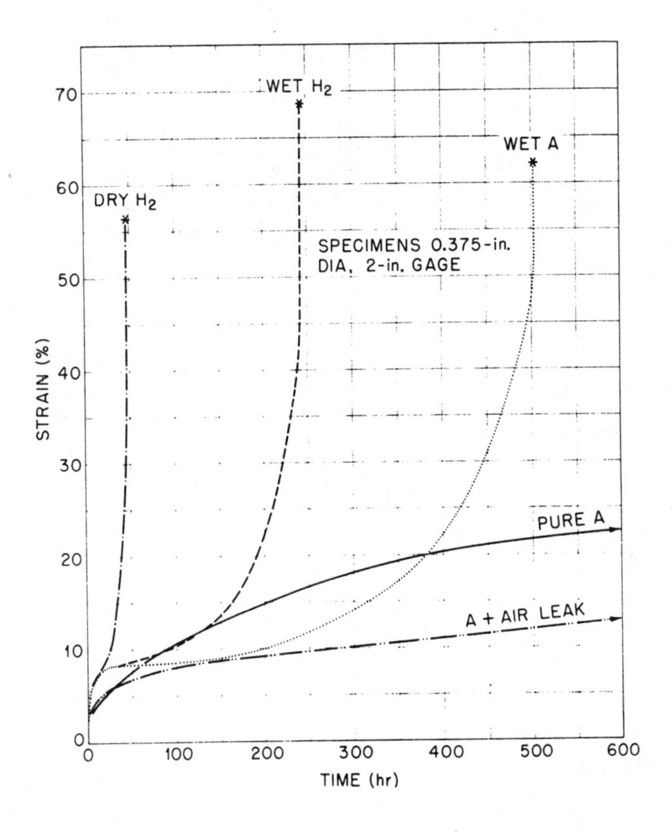

Effect of nitrogen on creep properties of niobium at 1850 °F and 3500 psi.

Analysis of niobium tested in various environments

Environment	Temp., °C	Exposure time, hr	Oxygen, wt-%	Nitrogen, wt-%	Hydrogen, wt-%
Nitrogen	1010	2321	0.019	0.045	0.0003
Argon	1010	56	0.037	0.012	0.0002
Argon	982	2538	0.024	0.0046	0.0001
Argon (air leak)	982	1121	0.041	0.0077	0.0003
Hydrogen (dry)	982	53	0.046	0.010	0.014
Hydrogen (wet)	982	244	0.510	0.014	0.017
Argon (wet)	982	503	0.330	0.0054	0.024

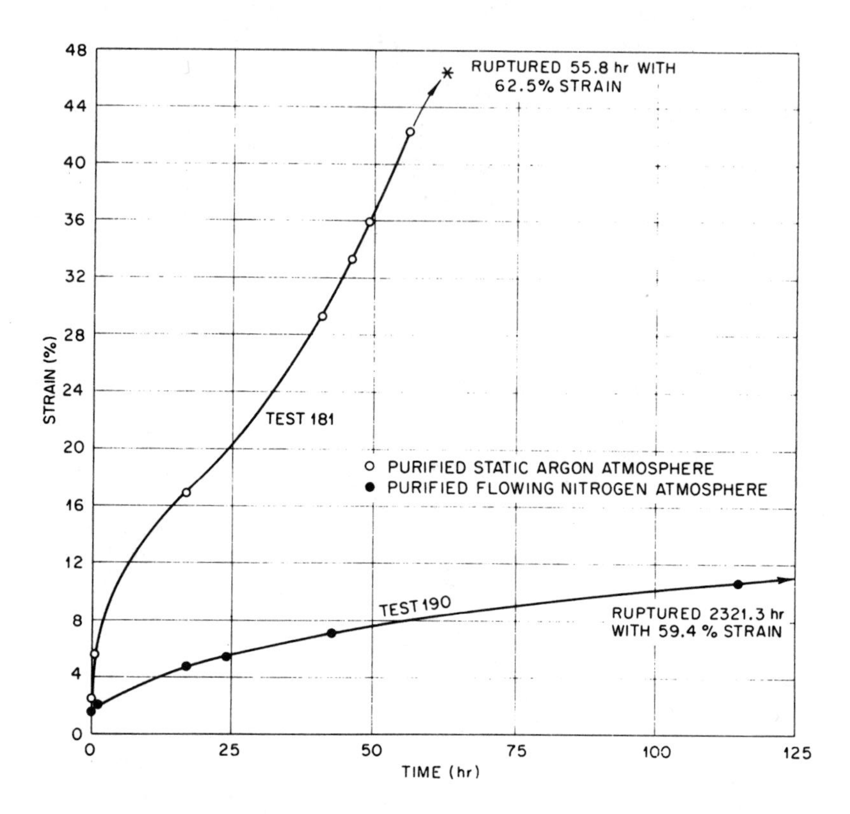

Effect of environment on the creep properties of niobium at 1800 °F and 3500 psi.

Source: H.E. McCoy and D.A. Douglas, "Effect of Various Gaseous Contaminants on the Strength and Formability of Columbium," in Columbium Metallurgy (Proceedings of a Symposium in Bolton Landing, NY), D.L. Douglas and F.W. Kunz, Eds., Interscience, New York, 1961, p 108-109

Recrystallized Niobium

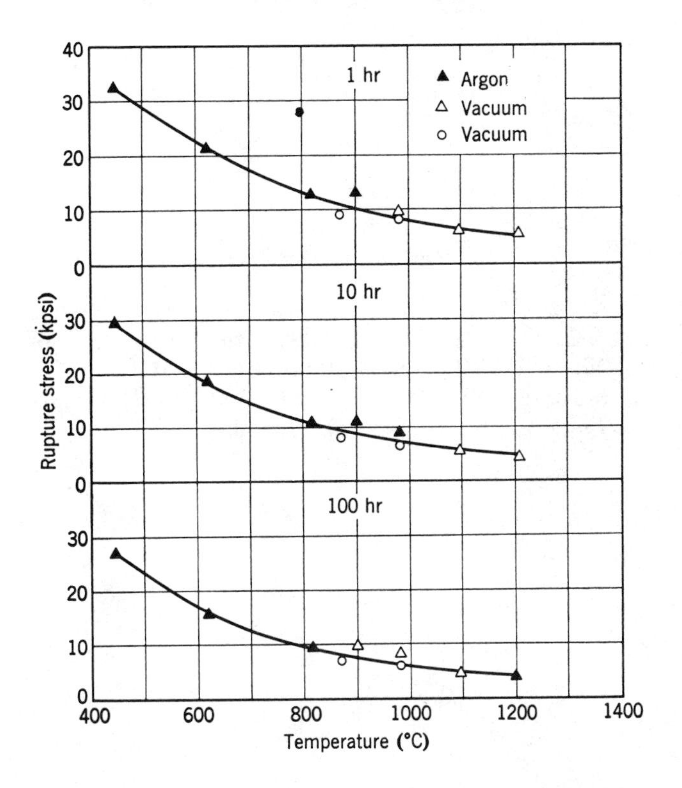

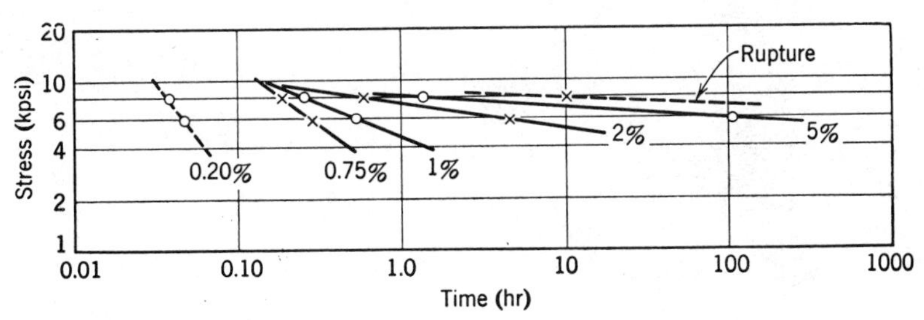

Rupture properties (top) and creep deformation at 650 °C (bottom) of recrystallized niobium.

Source: R.W. Fountain and C.R. McKinsey, "Physical and Mechanical Properties of Columbium and Tantalum and Their Alloys," in Columbium and Tantalum, F.T. Sisco and E. Epremian, Eds., John Wiley & Sons, New York, 1963, p 218, 220

8.50 65Nb-7Ti-28W, 67Nb-10Ti-20W-3V, and 70Nb-7Ti-20W-3Mo: Comparison of Stress-Rupture Properties in Vacuum

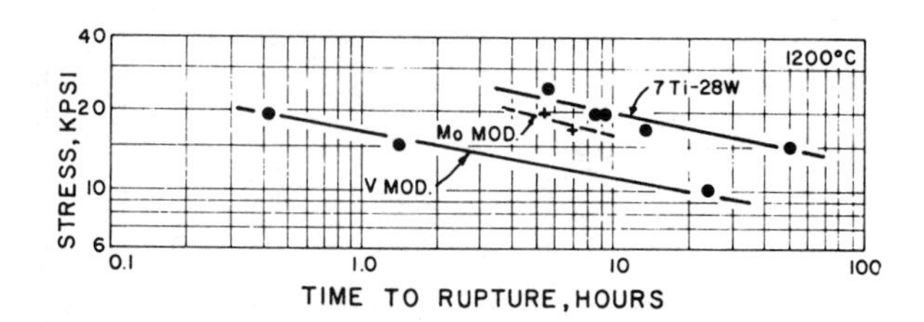

The 1200 °C vacuum stress-rupture properties of 65Nb-7Ti-28W, 67Nb-10Ti-20W-3V (V mod), and 70Nb-7Ti-20W-3Mo (Mo mod). All samples were in the as-worked condition.

Source: W.F. Sheely and J.L. Wilson, "The Properties of Columbium-Titanium-Tungsten Alloys. Part II. Structure and Mechanical Properties," in Columbium Metallurgy (Proceedings of a Symposium in Bolton Landing, NY), D.L. Douglass and F.W. Kunz, Eds., Interscience, New York, 1961, p 208

AS-55 and D-43: Comparison of Rupture Properties After 5000- and 10,000-h Exposure to Potassium

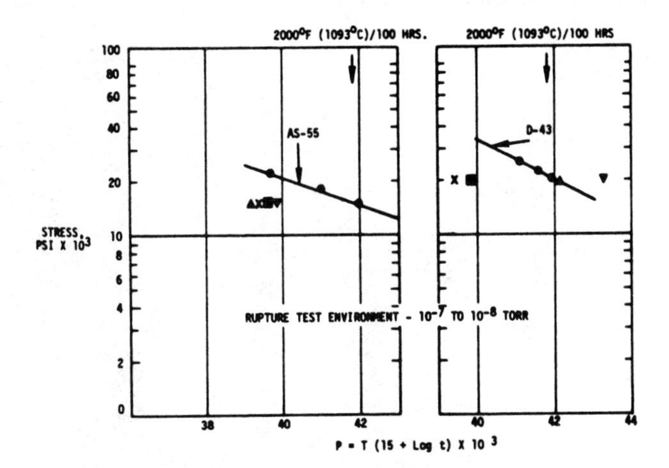

Effect of 5000- and 10,000-h exposure to potassium at 2000 °F (1093 °C) on the stress-rupture properties of AS-55 and D-43 alloys.

Source: R.G. Frank, "Recent Advances in Columbium Alloys," in Refractory Metal Alloys (Proceedings of a Symposium in Washington, D.C.), I. Machlin, R.T. Begley, and E.D. Weisert, Eds., Plenum, New York, 1968, p 353

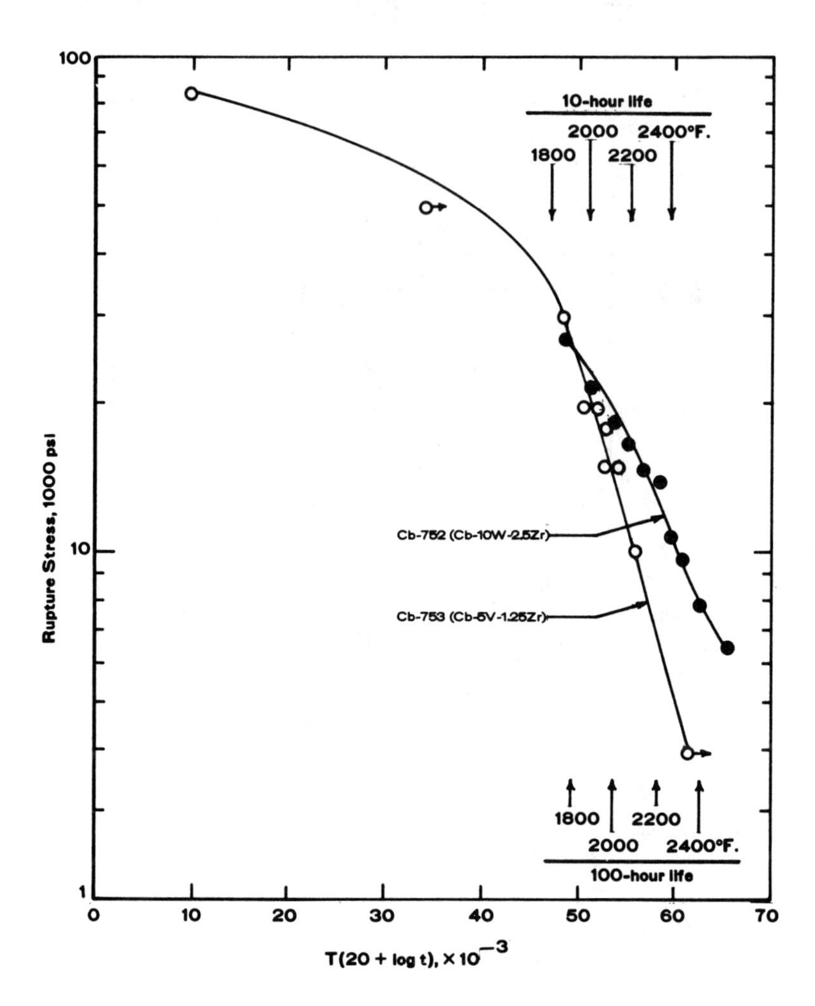

Larson-Miller plot of stress-rupture properties of alloys Cb-752 and Cb-753.

Tests were conducted on specimens machined from recrystallized 0.375-in.-diameter rod. These tests were conducted under vacuum of approximately 10^{-5} torr at temperatures of 700, 1100, 1500, 2000, and 2200 °F. Varying stress levels and a number of specimens were employed to establish the relationship of the rupture-stress vs time-to-rupture at each test temperature. The results are shown in the Larson-Miller plot above.

Source: M. Schussler, "Properties of Columbium Alloy Cb-753," in Refractory Metals and Alloys IV, Volume I (Proceedings of a Conference in French Lick, IN), R.I. Jaffee, G.M. Ault, J. Maltz, and M. Semchyshen, Eds., Gordon and Breach, New York, 1967, p 393, 395

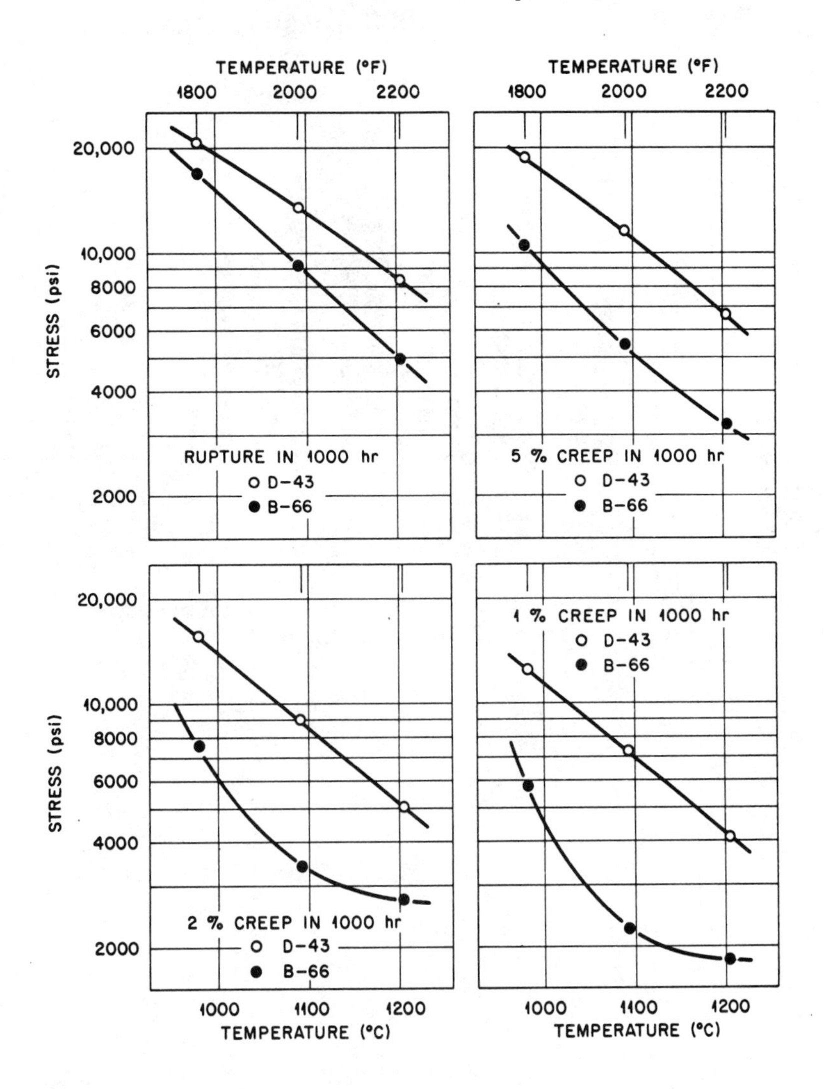

Comparative creep-rupture properties vs temperature for D-43 and B-66 alloys.

The fracture ductility of B-66 alloy is much greater than that of D-43 alloy. In addition, the secondary creep rate at a given temperature and stress is faster for B-66 alloy than for D-43 alloy. As a result, the B-66 alloy compares even less favorably on the basis of stresses to produce a specific strain in a specified time than on the basis of stress to produce rupture.

Source: R.L. Stephenson, "Comparative Creep-Rupture Properties of D-43 and B-66 Columbium Alloys," in Refractory Metals and Alloys IV, Volume II (Proceedings of a Conference in French Lick, IN), R.I. Jaffee, G.M. Ault, J. Maltz, and M. Semchyshen, Eds., Gordon and Breach, New York, 1967, p 804-806

Niobium Alloys: Comparison of Room-Temperature Ductilities Following 1000-h Aging Treatment

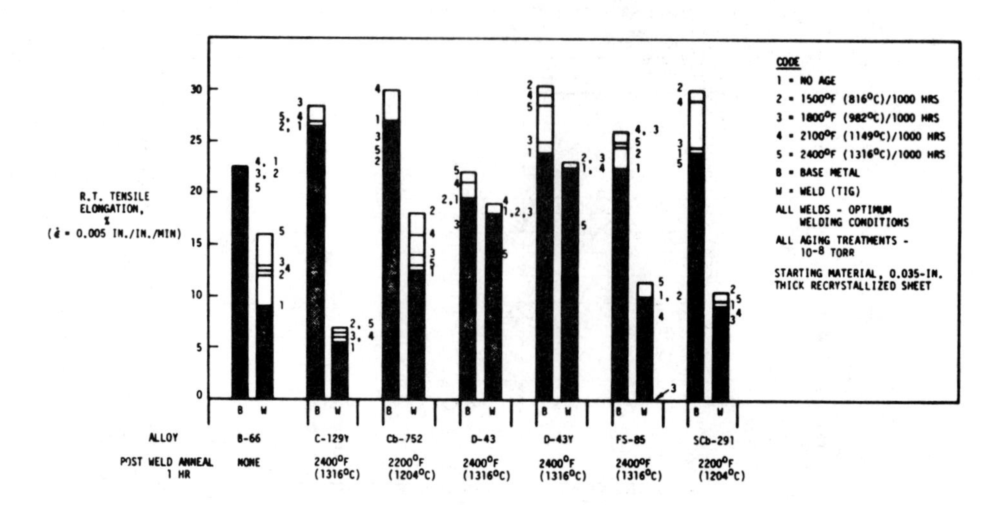

Effect of 1000-h aging treatment on room-temperature ductility of niobium alloys.

Source: R.G. Frank, "Recent Advances in Columbium Alloys," in Refractory Metal Alloys (Proceedings of a Symposium in Washington, D.C.), I. Machlin, R.T. Begley, and E.D. Weisert, Eds., Plenum, New York, 1968, p 353

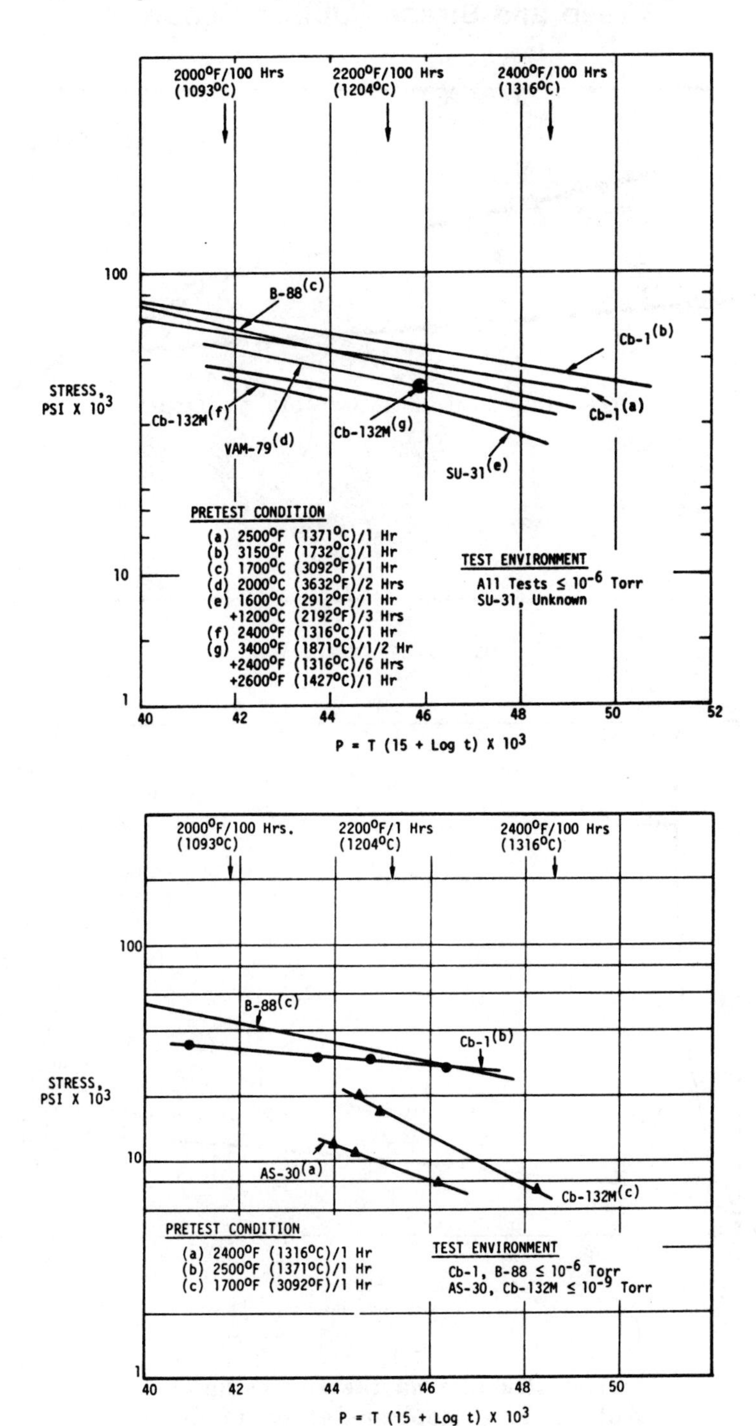

Stress-rupture properties (top) and creep properties under 1% strain (bottom) of high-strength niobium alloys.

Source: R.G. Frank, "Recent Advances in Columbium Alloys," in Refractory Metal Alloys (Proceedings of a Symposium in Washington, D.C.), I. Machlin, R.T. Begley, and E.D. Weisert, Eds., Plenum, New York, 1968, p 342

Niobium Alloys: Effect of Strain Hardening on Creep and Stress-Rupture Behavior

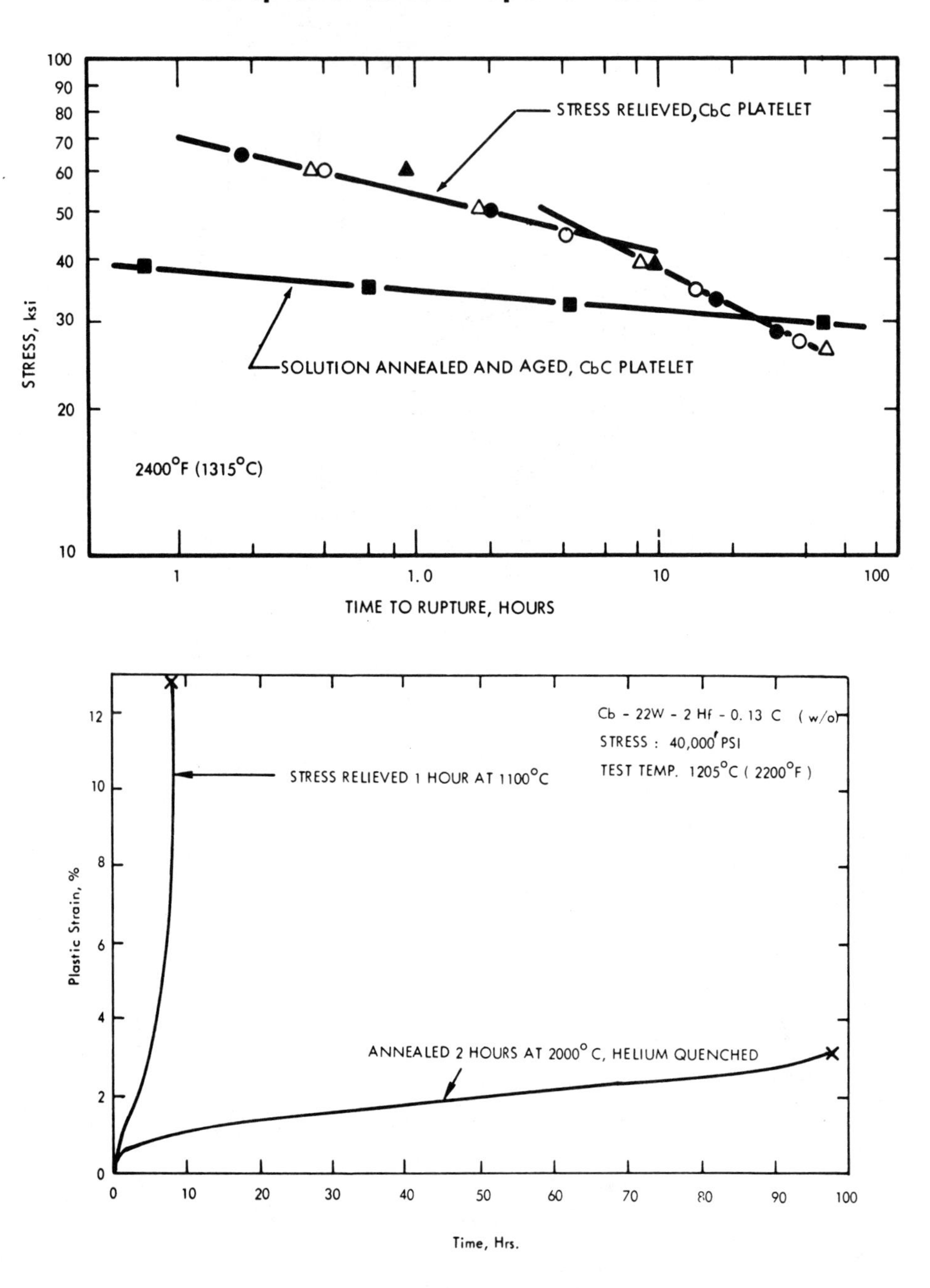

Effect of structure on the stress-rupture properties of Cb-TZM (top) and effect of thermal-mechanical treatment on creep behavior of a Nb-W-Hf-C alloy.

Niobium Alloys: Effect of Strain Hardening on Creep and Stress-Rupture Behavior (Continued)

Stress-rupture data for Cb-TZM (top) show that for short times at 1315 °C (0.55 Tm) strain-hardened material stabilized by platelet CbC precipitates had much superior rupture properties compared to unworked material. However, for times greater than about 20 h, the benefits of strain hardening were lost due to structural instability.

In the Group V_A metals, strain hardening appears to be considerably less effective, as shown by the data (bottom) for a strain-hardened and solution annealed Nb-W-Hf-C alloy. At 1200 °C (0.5 Tm), the warm worked plus stress-relieved condition resulted in a marked decrease in rupture life and an increase in creep rate compared to the solution annealed condition. Similar effects have been observed in carbide-strengthened Ta alloys.

Source: R.T. Begley, D.L. Harrod, and R.E. Gold, "High Temperature Creep and Fracture Behavior of the Refractory Metals," in Refractory Metal Alloys (Proceedings of a Symposium in Washington, D.C.), I. Machlin, R.T. Begley, and E.D. Weisert, Eds., Plenum, New York, 1968, p 73-74

8.58 **Niobium and a Niobium-Titanium Alloy: Comparison of Stress-Rupture Data for Cold Worked and Recrystallized Material**

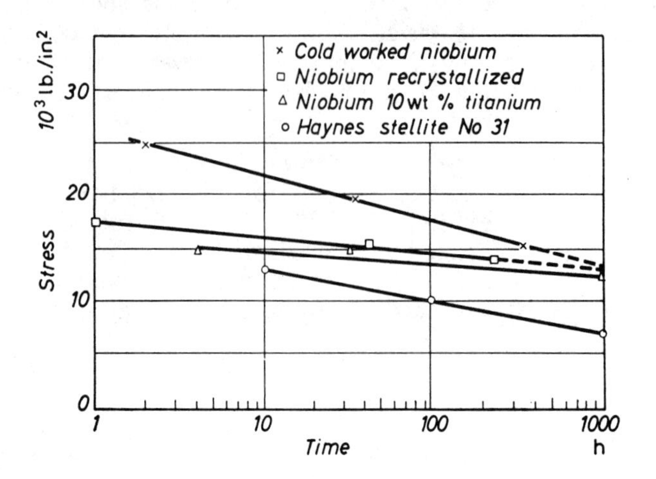

Stress-rupture curves for cold worked and recrystallized niobium and a niobium-titanium alloy at 1800 °F.

Source: G.L. Miller, Tantalum and Niobium, Academic Press, New York, 1959, p 426

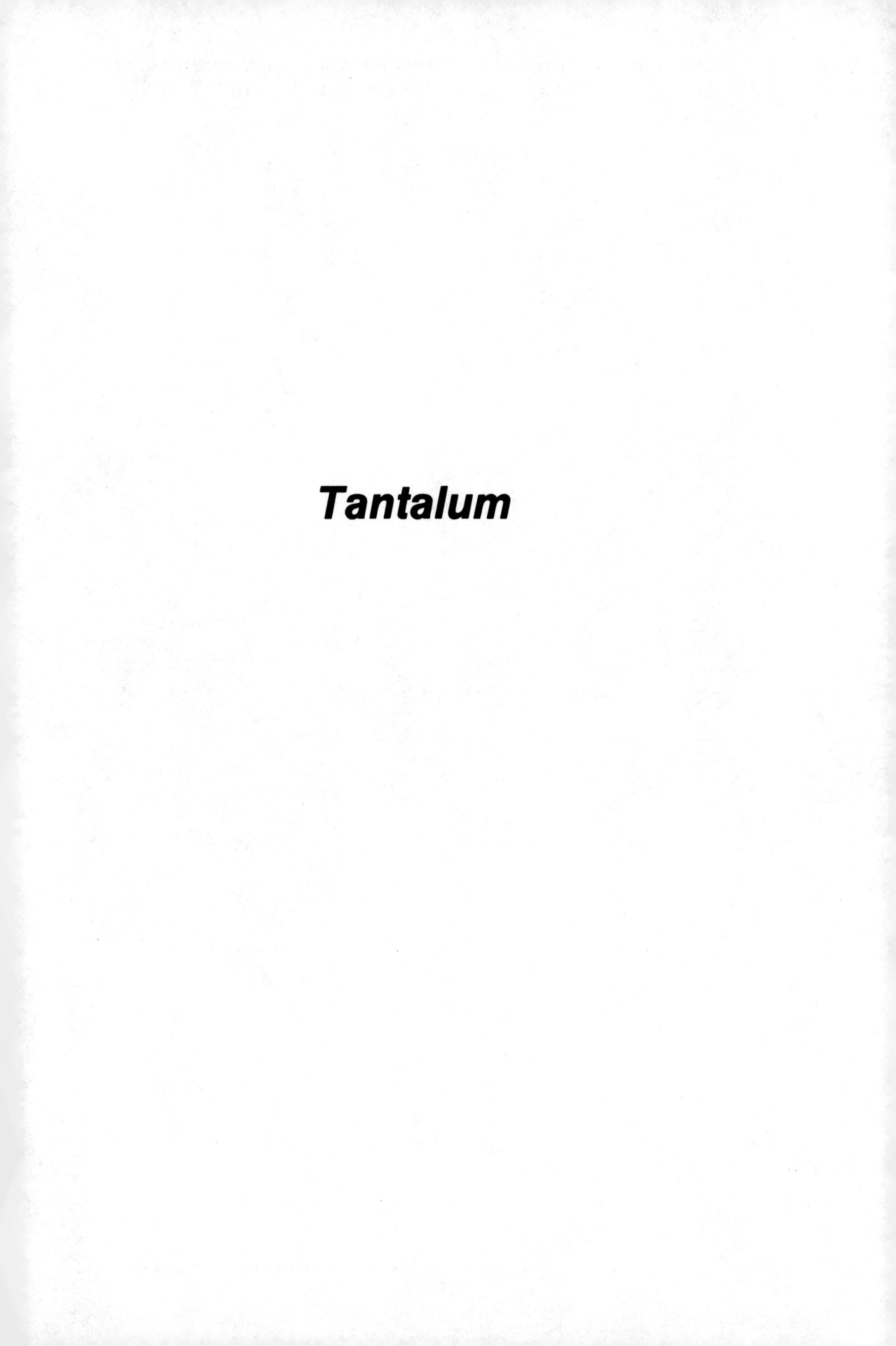

Tantalum

Recrystallized P/M Tantalum: Effect of Vacuum Degassing and Grain Size, in Helium

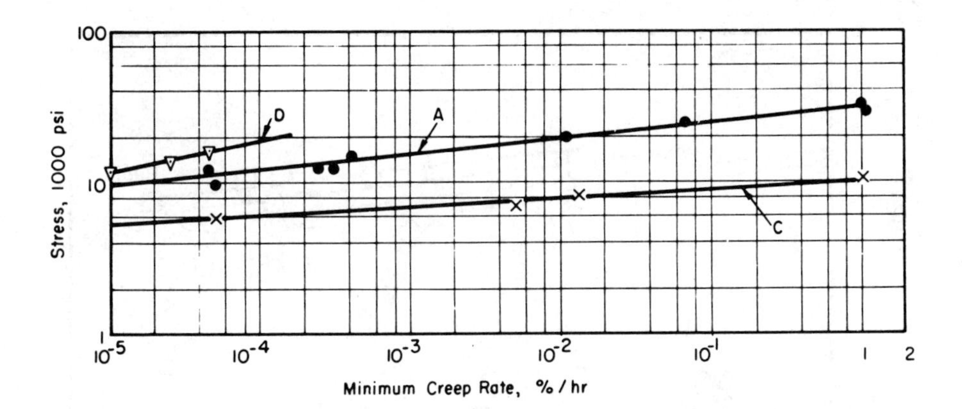

Stress vs minimum creep rate for recrystallized P/M tantalum sheet (processed under different conditions) tested at 1200 °F in helium. Closed circles represent recrystallized specimens; x's represent vacuum-degassed, coarse-grained specimens; triangles represent vacuum-degassed, fine-grained specimens.

Source: T.E. Tietz and J.W. Wilson, Behavior and Properties of Refractory Metals, Stanford University Press, Stanford, 1965, p 257

Sintered, Rolled, and Annealed Tantalum: Sheet Tested in Argon

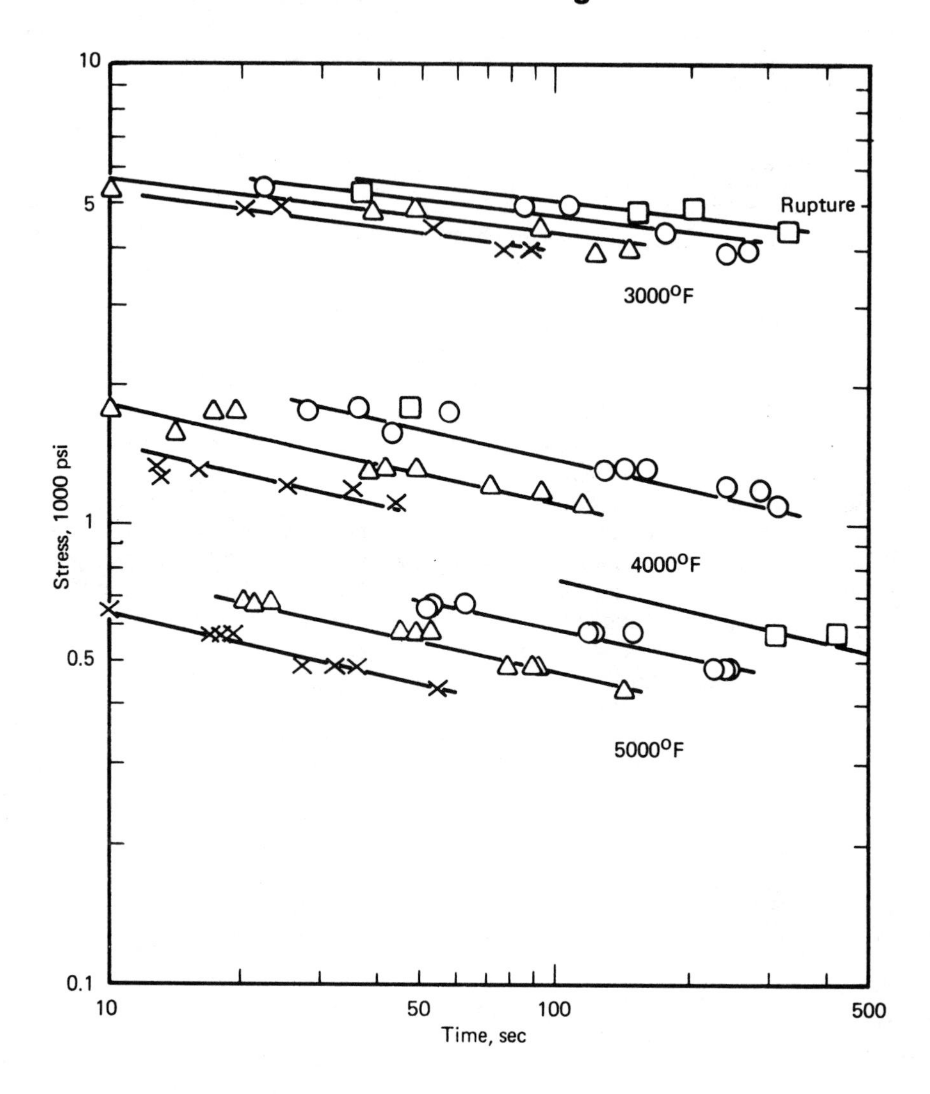

Creep-rupture data for sintered, rolled, and annealed tantalum sheet tested in argon.

Source: J.B. Conway and P.N. Flagella, Creep-Rupture Data for the Refractory Metals to High Temperatures, Gordon and Breach, New York, 1971, p 460

T-111: In Vacuum

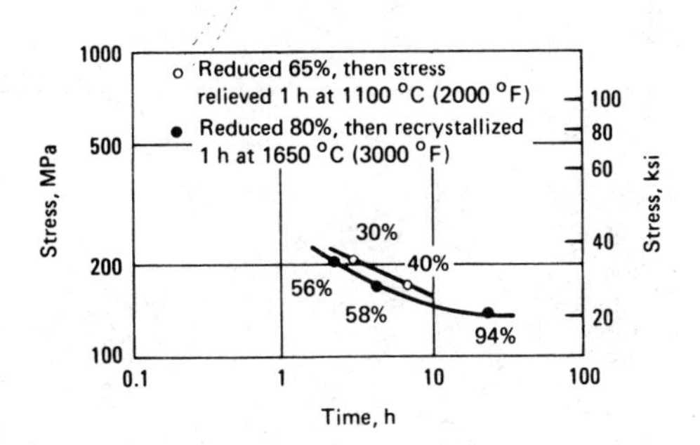

Stress-rupture properties of T-111 at 1200 °C (2400 °F) in vacuum.

Source: Metals Handbook, Ninth Edition, Volume 3, Properties and Selection: Stainless Steels, Tool Materials and Special-Purpose Metals, American Society for Metals, Metals Park, OH, 1980, p 346

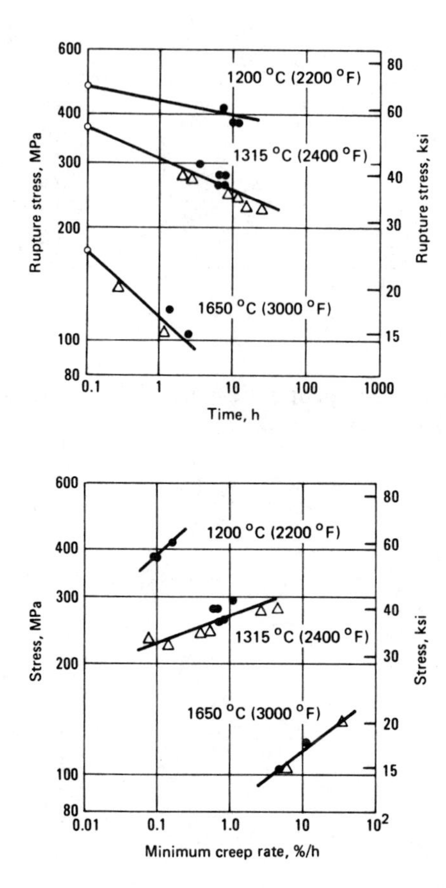

Creep-rupture curves (top) and minimum creep rate curves (bottom) for T-222 alloy at 1200, 1315, and 1650 °C (2200, 2400, and 3000 °F).

Source: Metals Handbook, Ninth Edition, Volume 3, Properties and Selection: Stainless Steels, Tool Materials and Special-Purpose Metals, American Society for Metals, Metals Park, OH, 1980, p 347

Ta-10W: Creep at Various Temperatures

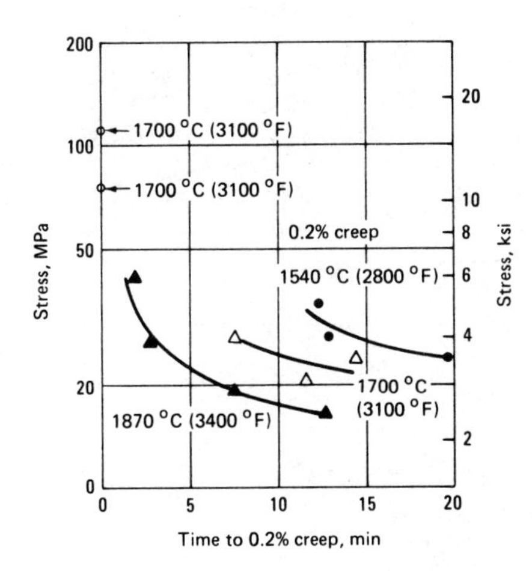

Time for 0.2% creep at various temperatures for Ta-10W tantalum.

Source: Metals Handbook, Ninth Edition, Volume 3, Properties and Selection: Stainless Steels, Tool Materials and Special-Purpose Metals, American Society for Metals, Metals Park, OH, 1980, p 345

Ta-10W: Sheet Tested in Vacuum

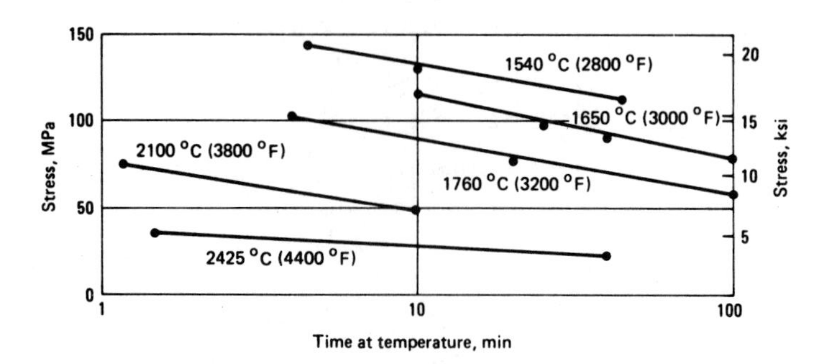

Typical stress-rupture data for Ta-10W sheet tested in vacuum.

Source: Metals Handbook, Ninth Edition, Volume 3, Properties and Selection: Stainless Steels, Tool Materials and Special-Purpose Metals, American Society for Metals, Metals Park, OH, 1980, p 346

Ta-W-Hf Alloys: Effect of Rhenium Content

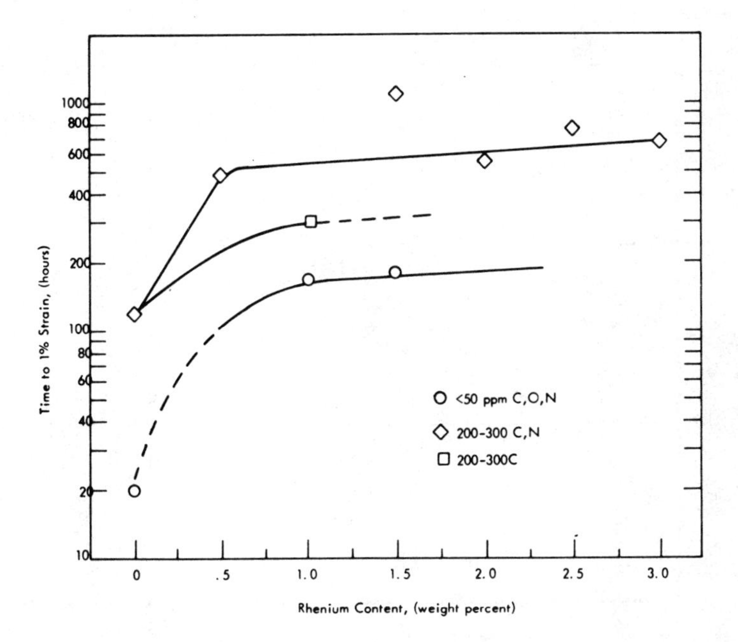

Effect of rhenium on creep properties of Ta-W-Hf alloys tested at 2400 °F and 15,000 psi. Specimens were annealed 1 h at 3000 °F prior to test.

Although the physical properties of W and Re are quite similar, the latter appears to exert a much more significant effect on the creep of Ta alloys than does an equivalent amount of W. The creep behavior of complex Ta alloys is sensitive to minor Re additions, as shown in the above figure.

Source: R.W. Buckman, Jr. and R.C. Goodspeed, "Considerations in the Development of Tantalum Base Alloys," in Refractory Metal Alloys (Proceedings of a Symposium in Washington, D.C.), Plenum, New York, 1968, p 380-381

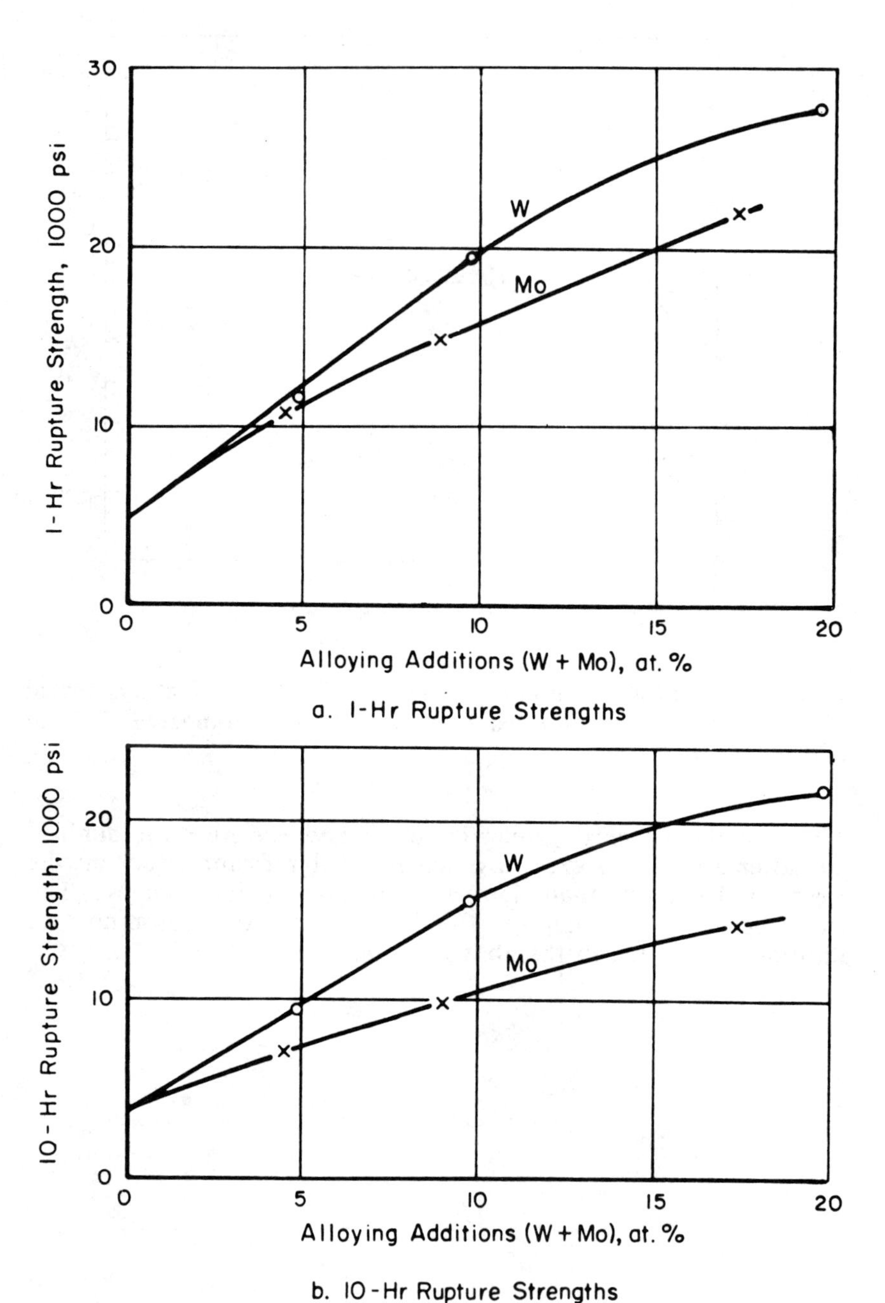

Effect of alloy content on rupture strength of Ta-W-Mo alloys at 2700 °F.

Ta-W-Mo Alloys: Effect of Alloy Content (Continued)

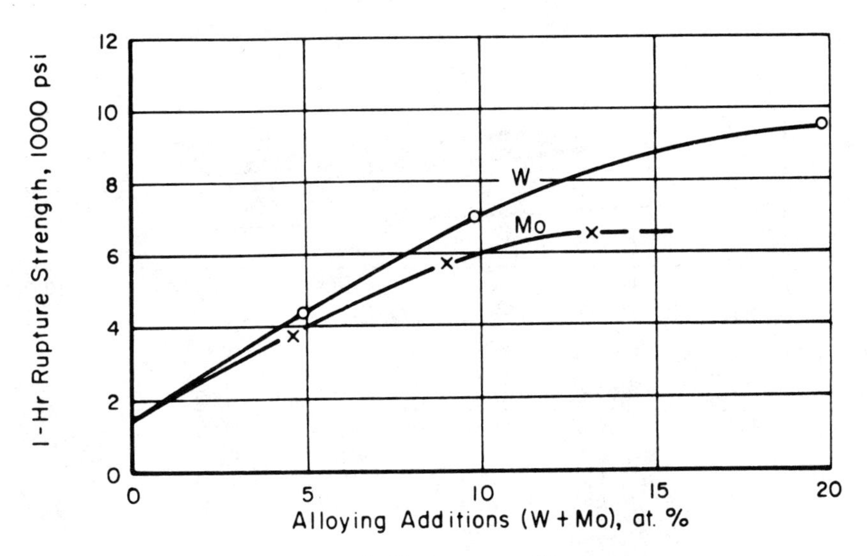

a. 1-Hr Rupture Strengths

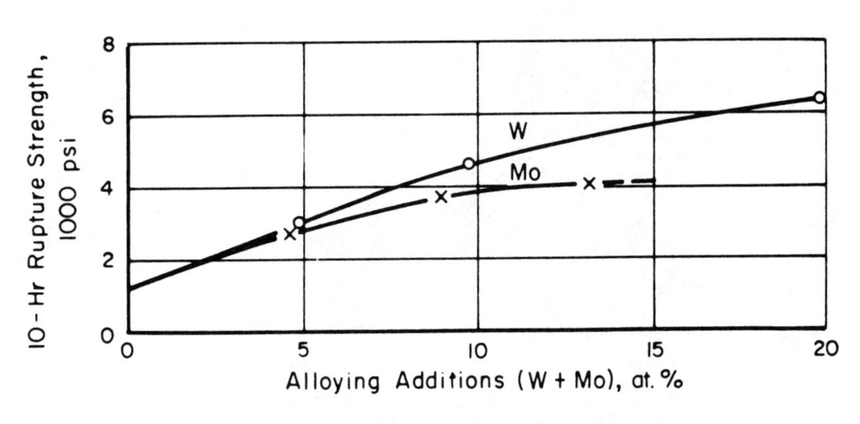

b. 10-Hr Rupture Strengths

Effect of alloy content on rupture strength of Ta-W-Mo alloys at 2700 °F.

Source: J.B. Conway and P.N. Flagella, Creep-Rupture Data for the Refractory Metals to High Temperatures, Gordon and Breach, New York, 1971, p 570-571

8.68 Tantalum: Creep-Stress-Time Curves for 0.2% Deformation in Materials Processed Under Different Conditions

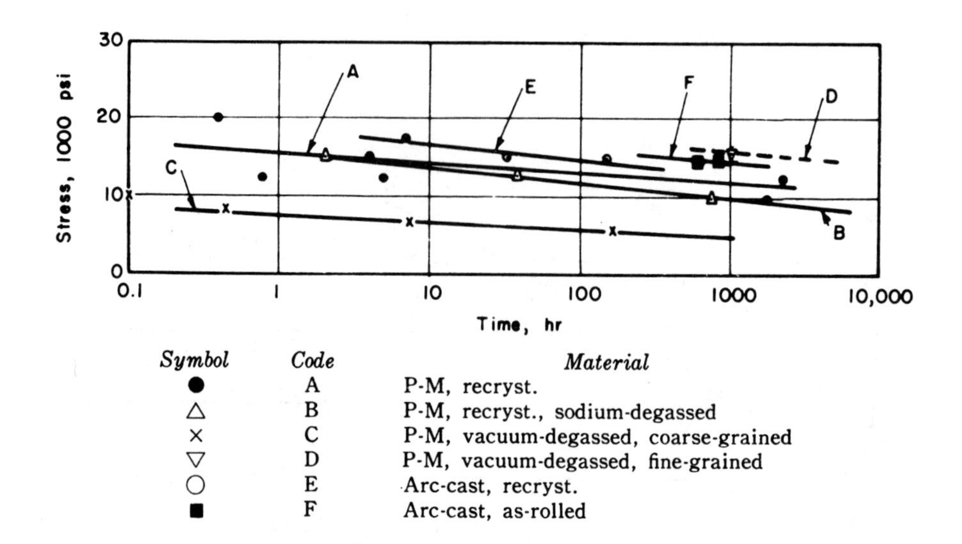

Creep stress vs time for 0.2% total deformation in helium at 650 °C (1200 °F) for tantalum processed under different conditions.

Source: T.E. Tietz and J.W. Wilson, Behavior and Properties of Refractory Metals, Stanford University Press, Stanford, 1965, p 256

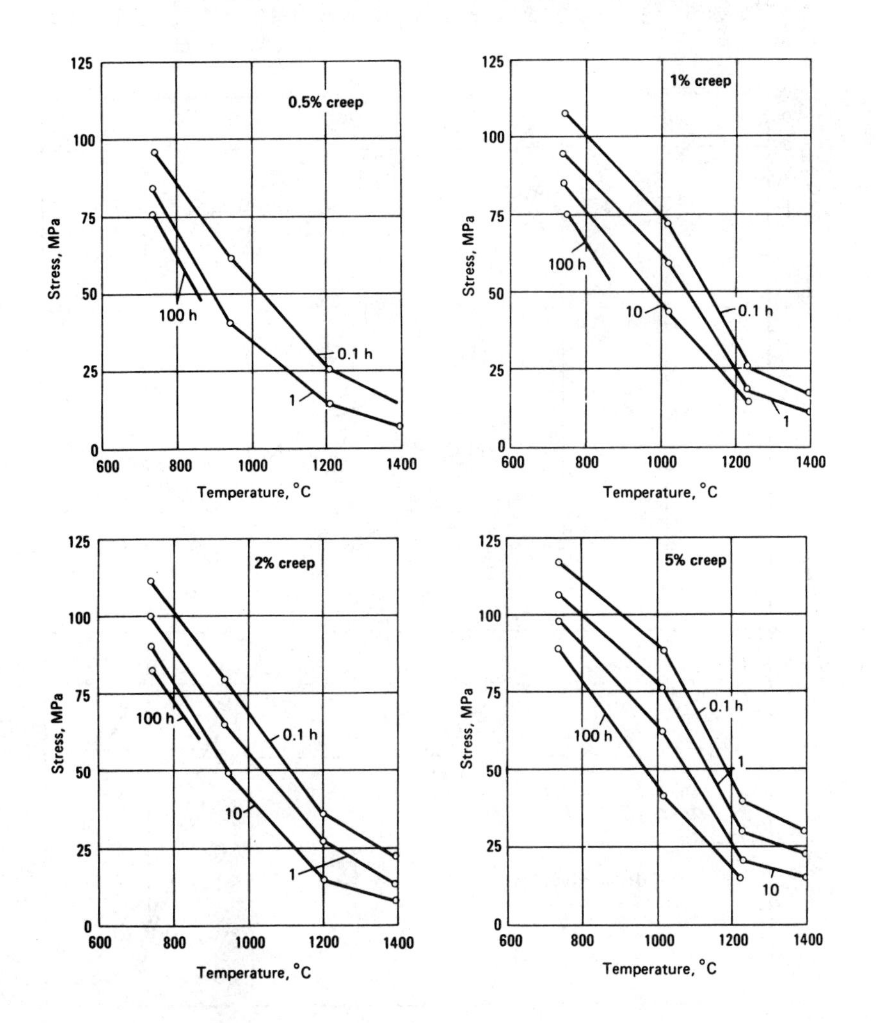

Creep characteristics of 1-mm-thick electron-beam-melted tantalum sheet.

Sample impurities: 0.0030% C, 0.0016% O, 0.0010% N, <0.040% others. Sheet was cold rolled 75% and recrystallized by heating for 1 h at 1200 °C (2190 °F).

Source: Metals Handbook, Ninth Edition, Volume 2, Properties and Selection: Nonferrous Alloys and Pure Metals, American Society for Metals, Metals Park, OH, 1979, p 803

High-Purity Tantalum and Ta-C, Ta-O, and Ta-N: A Comparison

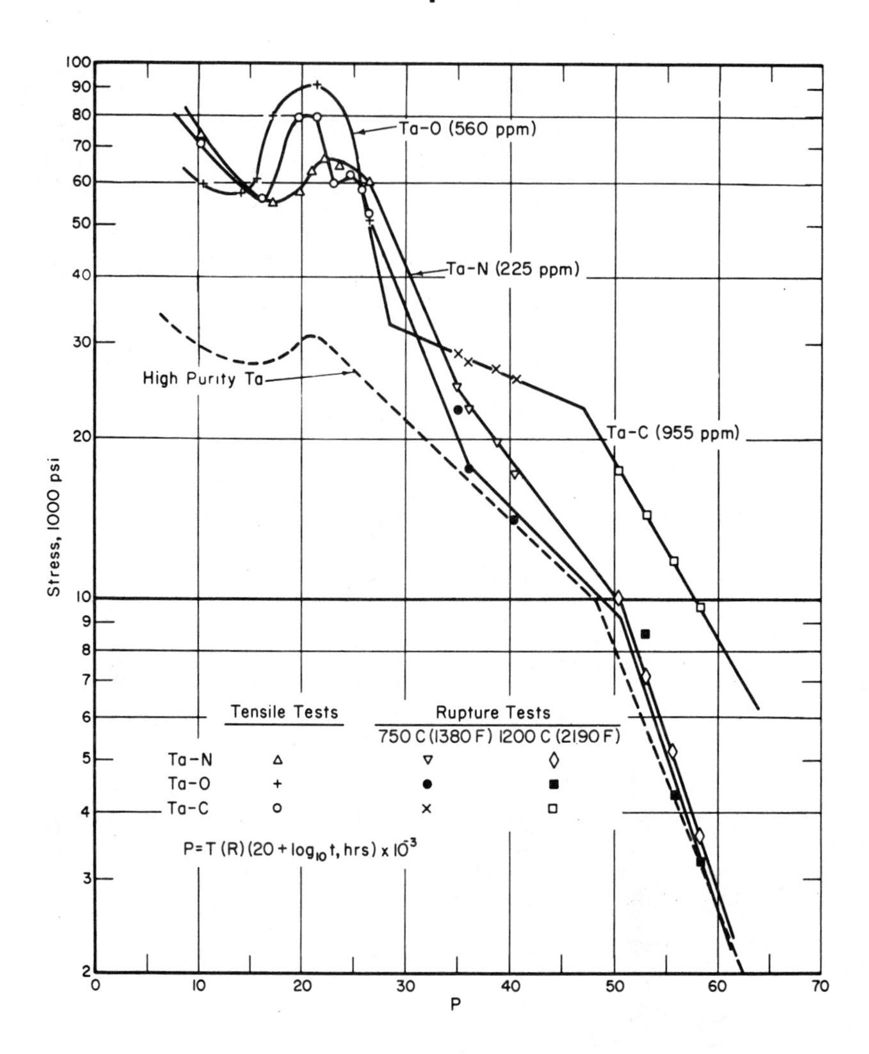

Larson-Miller plot comparing rupture properties of high-purity tantalum and Ta-C, Ta-O, and Ta-N alloys.

The figure above shows that the strengthening effects of oxygen and nitrogen are lost with increasing time and temperatures of testing. The strengthening effect of carbon, however, appears to increase with both time and temperature. The strength peaks shown in the curves at lower temperatures are the result of strain aging reactions.

Source: D.J. Maykuth and H.R. Ogden, Present and Future Status of Tantalum, Tungsten, and Their Alloys, in High Temperature Materials II, G.M. Ault, W.F. Barclay, and H.P. Munger, Eds., Interscience, New York, 1963, p 93, 95

Ta-10W vs Ta-8W-2Hf: A Comparison

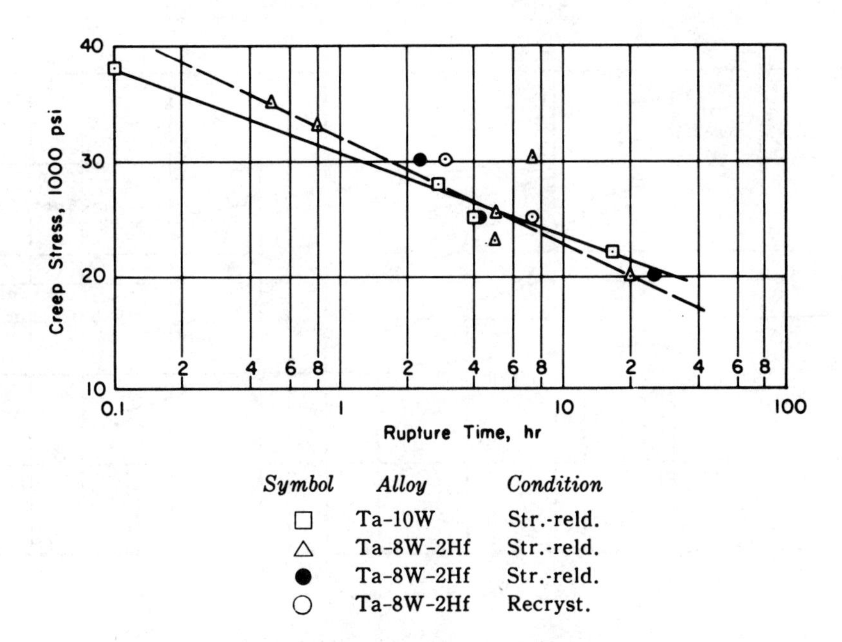

Symbol	*Alloy*	*Condition*
□	Ta-10W	Str.-reld.
△	Ta-8W-2Hf	Str.-reld.
●	Ta-8W-2Hf	Str.-reld.
○	Ta-8W-2Hf	Recryst.

Stress vs rupture time for Ta-10W and Ta-8W-2Hf alloys at 2400 °F.

Source: T.E. Tietz and J.W. Wilson, Behavior and Properties of Refractory Metals, Stanford University Press, Stanford, 1965, p 402

Tantalum Alloys: Comparison of Rupture Properties in Vacuum

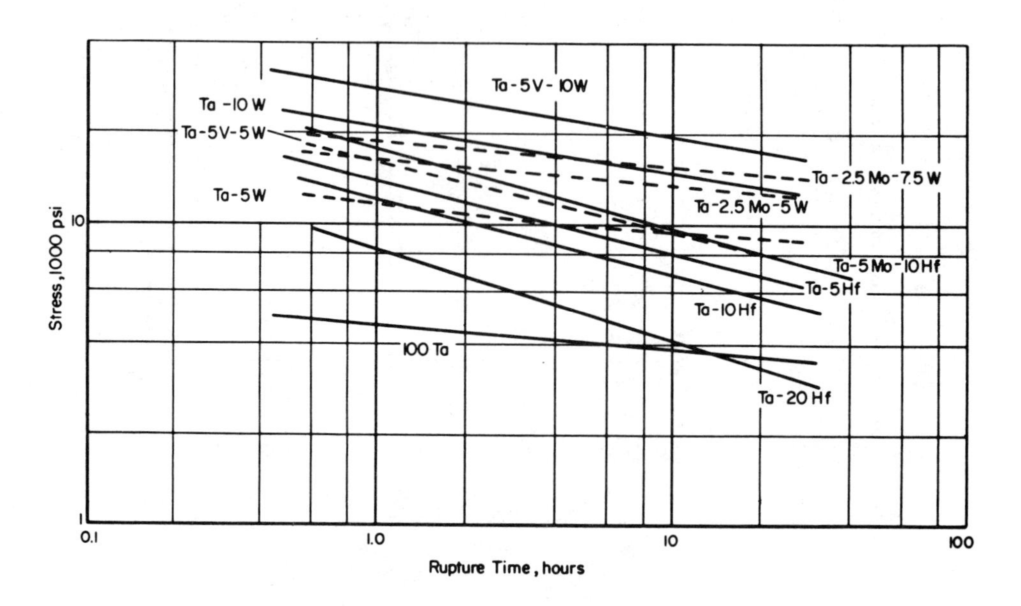

Stress-rupture data of various tantalum-base alloys tested in vacuum at 1480 °C.

Source: J.B. Conway and P.N. Flagella, Creep-Rupture Data for the Refractory Metals to High Temperatures, Gordon and Breach, New York, 1971, p 560

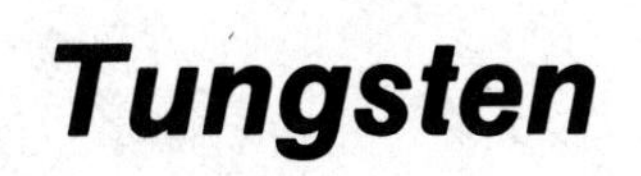

Tungsten

Arc- and Electron-Beam-Melted Tungsten: Relationship Between Transient and Steady-State Creep

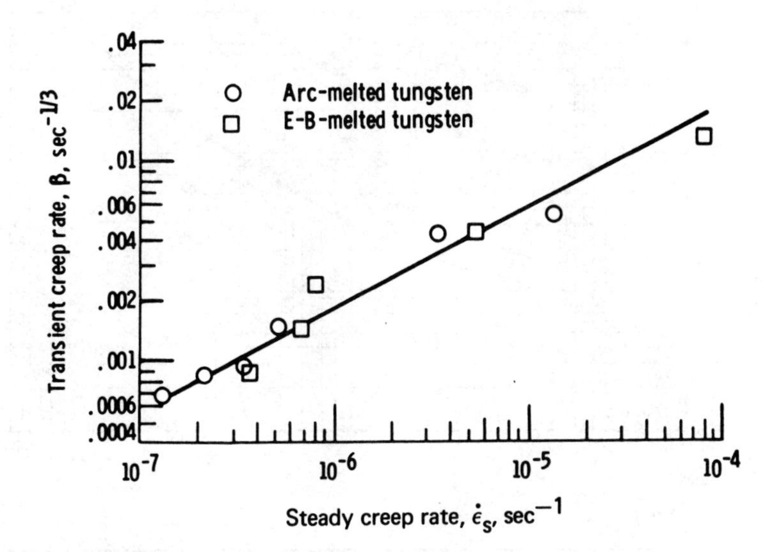

Relationship between transient and steady-state creep of tungsten at 2875 and 3500 °F.

Source: J.B. Conway and P.N. Flagella, Creep-Rupture Data for the Refractory Metals to High Temperatures, Gordon and Breach, New York, 1971, p 322

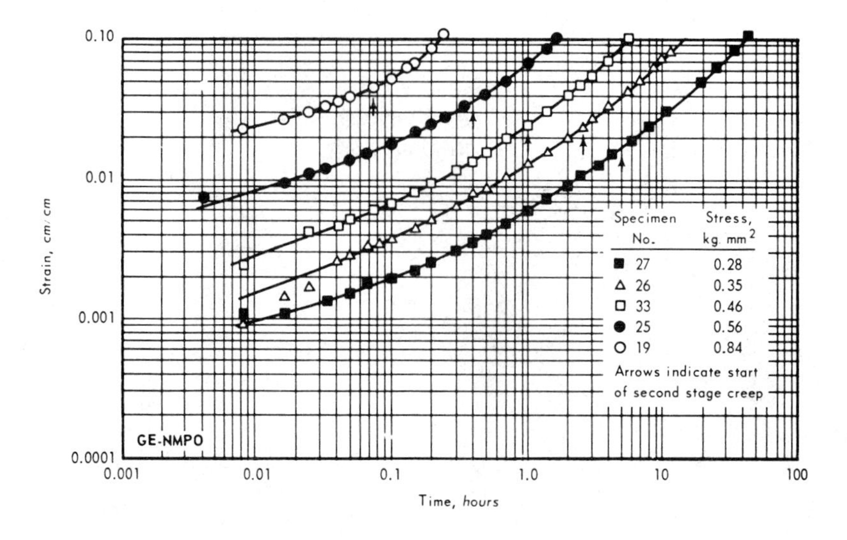

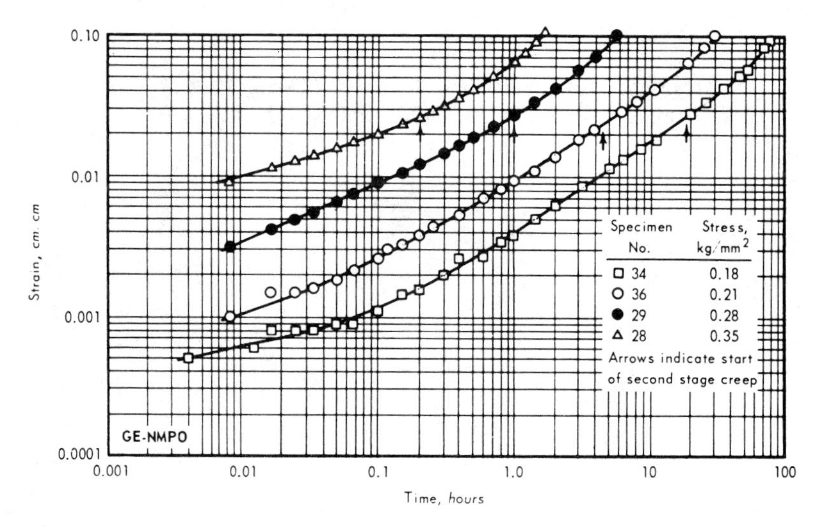

Creep curves for 0.15-cm-thick arc-cast tungsten at 2600 °C (top) and at 2800 °C (bottom).

A desirable feature of the type of plot shown in the top figure, aside from comparison of creep curves at various stresses for a given temperature, relates to first-stage creep analysis. Ordinarily a parabolic creep law of the form:

$$\varepsilon = bt^m + \varepsilon_0$$

Source: J.B. Conway and P.N. Flagella, Creep-Rupture Data for the Refractory Metals to High Temperatures, Gordon and Breach, New York, 1971, p 370-371

applies to first-stage creep behavior, where ε is the instantaneous strain, ε_0 is the strain immediately following load application, t is the time, and b and m are constants. Obviously, when ε_0 is negligible (it can never be zero), the parabolic law describes a linear relationship for first-stage creep behavior when strain and time are both plotted on logarithmic scales. Nonlinearity on such a plot in the first stage of creep results when ε_0 assumes values that are no longer negligible. In these cases, a replotting of the data in terms of $\varepsilon - \varepsilon_0$ vs time on logarithmic coordinates is dictated to obtain linearity. Then the slope of this line yields the exponent of time in the parabolic creep law. Such an analysis has been applied to various materials. Identification of the value of ε_0 is said to result from strain measurements in the early moments following load application, which permits an extrapolation to zero time to obtain ε_0. Because such data are not always obtainable, a trial and error approach was proposed, in which values of ε_0 were assumed until the logarithmic plot of $\varepsilon - \varepsilon_0$ vs time was linear.

P/M and Vacuum-Arc-Melted Tungsten: Annealed for 1 h

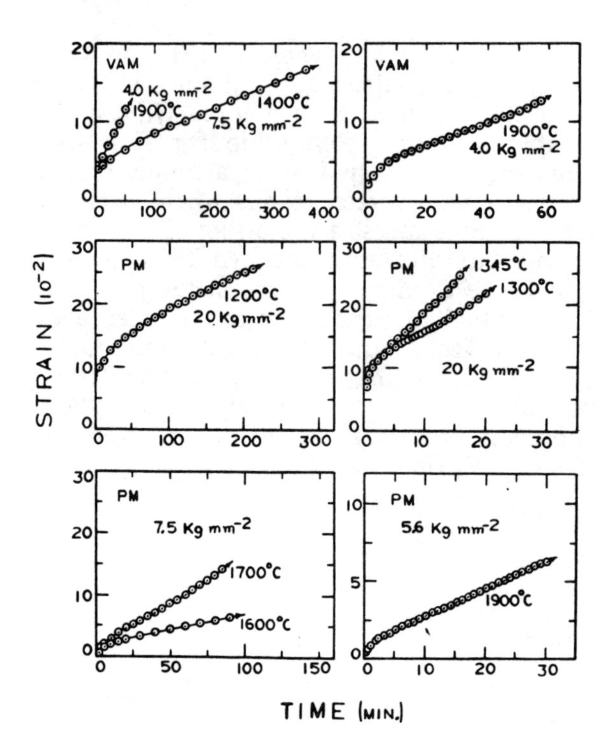

Strain vs time for vacuum-arc-melted and P/M tungsten annealed for 1 h at 2000 °C.

Typical strain vs time curves are shown above. Some tests were carried out to rupture to confirm the observed steady-state creep. The vacuum environment was between 2×10^{-6} and 3×10^{-8} mm of Hg.

Source: E.R. Gilbert, J.E. Flinn, and F.L. Yaggee, "Multimechanism Behavior in the Creep of Tungsten," in Refractory Metals and Alloys IV, Volume II (Proceedings of a Conference in French Lick, IN), R.I. Jaffee, G.M. Ault, J. Maltz, and M. Semchyshen, Eds., Gordon and Breach, New York, 1967, p 865-866

P/M Tungsten: Short-Time Rupture Properties of Stress-Relieved Sheet

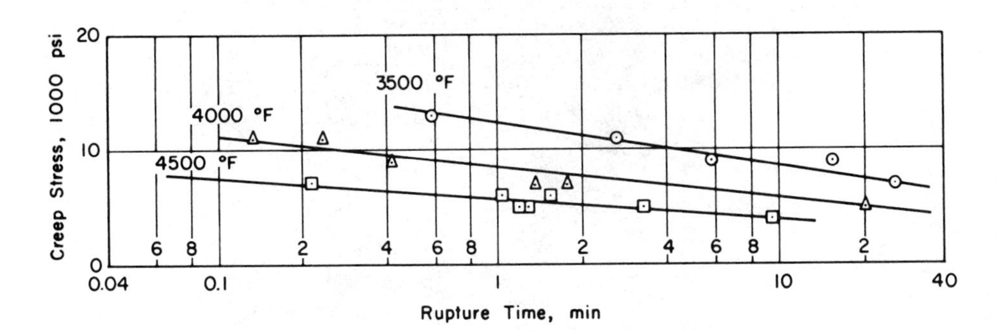

Short-time stress-rupture properties of stress-relieved 0.045-in. P/M tungsten sheet.

Source: T.E. Tietz and J.W. Wilson, Behavior and Properties of Refractory Metals, Stanford University Press, Stanford, 1965, p 405

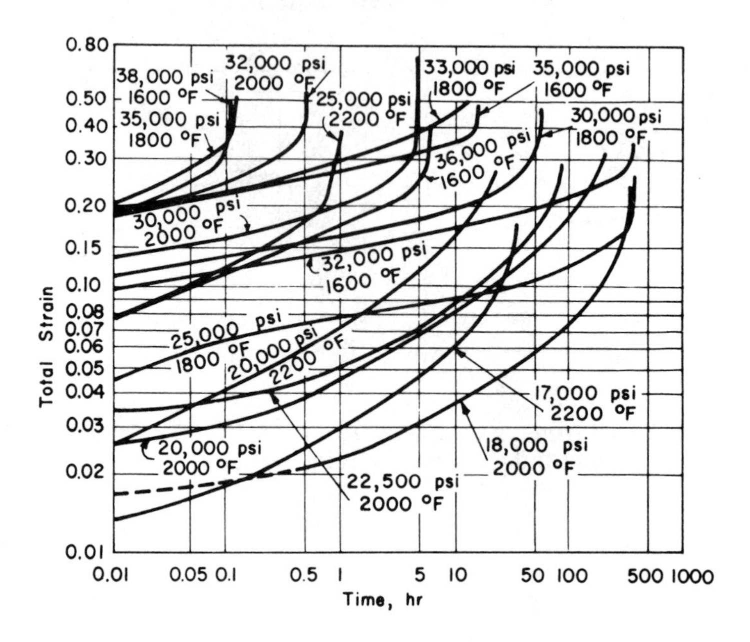

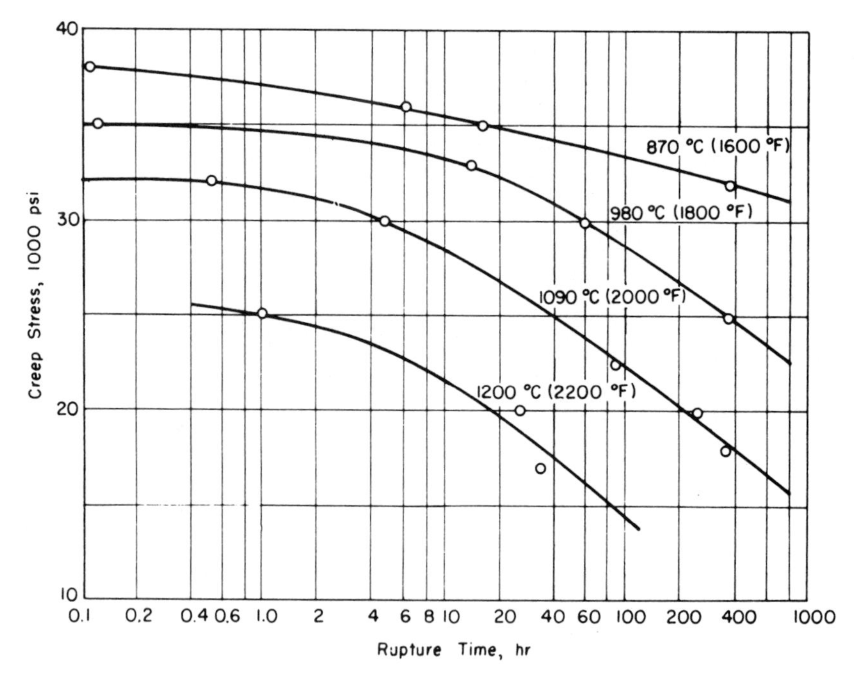

Creep curves for recrystallized tungsten for various test conditions of temperature and stress.

Source: T.E. Tietz and J.W. Wilson, Behavior and Properties of Refractory Metals, Stanford University Press, Stanford, 1965, p 314-315

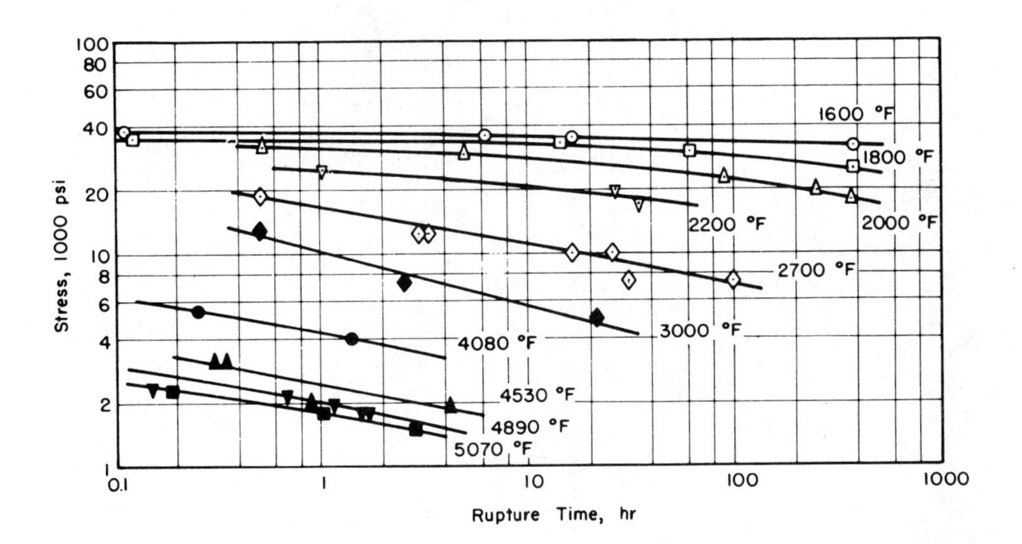

Creep stress vs rupture time for recrystallized tungsten tested in an inert atmosphere.

Source: T.E. Tietz and J.W. Wilson, Behavior and Properties of Refractory Metals, Stanford University Press, Stanford, 1965, p 315

Sintered Tungsten: Sheet Tested in Vacuum

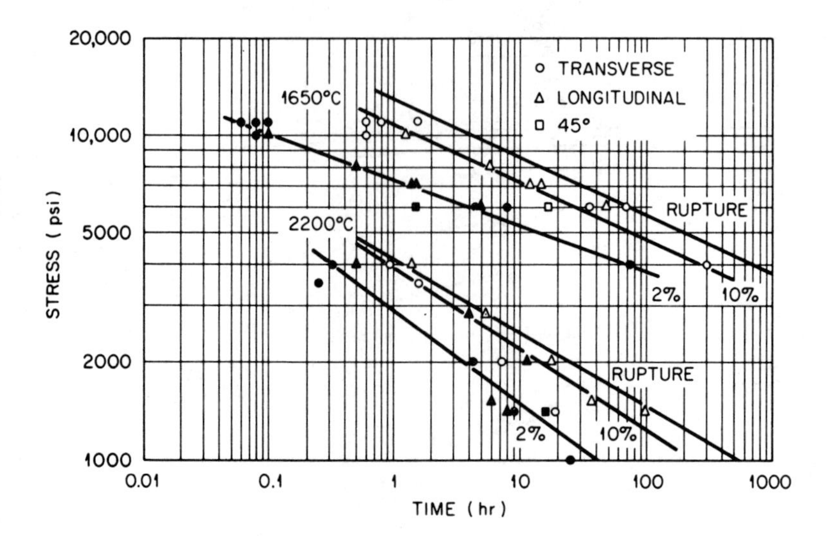

Creep-rupture data for sintered tungsten sheet tested in vacuum.

Source: J.B. Conway and P.N. Flagella, Creep-Rupture Data for the Refractory Metals to High Temperatures, Gordon and Breach, New York, 1971, p 341

Tungsten: Temperature Dependence of 1-h Rupture Strength

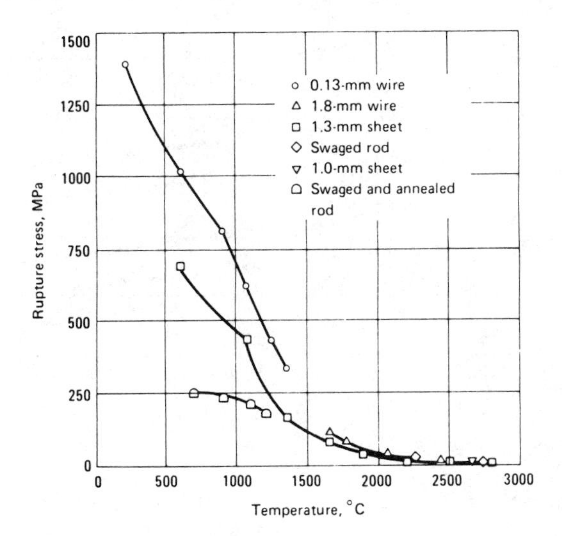

Temperature dependence of 1-h rupture strength of tungsten.

Source: **Metals Handbook, Ninth Edition, Volume 2, Properties and Selection: Nonferrous Alloys and Pure Metals, American Society for Metals, Metals Park, OH, 1979, p 819**

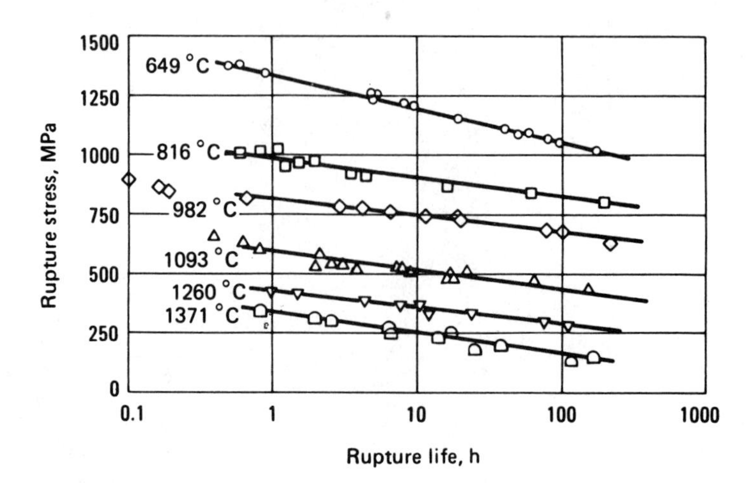

Typical stress-rupture behavior of tungsten wire 0.127 mm (0.005 in.) in diameter, as-drawn.

Source: Metals Handbook, Ninth Edition, Volume 2, Properties and Selection: Nonferrous Alloys and Pure Metals, American Society for Metals, Metals Park, OH, 1979, p 820

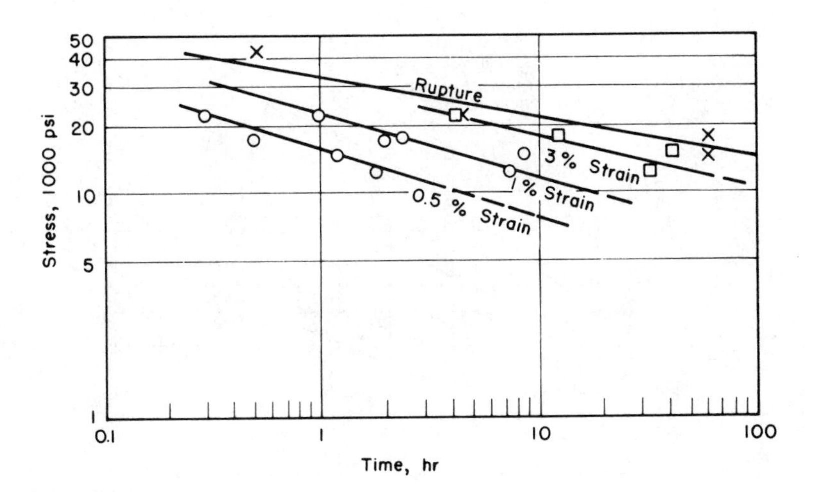

Creep properties of W-0.38TaC alloy at 2700 °F.

Source: T.E. Tietz and J.W. Wilson, Behavior and Properties of Refractory Metals, Stanford University Press, Stanford, 1965, p 316

W-2ThO$_2$: Rod Tested in Vacuum

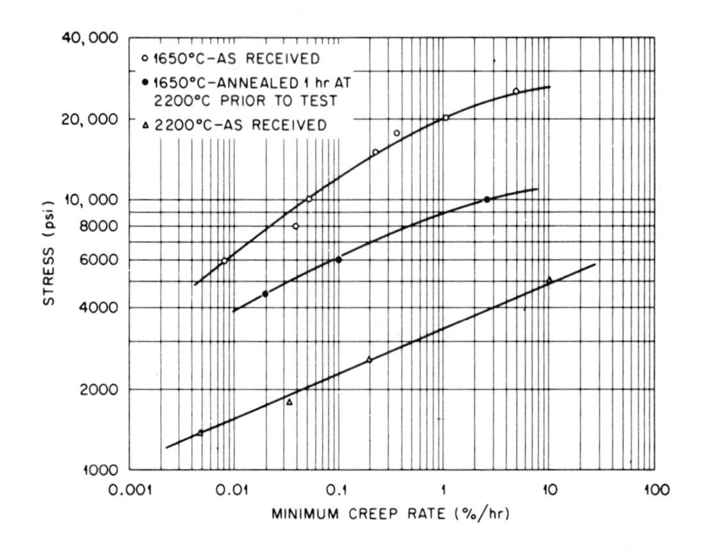

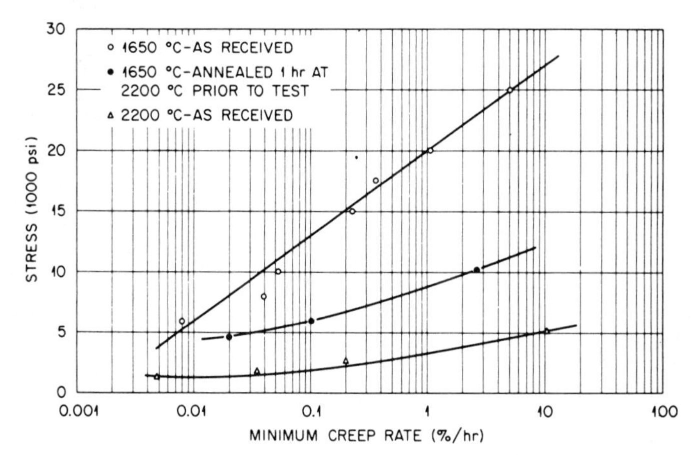

Creep rate data for W-2ThO$_2$ rod material tested in vacuum.

Source: J.B. Conway and P.N. Flagella, Creep-Rupture Data for the Refractory Metals to High Temperatures, Gordon and Breach, New York, 1971, p 383

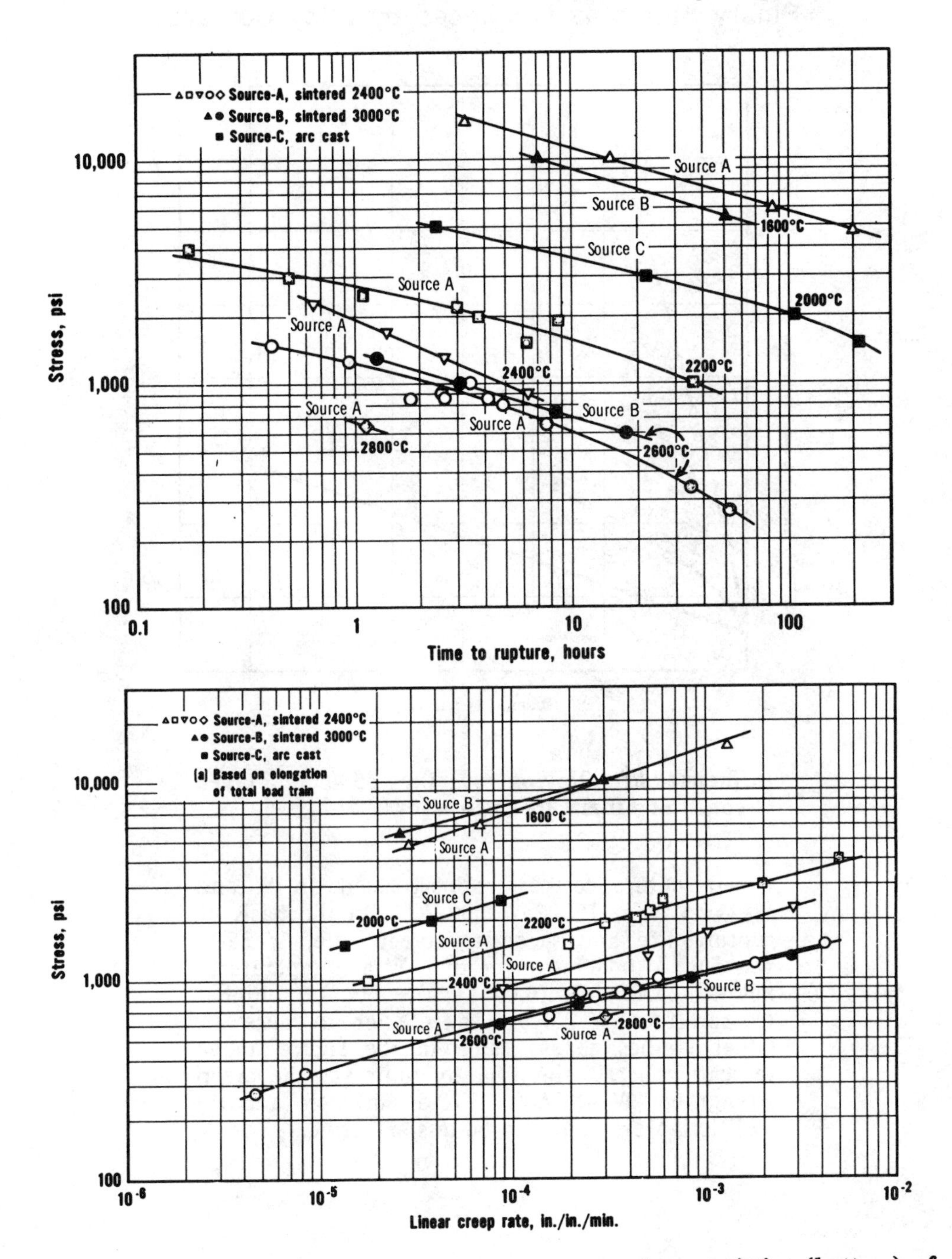

Stress-rupture characteristics (top) and creep rate characteristics (bottom) of W-25Re tested in hydrogen.

Stress-rupture and creep results obtained for W-25Re at 1600, 2000, 2200, 2400, 2600, and 2800 °C are presented above. The data were obtained from three sources of material and in general appear to be consistent, but some differences were observed.

Source: P.N. Flagella and C.O. Tarr, "Creep-Rupture Properties of Rhenium and Some Alloys of Rhenium at Elevated Temperature," in Refractory Metals and Alloys IV, Volume II (Proceedings of a Conference in French Lick, IN), R.I. Jaffee, G.M. Ault, J. Maltz, and M. Semchyshen, Eds., Gordon and Breach, New York, 1967, p 829, 832-833

Tungsten Alloys: Comparison of the Creep Strengths of Binary Alloys as Influenced by Alloy Content

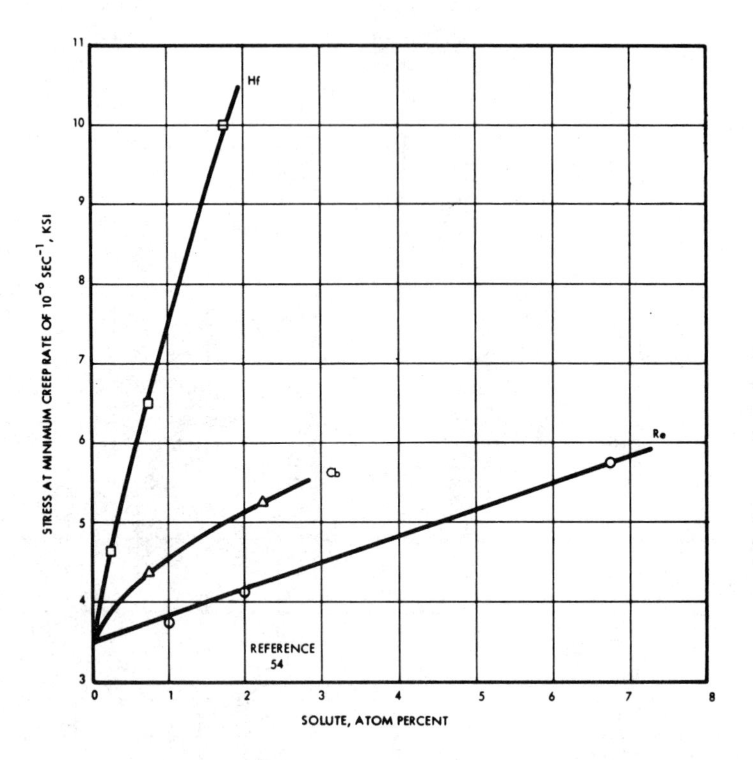

Influence of alloying on the 3500 °F creep strength of binary tungsten alloys.

Creep-rupture tests were performed on W and W-25Re from 1600 to 2600 °C. On the basis of rupture life and secondary creep rate, W-25Re was found to be stronger than W at temperatures below 2000 °C (0.62 Tm), but at higher temperatures W-25Re was inferior to pure W. Other studies as suggested in the figure above indicate that Hf significantly improves the creep strength of W at 3500 °F (0.6 Tm). Nb and Re additions were found to be less effective.

Source: R.T. Begley, D.L. Harrod, and R.E. Gold, "High Temperatue Creep and Fracture Behavior of the Refractory Metals," in Refractory Metal Alloys (Proceedings of a Symposium in Washington, D.C.), I. Machlin, R.T. Begley, and E.D. Weisert, Eds., Plenum, New York, 1968, p 66-67

Tungsten and Tungsten Alloys: Comparison of Swaged and Recrystallized Material

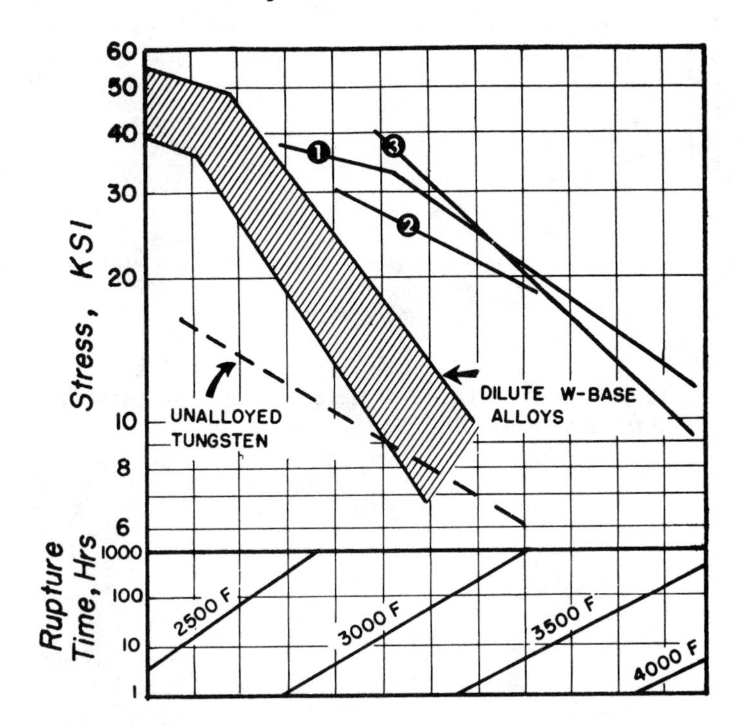

Stress vs time-temperature parameter for swaged tungsten and tungsten-base alloys:

(1) W + 0.48% Zr + 0.048% C (as-swaged)

(2) W + 0.48% Zr + 0.048% C (recrystallized)

(3) W + 1.18% Hf + 0.086% C (as-swaged)

W + 0.48% Zr + 0.048% C and W + 1.18% Hf + 0.086% C were tested at 3000, 3500, and 4000 °F in the as-swaged condition; the W + 0.48% Zr + 0.048% C alloy was also tested in the recrystallized state at 3000 and 3500 °F. The creep-rupture tests were conducted in vacuum equipment operating at pressures of less than 10^{-4} torr.

The test specimen configuration was essentially similar to that of the tensile test specimens. Grip failures were encountered in the first few creep-rupture tests; this problem was circumvented in subsequent tests by reducing the specimen gage diameter to 0.150 in.

Stress values as a function of time and temperature are plotted above; the abscissa is based on the Larson-Miller parameter:

$$T(C + \log t)$$

where T is the absolute temperature, t is the time in hours, and C is a constant.

Source: M. Semchyshen and E. Kalns, "Tungsten-Base Alloy Development," in Refractory Metals and Alloys IV, Volume I (Proceedings of a Conference in French Lick, IN), R.I. Jaffee, G.M. Ault, J. Maltz, and M. Semchyshen, Eds., Gordon and Breach, New York, 1967, p 492-493

W-5Mo, W-0.52Cb, and W-15Mo: A Comparison

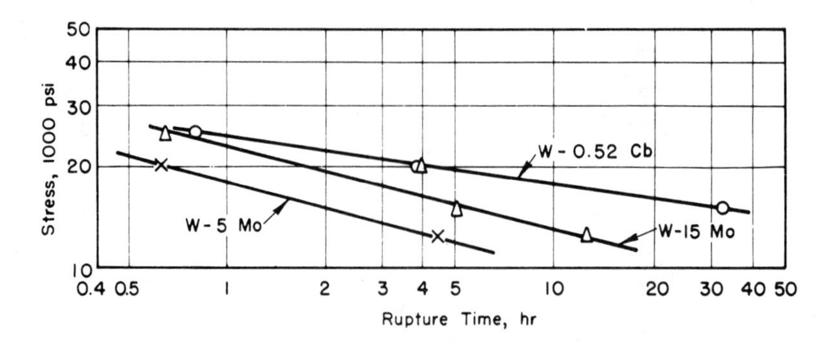

Creep stress vs rupture time for three tungsten binary alloys at 3000 °F.

Source: T.E. Tietz and J.W. Wilson, Behavior and Properties of Refractory Metals, Stanford University Press, Stanford, 1965, p 317

W-ThO_2 Alloys: Comparison of the Creep and Stress-Rupture Properties of Thoriated Bar

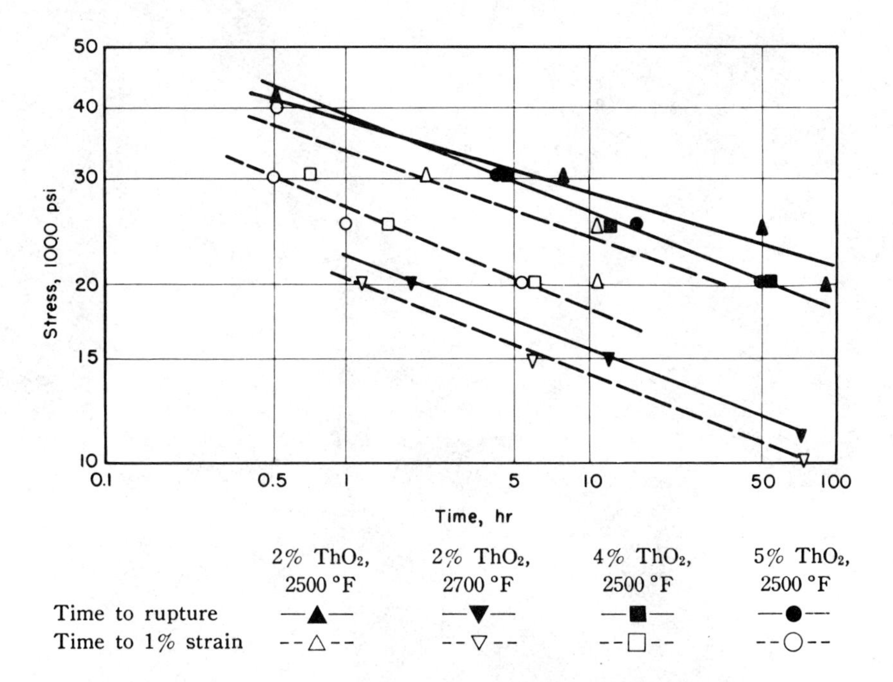

Creep stress vs time to 1% strain and time to rupture for thoriated tungsten bar alloys.

Source: T.E. Tietz and J.W. Wilson, Behavior and Properties of Refractory Metals, Stanford University Press, Stanford, 1965, p 316

9

Refractory Metals Comparisons

Molybdenum and Molybdenum Alloys vs Selected Superstrength Alloys: A Comparison

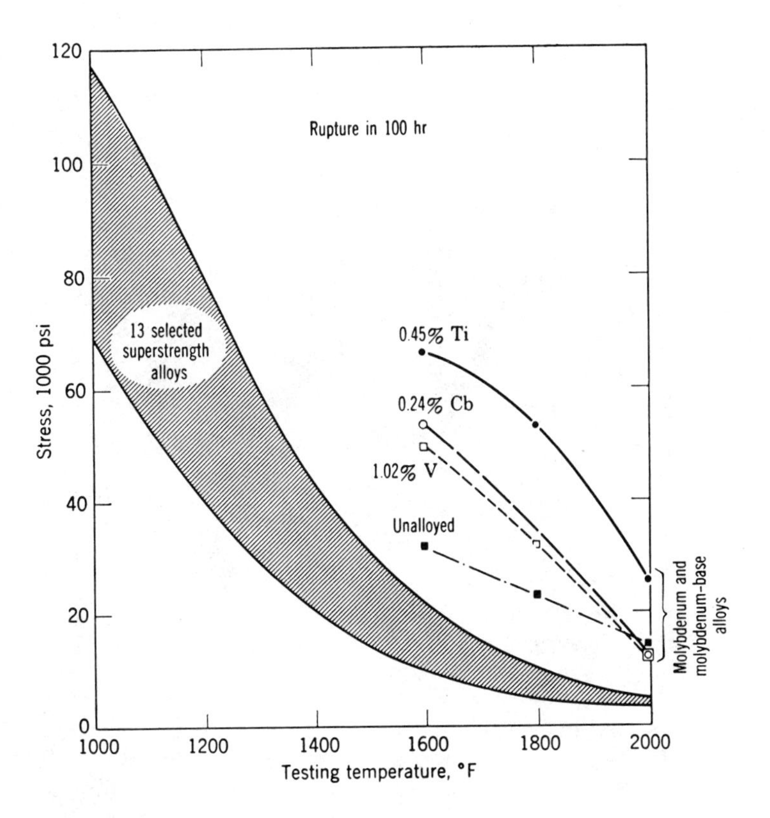

Stress-to-rupture properties of molybdenum and molybdenum alloys.

Within the temperature range of 1600-2000 °F, the creep-to-rupture strength of the unalloyed metal itself is superior to that of the best of Fe-base, Co-base, Ni-base, or other commercial superalloys. At 2000 °F, Mo-base alloys have strength properties superior to those of commercial high-temperature alloys at 1600 °F.

Source: J.J. Harwood and M. Semchyshen, "Molybdenum, Its Alloys and Its Protection," in High Temperature Materials, R.F. Hehemann and G.M. Ault, Eds., American Institute of Mining, Metallurgical, and Petroleum Engineers, Warrendale, PA, 1959, p 245-247

Molybdenum and Niobium Alloys: Comparison of 10-h Rupture Strengths

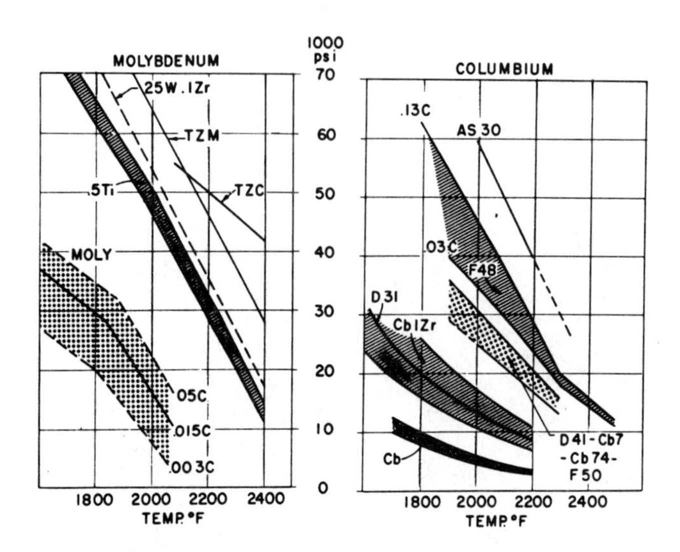

Ten-hour rupture strengths of select molybdenum and niobium alloys.

As the result of highly pure metals and modern electron beam technology, Mo can be at least two times as strong as Nb. From these base points, alloying has raised Mo about 8-fold in strength, and it has raised Nb about 12-fold.

Source: L.P. Jahnke, "Review of the Status and Future of Molybdenum and Columbium Alloys," in High Temperature Materials II, G.M. Ault, W.F. Barclay, and H.P. Munger, Eds., Interscience, New York, 1963, p 286

Niobium Alloys in the Stress-Relieved Condition vs Mo-0.5Ti Bar: A Comparison

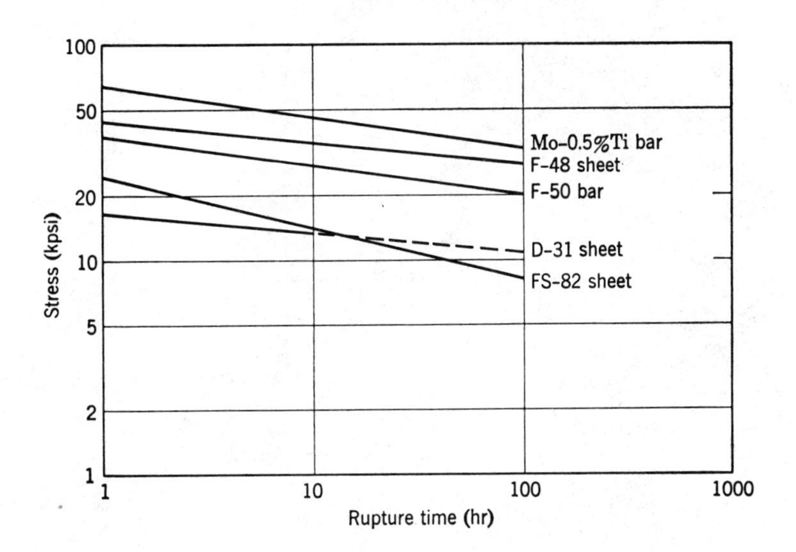

Stress-rupture properties of stress-relieved niobium-base alloys at 1095 °C compared to 0.5%Ti molybdenum bar.

F-48 has a 100-h rupture strength of 29,000 psi at this temperature, which is only slightly lower than that of the molybdenum alloy.

Source: R.W. Fountain and C.R. McKinsey, "Physical and Mechanical Properties of Columbium and Tantalum and Their Alloys," in Columbium and Tantalum, F.T. Sisco and E. Epremian, Eds., John Wiley & Sons, New York, 1963, p 277-279

Niobium, Vanadium, and Molybdenum Alloys: Comparison of Tests in Helium

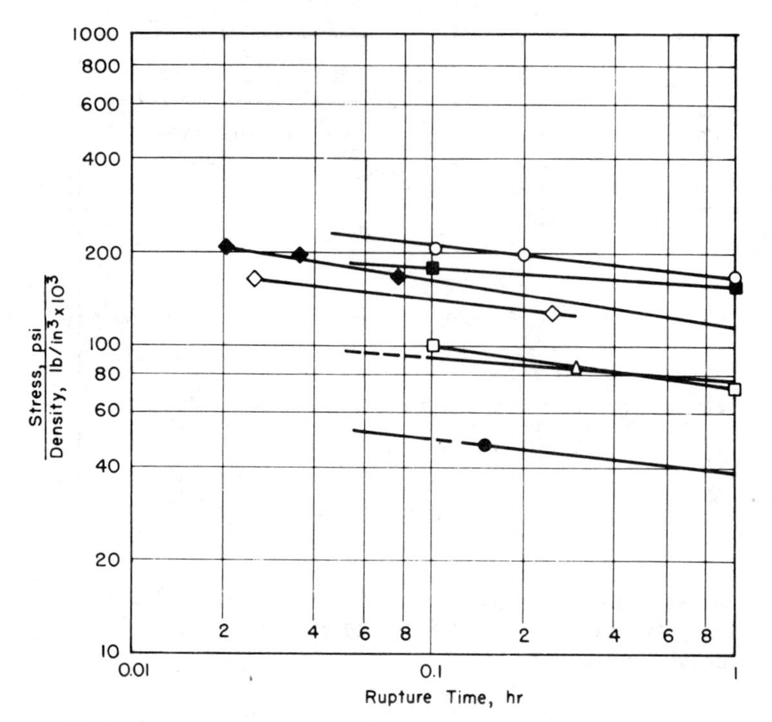

○ Mo-0.5Ti, stress-relieved
■ F-48 (Cb-15W-5Mo-1Zr), stress-relieved
◆ V-5Ti-20Cb, recrystallized
◇ V-5Ti-20Ta, recrystallized
□ Molybdenum, stress-relieved
△ Mo-0.5Ti, recrystallized
● Molybdenum, recrystallized

Stress-to-density ratio vs rupture time for selected vanadium, niobium, and molybdenum alloys tested in helium at 2000 °F.

Two V alloys are compared with commercial Mo and Nb alloys in the figure above on the basis of short-time rupture strength at 2000 °C. The V alloys were tested in the recrystallized state, and they were superior to the recrystallized Mo and Nb alloys on the basis of the strength-to-density ratio. The reverse is true, however, if the recrystallized V alloys are compared with the Mo and Nb alloys in the stress-relieved condition. The annealing treatment (1830 °F for 30 min) given the V alloys resulted in complete recrystallization of the cold worked structure.

Source: T.E. Tietz and J.W. Wilson, Behavior and Properties of Refractory Metals, Stanford University Press, Stanford, 1965, p 366-367

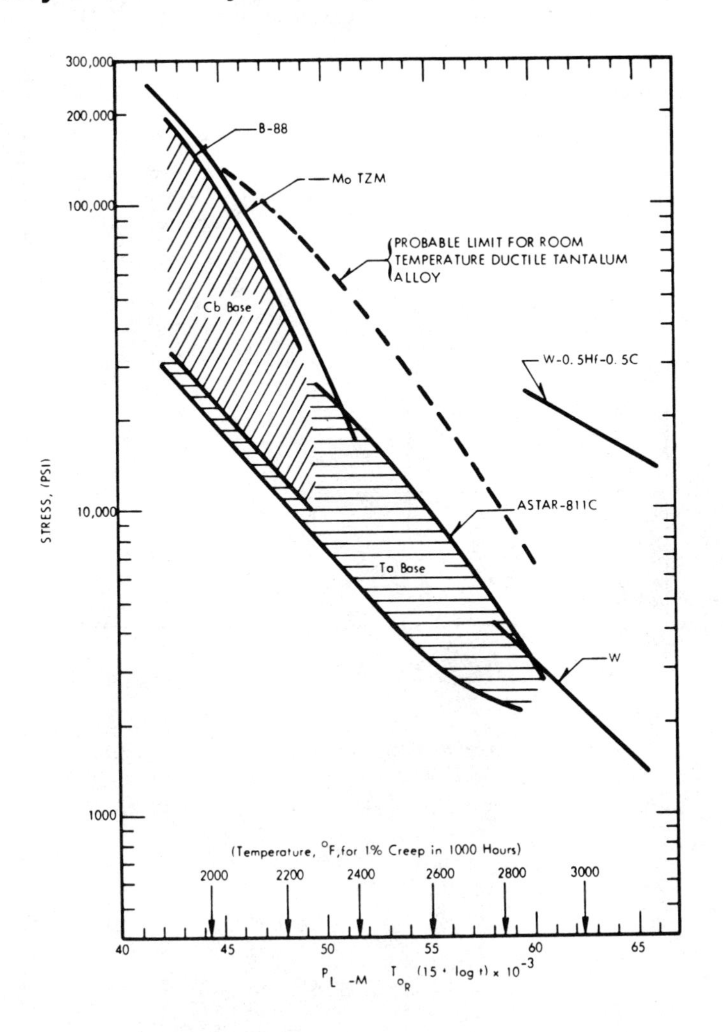

Larson-Miller parameter normalization for time to 1% creep for refractory metal alloys.

In this figure, time-dependent deformation is the criterion. It becomes obvious that Ta-base alloys have a distinct advantage over both Nb- and Mo-base alloys at temperatures above 2200 °F. The potential strength limit was estimated for an alloy composition that would exhibit base-metal ductility at room temperature. It would be of a type similar to ASTAR-811C, but would contain a total substitutional solute level on the order of 16-20 at%. It is most likely that Ta alloys of this solute level would find application primarily as bar and/or forging, because sheet fabrication will most likely be difficult.

Source: R.W. Buckman, Jr. and R.C. Goodspeed, "Considerations in the Development of Tantalum Base Alloys," in Refractory Metal Alloys (Proceedings of a Symposium in Washington, D.C.), Plenum, New York, 1968, p 387, 389-390

9.6 Refractory Metal Alloys and Unalloyed Tungsten: 10-min Stress-Rupture Strength-to-Density Ratio

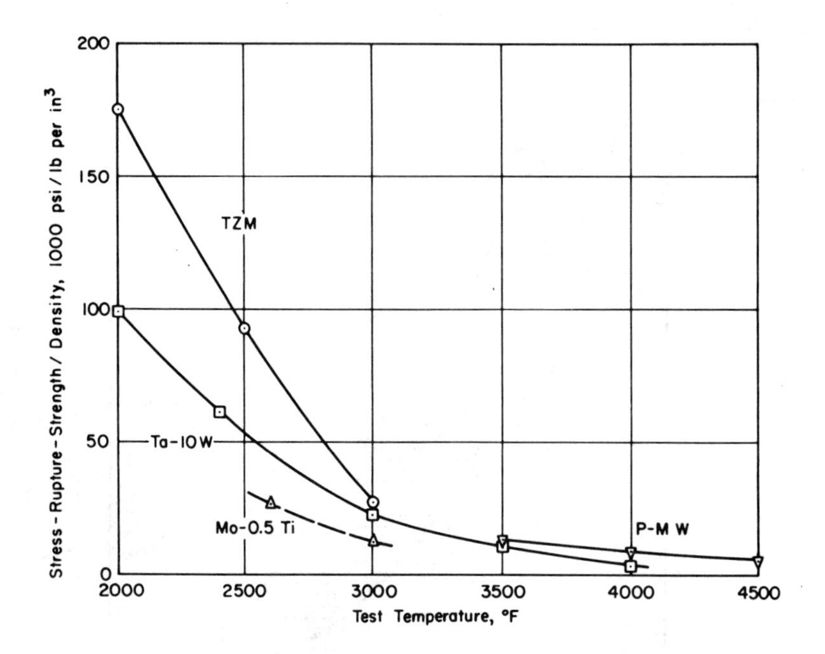

Ten-minute stress-rupture strength-to-density ratio for selected refractory metal alloys and unalloyed tungsten between 2000 and 4500 °F.

Source: T.E. Tietz and J.W. Wilson, Behavior and Properties of Refractory Metals, Stanford University Press, Stanford, 1965, p 407

Refractory Metals: Comparison of 10-h Rupture Strengths and Larson-Miller Curves

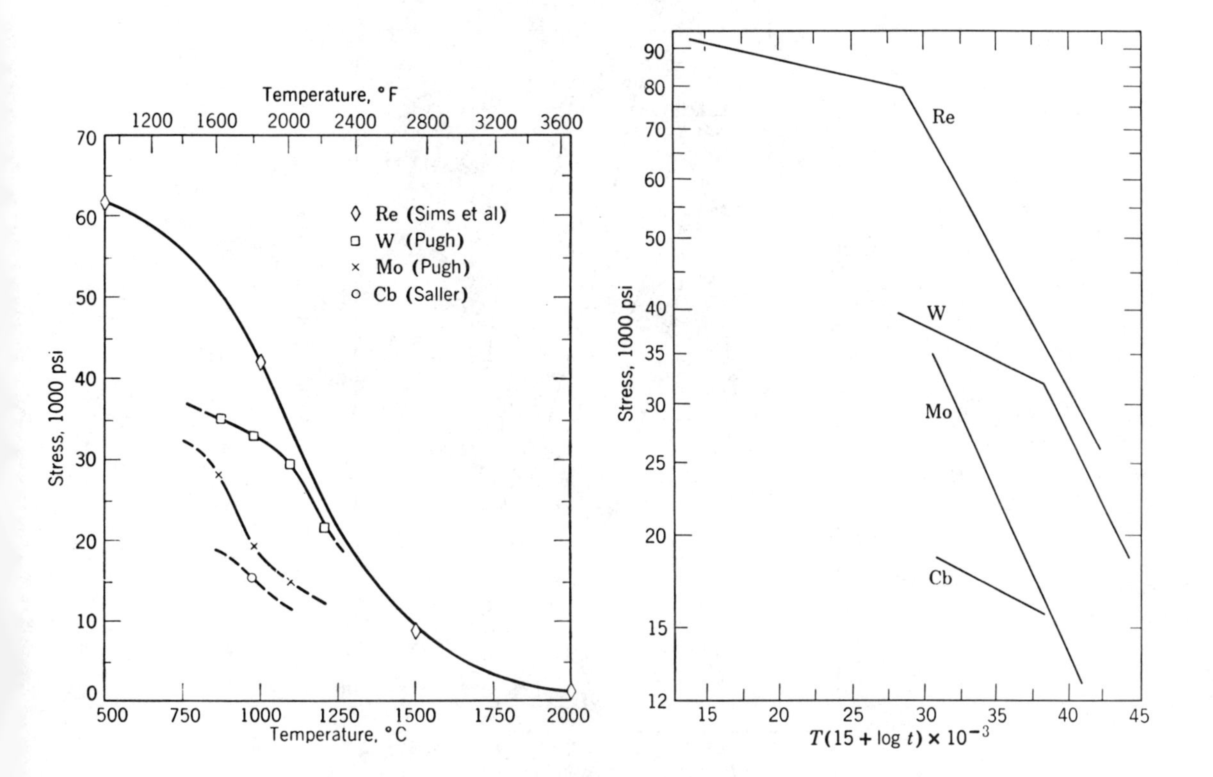

Stress-rupture strength for 10-h life.

Creep-rupture strength as a function of Larson-Miller parameter.

Re shows less superiority over W at high temperatures. Mo and Nb both appear considerably weaker. The bottom figure provides a comparison of stress-rupture properties based on the Larson-Miller parameter. Re is again most outstanding at lower temperatures and for shorter times. The order of superiority of body-centered cubic metals again appears to be related to their melting points.

Source: J.W. Pugh, "Refractory Metals: Tungsten, Tantalum, Columbium, and Rhenium," in High Temperature Materials, R.F. Hehemann and G.M. Ault, Eds., American Institute of Mining, Metallurgical, and Petroleum Engineers, Warrendale, PA, 1959, p 312-313

9.8 Refractory Metals: Comparison of 100-h Rupture Strengths for Recrystallized Material

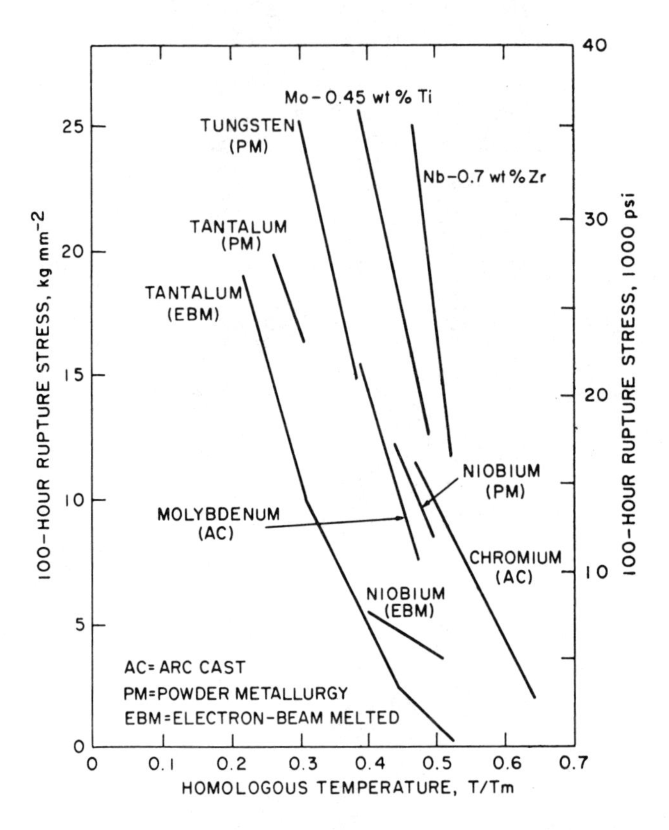

One-hundred-hour rupture strength versus homologous temperature for refractory metals in the recrystallized condition.

Source: W.D. Wilkinson, Properties of Refractory Metals, Gordon and Breach, New York, 1969, p 23

Refractory Metals and Their Alloys: Comparison of 1000-h Rupture Strengths

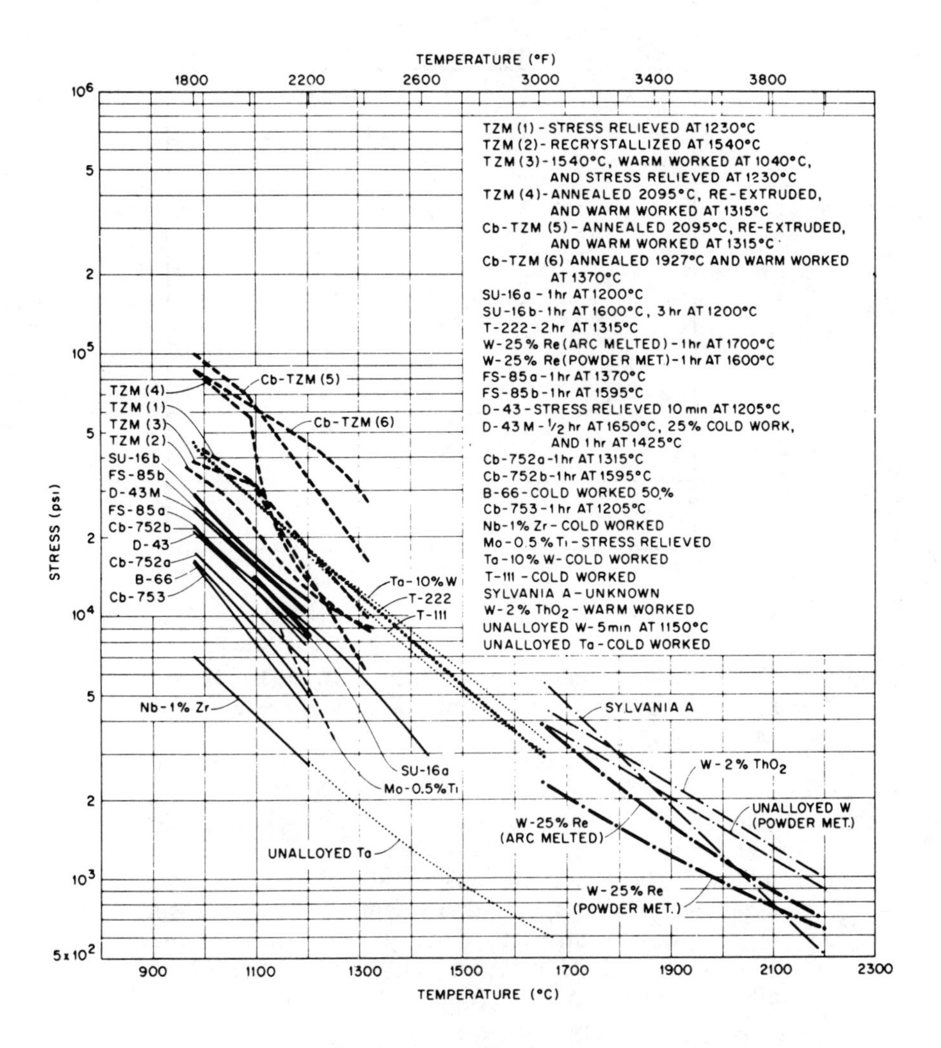

Stress to produce rupture in 1000 h for selected refractory metals and refractory metal alloys.

Source: J.B Conway and P.N. Flagella, Creep-Rupture Data for the Refractory Metals to High Temperatures, Gordon and Breach, New York, 1971, p 780

Vanadium-Base Alloys vs Commercial Alloys of Other Metals

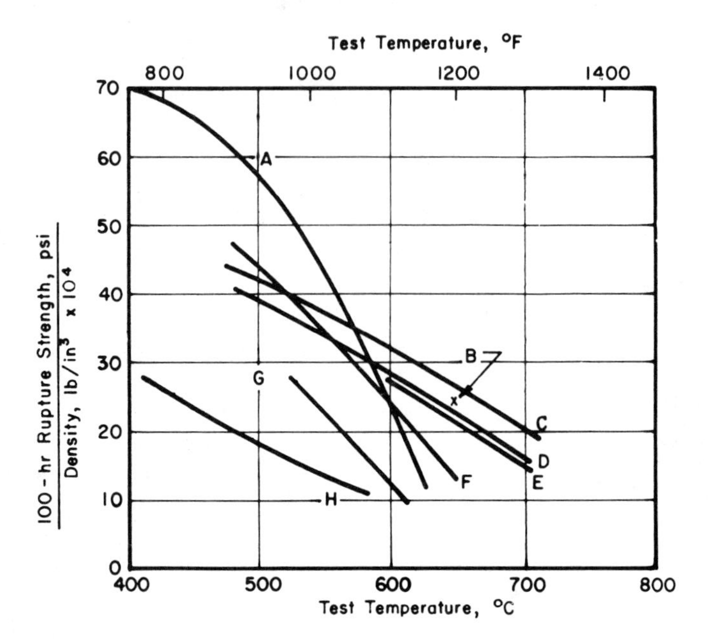

One-hundred-hour stress-rupture strength-density ratio as a function of test temperature for various alloys.

The figure shows comparative data relating the 100-h stress-rupture strength-density ratio vs temperature for various V-base alloys and commercial alloys of other metals. On the basis of the strength-density ratio, the V-40Ti-5Al-0.5C alloy is superior to the other alloys at 350 to 570 °C (660 to 1060 °F).

Source: T.E. Tietz and J.W. Wilson, Behavior and Properties of Refractory Metals, Stanford University Press, Stanford, 1965, p 366-367

10

ACI Casting Alloys

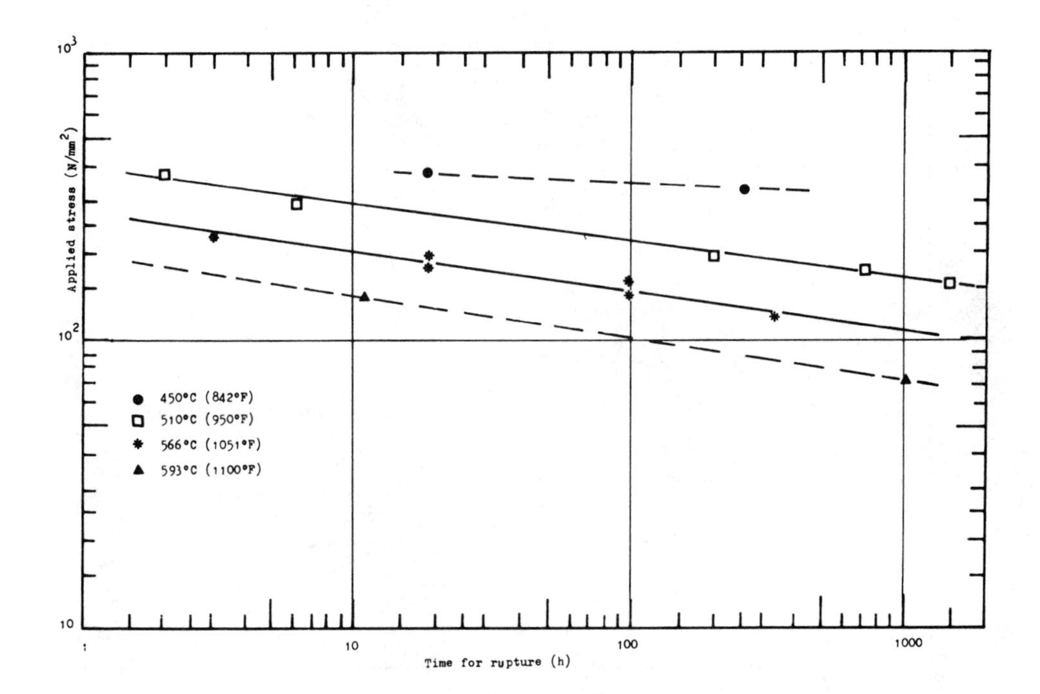

Isothermal creep curves.

Isothermal creep curves $\log_{10}\sigma$-$\log_{10}t$ determined by linear correlation

Test Temperature	Tests Carried out	Linear Correlation Coefficient	Equation Calculated
450°C (842°F)	2	−1	$\log_{10}\sigma = 2.632 - 0.0416 \log_{10}t$
510°C (950°F)	5	−0.993813	$\log_{10}\sigma = 2.596 - 0.1240 \log_{10}t$
566°C (1051°F)	6	−0.947503	$\log_{10}\sigma = 2.451 - 0.1474 \log_{10}t$
593°C (1100°F)	2	−1	$\log_{10}\sigma = 2.305 - 0.1530 \log_{10}t$

NOTES: σ = stress, N/mm^2; t = time to rupture, h.

For a given stress, the LM constant can be found in the table by plotting log *tr* versus $1/T(K)$, because this constant is intercepted by the straight line on the ordinate axis. The time to rupture at constant stress was obtained from the diagram shown above. The computer analysis of these data by linear correlation gave the equation of the straight line and the value of the LM constant.

Source: C. Santafè, M. Priante, G. Amici, and M. Finocchio, "Manufacturing and Properties of Large Castings in 13Cr-1Ni Low-Carbon Steel (ASTM A217)," in Stainless Steel Castings, ASTM Special Technical Publication No. 756, American Society for Testing and Materials, Philadelphia, 1982, p 373-375

CF-3 and CF-8

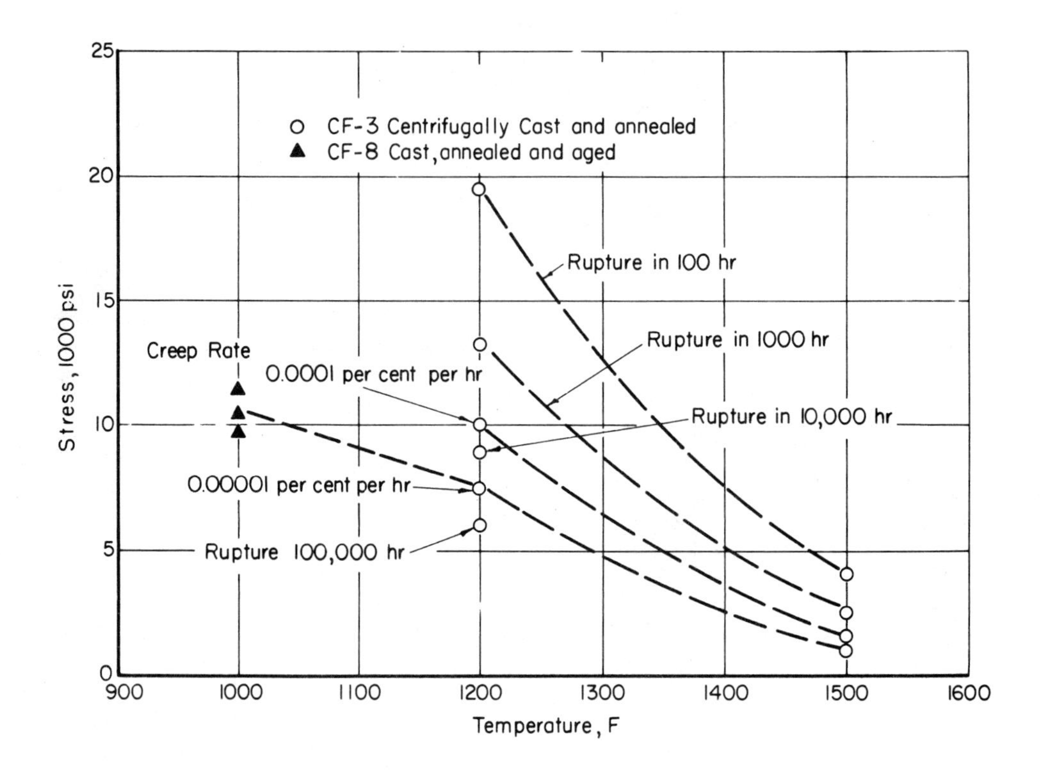

Rupture and creep strengths of CF alloy castings.

Source: The Elevated-Temperature Properties of Stainless Steels, ASTM Data Series Publication DS 5-S1, American Society for Testing and Materials, Philadelphia, 1965, p 57

CF-8: Effect of Nitrogen Content

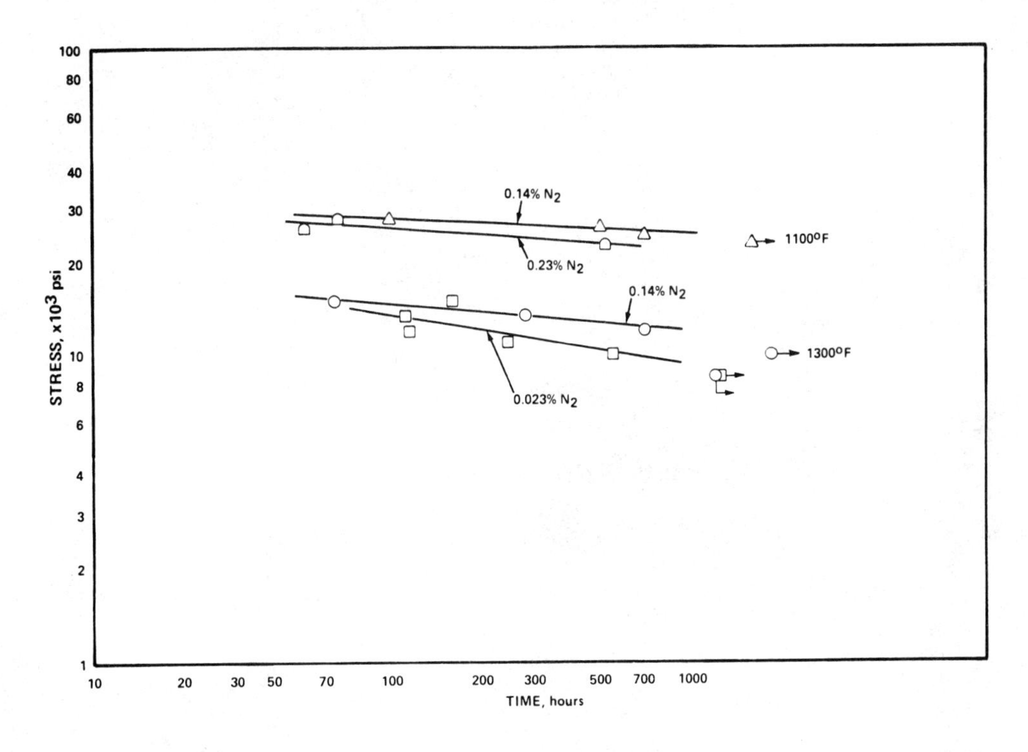

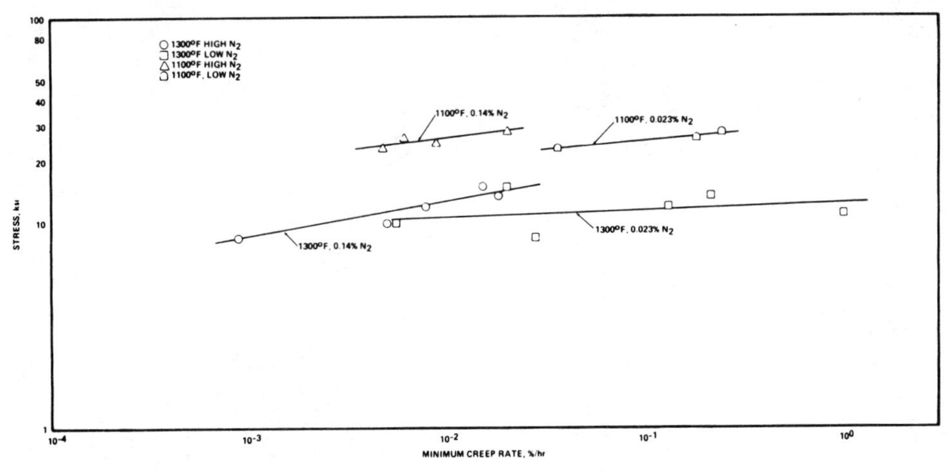

Time to rupture (top) and minimum creep rate (bottom) for CF-8 castings containing 0.023 and 0.14% N_2 at 1100 and 1300 °F.

Source: C.E. Bates, O.V. Rogers, and R.W. Monroe, "Effects of Nitrogen Additions on CF-3 and CF-8 Stainless Steel," in New Developments in Stainless Steel Technology, R.A. Lula, Ed., American Society for Metals, Metals Park, OH, 1985, p 218

HC

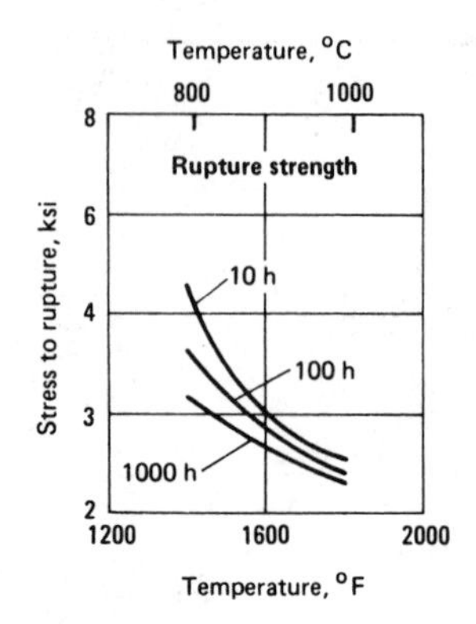

Rupture strength of alloy HC.

Source: Metals Handbook, Ninth Edition, Volume 3, Properties and Selection: Stainless Steels, Tool Materials and Special-Purpose Metals, American Society for Metals, Metals Park, OH, 1980, p 286

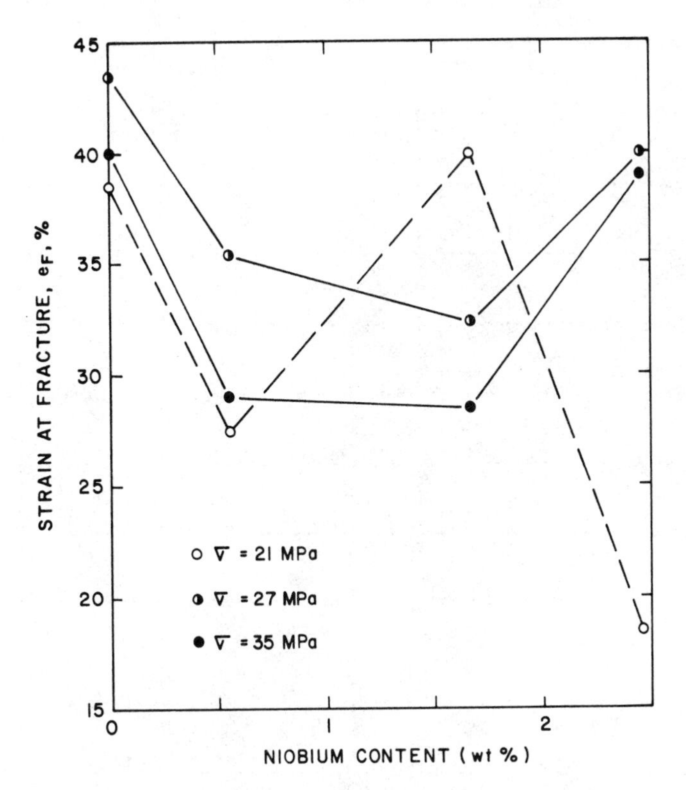

Strain to fracture (ductility) as a function of niobium content.

Source: E.A.A.G. Ribeiro, R. Papaléo, and J.R. Guimaraes, "Microstructure and Creep Behavior of a Niobium Alloyed Cast Heat-Resistant 26 Pct Cr Steel," Metallurgical Transactions A, Vol. 17A, April 1986, p 694

HD

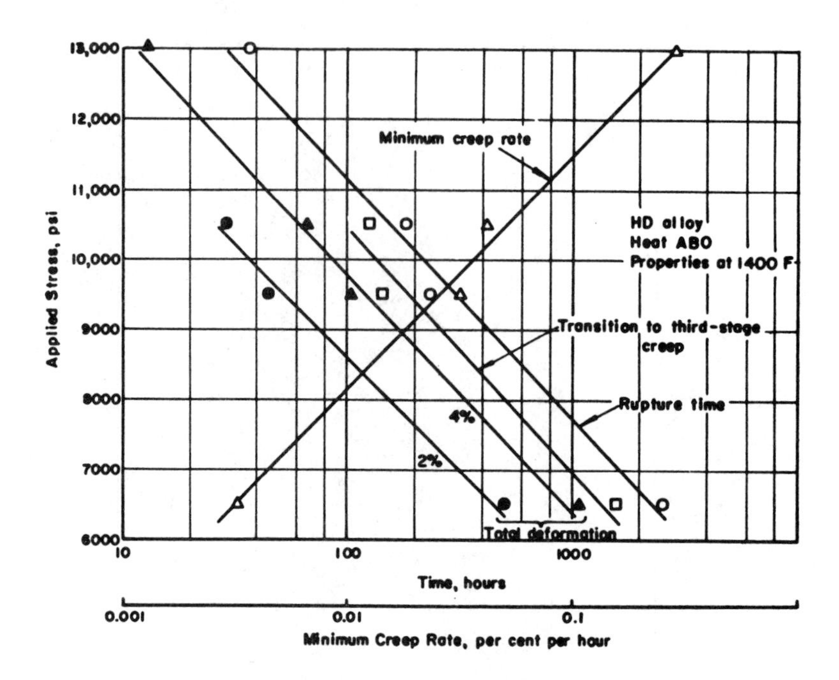

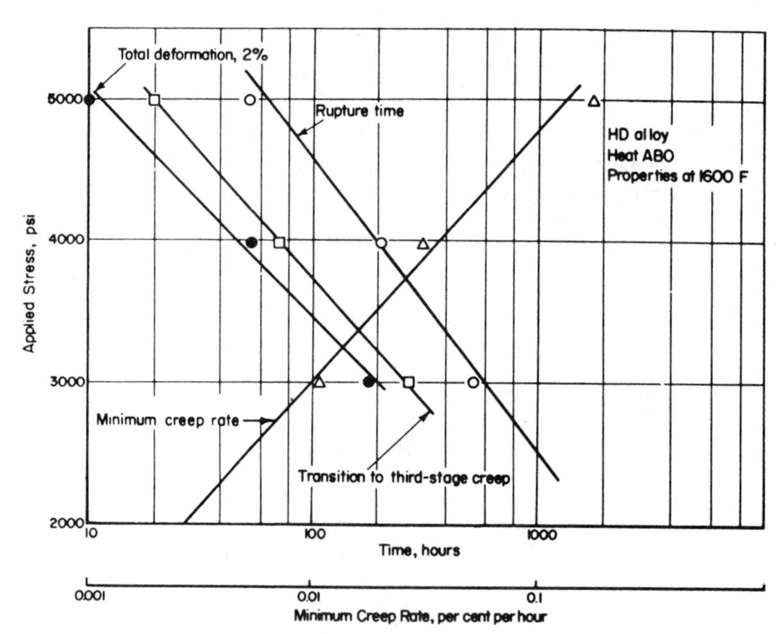

Stress-rupture and creep properties of HD alloy at 960 °C (1400 °F) (top) and 871 °C (1600 °F) (bottom).

Source: L.E. Finch, "Effect of Nitrogen in Cast Heat-Resistant Alloys for Service in Atmospheres Containing High Sulfur at High Temperature," in Stainless Steel Castings, ASTM Special Technical Publication No. 756, American Society for Testing and Materials, Philadelphia, 1982, p 272

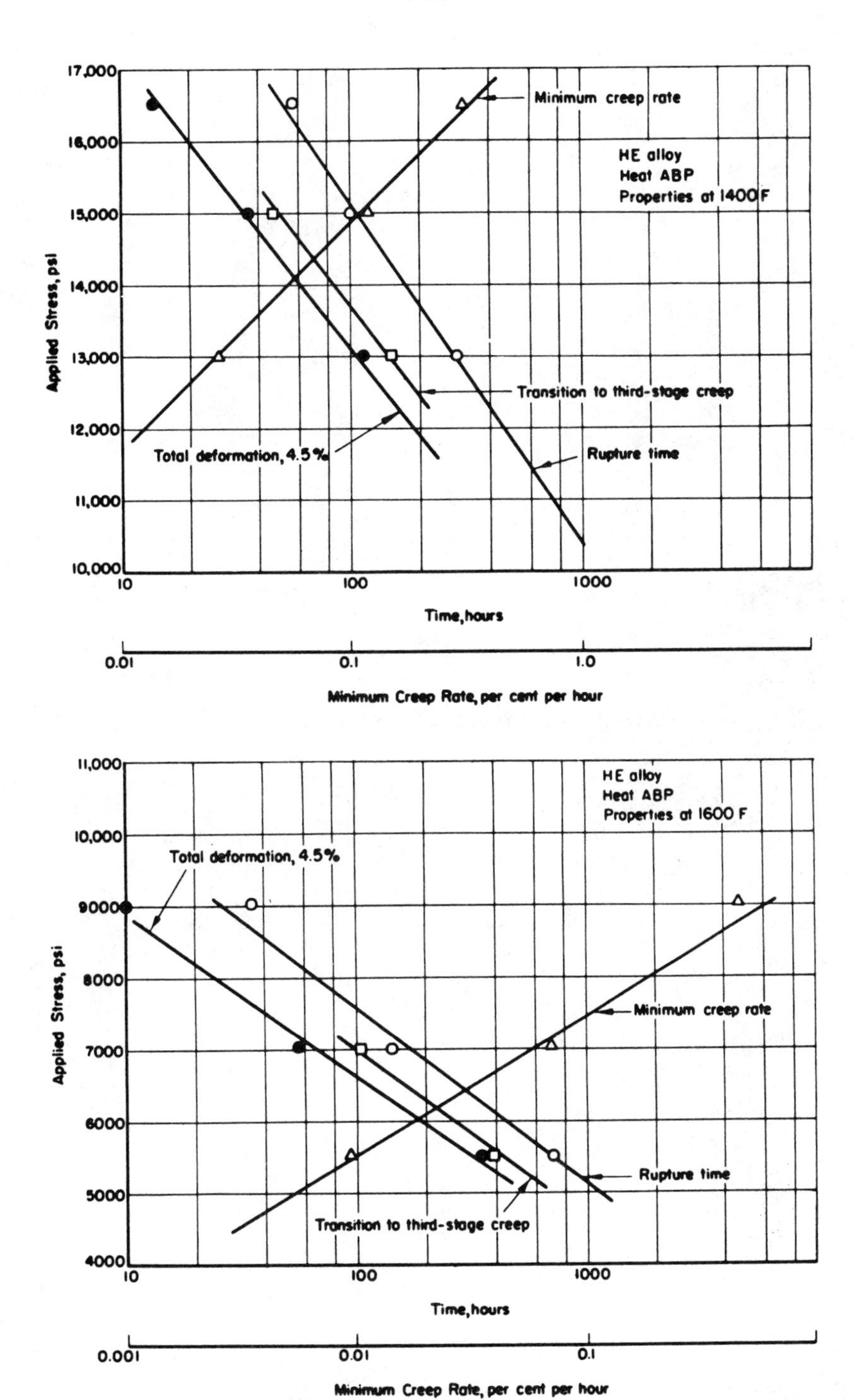

Stress-rupture and creep properties of HE alloy at 760 °C (1400 °F) (top) and 871 °C (1600 °F) (bottom).

Source: L.E. Finch, "Effect of Nitrogen in Cast Heat-Resistant Alloys for Service in Atmospheres Containing High Sulfur at High Temperature," in Stainless Steel Castings, ASTM Special Technical Publication No. 756, American Society for Testing and Materials, Philadelphia, 1982, p 273-274

HF

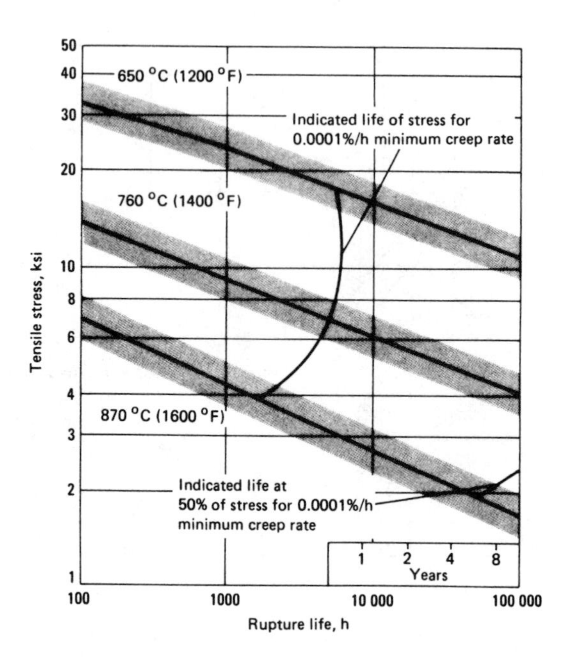

Creep-rupture properties of alloy HF.

The scatterbands are set arbitrarily at ±20% of the stress for the central tendency line. Such a range usually embraces test data for similar alloy compositions, but should not be considered statistically significant confidence limits. Scatter of values may be much wider, particularly at the longer times and higher temperatures.

Source: Metals Handbook, Ninth Edition, Volume 3, Properties and Selection: Stainless Steels, Tool Materials and Special-Purpose Metals, American Society for Metals, Metals Park, OH, 1980, p 292

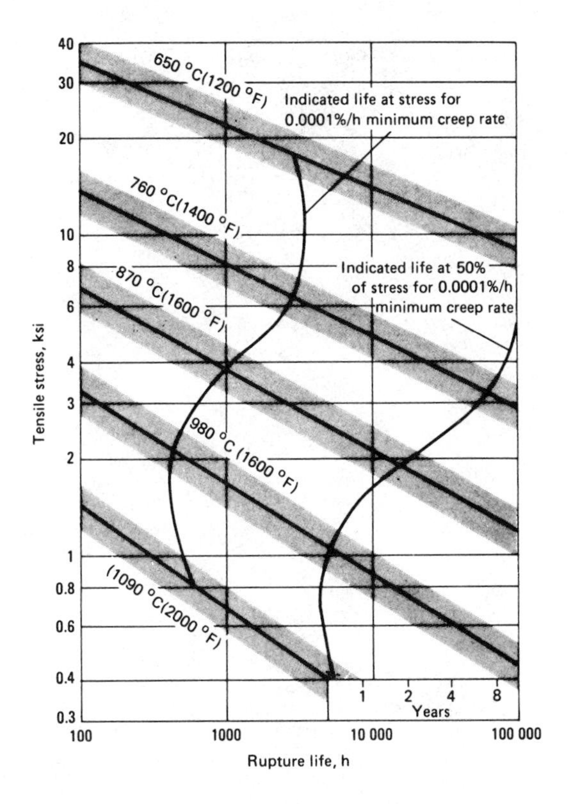

Creep-rupture properties of alloy HH.

The scatterbands are set arbitrarily at ±20% of the stress for the central tendency line. Such a range usually embraces test data for similar alloy compositions, but should not be considered statistically significant confidence limits. Scatter of values may be much wider, particularly at the longer times and higher temperatures.

Source: Metals Handbook, Ninth Edition, Volume 3, Properties and Selection: Stainless Steels, Tool Materials and Special-Purpose Metals, American Society for Metals, Metals Park, OH, 1980, p 295

HI

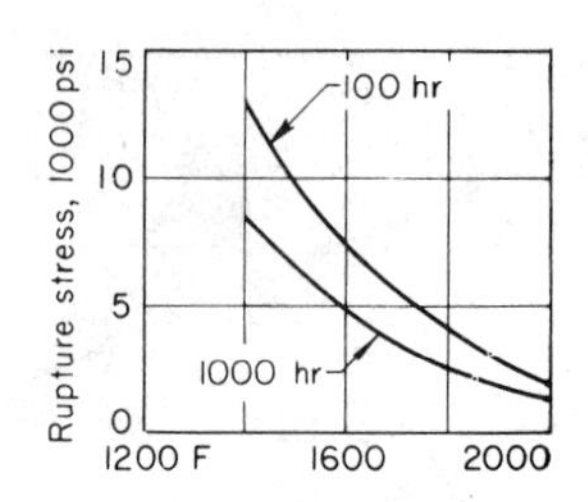

Stress-rupture curves for ACI casting alloy HI.

Source: "Heat-Resistant Alloy Castings," in Source Book on Materials for Elevated-Temperature Applications, E.F. Bradley, Ed., American Society for Metals, Metals Park, OH, 1979, p 230

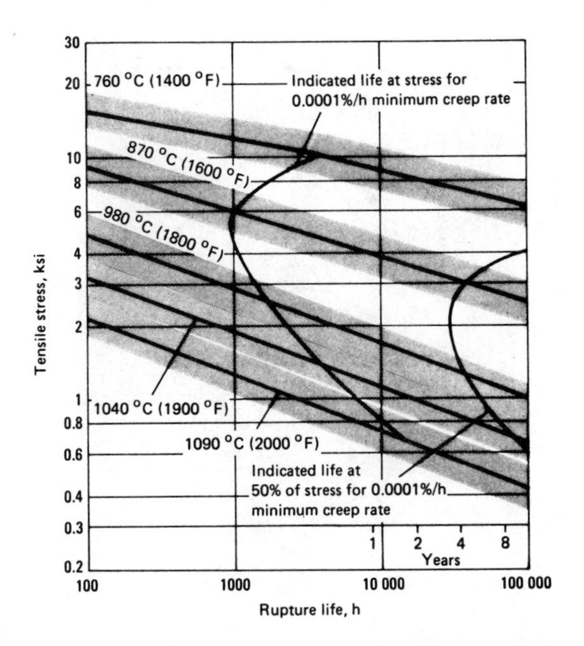

Creep-rupture properties of alloy HK-40.

The scatterbands are set arbitrarily at ±20% of the stress for the central tendency line. Such a range usually embraces test data for similar alloy compositions, but should not be considered statistically significant confidence limits. Scatter of values may be much wider, particularly at the longer times and higher temperatures.

Source: Metals Handbook, Ninth Edition, Volume 3, Properties and Selection: Stainless Steels, Tool Materials and Special-Purpose Metals, American Society for Metals, Metals Park, OH, 1980, p 300

HK-40: Effect of Titanium and Niobium Additions

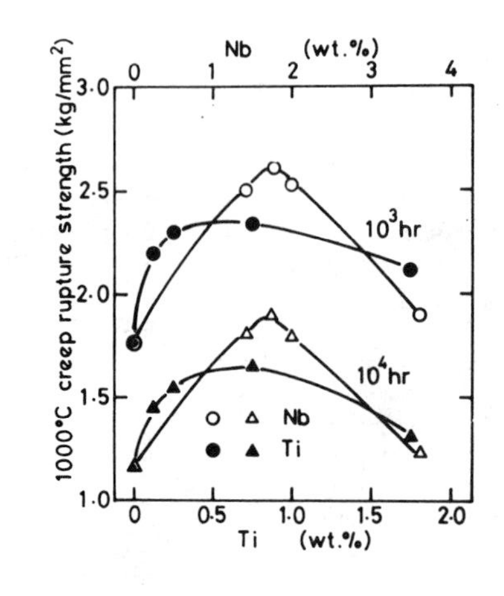

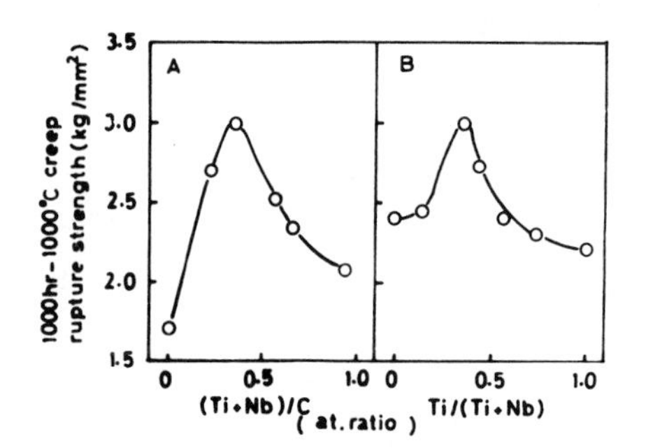

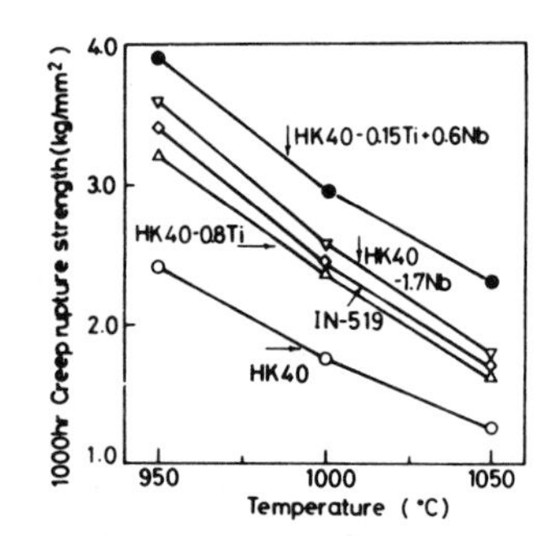

Creep-rupture properties of HK-40 steel as affected by titanium and niobium additions.

Although the constant "C" of the Larson-Miller parameter has been taken as 15 for plain HK-40 steel, this value is not suitable for steels with Ti and Nb additions. Through a regression analysis, a constant of 22 was found suitable for these steels. The 1000 °C 1000- and 10,000-h interpolated and extrapolated values from the master curves of Ti and Nb separately added steels are presented in the top figure in relation to the Ti or Nb addition. The remarkable increase in creep-rupture strength of HK-40 steel by Ti addition and in particular by Nb addition, irrespective of creep time, is obvious. As shown in figure A (bottom left), maximum creep-rupture strength is obtained at the atomic ratio of (Ti + Nb)/C $\simeq$ 0.3, and an increase in this ratio is followed by a decrease in creep-rupture strength. Increasing the ratio more

HK-40: Effect of Titanium and Niobium Additions (Continued)

than 0.3 leads to a decrease in the amount of the $M_{23}C_6$ secondary carbides, which are important for dispersion strengthening. As shown in figure B (bottom left), maximum creep-rupture strength is obtained at the atomic ratio of Ti/(Ti + Nb) $\simeq$ 0.3. Also, creep-rupture strength decreases with an increase in this ratio. Thus, creep-rupture strength is affected more by the Nb addition, which causes the lamellar eutectic; nevertheless, about one third of the total addition should be Ti to utilize its role in causing uniform precipitation of secondary carbides. The interpolated and extrapolated 1000-h creep-rupture strengths of all the steels in the present work at different temperatures in comparison with IN-519 alloy are shown in the bottom right figure. It is obvious that the 0.15% Ti + 0.16% Nb added steel is superior in strength to IN-519, and it has a strength about 1.7 times higher than the plain HK-40 steel at the same temperature. At an equal stress, it can be used at about 100 °C higher than the HK-40 steel.

Source: M.B. Zaghloul, T. Shinoda, and R. Tanaka, "On the Strengthening of the Centrifugally Cast HK-40 Tube," in Superalloys: Metallurgy and Manufacture (Proceedings of the Third International Symposium in Seven Springs, PA), Claitor's, 1976, p 270-272

HL

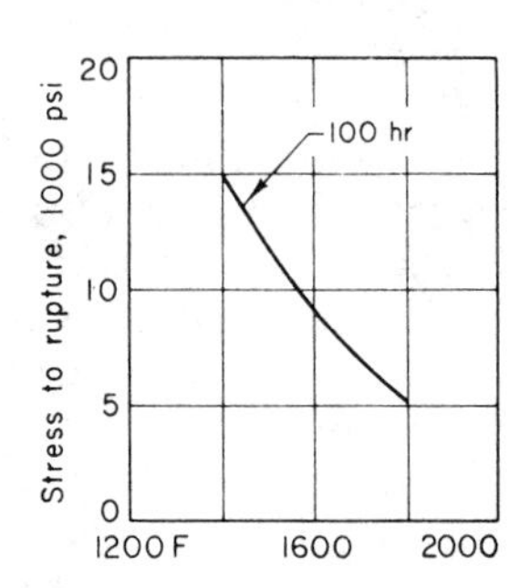

Stress-rupture curve for ACI casting alloy HL.

Source: "Heat-Resistant Alloy Castings," in Source Book on Materials for Elevated-Temperature Applications, E.F. Bradley, Ed., American Society for Metals, Metals Park, OH, 1979, p 231

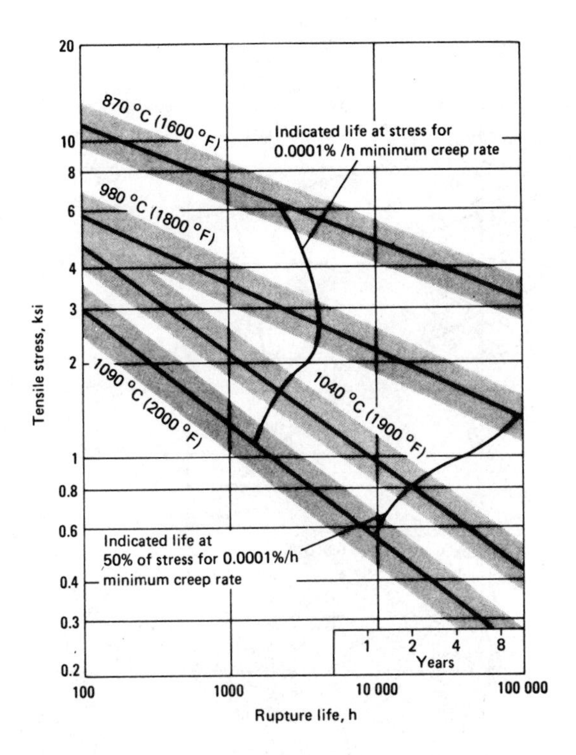

Creep-rupture properties of alloy HN.

The scatterbands are set arbitrarily at ±20% of the stress for the central tendency line. Such a range usually embraces test data for similar alloy compositions, but should not be considered statistically significant confidence limits. Scatter of values may be much wider, particularly at the longer times and higher temperatures.

Source: Metals Handbook, Ninth Edition, Volume 3, Properties and Selection: Stainless Steels, Tool Materials and Special-Purpose Metals, American Society for Metals, Metals Park, OH, 1980, p 303

HP

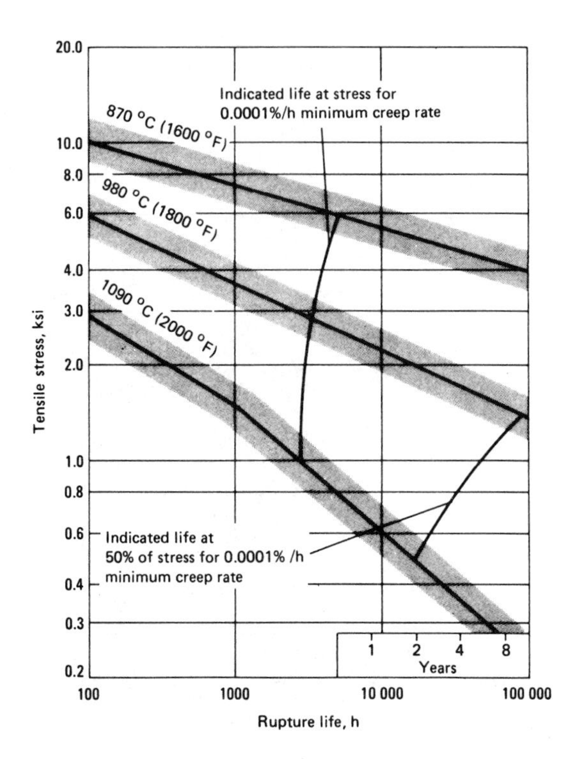

Creep-rupture properties of alloy HP.

The scatterbands are set arbitrarily at ±20% of the stress for the central tendency line. Such a range usually embraces test data for similar alloy compositions, but should not be considered statistically significant confidence limits. Scatter of values may be much wider, particularly at the longer times and higher temperatures.

Source: Metals Handbook, Ninth Edition, Volume 3, Properties and Selection: Stainless Steels, Tool Materials and Special-Purpose Metals, American Society for Metals, Metals Park, OH, 1980, p 306

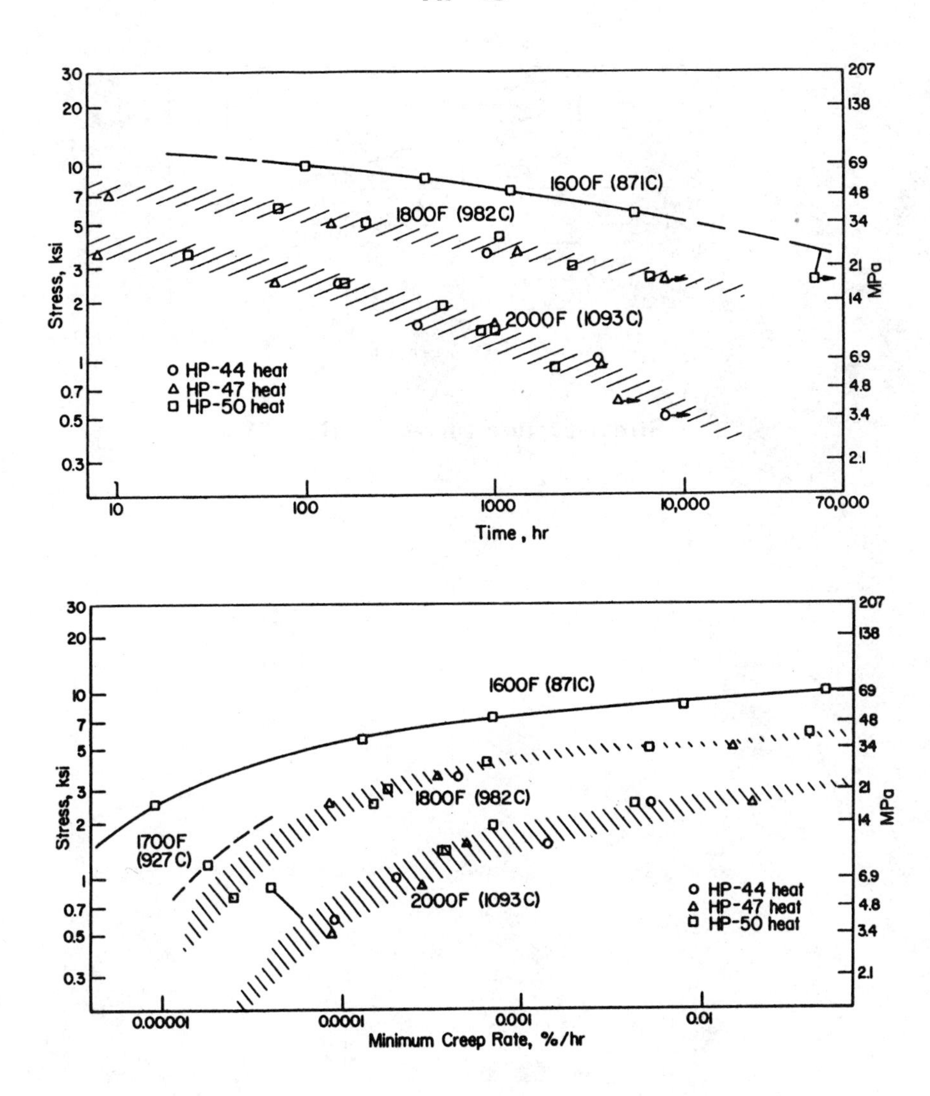

Stress-rupture properties (top) and stress vs minimum creep rate properties (bottom) of HP-45 alloy.

Source: D.B. Roach and J.A. Van Echo, "Comparison of the Properties of the HK-40 and HP-45 Cast Heat-Resistant Alloys," in Stainless Steel Castings, ASTM Special Technical Publication No. 756, American Society for Testing and Materials, Philadelphia, 1982, p 298-299

HP-50WZ: Stress-Rupture Properties

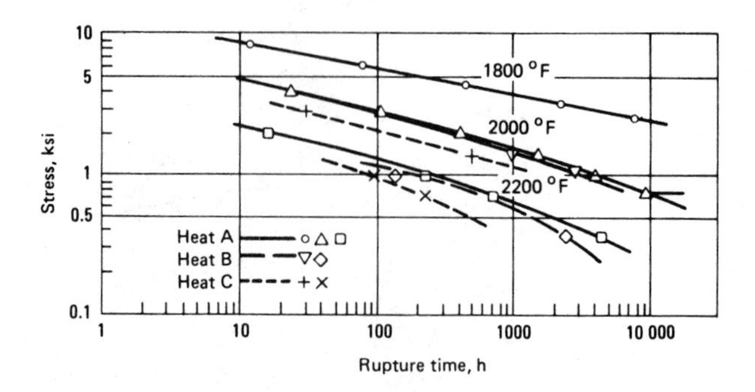

Stress-rupture curves for HP-50WZ.

Source: Metals Handbook, Ninth Edition, Volume 3, Properties and Selection: Stainless Steels, Tool Materials and Special-Purpose Metals, American Society for Metals, Metals Park, OH, 1980, p 307

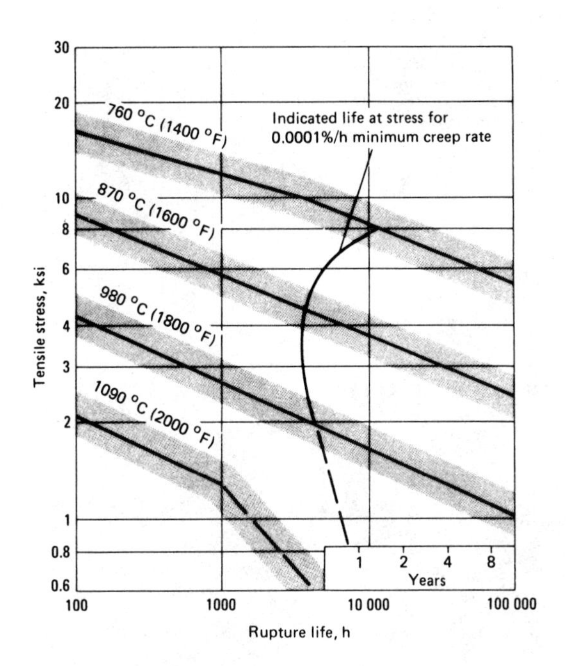

Creep-rupture properties of alloy HT.

The scatterbands are set arbitrarily at ±20% of the stress for the central tendency line. Such a range usually embraces test data for similar alloy compositions, but should not be considered statistically significant confidence limits. Scatter of values may be much wider, particularly at the longer times and higher temperatures.

Source: Metals Handbook, Ninth Edition, Volume 3, Properties and Selection: Stainless Steels, Tool Materials and Special-Purpose Metals, American Society for Metals, Metals Park, OH, 1980, p 308

HU

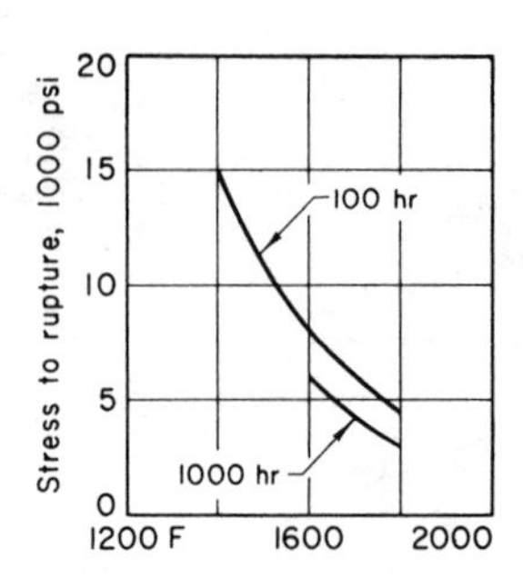

Stress-rupture curves for ACI casting alloy HU.

Source: "Heat-Resistant Alloy Castings," in Source Book on Materials for Elevated-Temperature Applications, E.F. Bradley, Ed., American Society for Metals, Metals Park, OH, 1979, p 233

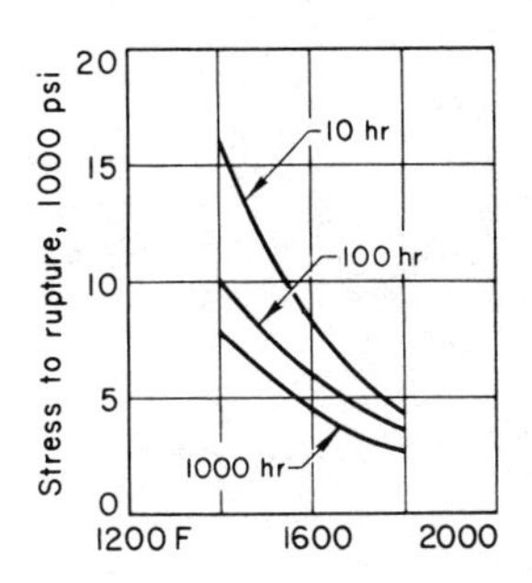

Stress-rupture curves for ACI casting alloy HW.

Source: "Heat-Resistant Alloy Castings," in Source Book on Materials for Elevated-Temperature Applications, E.F. Bradley, Ed., American Society for Metals, Metals Park, OH, 1979, p 233

HX

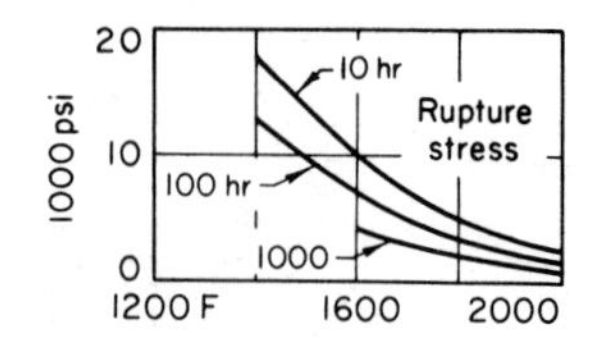

Stress-rupture curves for ACI casting alloy HX.

Source: "Heat-Resistant Alloy Castings," in Source Book on Materials for Elevated-Temperature Applications, E.F. Bradley, Ed., American Society for Metals, Metals Park, OH, 1979, p 234

HF, HH, HK-40, and HN: Comparison of Design Data

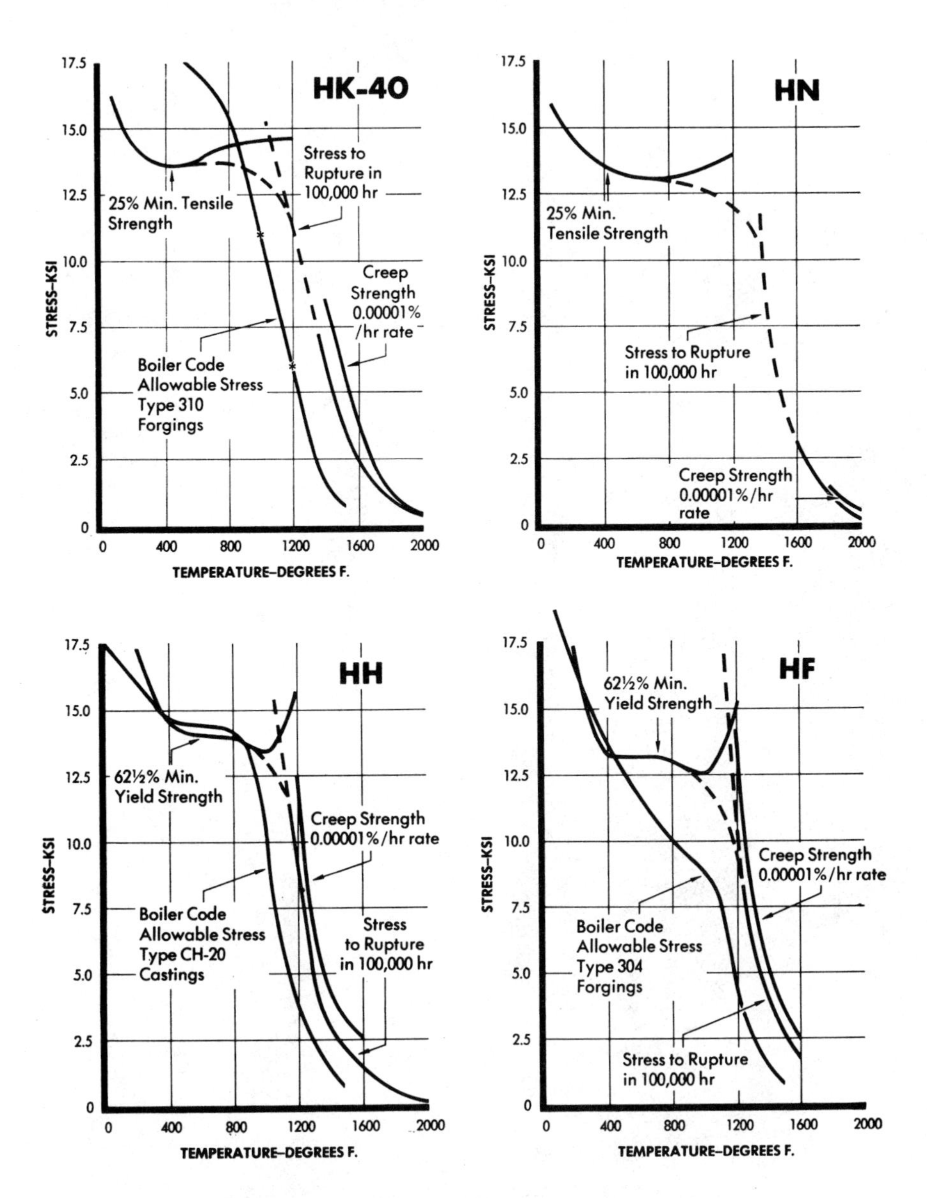

Design data for four heat-resistant cast steels.

These graphs show the values of allowable stress resulting from application of the ASME Boiler Code criteria to short-time tensile and creep-rupture properties of four alloys. There is an interesting distinction between the creep-rupture properties of the cast and wrought alloys: The creep strength curves for wrought steels (0.0001%/h rate) are lower than the rupture-stress curves for 100,000-h life. The reverse is true for the cast alloys. Thus, the cast alloys are not only stronger than the corresponding wrought grades, but they are much more resistant to deformation during their service life as well.

Source: R.W. Zillmann, "Elevated Temperature Properties of Cast Heat Resistant Alloys," Industrial Heating, Vol. 38, Sept. 1971, p 1720, 1722

11

Austenitic Stainless Steels

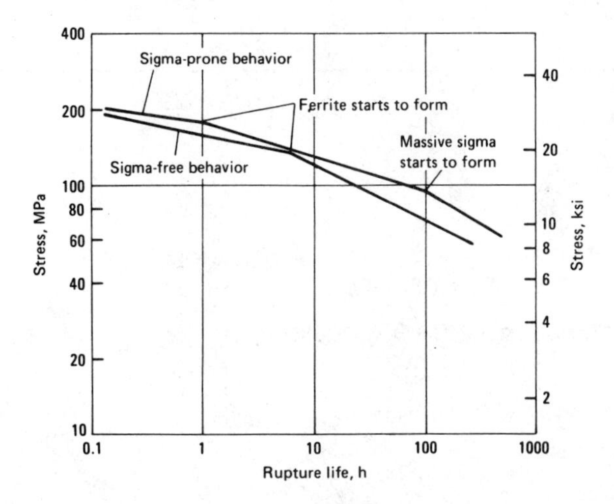

Log stress vs log rupture life at 700 °C (1300 °F) for 18-8 stainless steel.

Source: Metals Handbook, Ninth Edition, Volume 3, Properties and Selection: Stainless Steels, Tool Materials and Special-Purpose Metals, American Society for Metals, Metals Park, OH, 1980, p 229

11.2

21Cr-4Ni-9Mn Austenitic Stainless Steel: Effect of Grain Boundary Type

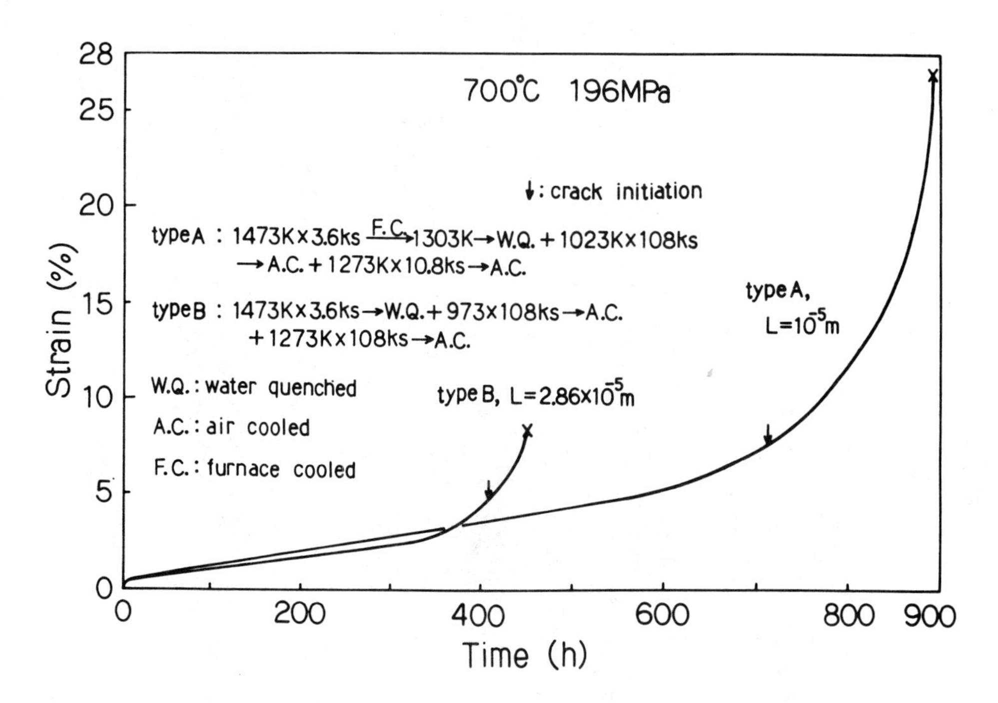

Creep curves of specimens with type A and type B grain boundaries of austenitic 21Cr-4Ni-9Mn stainless steel.

Source: M. Tanaka and H. Iizuka, "Grain-Boundary Crack Initiation and Toughening Mechanism in High-Temperature Creep of an Austenitic Cr-Mn-Ni Steel," in Creep (Proceedings of the International Conference in Tokyo), Japan Society of Mechanical Engineers, Tokyo, 1986, p 188

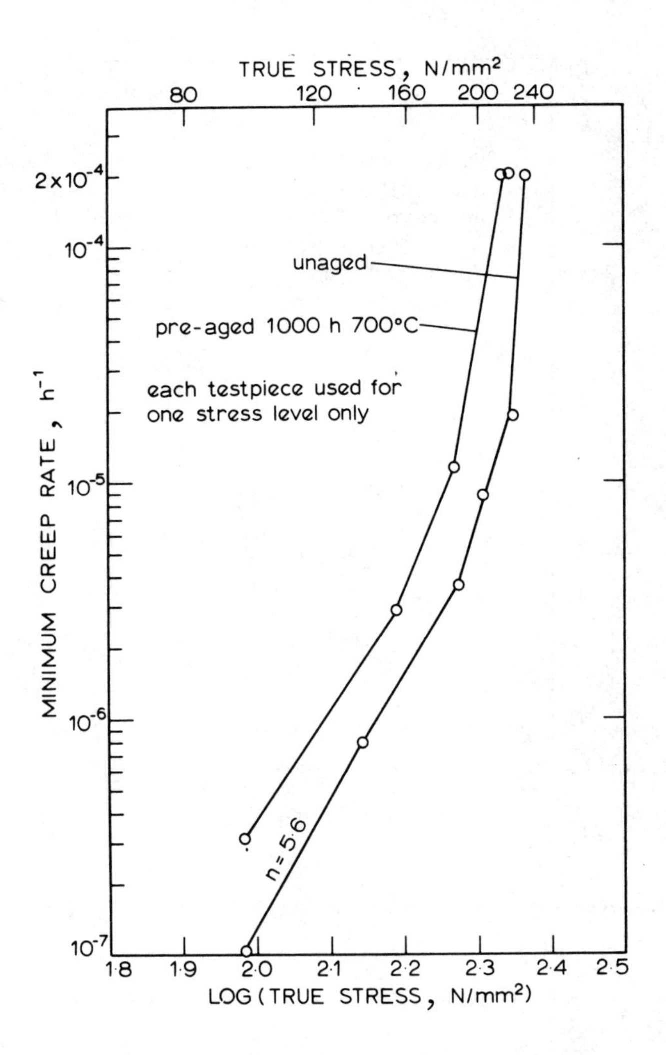

Stress dependence of creep rate for *AH* material aged up to 1000 h at 700 °C.

Initial treatments and conditions

Code	Initial treatment	Grain structure mean diameter, μm	Matrix precipitate dispersion (NbC)	Hardness, HV (30 kg)
AC	15 min 1 200°C WQ 40% cold reduction 1 h 1 000°C AC	Uniform, 18	Fine, fairly uniform	172
AH	30 min 1 100°C AC	Uniform, 15	Patches of fine precipitate	160
AE	20 min 1 250°C WQ 2 h 1 050°C AC	Uniform, 124	Medium	141
DE	20 min 1 250°C WQ 2 h 1 050°C AC	Uniform, 110	Medium	148
AS	30 min 1 150°C WQ	30–60% ~ 250 μm remainder 30–45 μm	Very sparse	160

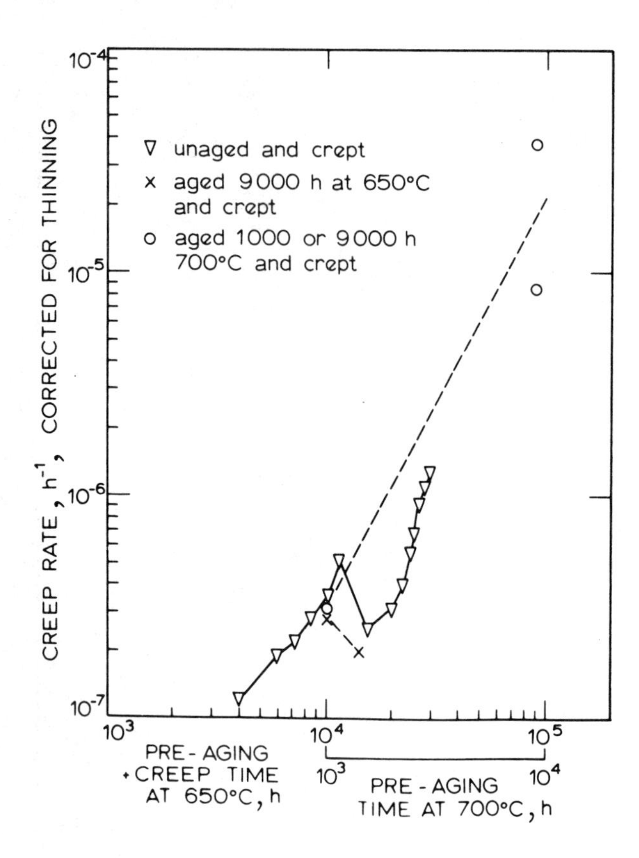

Actual and predicted creep behavior of *AH* at 650 °C and 96 N/mm².

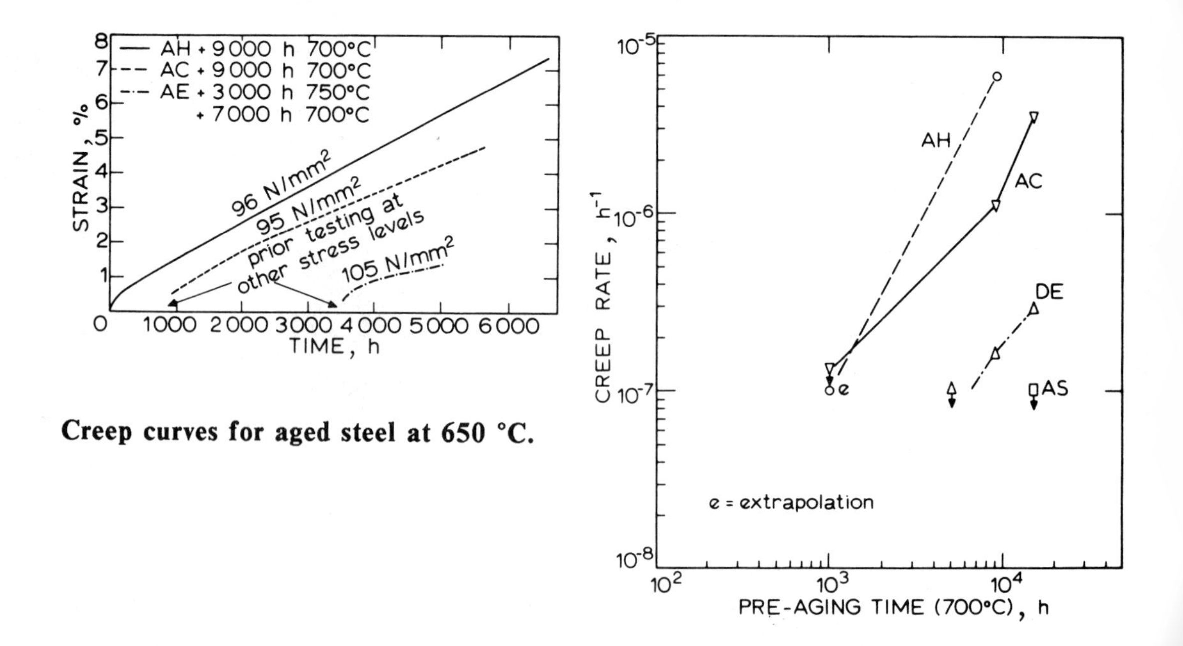

Creep curves for aged steel at 650 °C.

Effect of pre-aging at 700 °C on creep rate at 650 °C and 80 N/mm².

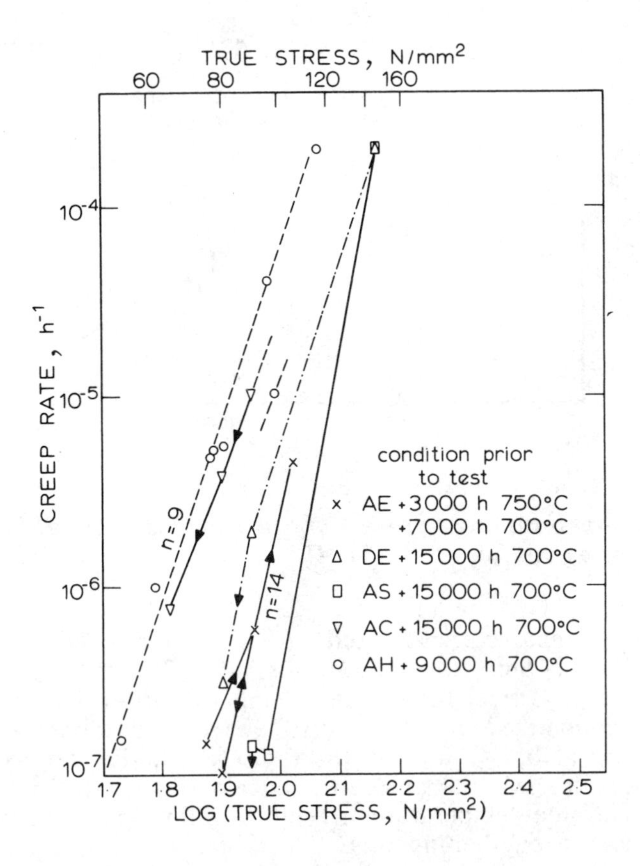

Stress dependence of creep rate at 650 °C after prolonged pre-aging.

The steel investigated contained 16% Cr, 10% Ni, and 6% Mn, with smaller amounts of Nb, Mo, V, and B, and is known commercially as Esshete 1250. It was developed for use as tubes or pipes in steam circuits at operating conditions in excess of 570 °C. The steel has the high flow strength of steels that precipitate NbC, together with good resistance to intergranular creep failure. The latter feature implies that the rupture life of this steel is governed by the creep flow rather than by intercrystalline creep fracture. The experimental data showed that the flow strength was very dependent on the initial level of supersaturation of Nb and C. Even with an initial solution treatment as low as 1050 °C, a substantial decrease in flow strength was caused by aging at 700 °C.

Source: F.E. Asbury and G. Willoughby, "Aging and Creep Behavior of a Cr-Ni-Mn Austenitic Steel," in Creep Strength in Steel and High-Temperature Alloys, The Metals Society, London, 1972, p 144, 149-150

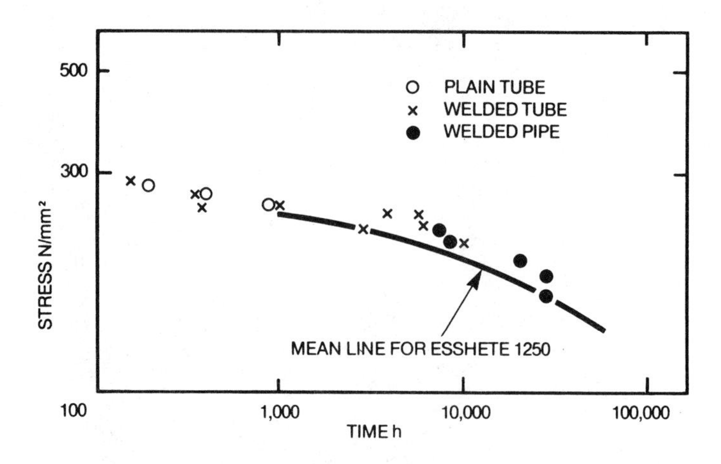

Stress-rupture properties of welded joints in tubes and pipes at 650 °C.

The characteristics and service performance of British Steel Corporation Esshete 1250 austenitic stainless steel for elevated temperature service are summarized in the figure above. Additions of Mo, Nb, V, and B to a Cr-Ni-Mn-Fe matrix result in an austenitic steel that has high elevated-temperature strength. Both the matrix and strengthening NbC precipitate dispersion are stable, yielding good creep-rupture ductility and a low susceptibility to embrittlement in the solution-treated, cold-worked, or welded conditions.

Source: J. Orr and V.B. Nileshwar, "Esshete 1250: An Advanced Austenitic Stainless Steel for Power Station Tubes, Pipes and Headers," in Stainless Steels '84 (Proceedings of a Conference in Göteborg), Institute of Metals, London, 1985, p 533, 539

Type 301 Stainless Steel: Effect of Cold Rolling and Recrystallization

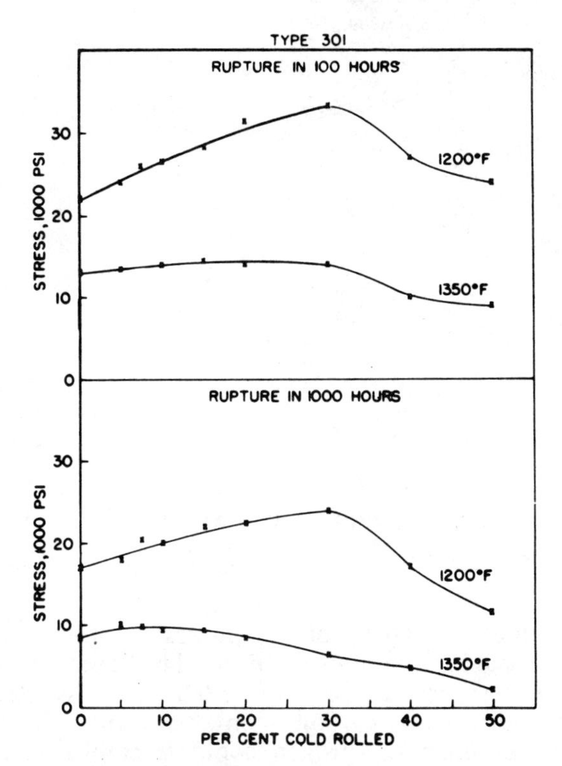

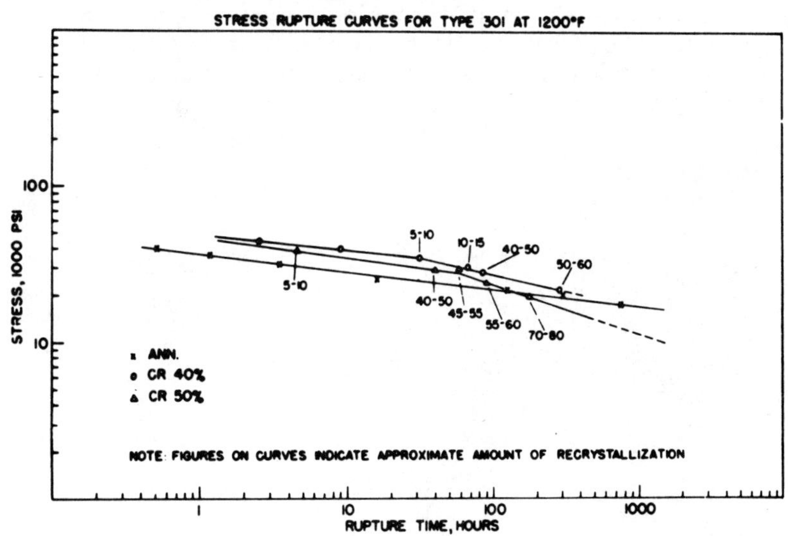

Variation in stress-rupture properties of type 301 stainless steel at 1200 °F with cold rolling (top) and recrystallization (bottom).

Source: A.J. Lena and F.A. Malagari, "Stainless Steels and Hot-Work Die Steels for High-Strength Applications," in Quality Requirements of Super-Duty Steels (Proceedings of a Conference in Pittsburgh), R.W. Lindsay, Ed., Interscience, New York, 1959, p 61

Type 302 Stainless Steel: Effect of tcp Phases

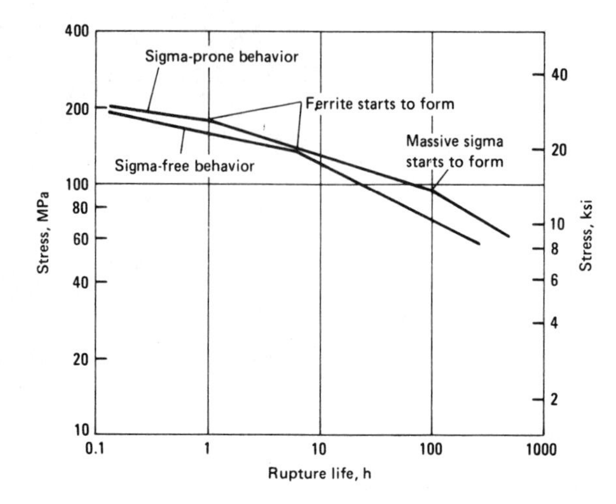

Log stress vs log rupture life at 700 °C (1300 °F) for type 302 stainless steel.

The potential effects of tcp phases on stress-rupture behavior are illustrated in the figure above, where test results for two 18-8 stainless steels (with differing carbon contents) are plotted. High carbon and nitrogen contents stabilized the alloy against sigma formation. Sigma formed in the alloy having lower carbon and nitrogen contents, with a resultant decrease in load-carrying capability.

Source: Metals Handbook, Ninth Edition, Volume 3, Properties and Selection: Stainless Steels, Tool Materials and Special-Purpose Metals, American Society for Metals, Metals Park, OH, 1980, p 226, 229

Type 302 Stainless Steel: Effect of Titanium

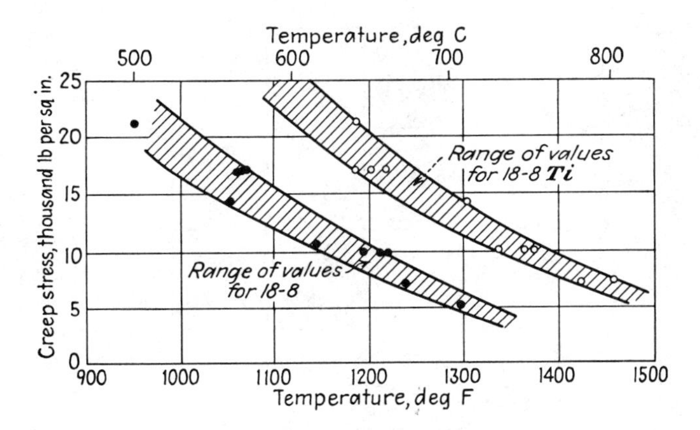

Variation of stress for 1% creep in 1000 h with temperature for 0.10% carbon, 18% chromium, 9% nickel steels with and without 0.4% titanium.

The improvement in high-temperature strength brought about by the addition of Ti to type 302 stainless steel is illustrated in the above figure. These data, indicating about double the strength for the Ti steel as for the steel without Ti at the same temperature, were obtained by means of the 35-h test commonly used in Europe.

Source: G.F. Comstock, Titanium in Iron and Steel, John Wiley & Sons, New York, 1955, p 220-221

Type 302 (Modified) Stainless Steel: Bar vs Forgings

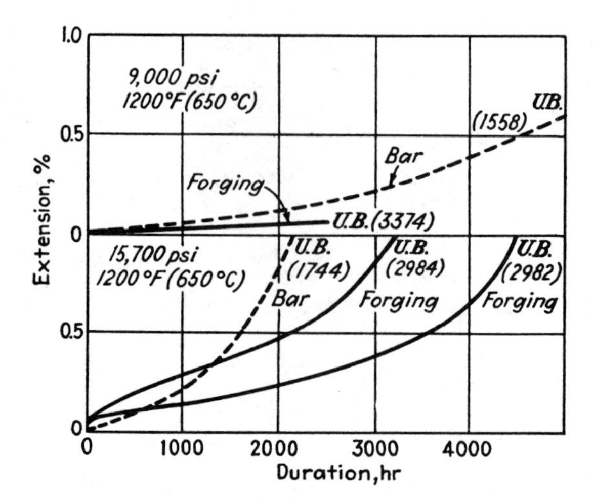

Creep curves at 1200 °F (650 °C) for modified type 302 (18-8-Cb) stainless steel bar compared to typical forgings. U.B. = unbroken after the number of hours shown in parentheses.

Source: R.A. Grange, F.J. Shortsleeve, D.C. Hilty, W.O. Binder, et. al., Boron, Calcium, Columbium, and Zirconium in Iron and Steel, John Wiley & Sons, New York, 1957, p 271

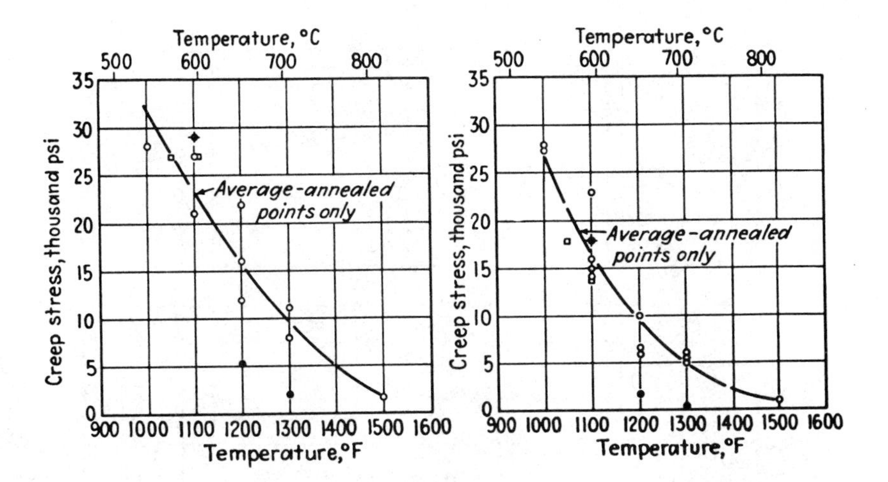

Creep of modified type 302 (18-8-Cb) stainless steel at 900 to 1600 °F (480 to 870 °C). Left: creep rate, 1%/10,000 h. Right: creep rate, 1%/100,000 h.

Source: R.A. Grange, F.J. Shortsleeve, D.C. Hilty, W.O. Binder, et. al., Boron, Calcium, Columbium, and Zirconium in Iron and Steel, John Wiley & Sons, New York, 1957, p 268

11.12 Type 302 (Modified) Stainless Steel: Effect of Tantalum and Niobium Additions

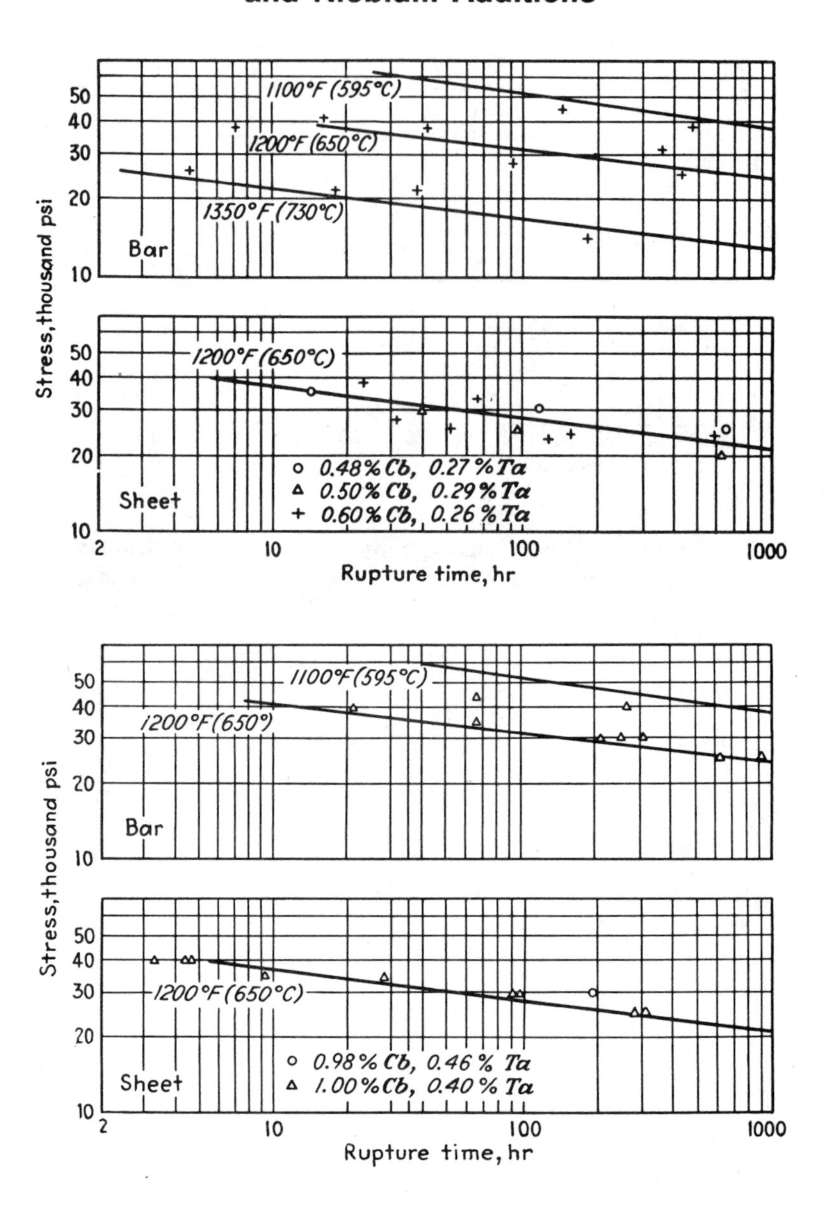

Stress-rupture data for type 302 stainless steel bar and sheet containing about 0.25% Ta and 0.50% Nb (top) and about 0.40% Ta and 1.0% Nb (bottom).

Source: R.A. Grange, F.J. Shortsleeve, D.C. Hilty, W.O. Binder, et. al., Boron, Calcium, Columbium, and Zirconium in Iron and Steel, John Wiley & Sons, New York, 1957, p 328-329

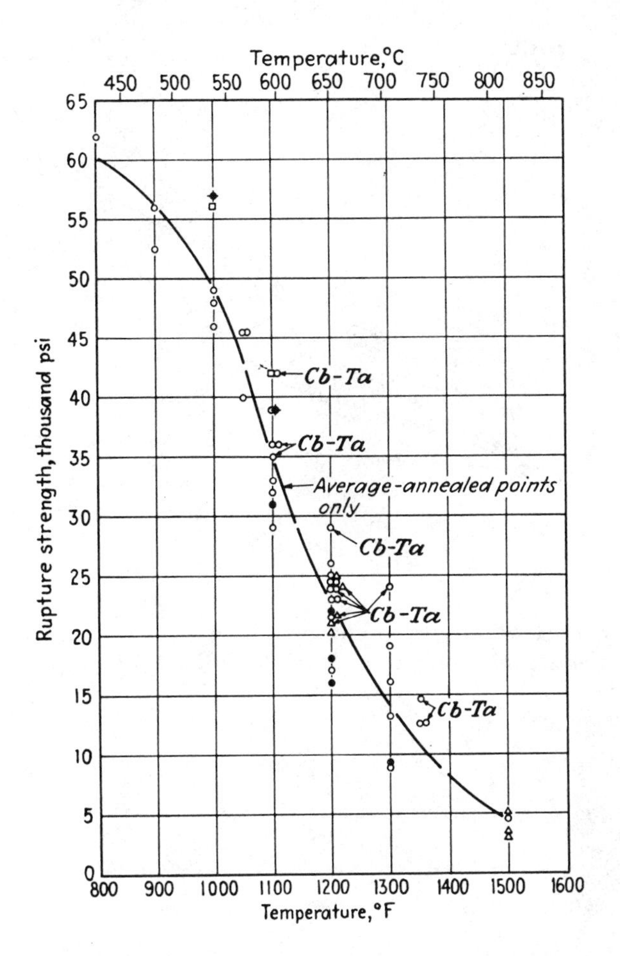

Rupture strength of modified type 302 (18-8-Cb) stainless steel at 800 to 1600 °F (425 to 870 °C) in 1000 h.

Source: R.A. Grange, F.J. Shortsleeve, D.C. Hilty, W.O. Binder, et. al., Boron, Calcium, Columbium, and Zirconium in Iron and Steel, John Wiley & Sons, New York, 1957, p 269

Type 304 Stainless Steel: Effect of Environment

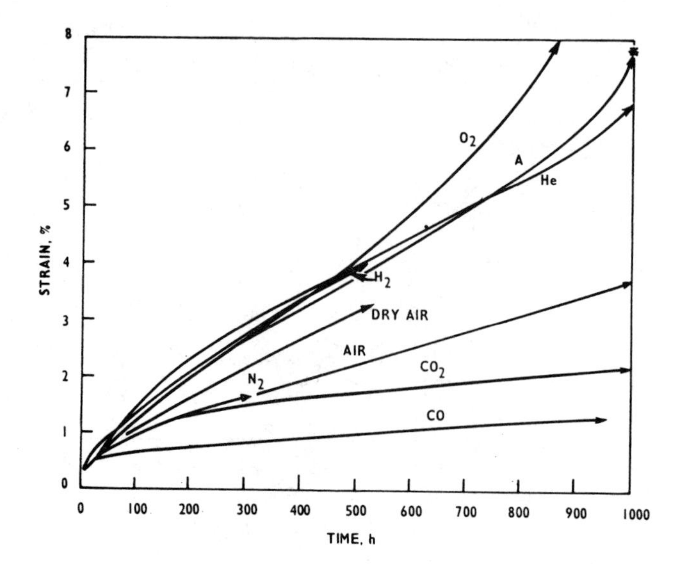

Effect of environment on creep behavior of type 304 stainless steel at 1090 K and 23.5 N/mm².

Source: R.H. Cook and R.P. Skelton, "Environment-Dependence of the Mechanical Properties of Metals at High Temperature," in Source Book on Materials for Elevated-Temperature Applications, E.F. Bradley, Ed., American Society for Metals, Metals Park, OH, 1979, p 75

Type 304 Stainless Steel: Experimental vs Predicted Creep-Rupture Life

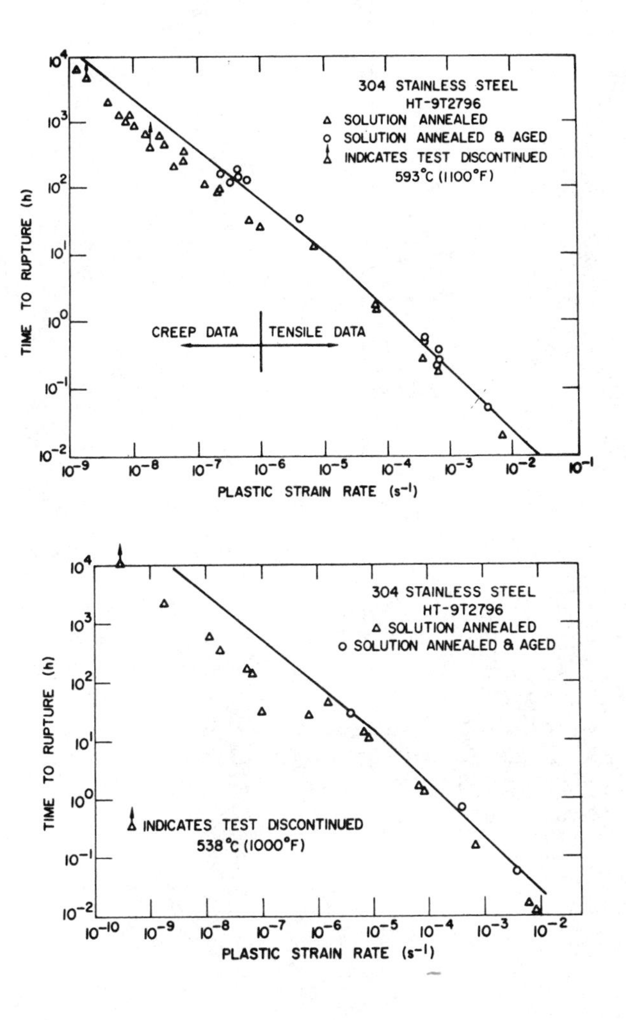

Experimental and predicted tensile and creep-rupture lives for type 304 stainless steel at 593 °C (1100 °F) (top) and 538 °C (1000 °F).

Source: S. Majumdar and P.S. Maiya, "A Damage Equation for Creep-Fatigue Interaction," in 1976 ASME-MPC Symposium on Creep-Fatigue Interaction, R.M. Curran, Ed., American Society of Mechanical Engineers, New York, 1976, p 331

11.16 Type 304 Stainless Steel: Fatigue and Stress-Rupture Data Correlations

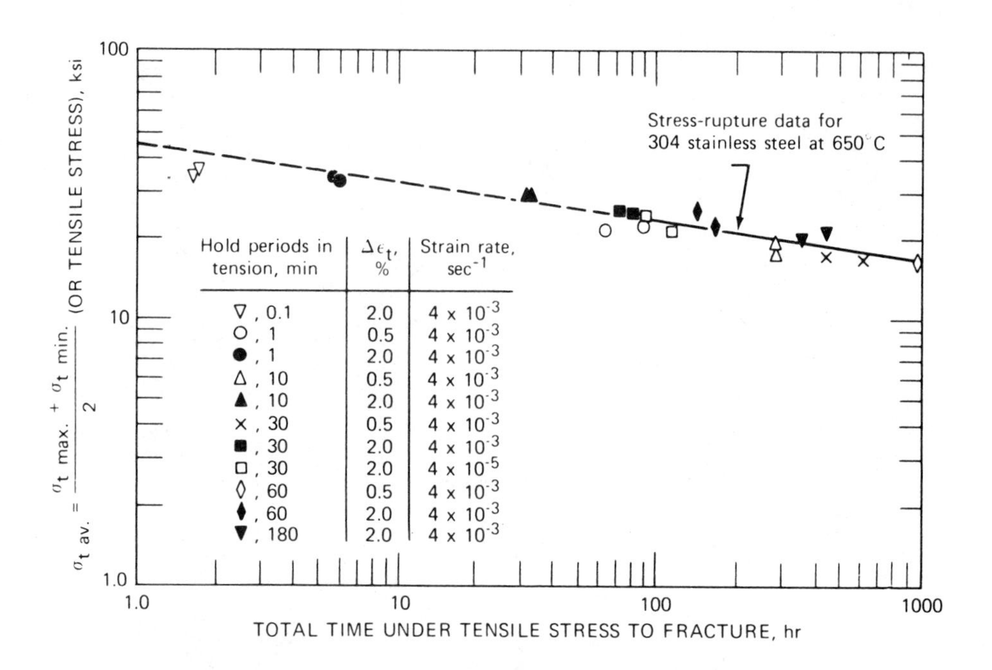

Correlation of fatigue data involving hold periods in tension only with typical stress-rupture data for type 304 stainless steel tested in air at a temperature of 650 °C.

Source: J.B. Conway, R.H. Stentz, and J.T. Berling, Fatigue, Tensile, and Relaxation Behavior of Stainless Steels, Technical Information Center, U.S. Atomic Energy Commission, Washington, D.C., 1975, p 57

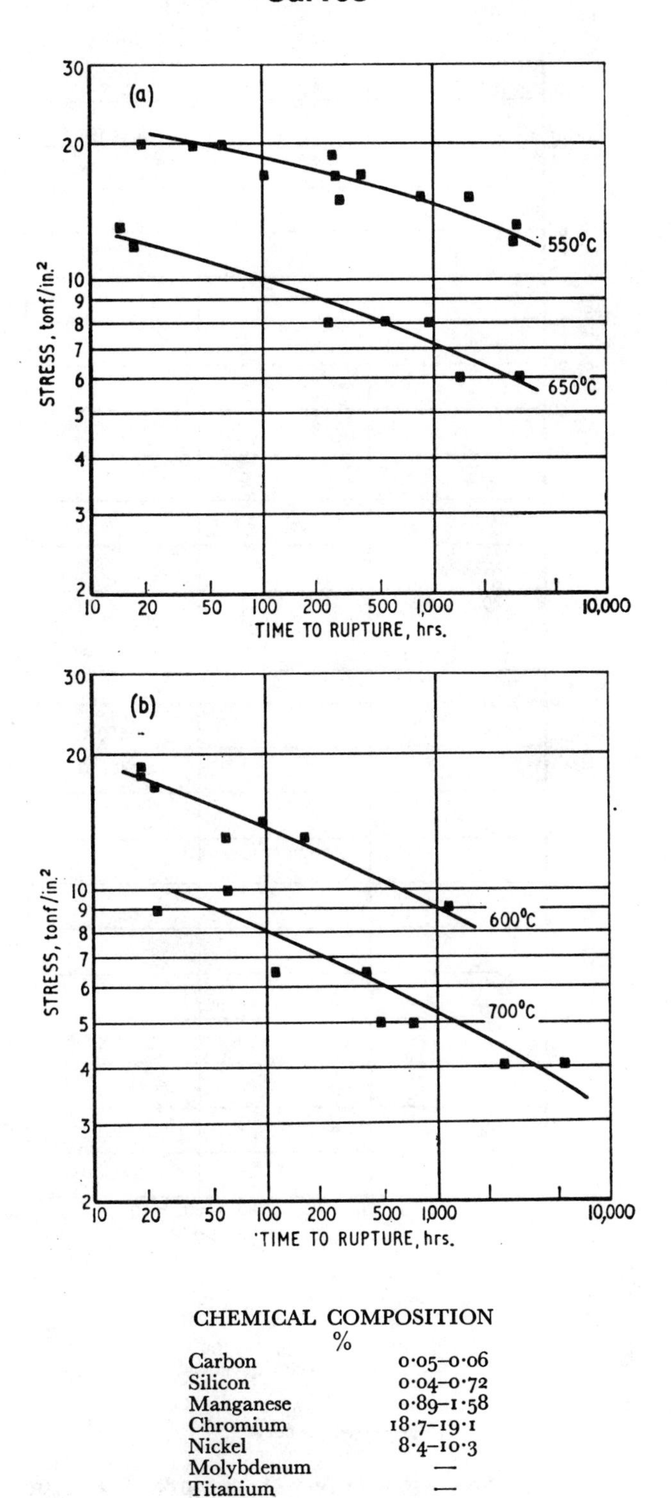

CHEMICAL COMPOSITION

	%
Carbon	0·05–0·06
Silicon	0·04–0·72
Manganese	0·89–1·58
Chromium	18·7–19·1
Nickel	8·4–10·3
Molybdenum	—
Titanium	—
Niobium	—
Sulphur	0·010–0·019
Phosphorus	0·018–0·026

Stress-rupture properties of type 304 stainless steel.

Type 304 Stainless Steel: Stress-Rupture and Elongation Curves (Continued)

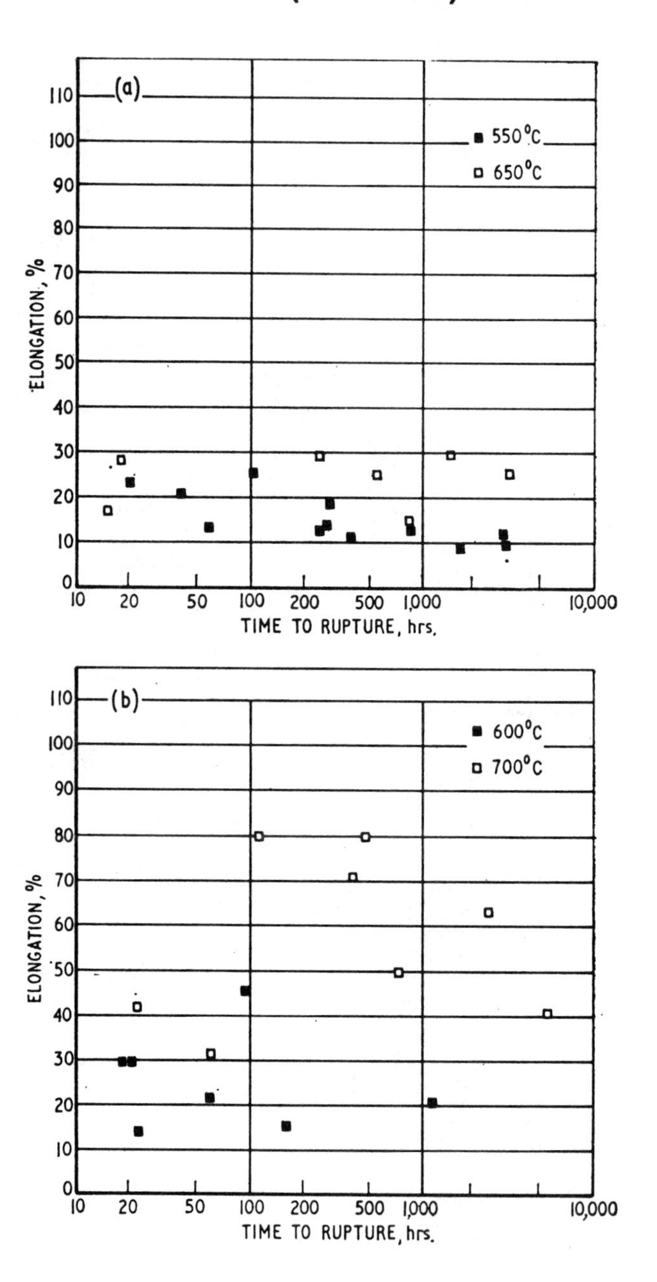

HEAT TREATMENT

Solution treated 1050–1100°C
Air cooled or water quenched
□ Bar
× Forging
——— Mean curve

Percentage elongation at rupture for type 304 stainless steel.

Source: R.J. Truman, R.P. Harvey, R.F. Johnson, and J.N. Smith, "Elevated-Temperature Tensile, Creep, and Rupture Properties of 18%Cr-8%Ni, 18%Cr-12%Ni-Mo, 18%Cr-10%Ni-Ti, and 18%Cr-12%Ni-Nb Steels," in High-Temperature Properties of Steels, Iron and Steel Institute, London, 1967, p 291

Type 304 Stainless Steel: Stress-Rupture Properties Between 1050 and 1500 °F

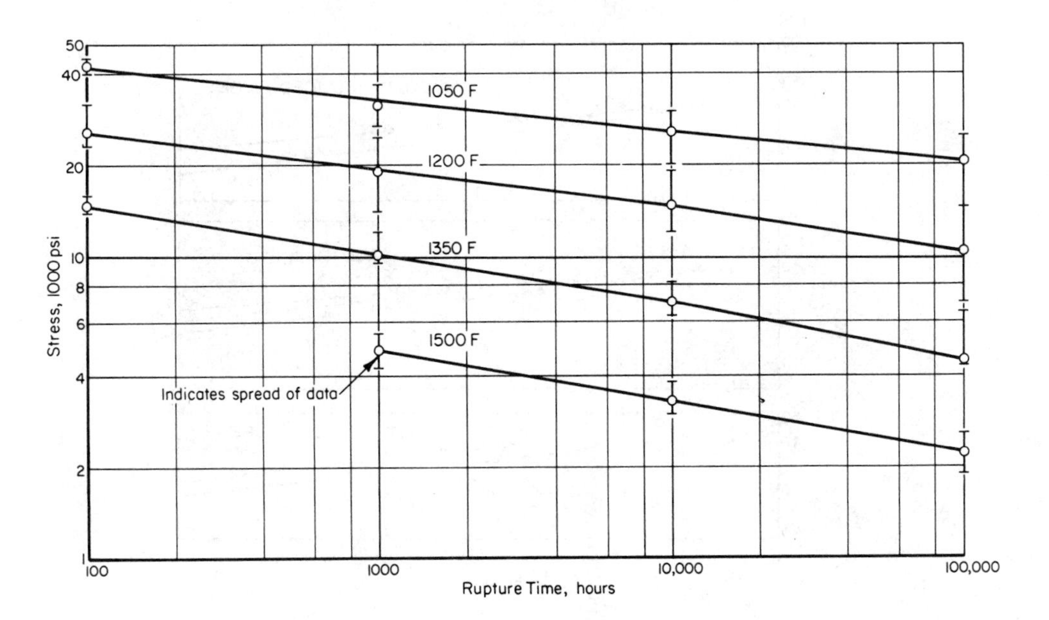

Stress vs rupture-time for type 304 bar and plate. Data are based on averages from stress-temperature curves.

Source: The Elevated-Temperature Properties of Stainless Steels, ASTM Data Series DS 5-S1, American Society for Testing and Materials, Philadelphia, 1965, p 11

Type 304H Stainless Steel: Effect of Production Method on Rupture Strength of Tubing

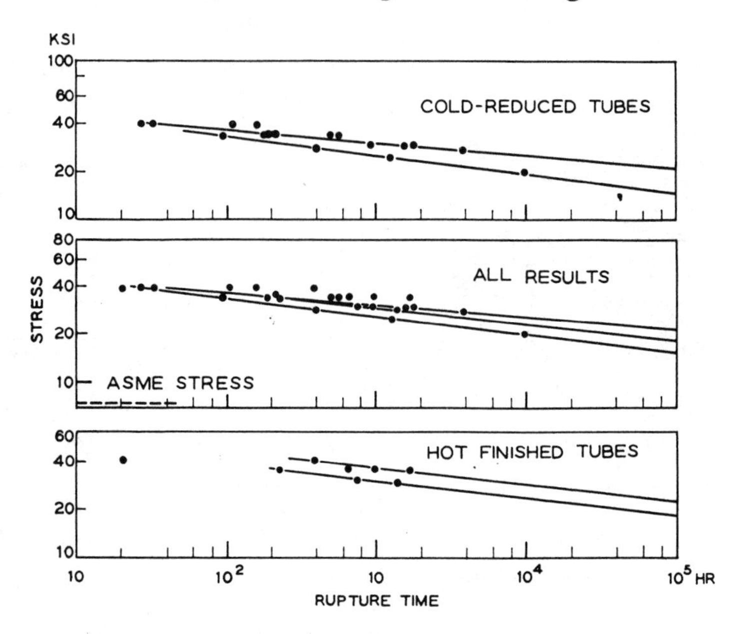

Effect of production method on the 1100 °F rupture strength of type 304H stainless steel tubing. Tubes, which ranged in size from 1.900 by 0.400 in. to 7$^3/_4$ by 1$^3/_4$ and 8.766 by 1.030 in., were water quenched from 1950 °F after hot finishing or cold reducing. Specimens were taken from tube wall.

Source: T.D. Parker, "Strength of Stainless Steels at Elevated Temperature," in Selection of Stainless Steels, American Society for Metals, Metals Park, OH, 1968, p 52

Type 304L Stainless Steel: Effect of Nitrogen Addition on Creep Strength

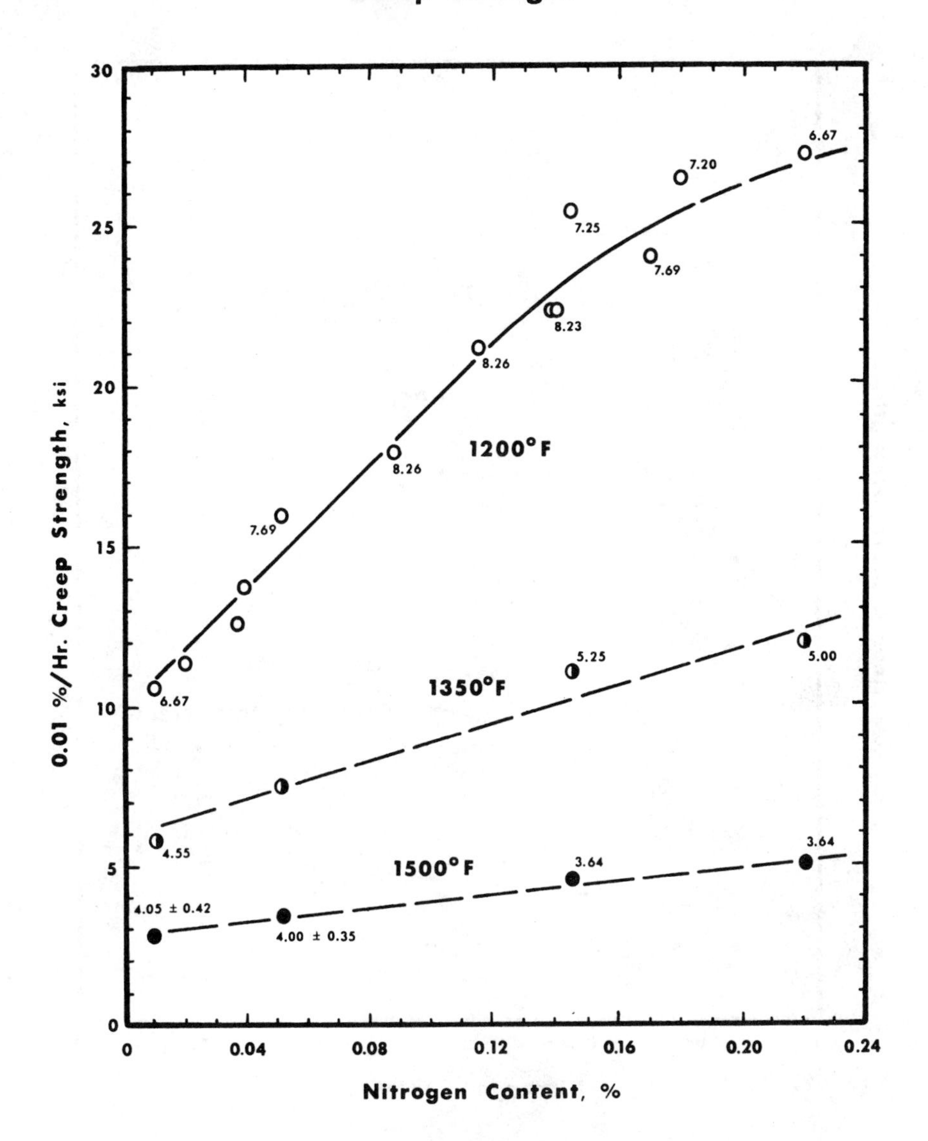

Influence of nitrogen on the 0.01%/h creep strength (based on minimum creep rate) of the nitrogen series heats at 1200, 1350, and 1500 °F (649, 732, and 816 °C). The numbers adjacent to the symbols are the stress exponent (n) for the relation $e = A\sigma^n$ at a minimum creep rate of 0.01%/h.

Source: P.D. Goodell and J.W. Freeman, "Elevated Temperature Properties of Nitrogen-Containing Type 304L Austenitic Stainless Steel," in Elevated Temperature Properties as Influenced by Nitrogen Additions to Types 304 and 316 Austenitic Stainless Steels (Proceedings of a Symposium in Atlantic City), ASTM Special Technical Publication No. 522, American Society for Testing and Materials, Philadelphia, 1973, p 52

Type 304L Stainless Steel: Effect of Nitrogen Addition on Rupture Strength

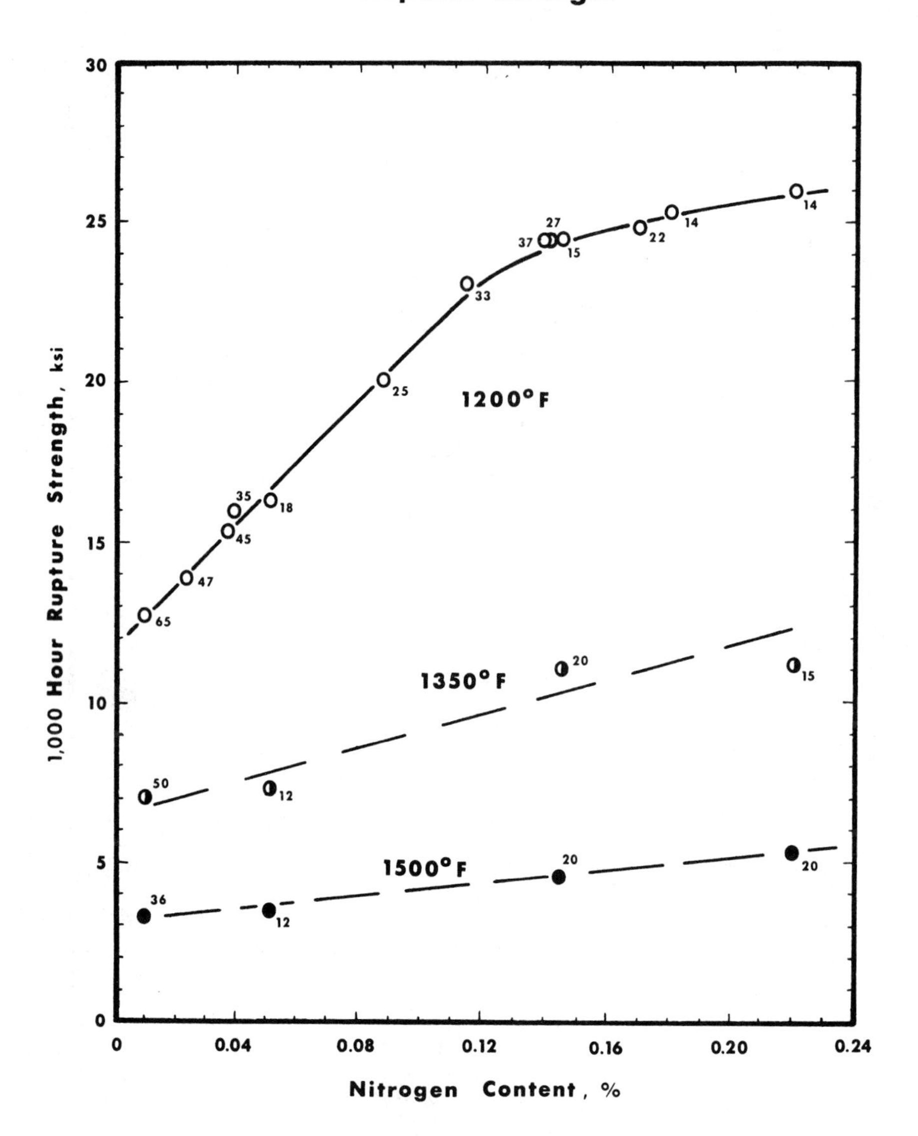

Influence of nitrogen on the 1000-h rupture strength of the nitrogen series heats at 1200, 1350, and 1500 °F (649, 732, and 816 °C). The numbers adjacent to the symbols are the estimated elongation for rupture in 1000 h.

Source: P.D. Goodell and J.W. Freeman, "Elevated Temperature Properties of Nitrogen-Containing Type 304L Austenitic Stainless Steel," in Elevated Temperature Properties as Influenced by Nitrogen Additions to Types 304 and 316 Austenitic Stainless Steels (Proceedings of a Symposium in Atlantic City), ASTM Special Technical Publication No. 522, American Society for Testing and Materials, Philadelphia, 1973, p 53

Type 304L Stainless Steel: Relationship Between Minimum Creep Rate and Rupture Time for Nitrogen-Containing Material

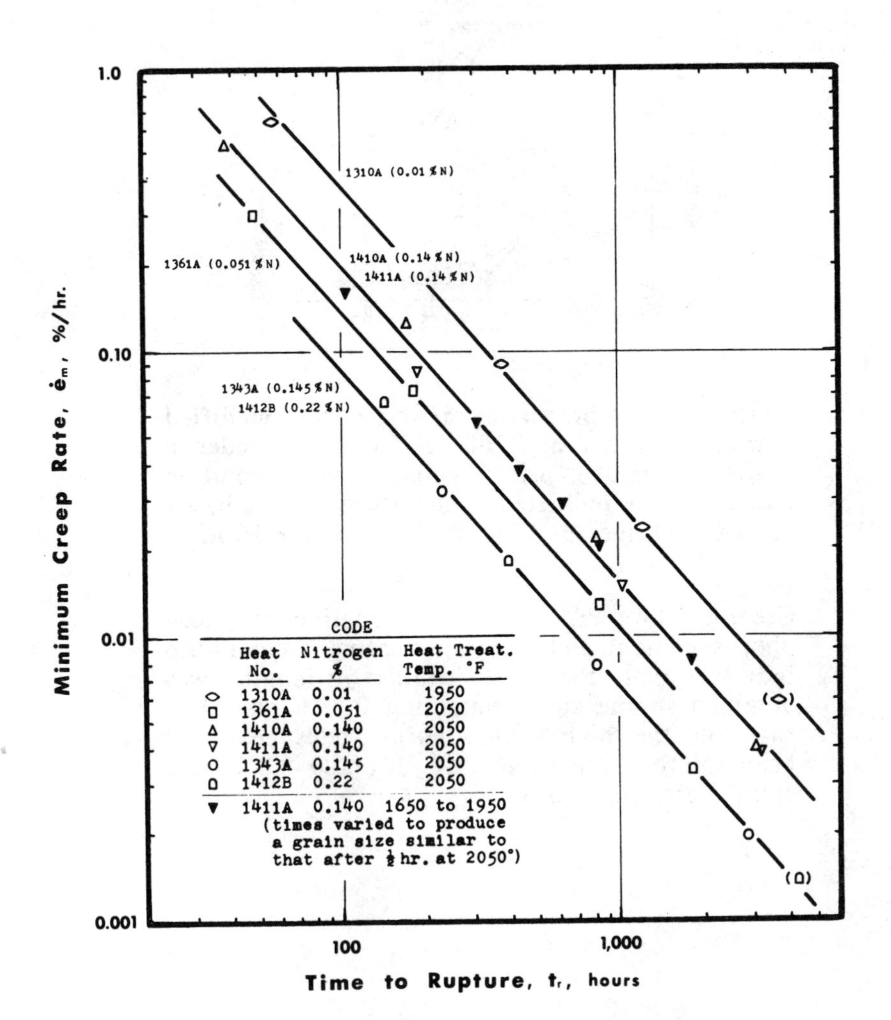

Relationship between minimum creep rate and time to rupture of the nitrogen series heats at 1200 °F (649 °C).

Source: P.D. Goodell and J.W. Freeman, "Elevated Temperature Properties of Nitrogen-Containing Type 304L Austenitic Stainless Steel," in Elevated Temperature Properties as Influenced by Nitrogen Additions to Types 304 and 316 Austenitic Stainless Steels (Proceedings of a Symposium in Atlantic City), ASTM Special Technical Publication No. 522, American Society for Testing and Materials, Philadelphia, 1973, p 54

Type 308 (Modified) Stainless Steel

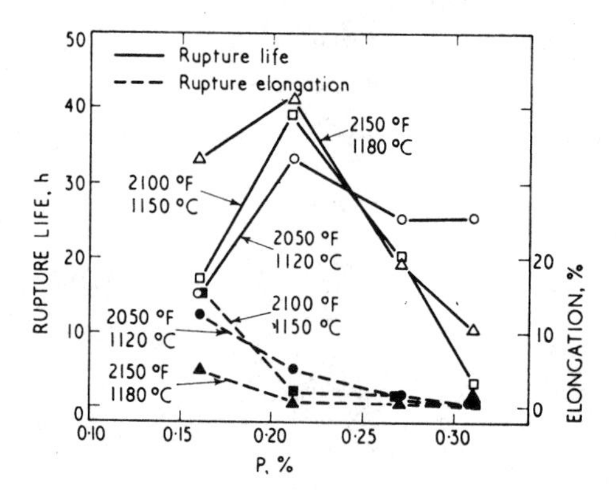

Creep-rupture properties of type 308 modified (Crucible HNM) at 1200 °F (650 °C) under a stress of 45,000 psi. Specimens were solution treated at the indicated temperature for 1/2 h, air cooled, and aged at 1300 °F (700 °C) for 16 h.

Creep-rupture properties as a function of phosphorus content and solution-treating temperature indicate that the best rupture strength was obtained in the steel containing 0.21% phosphorus, but the higher phosphorus steels that had been solution treated at 2050 °F (1120 °C) before aging were almost as good.

Source: E.J. Dulis, "Age-Hardening Austenitic Stainless Steels," in Metallurgical Developments in High Alloy Steels (Proceedings of a Conference in Scarborough, England), Iron and Steel Institute, London, 1964, p 169

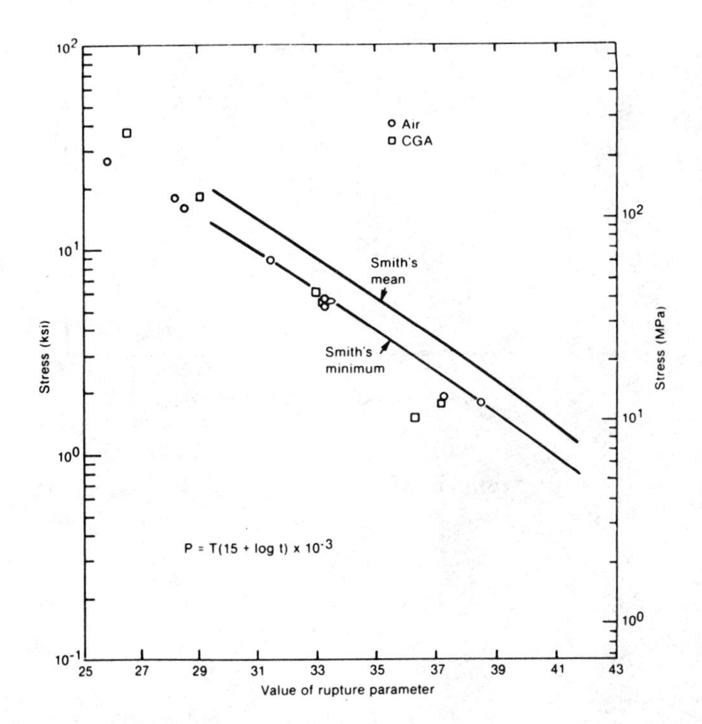

Variation of rupture parameter for type 310 stainless steel.

Source: R.M. Horton, "Effect of a Simulated Coal Gasifier Atmosphere on the Biaxial Stress Rupture Strength and Ductility of Selected Candidate Coal Gasifier Alloys," in Behavior of High Temperature Alloys in Aggressive Environments (Proceedings of the Petten International Conference, Petten, The Netherlands), I. Kirman, J.B. Marriott, M. Merz, P.R. Sahm, and D.P. Whittle, Eds., The Metals Society, London, 1980, p 907

Type 314 Stainless Steel: 1% Creep Strength

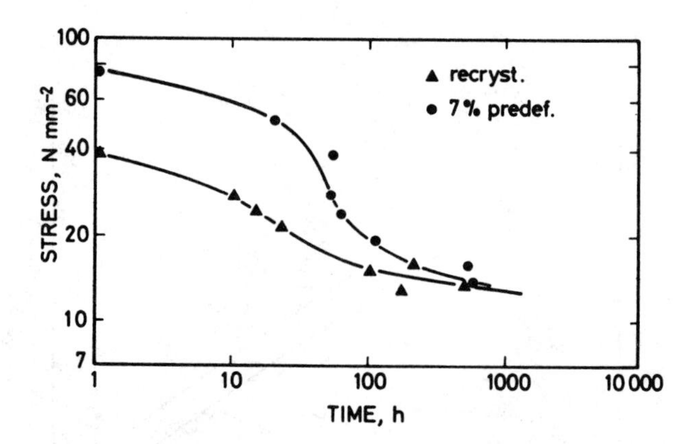

1% creep strength of type 314 stainless steel at 900 °C.

Source: V. Guttmann and J. Timm, "Relationship between Structure and Creep Properties of a Predeformed Austenitic Steel," in Mechanical Behaviour and Nuclear Applications of Stainless Steel at Elevated Temperatures (Proceedings of the International Conference in Varese, Italy), The Metals Society, London, 1981, p 107

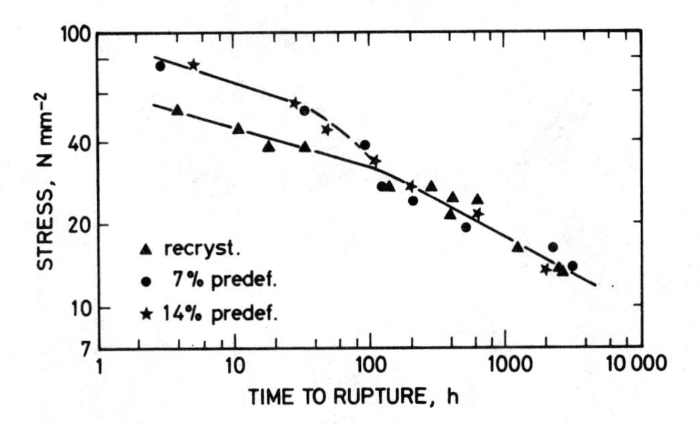

Creep-rupture strength of type 314 stainless steel at 900 °C.

Source: V. Guttmann and J. Timm, "Relationship between Structure and Creep Properties of a Predeformed Austenitic Steel," in Mechanical Behaviour and Nuclear Applications of Stainless Steel at Elevated Temperatures (Proceedings of the International Conference in Varese, Italy), The Metals Society, London, 1981, p 107

Type 314 Stainless Steel: Minimum Creep Rate vs Stress

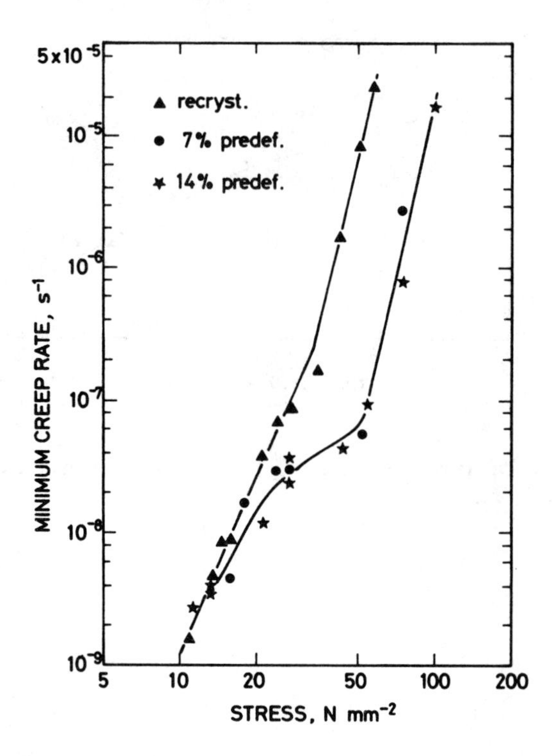

Variation of minimum creep rate with stress for type 314 stainless steel at 900 °C.

Source: V. Guttmann and J. Timm, "Relationship between Structure and Creep Properties of a Predeformed Austenitic Steel," in Mechanical Behaviour and Nuclear Applications of Stainless Steel at Elevated Temperatures (Proceedings of the International Conference in Varese, Italy), The Metals Society, London, 1981, p 107

Type 314 Stainless Steel: Minimum Creep Rate vs Time to Rupture

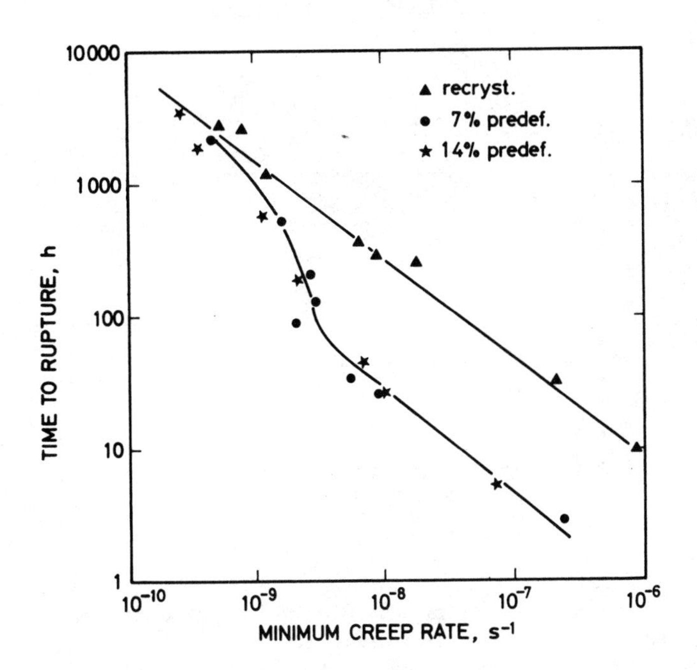

Minimum creep rate vs time to rupture for type 314 stainless steel at 900 °C.

Source: V. Guttmann and J. Timm, "Relationship between Structure and Creep Properties of a Predeformed Austenitic Steel," in Mechanical Behaviour and Nuclear Applications of Stainless Steel at Elevated Temperatures (Proceedings of the International Conference in Varese, Italy), The Metals Society, London, 1981, p 107

Type 316 Stainless Steel: 1% Total Plastic Strain Data

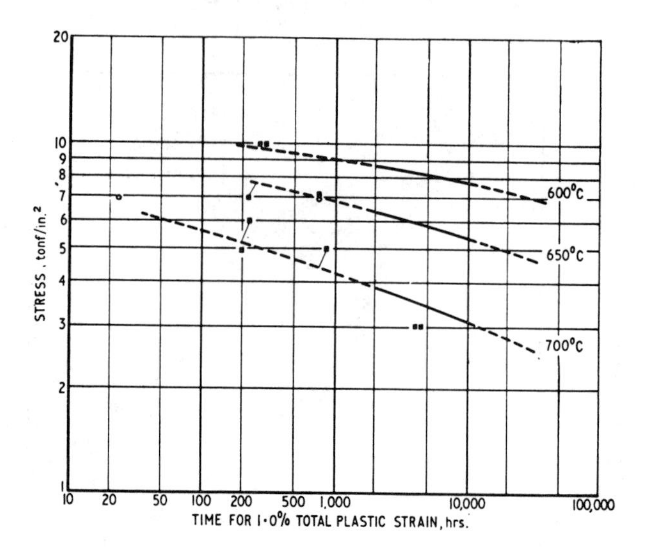

Typical 1.0% total plastic strain data for type 316 stainless steel. Solid line represents mean curve based on preliminary assessment by Electrical Research Association (Great Britain); broken line represents expected curve.

Source: R.J. Truman, R.P. Harvey, R.F. Johnson, and J.N. Smith, "Elevated-Temperature Tensile, Creep, and Rupture Properties of 18Cr-8Ni, 18Cr-12Ni-Mo, 18Cr-10Ni-Ti, and 18Cr-12Ni-Nb Steels," in High-Temperature Properties of Steels (Proceedings of a Conference in Eastbourne, England), Iron and Steel Institute, London, 1967, p 293

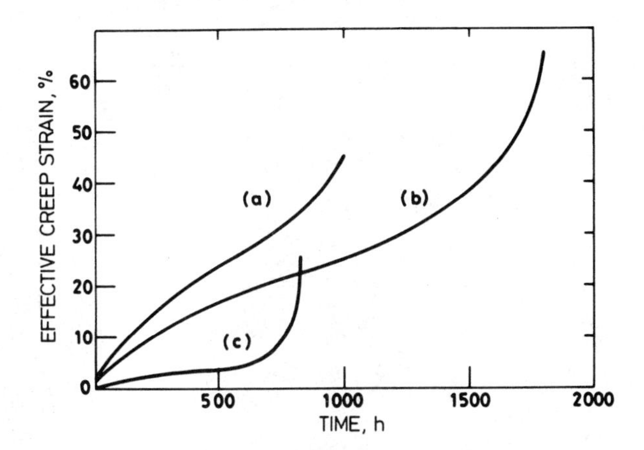

Creep properties of type 316 stainless steel at 600 °C for various initial hardened conditions. (a) 0% prestrain, σ_c = 290 MPa. (b) $\Delta\epsilon_p$ = 0.5%, cycled to N = 50 before creep test, σ_c = 290 MPa. (c) 25% prestrain, uniaxial σ_c = 300 MPa.

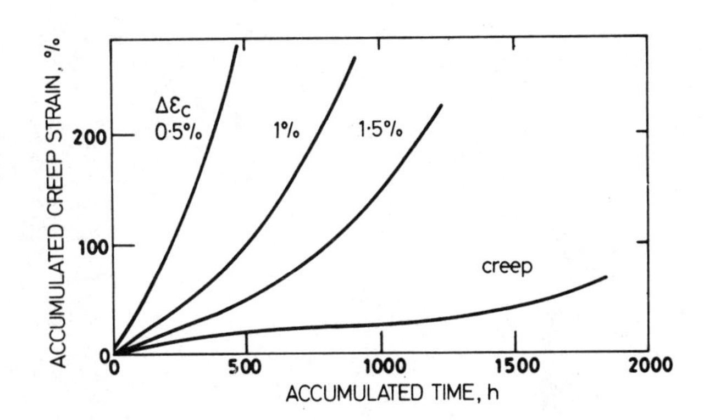

Accumulated creep strain vs accumulated dwell-time curves for different $\Delta\epsilon_c$ values.

Because material was placed in a cyclic hardened condition, its creep properties were not thought to be the same as those of solution-treated material, as shown in the top figure. Curve (a) is for solution-treated material and is weaker than material fatigue hardened to a steady cycle shape (curve (b)). For comparison, curve (c) shows that monotonic prestraining hardens the material to a far greater extent than fatigue cycling, although the former testing was in tension and is not strictly comparable. Failure in (a) and (b) was caused by a buckling instability, and thus the strain-to-failure values are more a reflection of specimen geometry than of material degradation. Accumulated creep strain/accumulated time plots in the bottom figure illustrate the systematic way in which average creep strain rates are increased as $\Delta\epsilon_c$ is reduced.

Source: K. Wei and B.F. Dyson, "Creep-Fatigue Interactions in 316 Stainless Steel under Torsional Loading," in Mechanical Behaviour and Nuclear Applications of Stainless Steel at Elevated Temperatures (Proceedings of the International Conference in Varese, Italy), The Metals Society, London, 1982, p 138

11.32 Type 316 Stainless Steel: Creep-Rupture of Partially and Fully Annealed Specimens

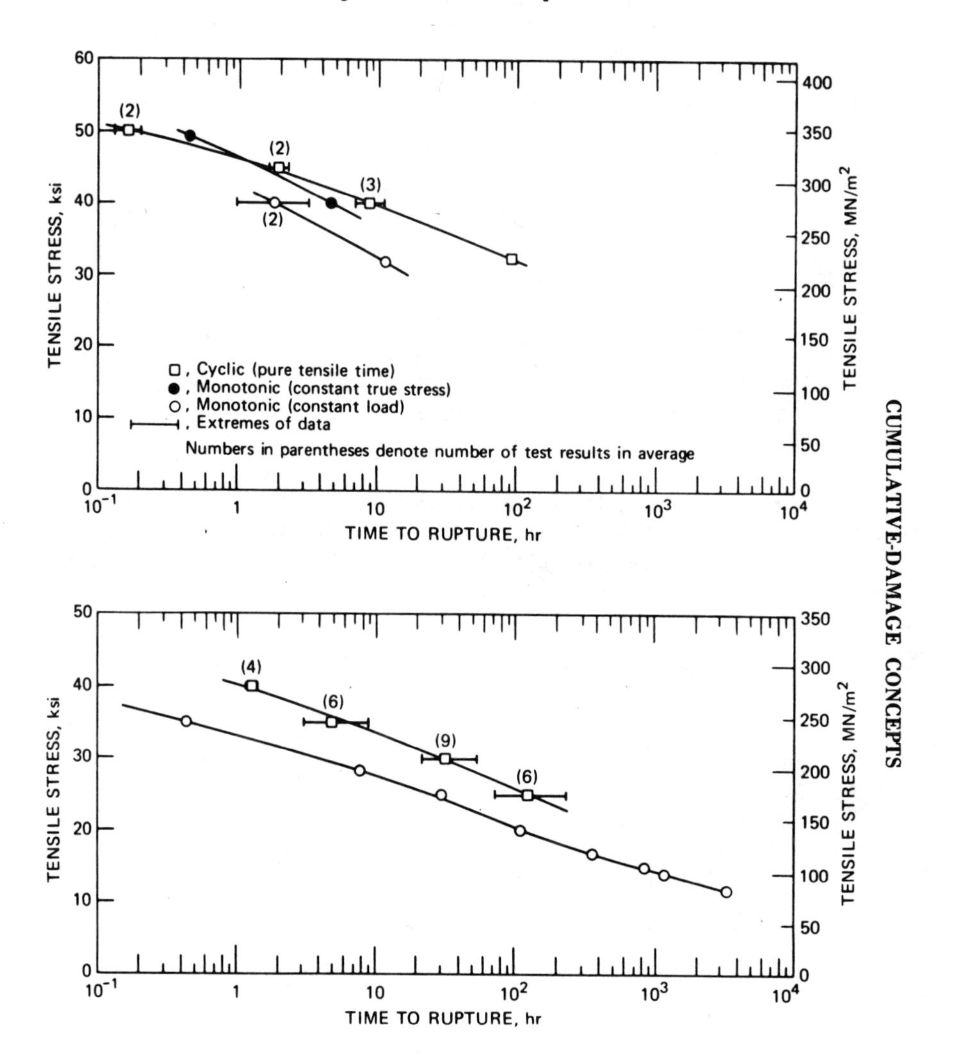

Creep-rupture properties of partially annealed (top) and fully annealed (bottom) type 316 stainless steel specimens tested at 1300 °F.

Source: J.B. Conway, R.H. Stentz, and J.T. Berling, Fatigue, Tensile, and Relaxation Behavior of Stainless Steels, Technical Information Center, U.S. Atomic Energy Commission, Washington, D.C., 1975, p 145

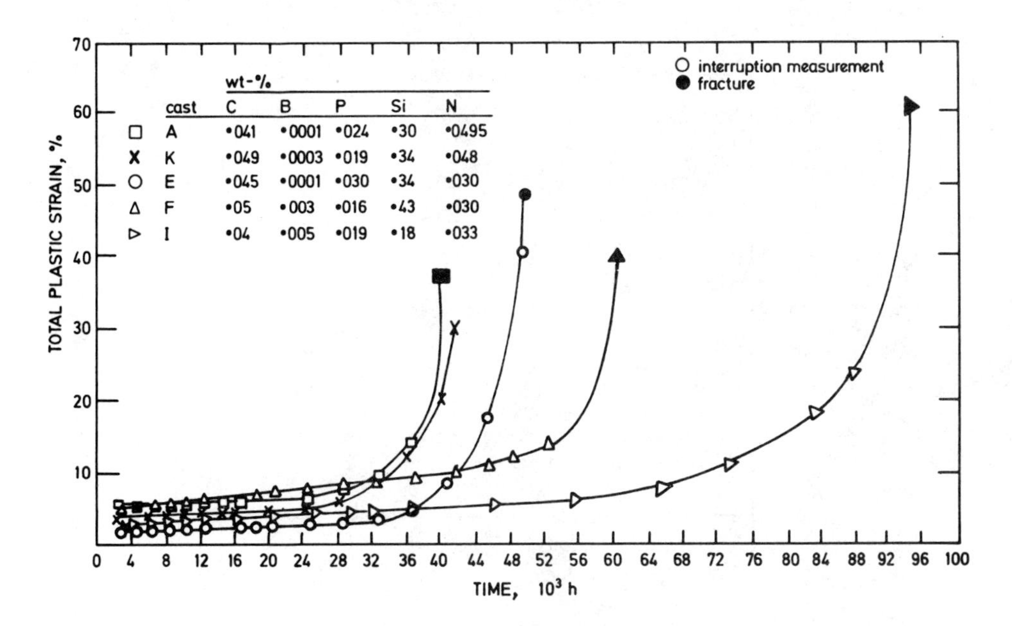

Interruption strain/time plots showing effect of boron on creep life of type 316 stainless steel casts tested at 600 °C and 154 MN/m².

It would appear that B increases creep-rupture life by delaying the onset of the tertiary stage, rather than increasing the creep strength by decreasing the creep rate. This phenomenon is highlighted when comparing the interruption strain data generated on five casts tested at a temperature of 600 °C. The data illustrated above show only marginal effect on the minimum creep rate yet a significant effect on the rupture life of higher B casts, as evidenced by the shifting of the tertiary stage to longer durations. A similar effect of B on rupture life was observed at 650 and 700 °C, as well as on rupture properties of type 316 welds.

Source: H.K. Grover and A. Wickens, "Effects of Minor Elements on Long-Term Creep-Rupture Properties of 316 Austenitic Stainless Steel at 550-700 °C," in Mechanical Behaviour and Nuclear Applications of Stainless Steel at Elevated Temperatures (Proceedings of the International Conference in Varese, Italy), The Metals Society, London, 1982, p 84

Type 316 Stainless Steel: Effect of Cold Strain and Heat Treatment

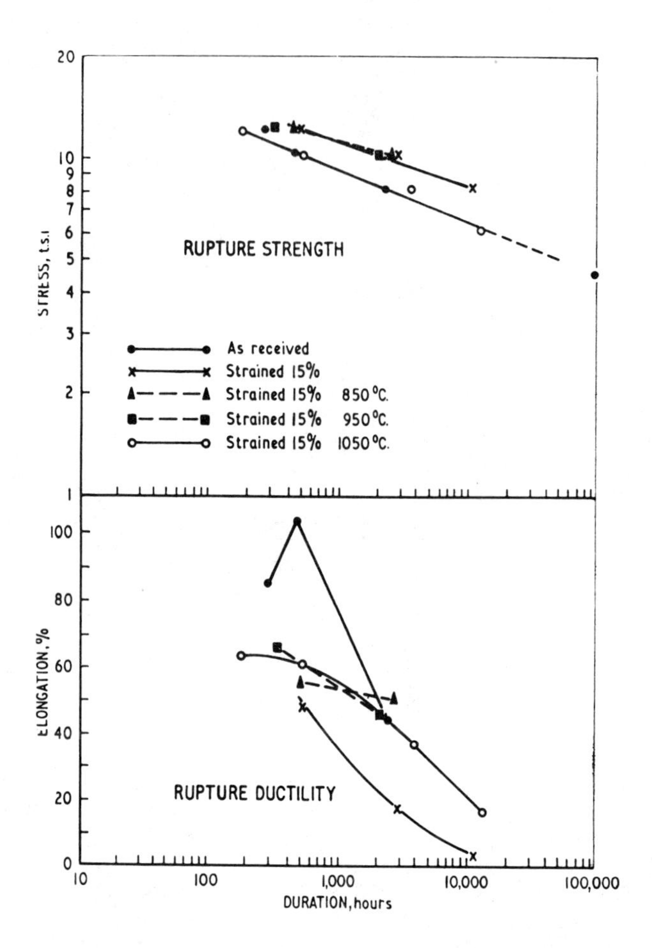

Effect of cold strain and heat treatment on the rupture strength (top) and rupture ductility (bottom) of type 316 stainless steel at 650 °C.

Source: B. Mitchell, "High-Temperature Materials for Electrical Power Generation," in High-Temperature Properties of Steels (Proceedings of a Conference in Eastbourne, England), Iron and Steel Institute, London, 1967, p 548

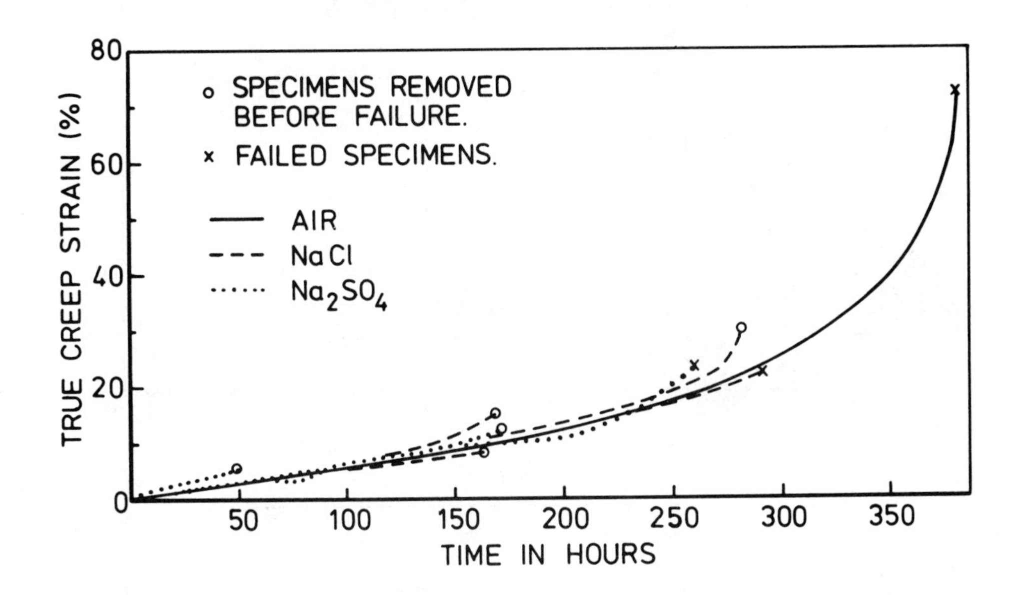

Creep curves for type 316 stainless steel in various environments at 800 °C and 50 MN/m².

As has been found for Nimonic 108, deposited layers of NaCl or Na_2SO_4 also severely degrade the creep properties of type 316 stainless steel (see figure), reducing creep lives by 25-30% and reducing creep ductility to a third of that for air tests. This degree of degradation is close to that observed for Nimonic 108 at 850 °C, except that specimens of type 316 stainless steel show lower ductilities in these corrosive environments than Nimonic 108.

Source: J.R. Nicholls, J. Samuel, R.C. Hurst, and P. Hancock, "The Influence of Hot Corrosion and Oxidation on the High Temperature Creep Behaviour of 316 Stainless Steel and Nimonic 108," in Behavior of High Temperature Alloys in Aggressive Environments (Proceedings of the Petten International Conference, Petten, The Netherlands) I. Kirman, J.B. Marriott, M. Merz, P.R. Sahm, and D.P. Whittle, Eds., The Metals Society, London, 1980, p 916, 924

Type 316 Stainless Steel: Effect of Nitrogen Addition on Creep and Rupture Properties

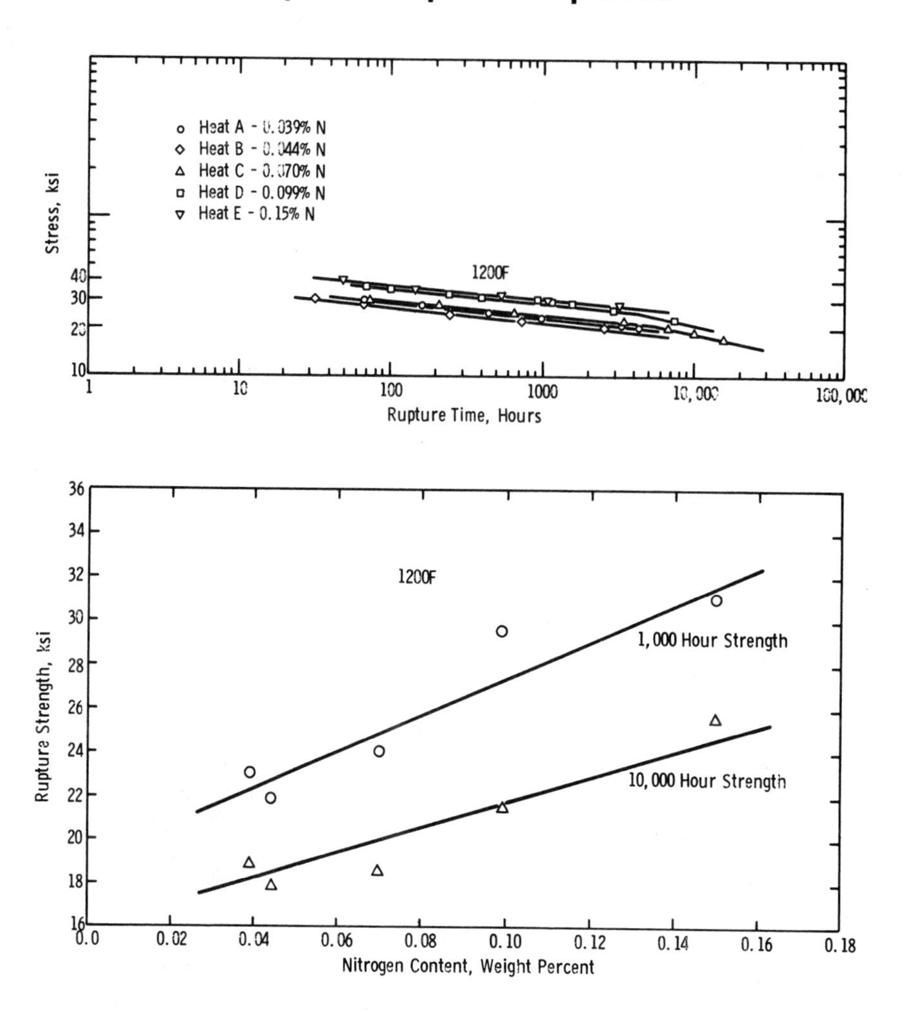

Effect of nitrogen content on rupture properties (top) and variation of 1000- and 10,000-h rupture strength at 1200 °F (649 °C) (bottom) of type 316 stainless steel.

The 1200 °F (649 °C) rupture results for five alloys with varying N contents are given in the top figure. This comparison clearly shows the beneficial effect of N on rupture properties of type 316 stainless steel. The 1000- and 10,000-h rupture strengths of these materials are plotted as a function of N content in the bottom figure. Again, the strong dependence of rupture strength on N content is clearly evident.

Source: T.M. Cullen and M.W. Davis, "Influence of Nitrogen on the Creep-Rupture Properties of Type 316 Steel," in Elevated Temperature Properties as Influenced by Nitrogen Additions to Types 304 and 316 Austenitic Stainless Steels (Proceedings of a Symposium in Atlantic City), ASTM Special Technical Publication No. 522, American Society for Testing and Materials, Philadelphia, 1973, p 61, 67-69

Type 316 Stainless Steel: Effect of Nitrogen Addition on Creep Resistance

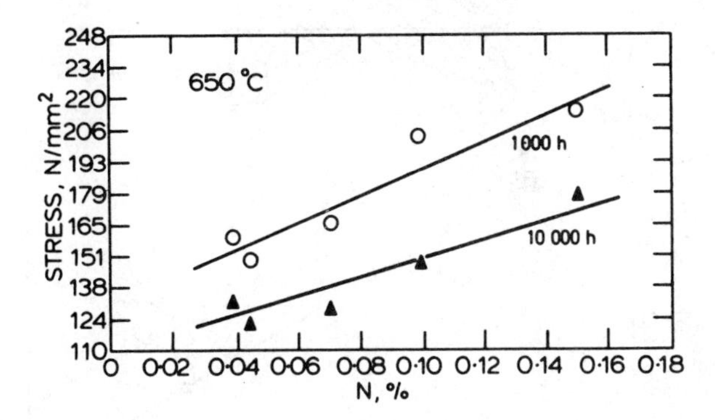

Effect of nitrogen addition on creep resistance of type 316 stainless steel.

Source: M. Ceccarelli, R. Santucci, and A. Bennani, "Hot Mechanical Properties of 316L Stainless Steel with Boron and Nitrogen Additions," in Mechanical Behaviour and Nuclear Applications of Stainless Steel at Elevated Temperatures (Proceedings of the International Conference in Varese, Italy), The Metals Society, London, 1982, p 40

11.38 Type 316 Stainless Steel: Effect of Nitrogen Addition on Stress-Rupture Properties

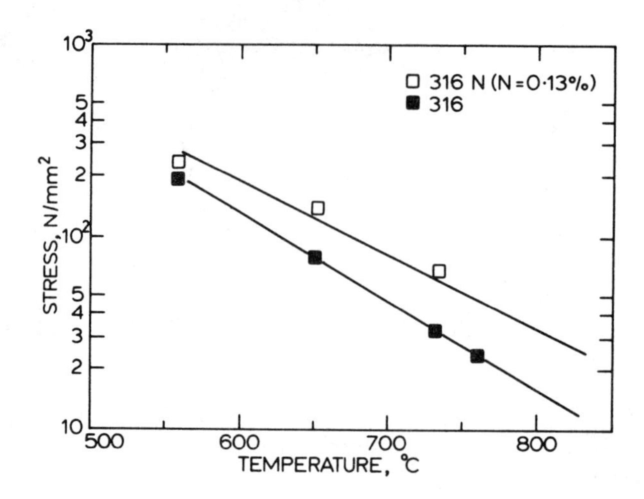

Effect of nitrogen on 10^5 h stress-rupture properties of type 316 stainless steel.

Source: M. Ceccarelli, R. Santucci, and A. Bennani, "Hot Mechanical Properties of 316L Stainless Steel with Boron and Nitrogen Additions," in Mechanical Behaviour and Nuclear Applications of Stainless Steel at Elevated Temperatures (Proceedings of the International Conference in Varese, Italy), The Metals Society, London, 1982, p 40

Type 316 Stainless Steel: Long-Term Creep and Rupture Strengths

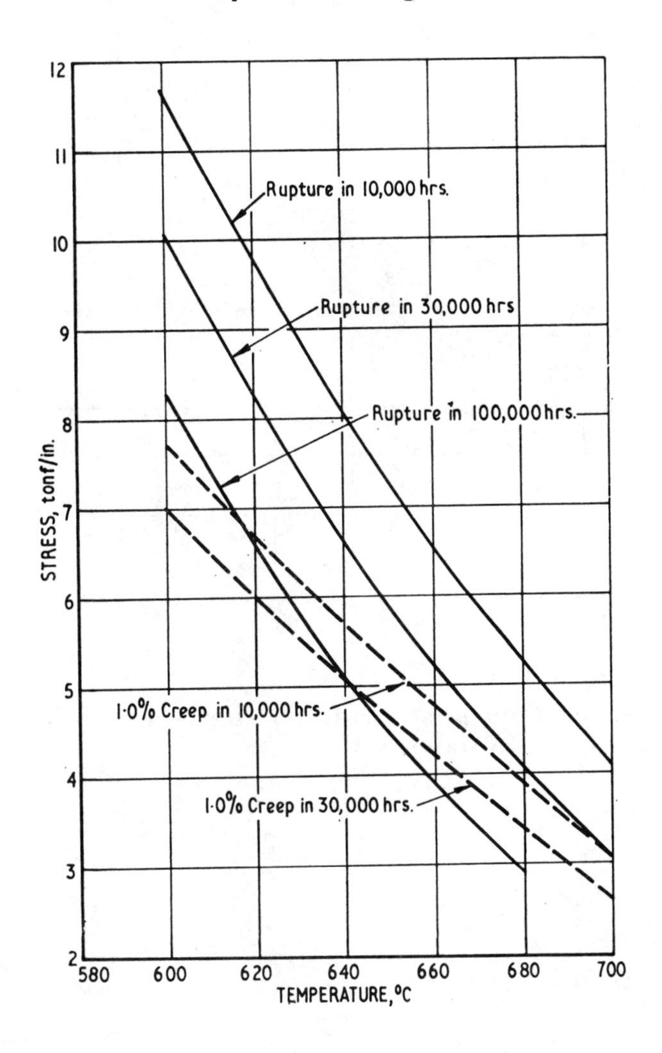

Estimated long-term 1.0% creep and rupture strengths of type 316 stainless steel.

Source: R.J. Truman, R.P. Harvey, R.F. Johnson, and J.N. Smith, "Elevated-Temperature Tensile, Creep, and Rupture Properties of 18Cr-8Ni, 18Cr-12Ni-Mo, 18Cr-10Ni-Ti, and 18Cr-12Ni-Nb Steels," in High-Temperature Properties of Steels (Proceedings of a Conference in Eastbourne, England), Iron and Steel Institute, London, 1967, p 293

Type 316 Stainless Steel: Sherby-Dorn Master Curve

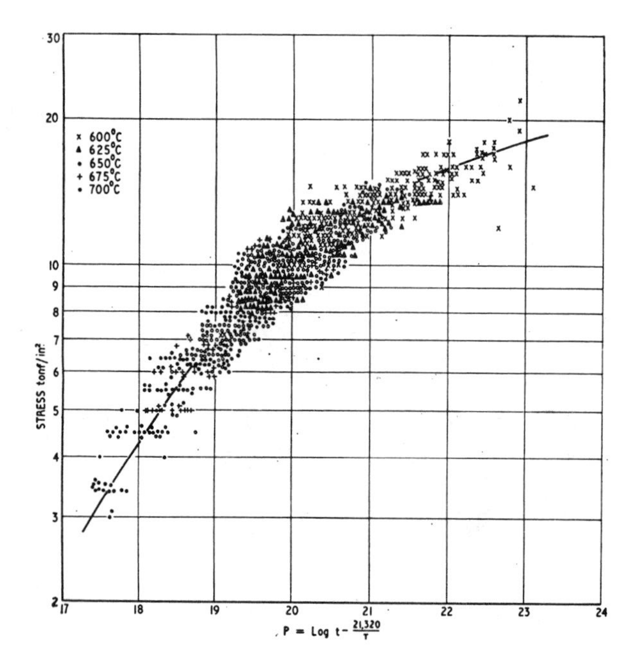

Sherby-Dorn master curve of stress-rupture data for type 316 stainless steel.

Source: R.J. Truman, R.P. Harvey, R.F. Johnson, and J.N. Smith, "Elevated-Temperature Tensile, Creep, and Rupture Properties of 18Cr-8Ni, 18Cr-12Ni-Mo, 18Cr-10Ni-Ti, and 18Cr-12Ni-Nb Steels," in High-Temperature Properties of Steels (Proceedings of a Conference in Eastbourne, England), Iron and Steel Institute, London, 1967, p 293

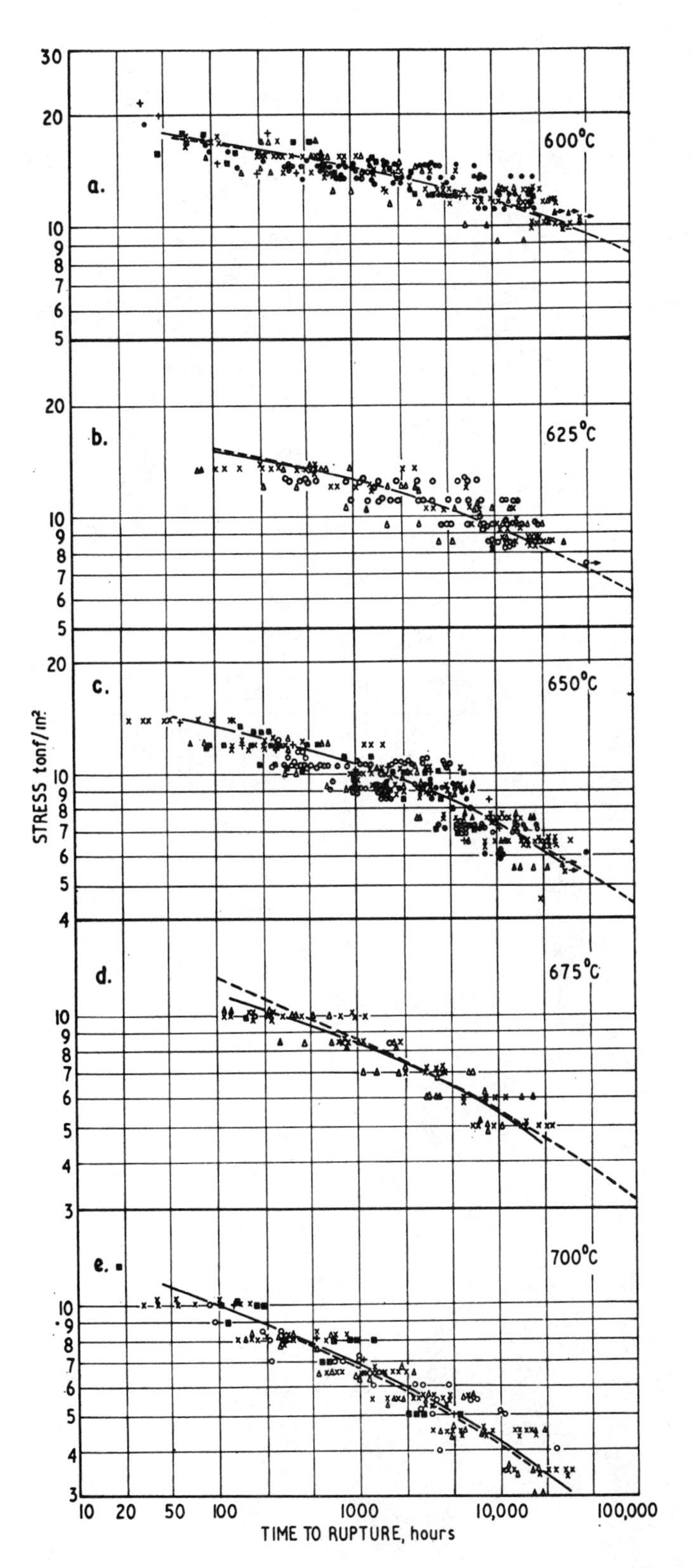

Stress-rupture properties of type 316 stainless steel. Solid lines represent calculated mean curve; broken lines represent curve predicted by Sherby-Dorn parametric equation.

Type 316 Stainless Steel: Stress-Rupture Properties (Continued)

Key to symbols in preceding figure

CHEMICAL COMPOSITION	%
Carbon	0·03–0·10
Silicon	0·18–0·76
Manganese	0·59–1·78
Chromium	15·4–18·0
Nickel	10·1–12·9
Molybdenum	2·11–2·91
Sulphur	0·005–0·035
Phosphorus	0·014–0·035

HEAT TREATMENT

Solution treated 1050–1100°C
Air cooled or water quenched
■ Bar
○ Tube
+ Forging
ERA data:—
× Tube
△ Bar

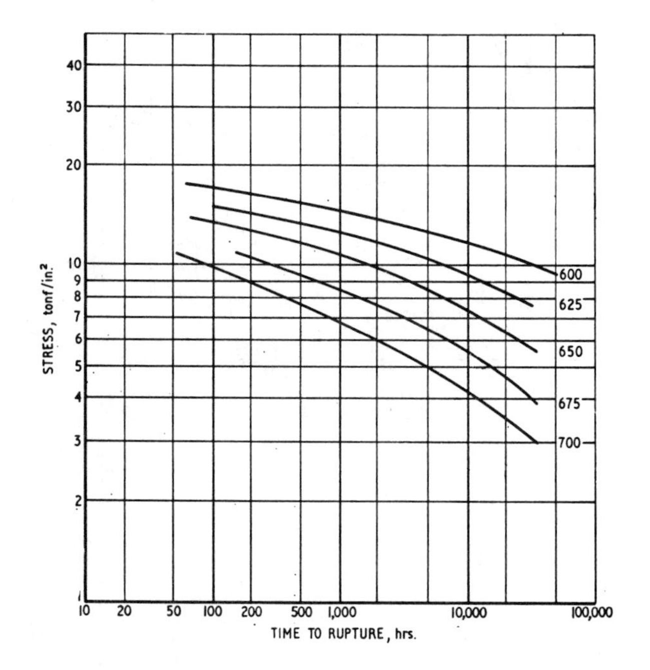

Summary of mean curves for stress-rupture data on type 316 stainless steel.

Source: R.J. Truman, R.P. Harvey, R.F. Johnson, and J.N. Smith, "Elevated-Temperature Tensile, Creep, and Rupture Properties of 18Cr-8Ni, 18Cr-12Ni-Mo, 18Cr-10Ni-Ti, and 18Cr-12Ni-Nb Steels," in High-Temperature Properties of Steels (Proceedings of a Conference in Eastbourne, England), Iron and Steel Institute, London, 1967, p 292-293

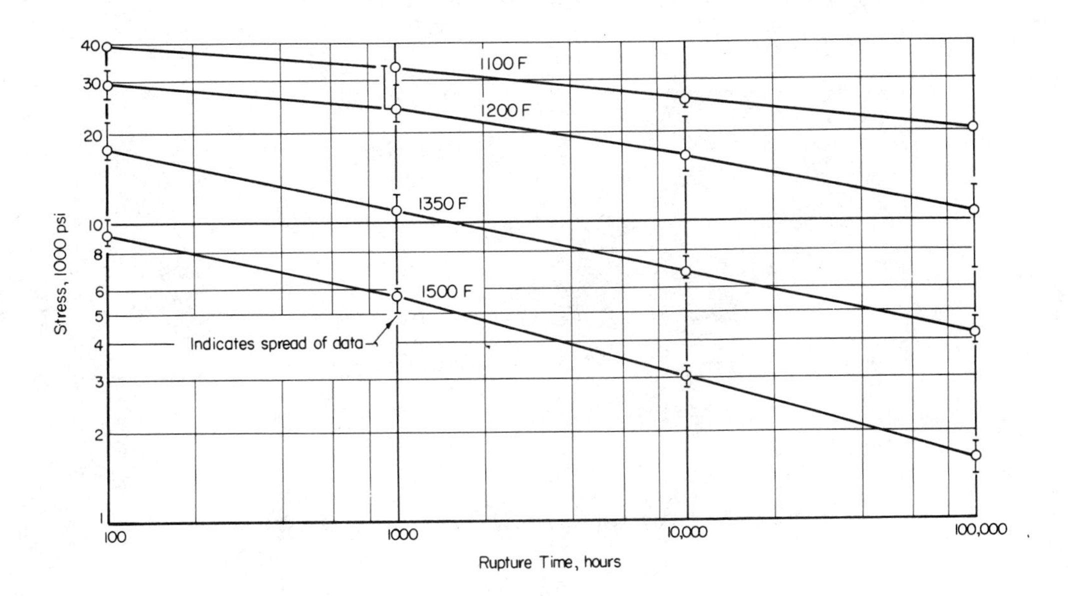

Stress vs rupture-time for type 316 stainless steel bar. Data are based on averages from stress-temperature curves.

Source: The Elevated-Temperature Properties of Stainless Steels, ASTM Data Series DS 5-S1, American Society for Testing and Materials, Philadelphia, 1965, p 25

Type 316 Stainless Steel: Typical Constant-Stress Creep Curves

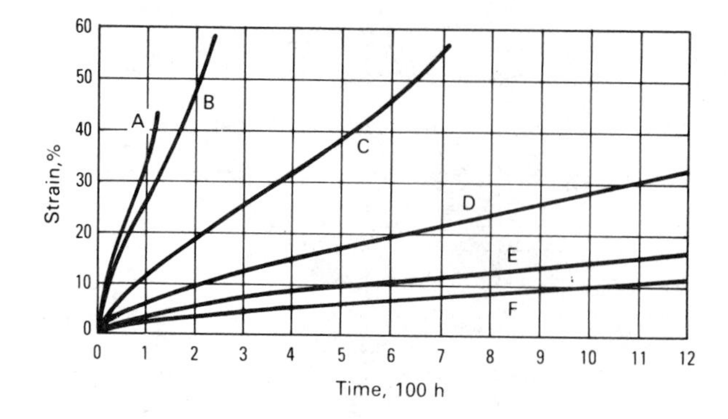

Specimen	Stress MPa	Stress ksi
A	160	23.2
B	147.5	21.4
C	128.2	18.6
D	106.9	15.5
E	91.0 (continued to 2100 h)	13.2
F	77.2 (continued to 2900 h)	11.2

Typical constant-stress creep curves obtained at 705 °C (1300 °F) for type 316 austenitic stainless steel.

Source: Metals Handbook, Ninth Edition, Volume 8, Properties and Selection: Mechanical Testing, American Society for Metals, Metals Park, OH, 1985, p 321

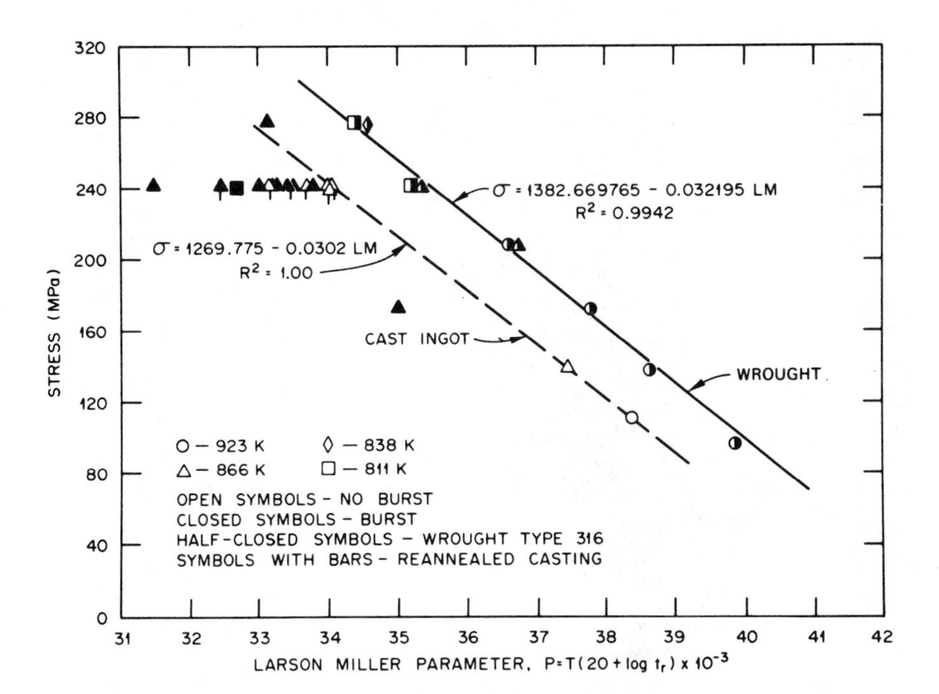

Comparison of Larson-Miller parameter for cast type 316 with the wrought type 316 stainless steel. The cast rupture strength is at least 40 MPa less than the wrought material.

Source: V.K. Sikka and S.A. David, "Discontinuous Creep Deformation in a Type 316 Stainless Steel Casting," Metallurgical Transactions A, Vol. 12A, May 1981, p 887

Type 316L Stainless Steel: Creep and Stress-Rupture Properties (10,000 h)

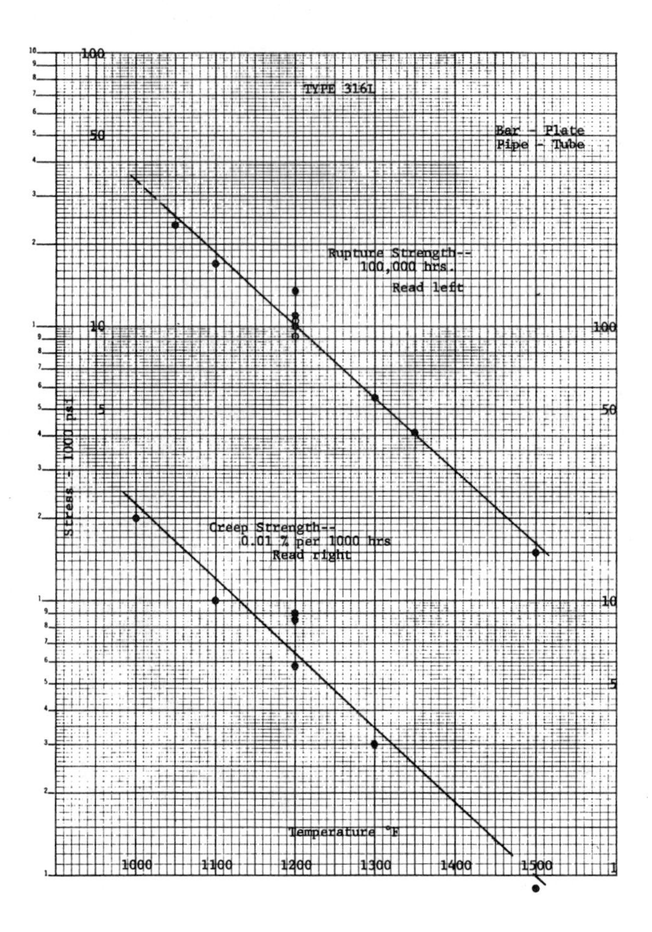

Variation of rupture strength (10,000 h) and creep strength (0.1%/1000 h) of type 316L stainless steel with temperature.

Source: An Evaluation of the Yield, Tensile, Creep, and Rupture Strengths of Wrought 304, 316, 321, and 347 Stainless Steels at Elevated Temperatures, ASTM Data Series DS 552, American Society for Testing and Materials, Philadelphia, 1969, p 76

Type 316L Stainless Steel: Creep and Stress-Rupture Properties (100,000 h)

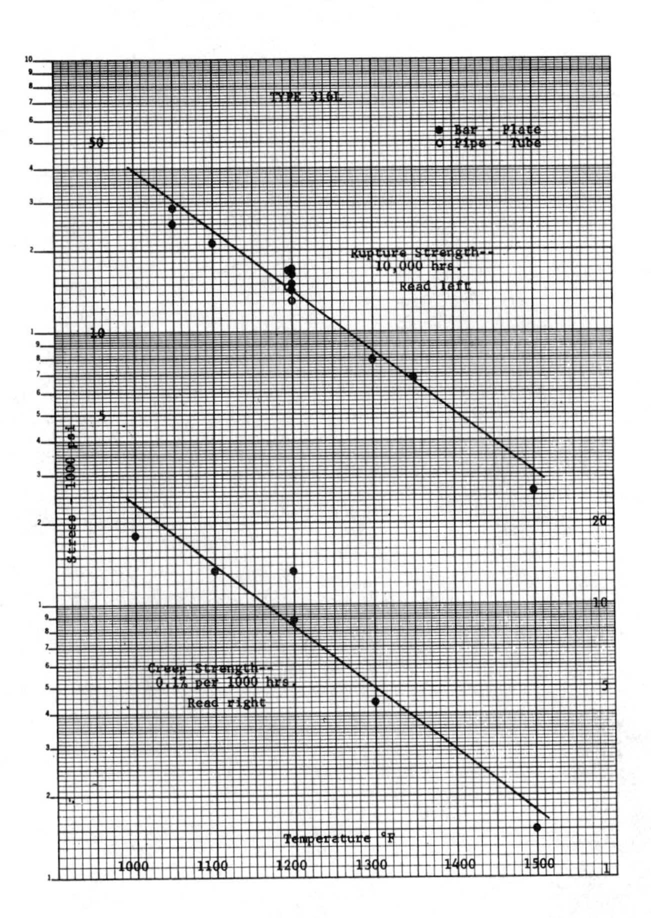

Variation of rupture strength (100,000 h) and creep strength (0.01%/1000 h) of type 316L stainless steel with temperature.

Source: An Evaluation of the Yield, Tensile, Creep, and Rupture Strengths of Wrought 304, 316, 321, and 347 Stainless Steels at Elevated Temperatures, ASTM Data Series DS 552, American Society for Testing and Materials, Philadelphia, 1969, p 80

Type 316L Stainless Steel: Effect of Neutron Irradiation

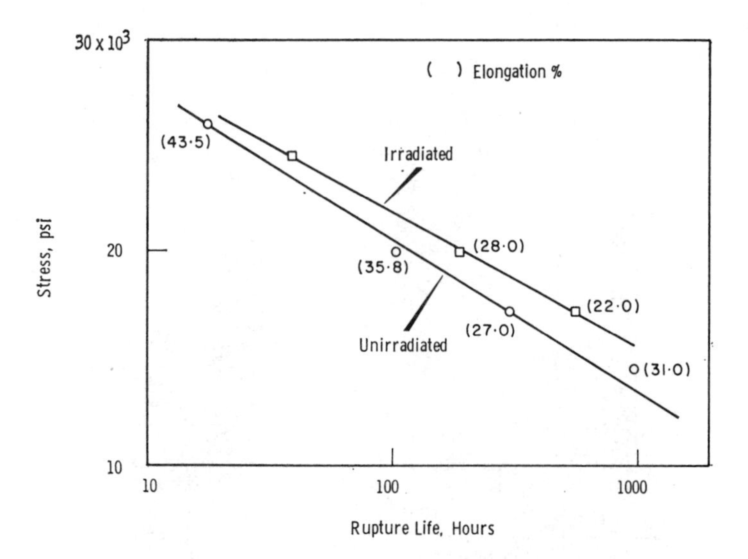

Effects of neutron irradiation under thermal conditions (1.7 × 10^{19} n/cm^2) and with fission (7.5 × 10^{18} n/cm^2 at 45 °C) on the 650 °C stress-rupture lives and ductilities of type 316L austenitic stainless steel.

Source: A.L. Bement, "Effects of Minor Constituents on Irradiation Damage to Austenitic Stainless Steels," in Effects of Residual Elements on Properties of Austenitic Stainless Steels (Proceedings of a Symposium in Atlantic City), ASTM Special Technical Publication No. 418, American Society for Testing and Materials, Philadelphia, 1967, p 63

Type 316L Stainless Steel: Effect of Nitrogen Content on Creep-Rupture Elongation

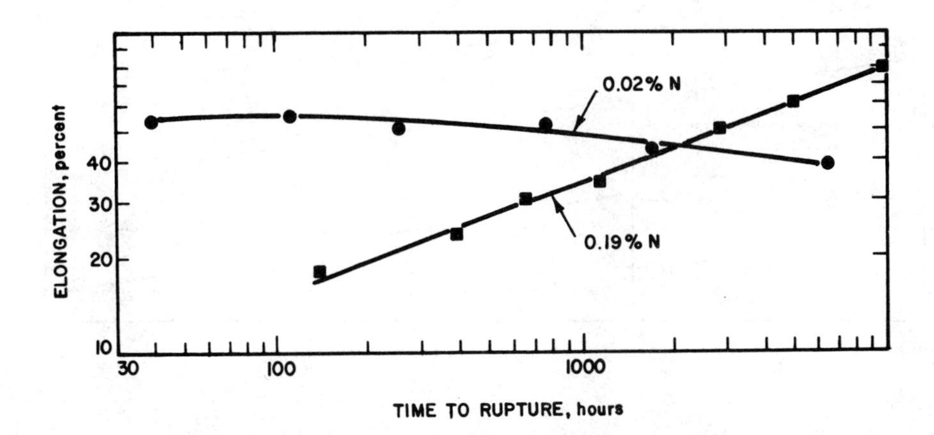

Creep-rupture elongation at 1300 °F (704 °C) for type 316L stainless steel containing 0.02 and 0.19% nitrogen.

Elongation of 0.19% N steel is initially lower than the elongation of the 0.02% N steel. However, the elongation of the 0.19% N steel increased with time, whereas the elongation of the 0.02% N steel decreased slightly with time, such that for times greater than about 2200 h at 1300 °F (704 °C), the steel with the 0.19% N content exhibited higher elongation compared with the 0.02% N steel. The reason for this divergent behavior is not understood.

Source: C.E. Spaeder, W. F. Domis, and K.G. Brickner, "High-Nitrogen Austenitic Stainless Steels," in Elevated Temperature Properties as Influenced by Nitrogen Additions to Types 304 and 316 Austenitic Stainless Steels (Proceedings of a Symposium in Atlantic City), ASTM Special Technical Publication No. 522, American Society for Testing and Materials, Philadelphia, 1973, p 43-45

Type 316L Stainless Steel: Stress vs Rupture Time

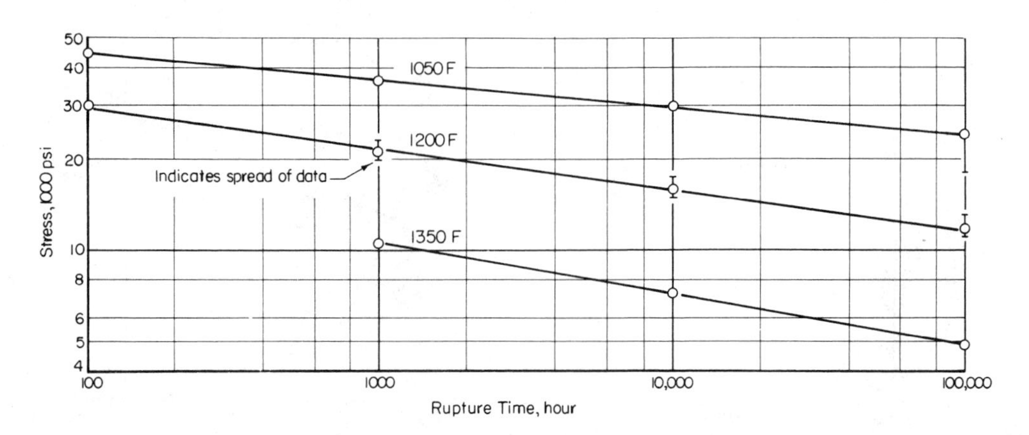

Stress vs rupture-time for type 316L stainless steel. Data are based on averages from stress-temperature curves.

Source: The Elevated-Temperature Properties of Stainless Steels, ASTM Data Series DS 5-S1, American Society for Testing and Materials, Philadelphia, 1965, p 32

Type 316N Stainless Steel: Minimum Creep Rate

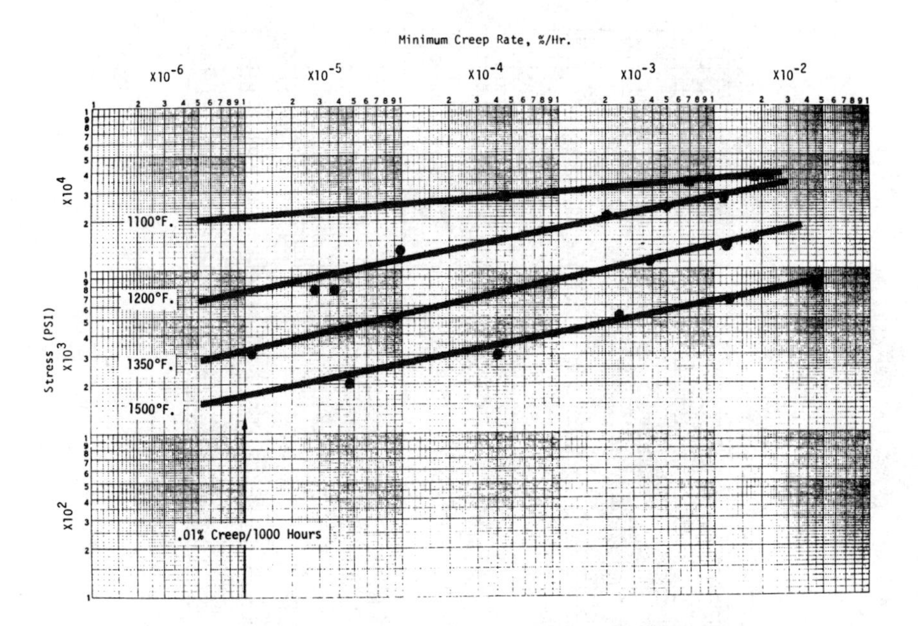

Minimum-creep rate data for type 316N stainless steel.

Source: P. Kadlecek, "Mechanical Property Data on Hot-Extruded 304N and 316N Stainless Steel Pipe," in Elevated Temperature Properties as Influenced by Nitrogen Additions to Types 304 and 316 Austenitic Stainless Steels (Proceedings of a Symposium in Atlantic City), ASTM Special Technical Publication No. 522, American Society for Testing and Materials, Philadelphia, 1973, p 8

Type 316N Stainless Steel: Stress-Rupture Properties

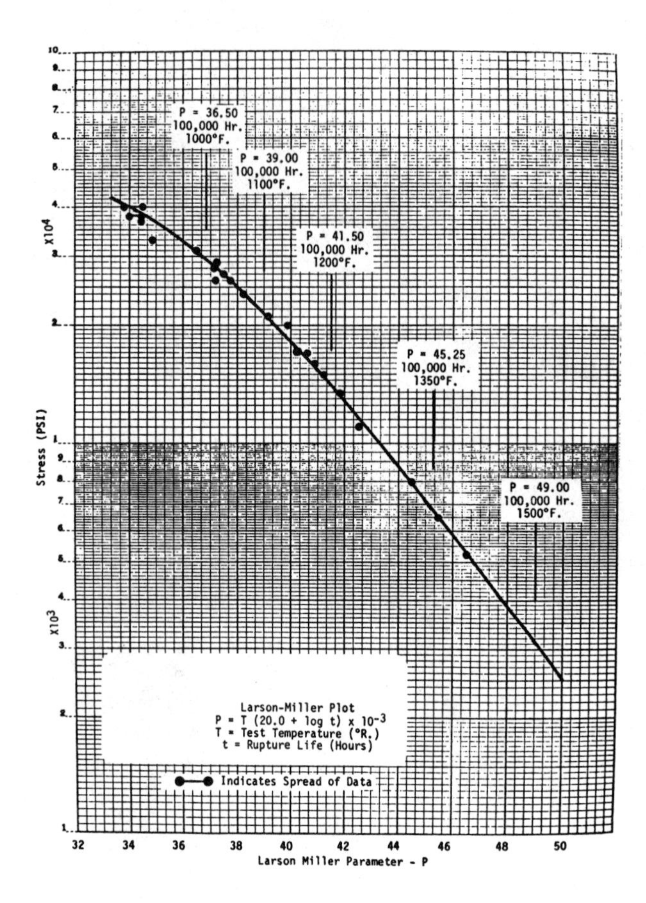

Stress-rupture data for type 316N stainless steel.

Source: P. Kadlecek, "Mechanical Property Data on Hot-Extruded 304N and 316N Stainless Steel Pipe," in Elevated Temperature Properties as Influenced by Nitrogen Additions to Types 304 and 316 Austenitic Stainless Steels (Proceedings of a Symposium in Atlantic City), ASTM Special Technical Publication No. 522, American Society for Testing and Materials, Philadelphia, 1973, p 7

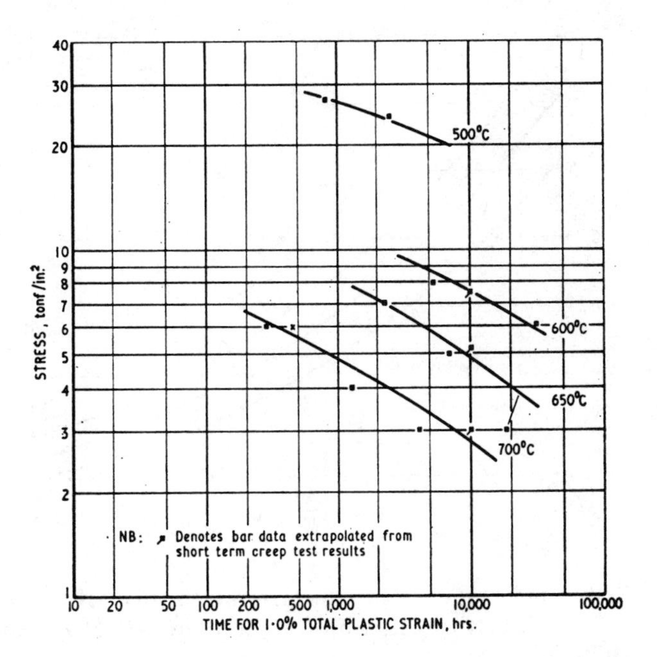

1.0% total plastic strain data for type 321 stainless steel that was solution treated at 1050 to 1100 °C.

Source: R.J. Truman, R.P. Harvey, R.F. Johnson, and J.N. Smith, "Elevated-Temperature Tensile, Creep, and Rupture Properties of 18%Cr-8%Ni, 18%Cr-12%Ni-Mo, 18%Cr-10%Ni-Ti, and 18%Cr-12%Ni-Nb Steels," in High-Temperature Properties of Steels, Iron and Steel Institute, London, 1967, p 295

Type 321 Stainless Steel: Long-Term Rupture Strength

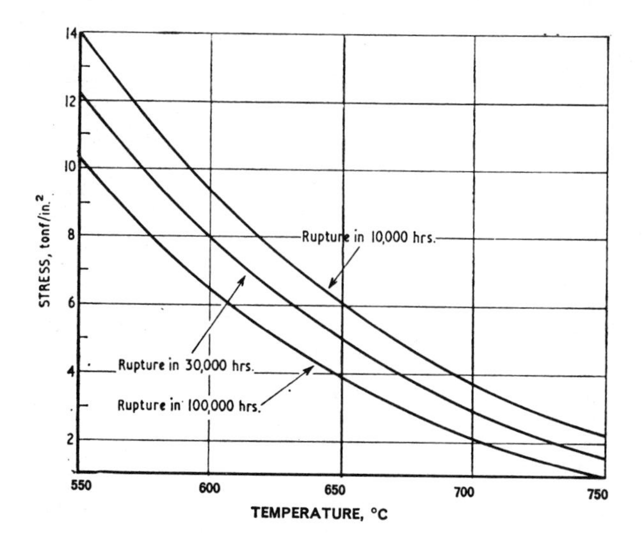

Estimated long-term rupture strengths of type 321 stainless steel solution treated at 1020 to 1050 °C.

Source: R.J. Truman, R.P. Harvey, R.F. Johnson, and J.N. Smith, "Elevated-Temperature Tensile, Creep, and Rupture Properties of 18%Cr-8%Ni, 18%Cr-12%Ni-Mo, 18%Cr-10%Ni-Ti, and 18%Cr-12%Ni-Nb Steels," in High-Temperature Properties of Steels, Iron and Steel Institute, London, 1967, p 295

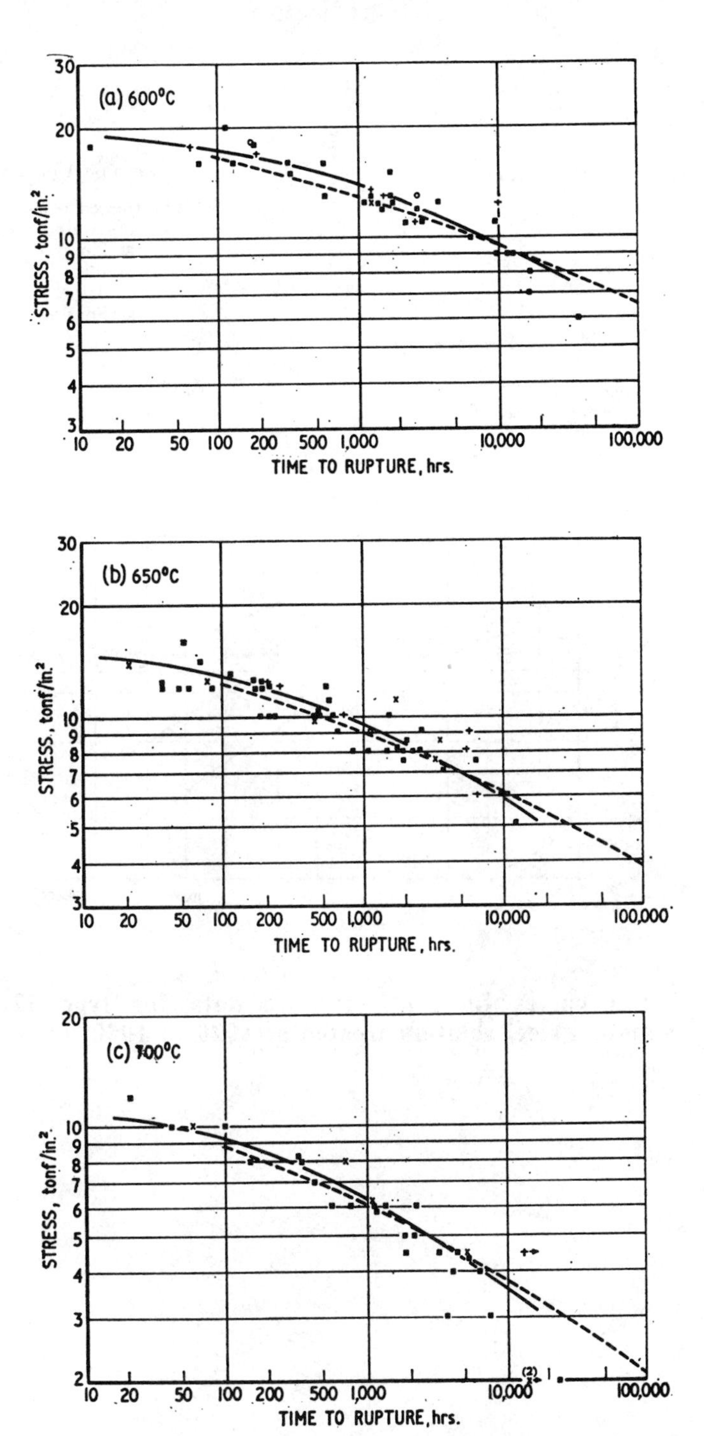

Stress-rupture properties of type 321 stainless steel solution treated at 1020 to 1050 °C. Solid lines represent calculated mean curve; broken lines represent curve predicted by Sherby-Dorn parametric equation.

Type 321 Stainless Steel: Stress-Rupture Properties (Continued)

CHEMICAL COMPOSITION

	%
Carbon	0·04–0·10
Silicon	0·32–1·15
Manganese	0·55–1·65
Chromium	17·5–18·6
Nickel	8·2–12·0
Molybdenum	—
Titanium	0·32–0·59
Niobium	—
Sulphur	0·007–0·030
Phosphorus	0·014–0·033

HEAT TREATMENT

Solution treated 1020–1050°C
Air cooled or water quenched
■ Bar
+ Plate
○ Tube
× Forging

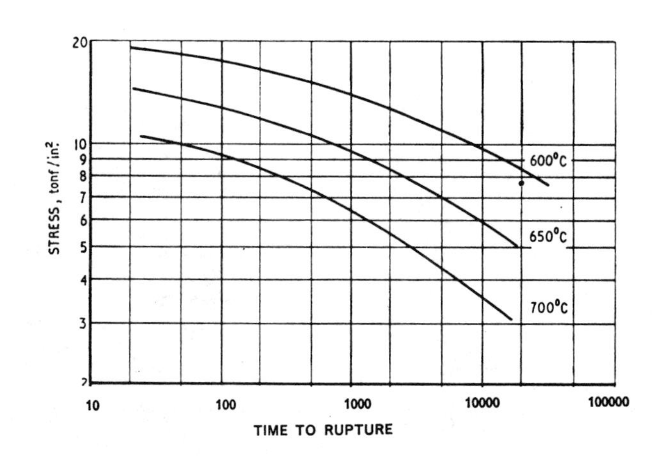

Mean curves for stress-rupture data for type 321 stainless steel solution treated at 1020 to 1050 °C.

Source: R.J. Truman, R.P. Harvey, R.F. Johnson, and J.N. Smith, "Elevated-Temperature Tensile, Creep, and Rupture Properties of 18%Cr-8%Ni, 18%Cr-12%Ni-Mo, 18%Cr-10%Ni-Ti, and 18%Cr-12%Ni-Nb Steels," in High-Temperature Properties of Steels, Iron and Steel Institute, London, 1967, p 294-295

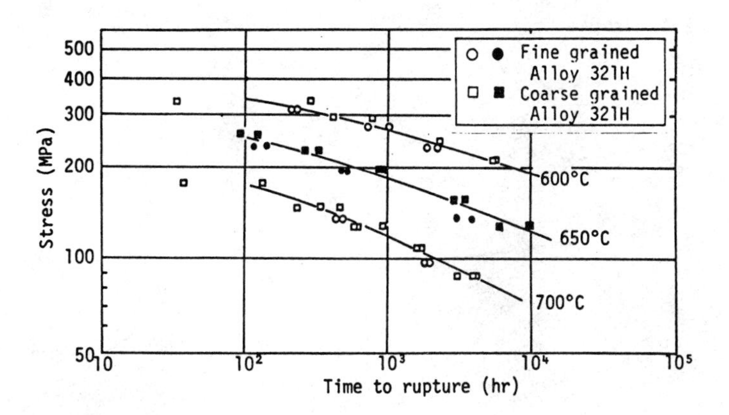

Stress vs time to rupture for fine- and coarse-grained type 321H stainless steel.

Source: H. Teranishi, K. Yoshikawa, and Y. Sawargi, "Application of Fine-Grained Austenitic Steel Tubes to Superheaters in Power Boiler," in Creep (Proceedings of the International Conference in Tokyo), Japan Society of Mechanical Engineers, Tokyo, 1986, p 235

Type 347 Stainless Steel: Annealed vs Cold Worked

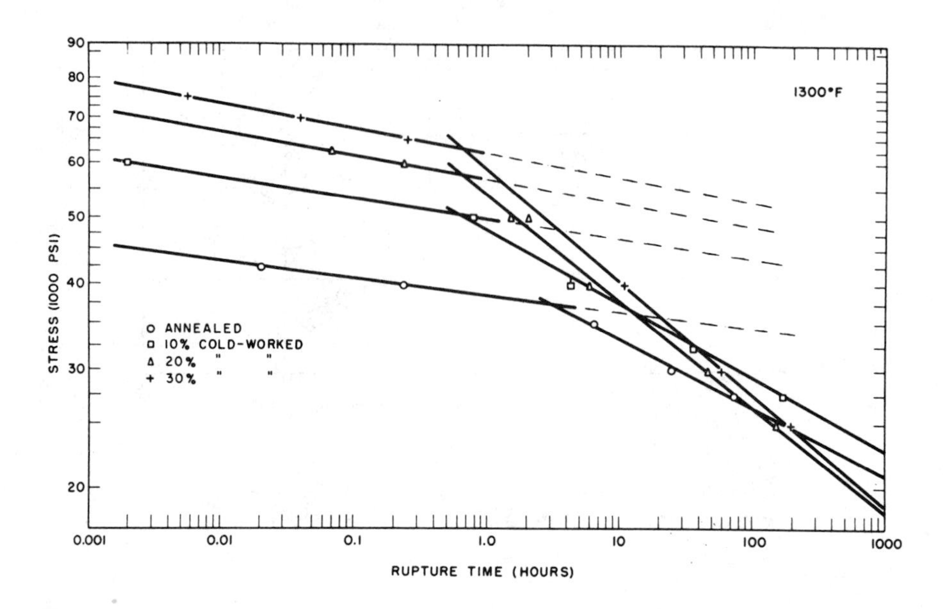

Log stress vs log rupture-time for annealed and cold-worked type 347 stainless steel at 1300 °F. Dashed curves indicate extrapolated strength values if grain boundary shear did not occur.

When examining the creep-rupture behavior of a conventional, wrought, solid-solution alloy in both the annealed and cold-worked states at elevated temperatures, the potential role of cold working can be seen in terms of strengthening metals at elevated temperatures, provided the increased strength can be retained in long-time tests. The figure above is a log stress vs log rupture-time plot for type 347 stainless steel at 1300 °F. About 30% cold work increases the short-time (less than 1 h) rupture stress by almost 2-to-1. These short-time tests fail in a ductile, transgranular manner. Unfortunately, at 1300 °F and above, grain boundary sliding and migration as well as recrystallization occur in the longer time tests, and marked weakening of the alloy results. Increasing cold work speeds up the weakening process and results in greater weakening in the longer time tests. If the flat slope obtained in short-time tests could be preserved, unusual strength values would be obtained at 100- and 1000-h life, as indicated by the dashed curves.

Source: N.J. Grant, "Dispersed Phase Strengthening," in The Strengthening of Metals, D. Peckner, Ed., Reinhold, New York, 1964, p 184-185

Type 347H Stainless Steel: Rupture Ductilities of Fine- vs Coarse-Grained Material

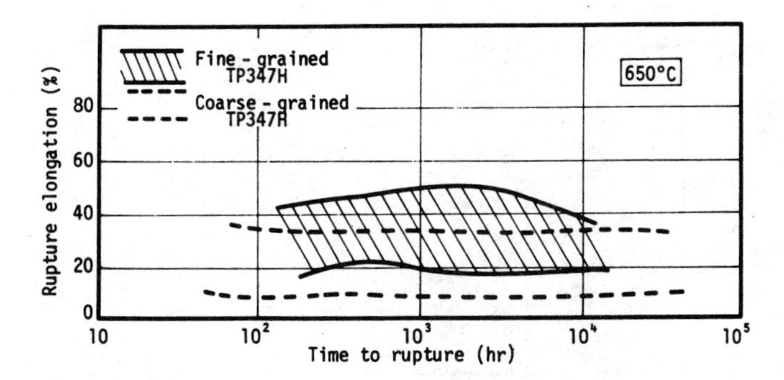

Creep-rupture ductility of fine- and coarse-grained type 347H stainless steel.

Source: H. Teranishi, K. Yoshikawa, and Y. Sawargi, "Application of Fine-Grained Austenitic Steel Tubes to Superheaters in Power Boiler," in Creep (Proceedings of the International Conference in Tokyo), Japan Society of Mechanical Engineers, Tokyo, 1986, p 235

Type 347H Stainless Steel: Stress-Rupture Properties of Fine- vs Coarse-Grained Material

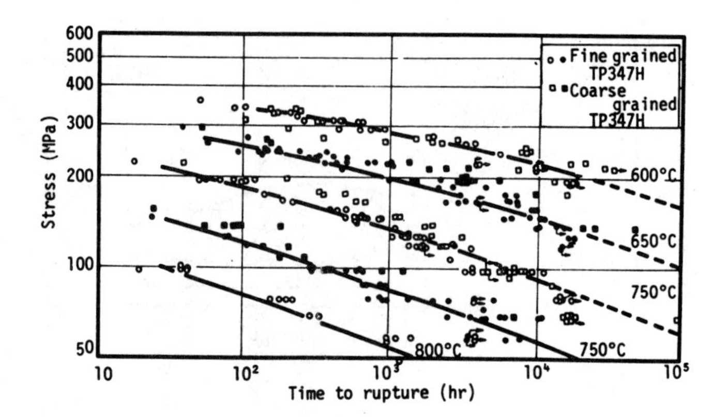

Stress vs time to rupture for fine- and coarse-grained type 347H stainless steel.

Source: H. Teranishi, K. Yoshikawa, and Y. Sawargi, "Application of Fine-Grained Austenitic Steel Tubes to Superheaters in Power Boiler," in Creep (Proceedings of the International Conference in Tokyo), Japan Society of Mechanical Engineers, Tokyo, 1986, p 235

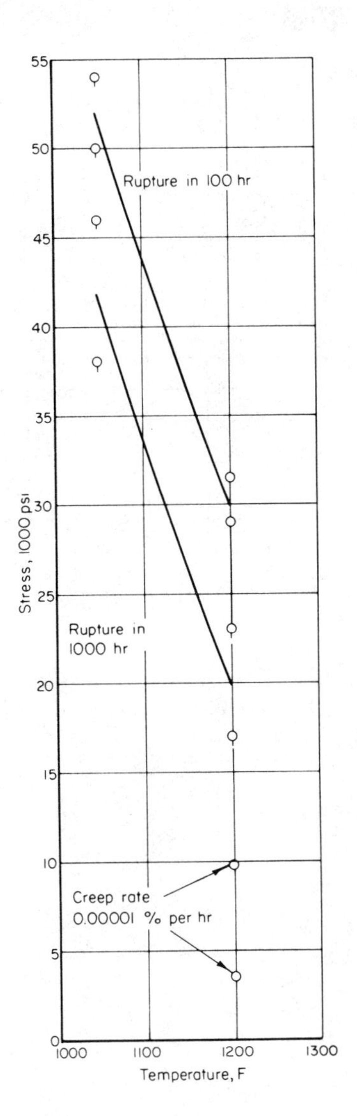

Creep and stress-rupture properties of type 347L stainless steel hot forged bar.

Source: The Elevated-Temperature Properties of Stainless Steels, ASTM Data Series DS 5-S1, American Society for Testing and Materials, Philadelphia, 1965, p 55

Types 302 vs Modified 302 (18-12-Cb): Comparison of Rupture Strengths, as Influenced by Grain Size

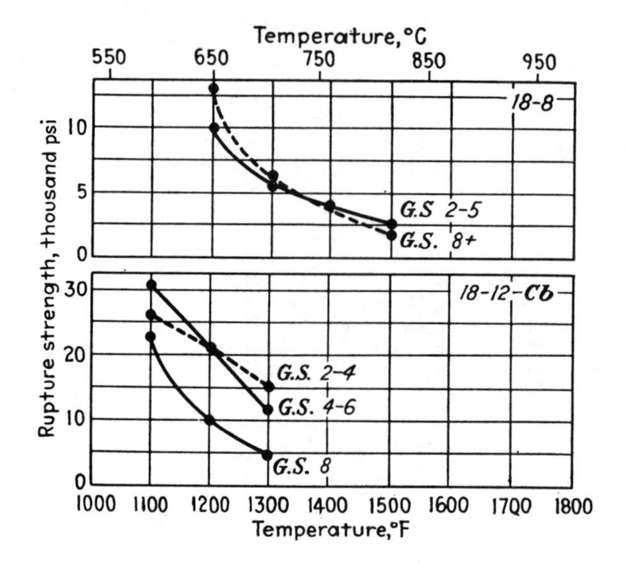

Effect of grain size on the rupture strength of austenitic steels at 1300 °F (705 °C) (10,000 h).

Source: R.A. Grange, F.J. Shortsleeve, D.C. Hilty, and W.O. Binder, et. al., Boron, Calcium, Columbium, and Zirconium in Iron and Steel, John Wiley & Sons, New York, 1957, p 271

Types 304 and 316 Stainless Steels: Comparison of Creep Strengths, as Influenced by Molybdenum

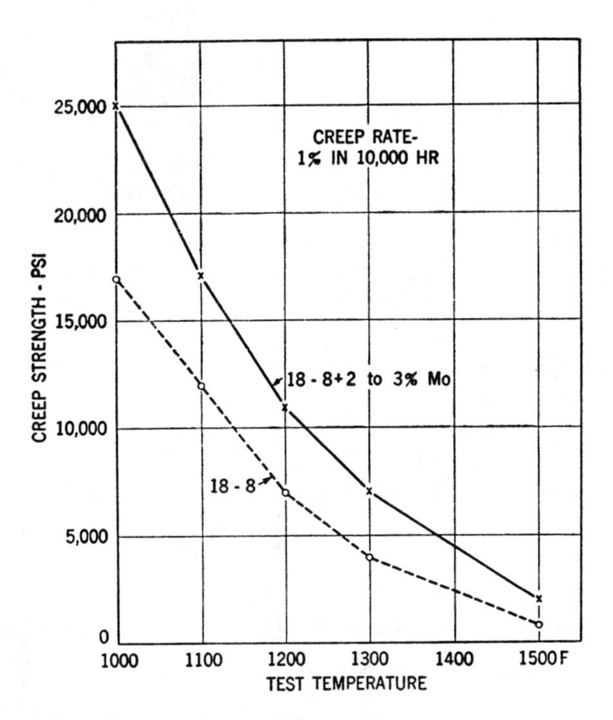

Effect of molybdenum on the creep strengths of types 304 and 316 stainless steels.

Source: R.S. Archer, J.Z. Briggs, and C.M. Loeb, Jr., Molybdenum Steels, Irons, Alloys, Climax Molybdenum Co., New York, 1948, p 11

Types 304 and 316 Stainless Steels: Comparison of Stress-Rupture Strengths

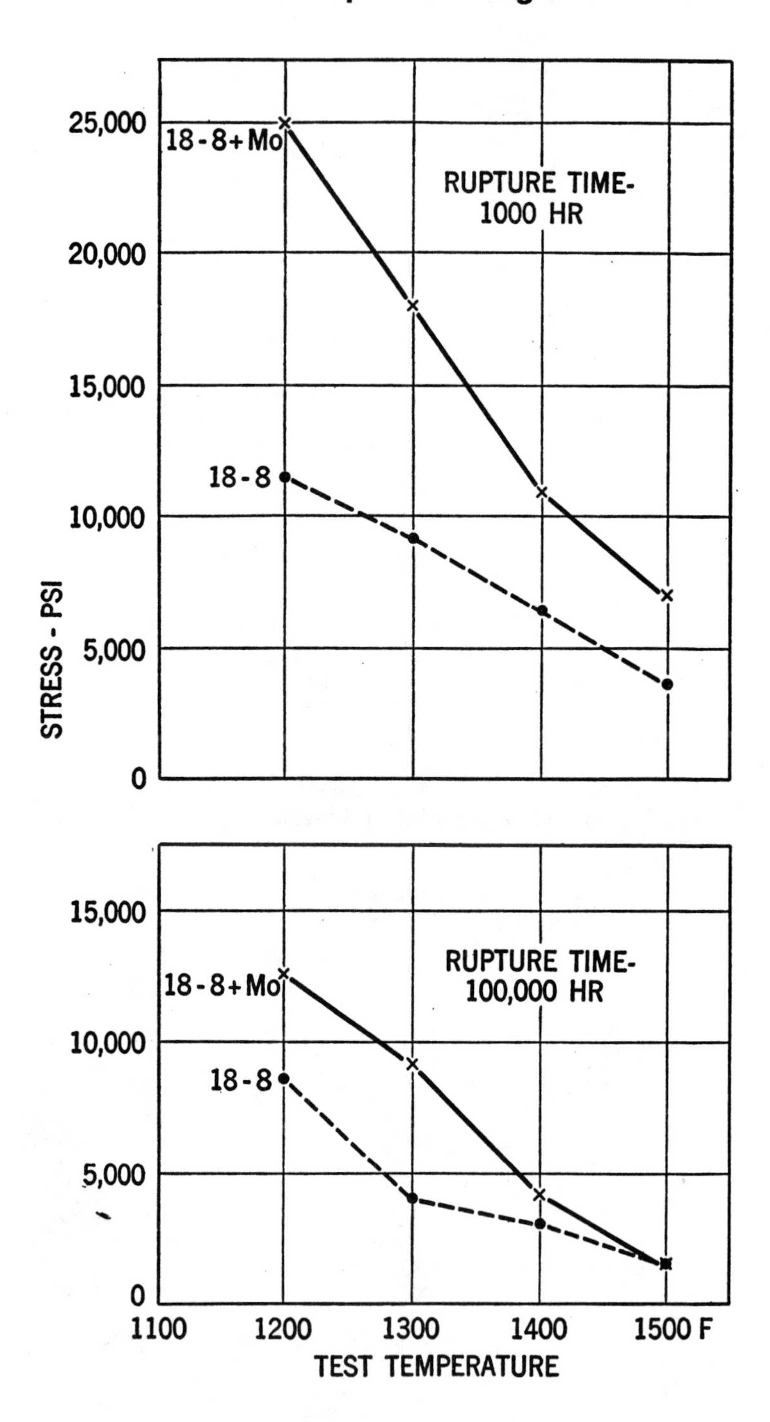

Stress-rupture strengths of types 304 and 316 stainless steels.

Source: R.S. Archer, J.Z. Briggs, and C.M. Loeb, Jr., Molybdenum Steels, Irons, Alloys, Climax Molybdenum Co., New York, 1948, p 165

Type 304 vs Cr-Mn-N Stainless Steel: Comparison of 10,000-h Rupture Strengths

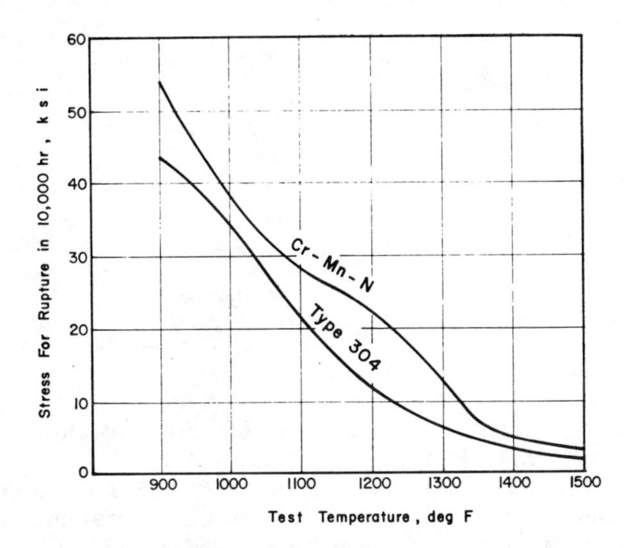

Stress for rupture in 10,000 h for type 304 and Cr-Mn-N austenitic stainless steels.

The creep strength and the creep-rupture strength of Cr-Mn-N stainless steel are also superior to those of the normal Cr-Ni stainless steels. Comparison of rupture strength in 10,000 h is shown in the figure above.

Source: J.J. Heger, "Mechanical Properties and Corrosion Resistance of a High-Strength Chromium-Manganese-Nitrogen Stainless Steel," in Advances in the Technology of Stainless Steels and Related Alloys, ASTM Special Technical Publication No. 369, American Society for Testing and Materials, Philadelphia, 1965, p 58-59

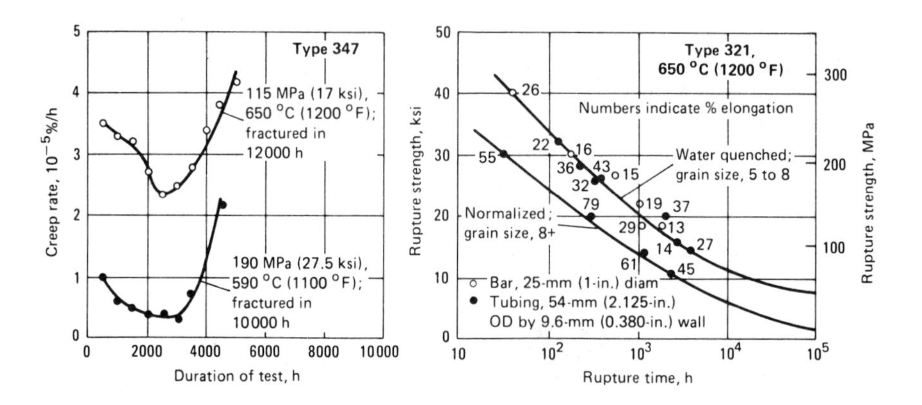

Creep and rupture characteristics of two austenitic stainless steels. (Left) Effect of time on creep rate of type 347 stainless steel (ASTM STP 124). (Right) Rupture characteristics of type 321 stainless steel, cold worked and solution treated. Specimens indicated by lower curve were normalized at 955 °C (1750 °F), which resulted in grain sizes of ASTM No. 8 or finer. Specimens water quenched from between 1040 and 1120 °C (1900 and 2050 °F) exhibited grain sizes of ASTM No. 5 to 8.

Stresses allowed by the ASME Boiler Code are determined by, and vary considerably with, the type of steel. Allowable stresses for two heat-resistant steels are compared with those for carbon steel. Time has a greater effect on creep rate than on rupture strength for type 347 steel at 590 and 650 °C (1100 and 1200 °F), as shown in the left figure. The fine-grain tubes were solution treated at 950 °C (1750 °F) and cooled in air, whereas the coarser-grain tubes were solution treated at 1040 °C (1900 °F) or higher and water quenched. The rupture characteristics of both fine-grain and coarse-grain tubes are shown in the right figure.

Types 347 and 321 Stainless Steel: A Comparison (Continued)

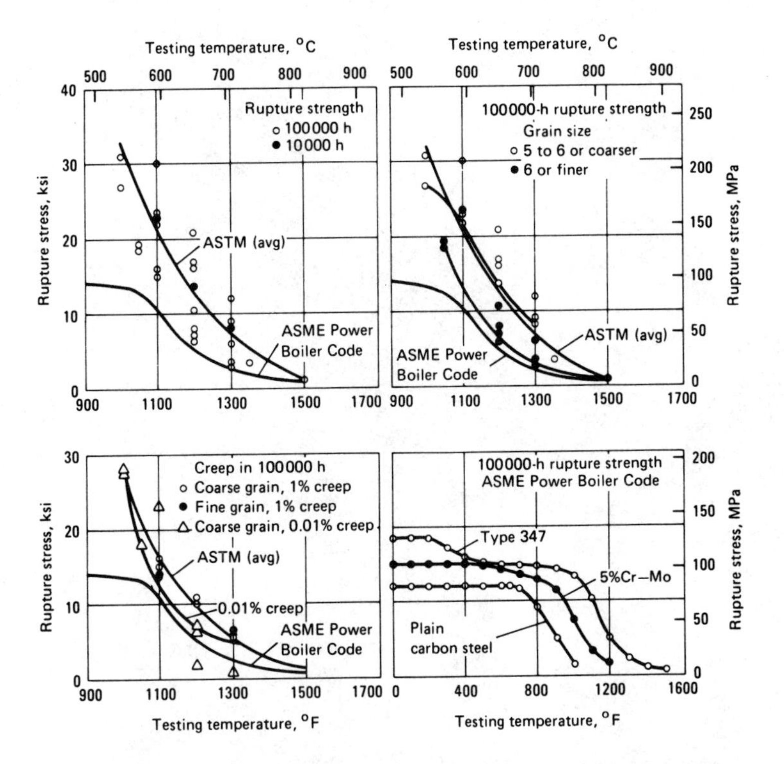

Rupture strength and creep properties of type 347 stainless steel compared with stresses permitted by the ASME Power Boiler Code.

Source: Metals Handbook, Ninth Edition, Volume 3, Properties and Selection: Stainless Steels, Tool Materials and Special-Purpose Metals, American Society for Metals, Metals Park, OH, 1980, p 193

12

Ferritic Stainless Steels

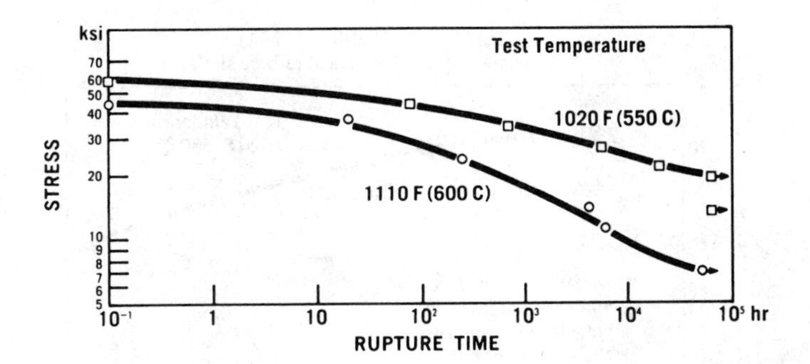

Stress vs rupture time for a class II (×15CrMoWV 12 1) nonstandard ferritic stainless steel, oil quenched and tempered at 1290 °F (700 °C) and aged at 1380 to 1435 °F (750 to 780 °C).

Source: J.Z. Briggs and T.D. Parker, "The Super 12% Cr Steels," in Source Book on Materials for Elevated-Temperature Applications, E.F. Bradley, Ed., American Society for Metals, Metals Park, OH, 1979, p 147

12.2 Class II (×20CrMoWV 12 1) Ferritic Stainless Steel

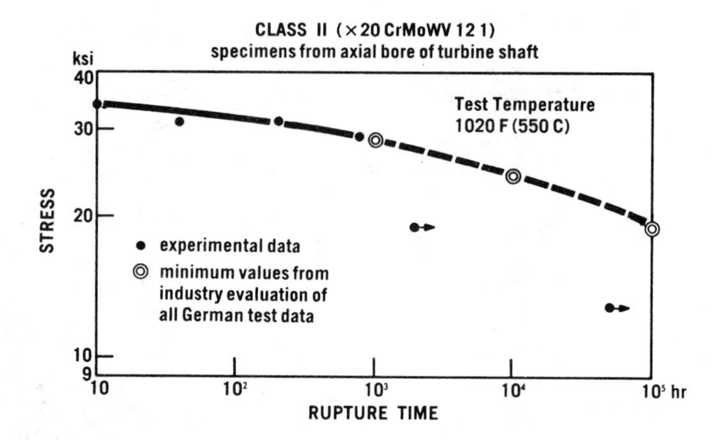

Stress-rupture data for specimens taken from the axial bore of a turbine shaft.

Source: J.Z. Briggs and T.D. Parker, "The Super 12% Cr Steels," in Source Book on Materials for Elevated-Temperature Applications, E.F. Bradley, Ed., American Society for Metals, Metals Park, OH, 1979, p 156

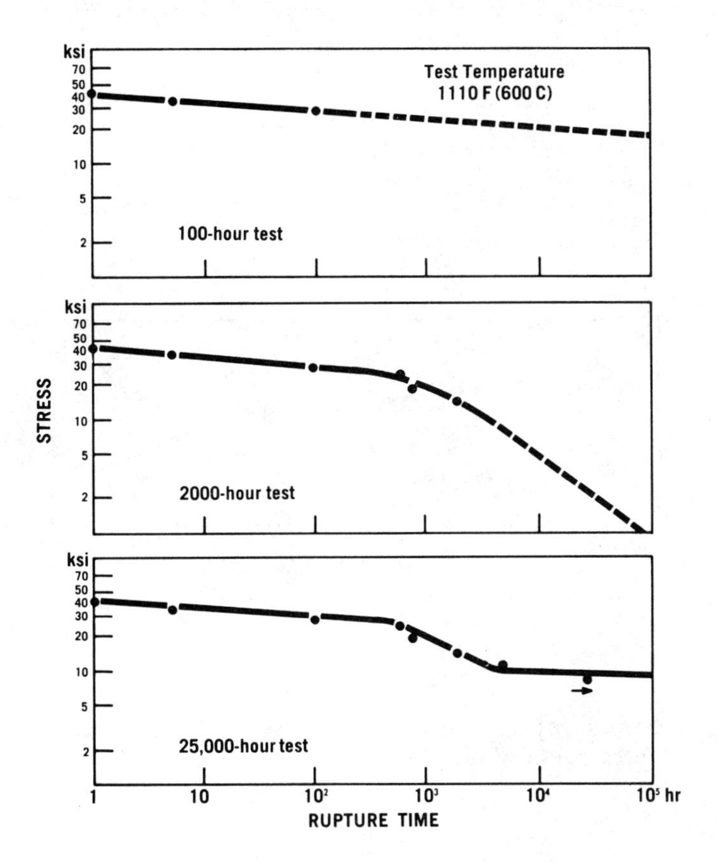

Stress vs rupture time for a class II (×20CrMoWV 12 1) nonstandard ferritic stainless steel heat treated to a tensile strength of about 128 ksi.

Source: J.Z. Briggs and T.D. Parker, "The Super 12% Cr Steels," in Source Book on Materials for Elevated-Temperature Applications, E.F. Bradley, Ed., American Society for Metals, Metals Park, OH, 1979, p 147

12.4 Class II (×20CrMoWV 12 1) Ferritic Stainless Steel: Stress-Rupture Time for Weld Metal, Weldment, and Parent Metal

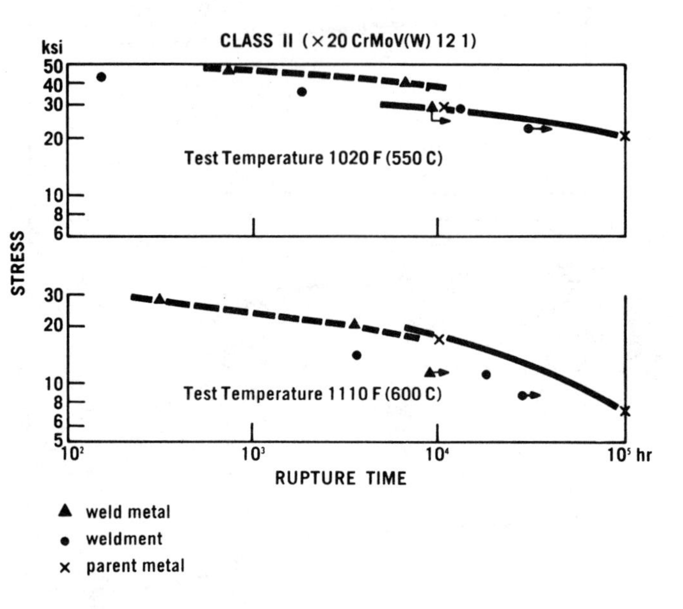

Stress-rupture data for a class II nonstandard ferritic stainless steel.

Source: J.Z. Briggs and T.D. Parker, "The Super 12% Cr Steels," in Source Book on Materials for Elevated-Temperature Applications, E.F. Bradley, Ed., American Society for Metals, Metals Park, OH, 1979, p 156

Type 422 Stainless Steel

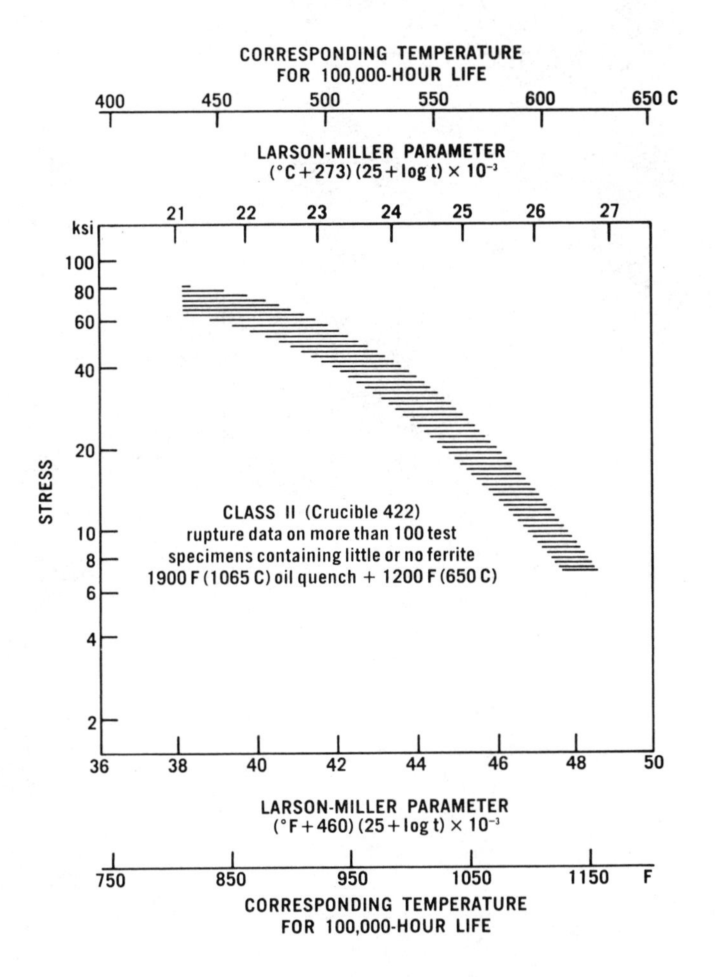

Stress-rupture data for type 422 stainless steel.

Source: J.Z. Briggs and T.D. Parker, "The Super 12% Cr Steels," in Source Book on Materials for Elevated-Temperature Applications, E.F. Bradley, Ed., American Society for Metals, Metals Park, OH, 1979, p 151

Type 422 Stainless Steel

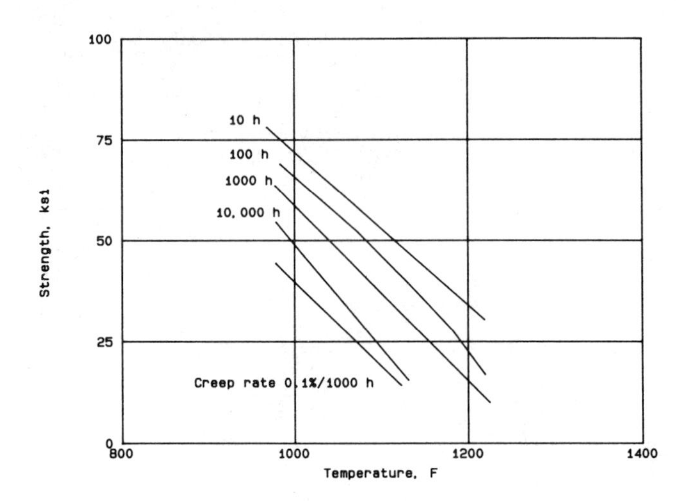

Creep and stress-rupture properties of bar. Specimens were tempered at 1200 °F to a room-temperature tensile strength of 150 ksi.

Source: High-Temperature Property Data: Ferrous Alloys, M.F. Rothman, Ed., ASM International, Metals Park, OH, 1988, p 9.90

Type 430 Stainless Steel

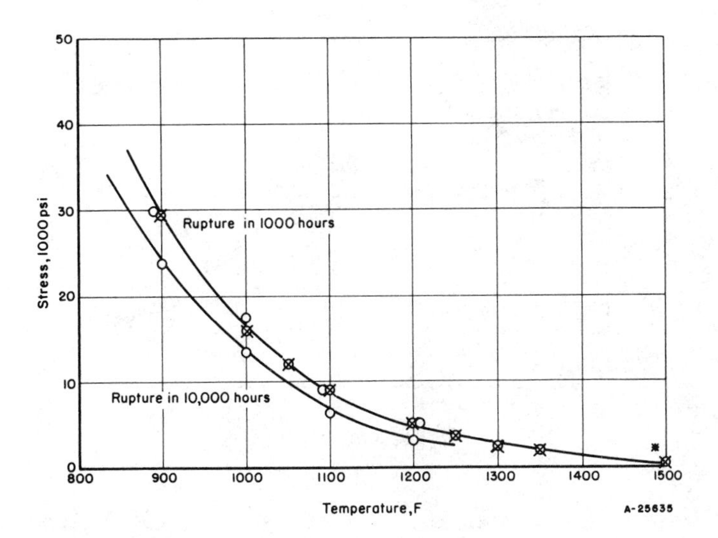

Rupture strength of type 430 stainless steel.

Source: W.F. Simmons and H.C. Cross, Report on Elevated-Temperature Properties of Chromium Steels, ASTM Special Technical Publication No. 228, American Society for Testing and Materials, Philadelphia, 1958, p 94

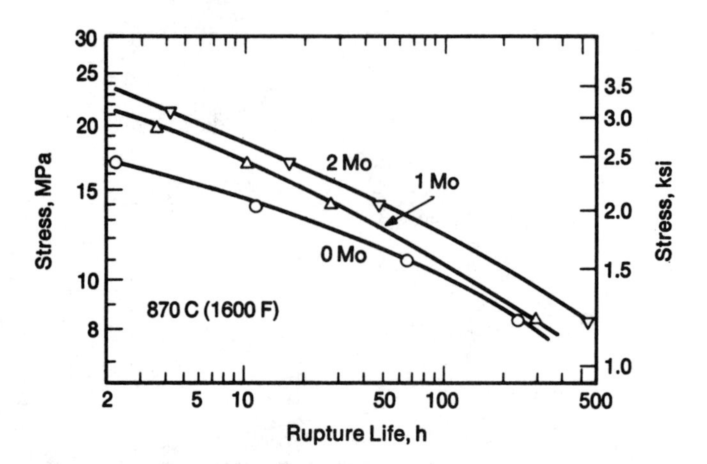

Creep-rupture life of low-interstitial 18Cr ferritic stainless steels with 0, 1, and 2% molybdenum. Prior to creep testing at 870 °C (1600 °F), each steel was recrystallized at 1095 °C (2000 °F) for 5 min and water quenched.

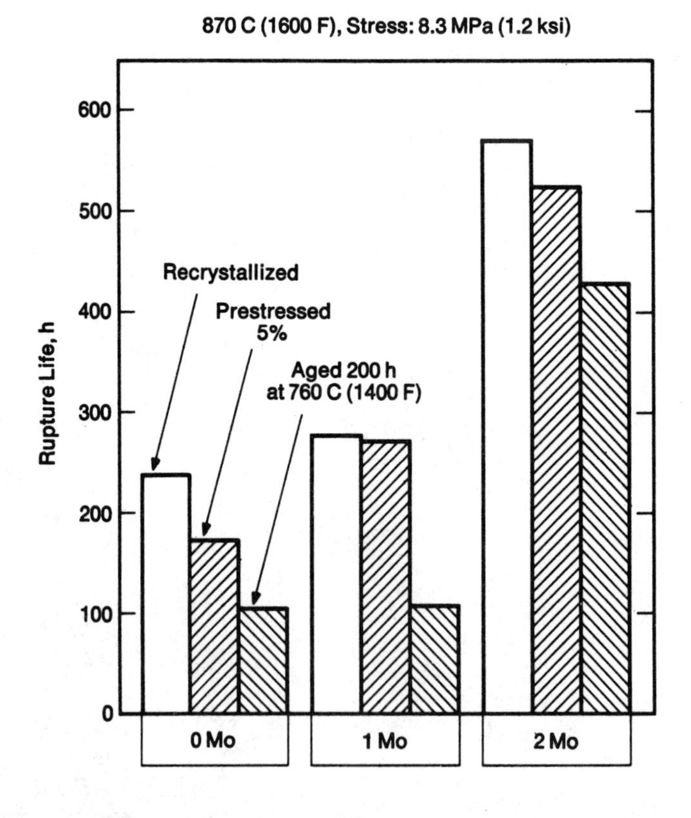

Creep-rupture life of low-interstitial 18Cr ferritic stainless steels with 0, 1, and 2% molybdenum at 870 °C (1600 °F) and 8.3 MPa (1.2 ksi).

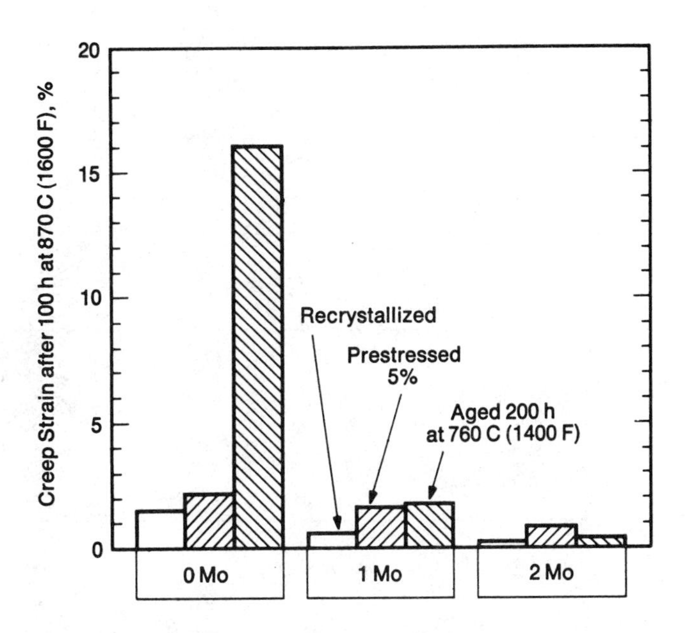

Total creep strain for low-interstitial 18Cr ferritic stainless steels with 0, 1, and 2% molybdenum after 100 h at 870 °C (1600 °F) and 8.3 MPa (1.2 ksi).

Mo additions had the predictable effect of increasing yield stress and tensile strength and slightly lowering ductility. Creep-rupture life at 870 °C (1600 °F) was determined for each steel in the recrystallized condition, and the results are depicted above. Each specimen was annealed at 1095 °C (2000 °F) for 5 min and water quenched. Increasing the Mo content improved the rupture life at all stress levels tested.

The creep-rupture life at 870 °C and 8.3 MPa (1600 °F, 1.2 ksi) was determined for each steel in the following conditions: (a) annealed at 1095 °C (2000 °F) for 5 min, water quenched; (b) annealed as in (a) + prestrained 5%; (c) annealed as in (a) + aged at 760 °C (1400 °F) for 200 h, water quenched.

The 5% prestrain was chosen to simulate the plastic deformation that occurs in the fabrication of the catalyst bead support plate. The 760 °C (1400 °F) aging treatment corresponds to the normal operating conditions of a catalytic converter; this heat treatment produced extensive precipitation of Laves phase. In all cases, increasing the Mo content improved the creep properties. Total creep strain after 100 h at 870 °C, 8.3 MPa (1600 °F, 1.2 ksi) is shown for each steel in all three conditions. The Mo-free steel after the 200-h aging treatment exhibited a strain of more than 16%, whereas the 18Cr-2Mo steel had less than 1% creep.

Source: R.K. Pitler, "Stainless Steels for Environmental Control Systems and Special Process Streams," in Alloys for the Eighties, R.Q. Barr, Ed., Climax Molybdenum Co., Greenwich, CT, 1981, p 327

Type 446 Stainless Steel: Rupture Strength

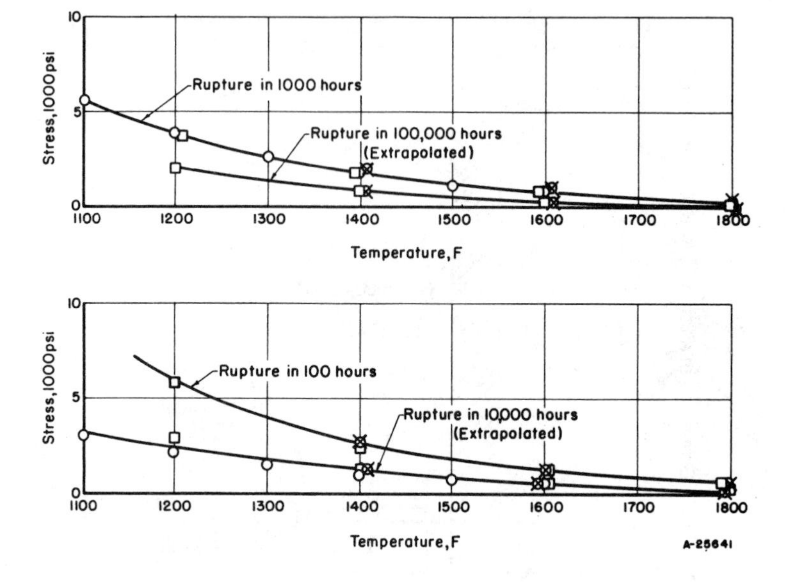

Rupture strength of type 446 stainless steel.

Source: W.F. Simmons and H.C. Cross, Elevated-Temperature Properties of Chromium Steels, ASTM Special Technical Publication No. 228, American Society for Testing Materials, Philadelphia, 1958, p 103

Type 446 Stainless Steel: Stress-Rupture Data

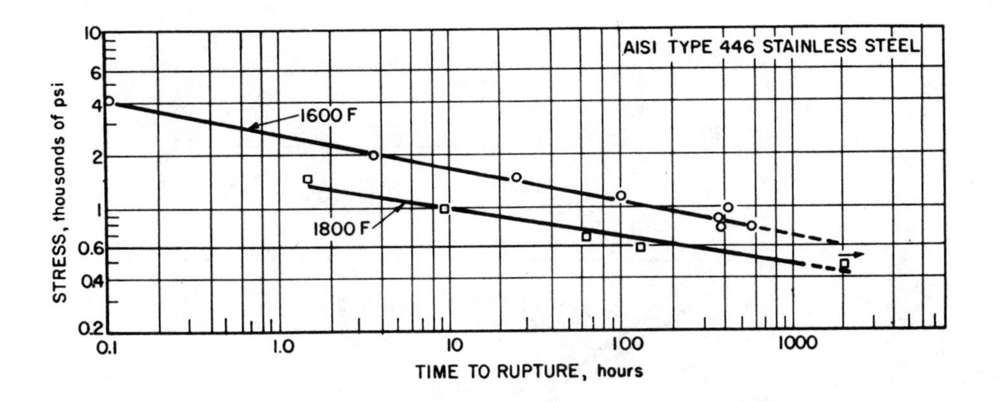

Relationship between stress and time to rupture for type 446 stainless steel.

Source: K.G. Brickner, G.A. Ratz, and R.F. Domagala, "Creep-Rupture Properties of Stainless Steels at 1600, 1800, and 2000 F," in Advances in the Technology of Stainless Steels and Related Alloys, ASTM Special Technical Publication No. 369, American Society for Testing and Materials, Philadelphia, 1965, p 106

Type 422 vs AM-350 and 17-7 PH Stainless Steels: A Comparison

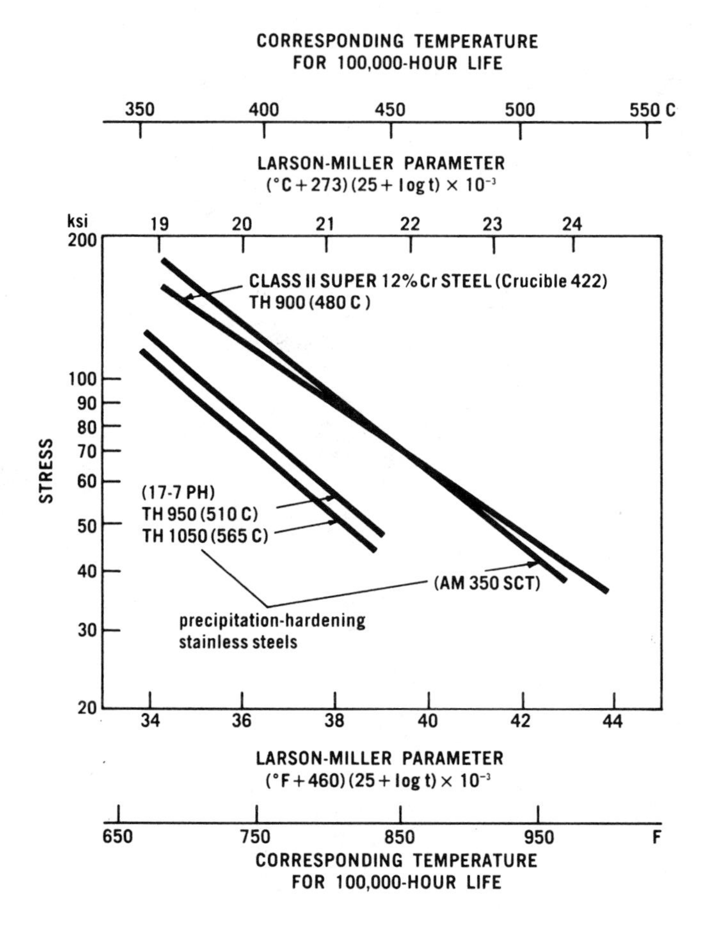

Stress-rupture data for type 422 martensitic stainless steel and two precipitation-hardening stainless steels.

Source: J.Z. Briggs and T.D. Parker, "The Super 12% Cr Steels," in Source Book on Materials for Elevated-Temperature Applications, E.F. Bradley, Ed., American Society for Metals, Metals Park, OH, 1979, p 154

Type 422 Stainless Steel vs Low-Alloy Steel and Discaloy: A Comparison

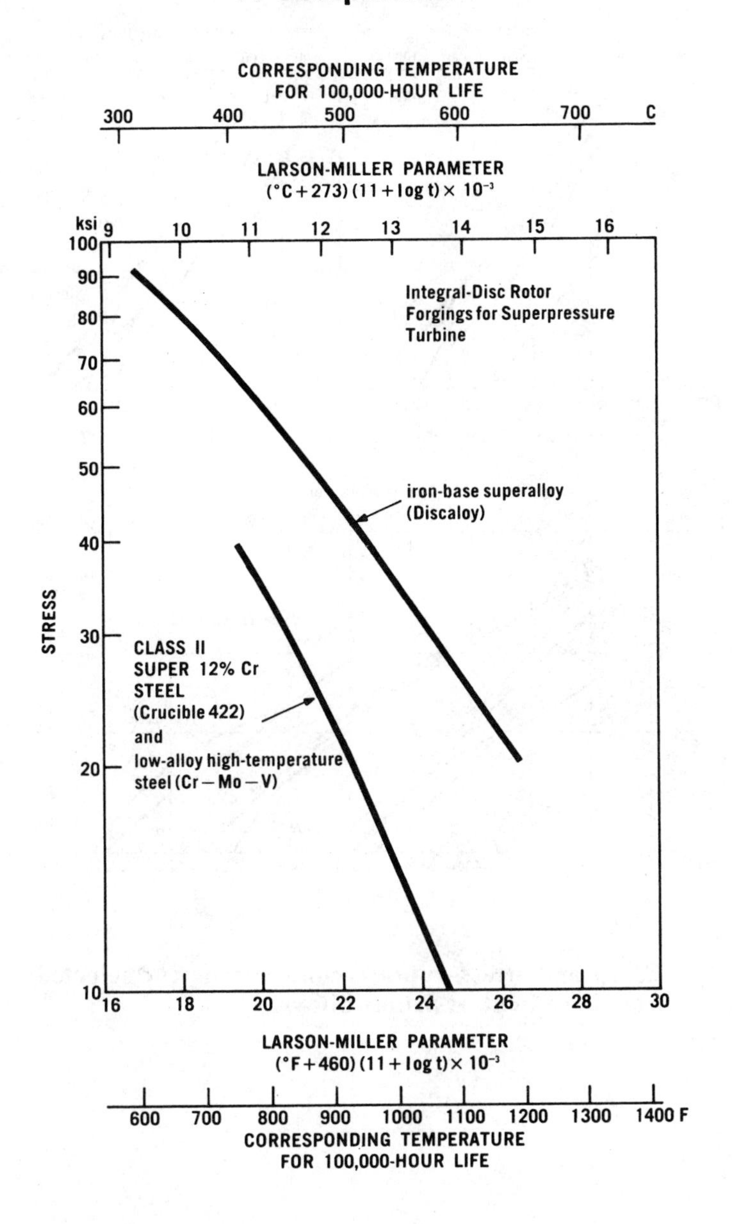

Stress-rupture data for type 422 stainless steel, low-alloy high-temperature steel, and the iron-base superalloy Discaloy.

Source: J.Z. Briggs and T.D. Parker, "The Super 12% Cr Steels," in Source Book on Materials for Elevated-Temperature Applications, E.F. Bradley, Ed., American Society for Metals, Metals Park, OH, 1979, p 156

Types 446, 430, and 316: A Comparison

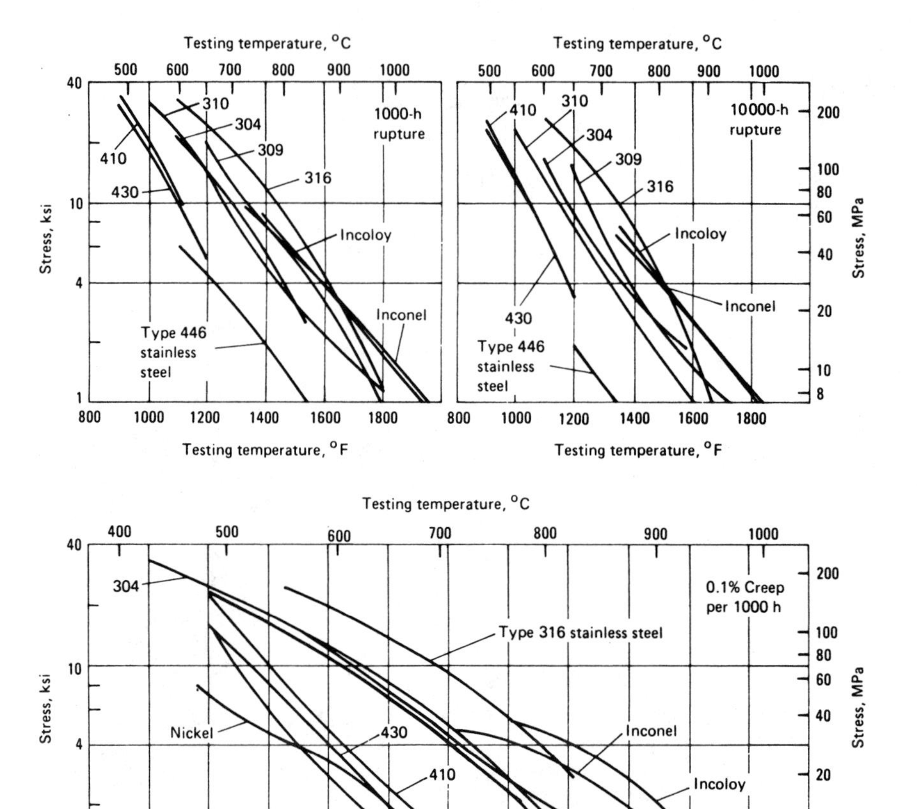

Creep and stress-rupture comparisons for selected iron-base heat-resistant alloys.

Source: Metals Handbook, Ninth Edition, Volume 3, Properties and Selection: Stainless Steels, Tool Materials and Special-Purpose Metals, American Society for Metals, Metals Park, OH, 1980, p 195

13

Martensitic Stainless Steels

Greek Ascoloy: Air-Cooled Bar

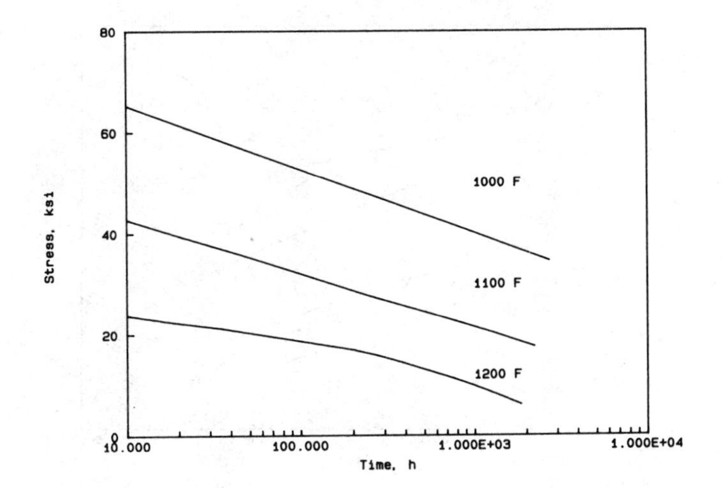

Creep-rupture curves for bar; specimens were heat treated at 1800 °F, air cooled, tempered at 1050 °F for 2 h, and air cooled.

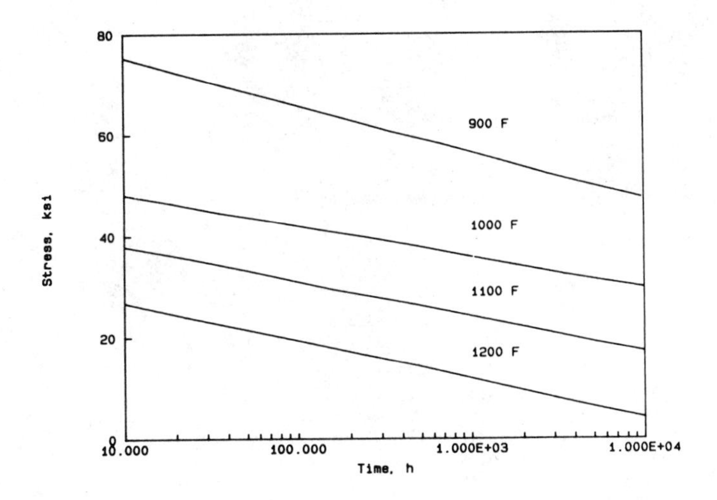

Creep-rupture curves for bar; specimens were heat treated at 1800 °F, air cooled, tempered at 1260 °F for 1$^1/_2$ h, and air cooled.

Source: High-Temperature Property Data: Ferrous Alloys, M.F. Rothman, Ed., ASM International, Metals Park, OH, 1988, p 9.147-148

Greek Ascoloy: Oil-Quenched Bar

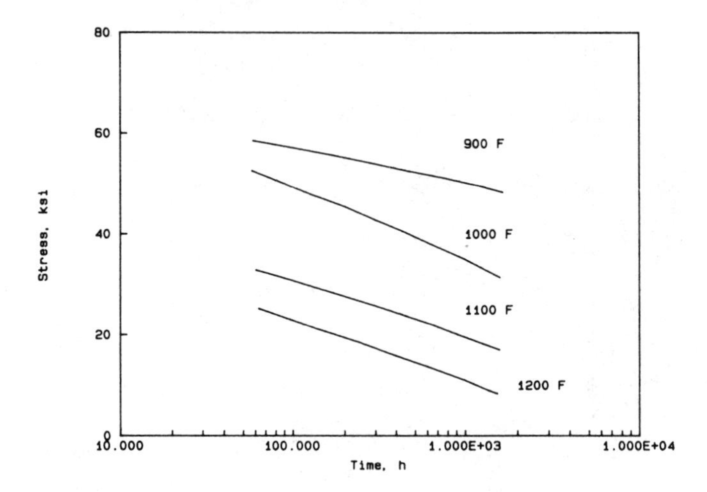

Creep-rupture curves for bar; specimens were heat treated at 1750 °F, oil quenched, tempered at 1200 °F for 1/2 h, and air cooled.

Source: High-Temperature Property Data: Ferrous Alloys, M.F. Rothman, Ed., ASM International, Metals Park, OH, 1988, p 9.147

Type 403 Stainless Steel: Bar

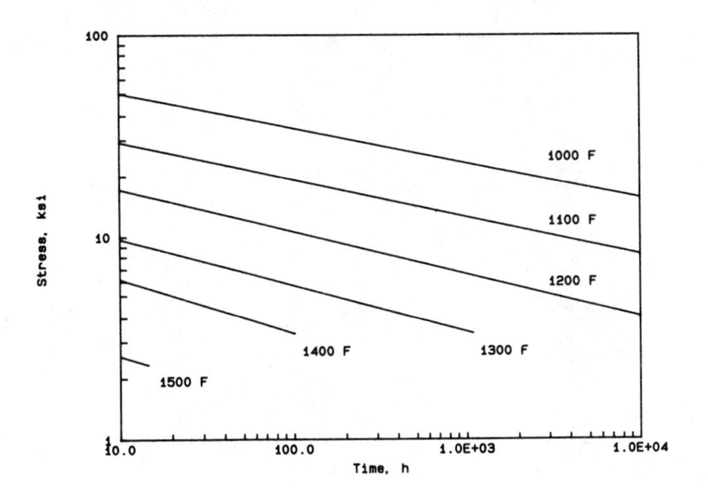

Type 403 (AISI 403) creep-rupture strength of bar at 1000 to 1500 °F. Specimens were heat treated at 1800 °F for 1/2 h, oil quenched, and held at 1000 °F for 2 h.

Source: High-Temperature Property Data: Ferrous Alloys, M.F. Rothman, Ed., ASM International, Metals Park, OH, 1988, p 9.69

Type 410 Stainless Steel: Stress-Rupture Strength

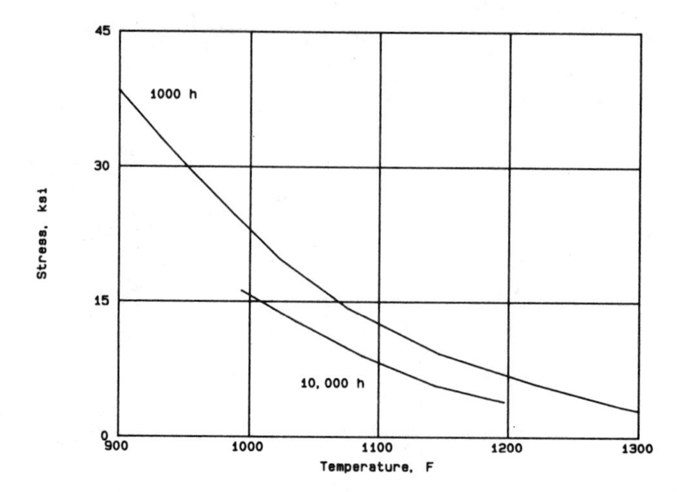

Type 410 (ANSI 410) stress-rupture strength; specimens were heat treated at 1800 °F for 1/2 h, oil quenched, aged at 1200 °F for 2 h, and air cooled.

Type 410 stainless steel: stress-rupture property data for bar

TEMP C	TEMP F	RPSTR MPa	RPSTR ksi	RPT h
Smooth				
316	600	1000	145	100
316	600	1069	155	100
316	600	1138	165	(a)
316	600	1103	160	(a)
371	700	1034	150	100
371	700	1034	150	500
427	800	1000	145	163
427	800	1000	145	8.7
482	900	931	135	(a)
Notched				
316	600	1138	165	100
316	600	1207	175	500
371	700	1138	165	200
371	700	1172	170	1000
427	800	1069	155	20.3
482	900	1000	145	12.5

Note: Room-temperature material hardness was 38 HRC
(a) Failed on loading

Source: High-Temperature Property Data: Ferrous Alloys, M.F. Rothman, Ed., ASM International, Metals Park, OH, 1988, p 9.77

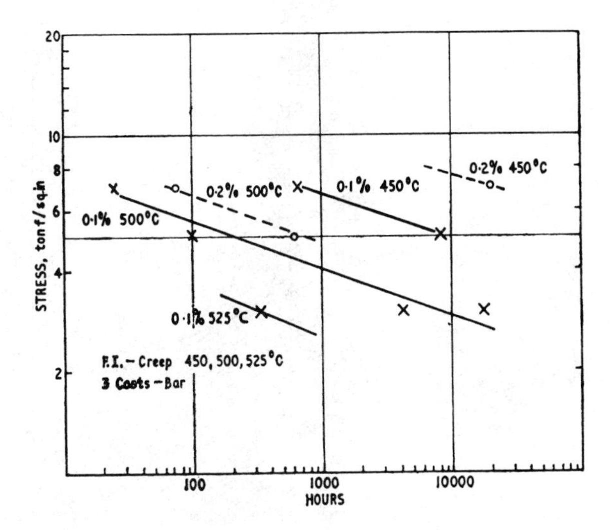

Creep data at 450 and 525 °C for modified type 410 (12Cr-0.1C, FI) stainless steel.

Type 410 (Modified) Stainless Steel: 12Cr-0.1C (FI) (Continued)

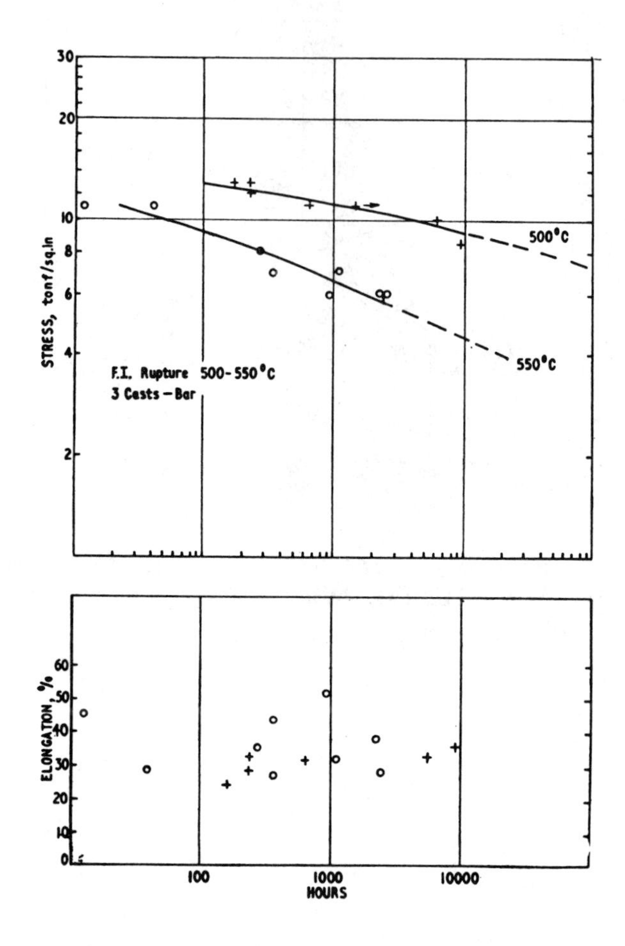

Rupture data at 500 and 550 °C for modified type 410 (12Cr-0.1C, FI) stainless steel.

Source: H.W. Kirkby and R.J. Truman, "12%Cr Steels: Creep-Resisting and High-Strength Variants," in High Temperature Properties of Steels (Proceedings of a Conference in Eastbourne, England), Iron and Steel Institute, London, 1967, p 368

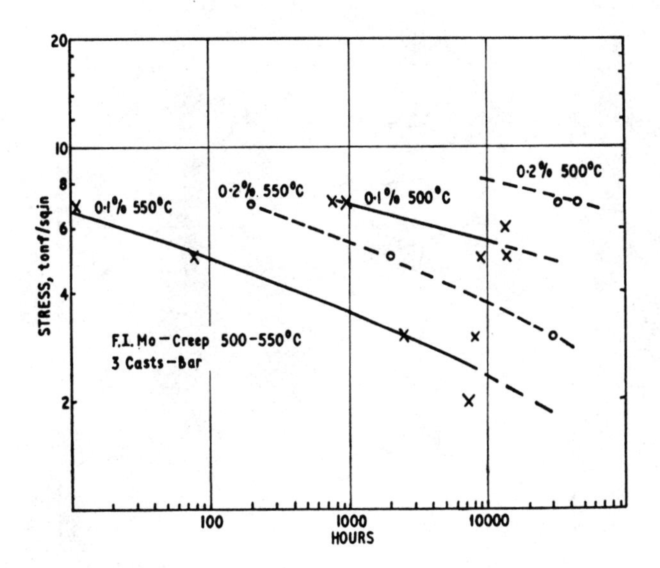

Creep data at 500 and 550 °C for modified type 410 (12Cr-Mo, FIMo) stainless steel.

Type 410 (Modified) Stainless Steel: 12Cr-Mo (FIMo) (Continued)

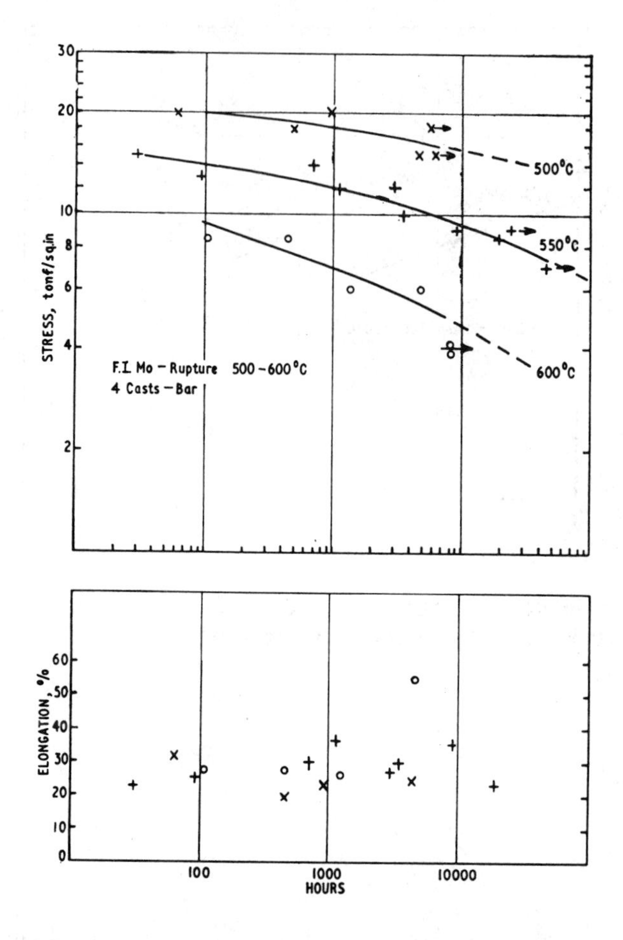

Rupture data at 500, 550, and 600 °C for modified type 410 (12Cr-Mo, FIMo) stainless steel.

Source: H.W. Kirkby and R.J. Truman, "12%Cr Steels: Creep-Resisting and High-Strength Variants," in High Temperature Properties of Steels (Proceedings of a Conference in Eastbourne, England), Iron and Steel Institute, London, 1967, p 369

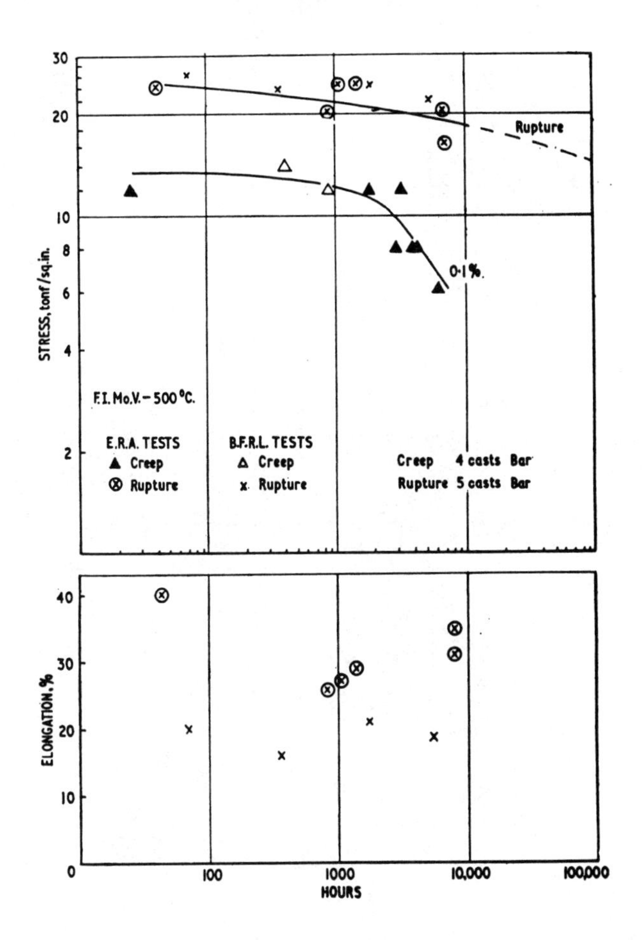

Creep and rupture data at 500 °C for modified type 410 (12Cr-Mo-V, FIMoV) stainless steel.

Type 410 (Modified) Stainless Steel: 12Cr-Mo-V (FIMoV) (Continued)

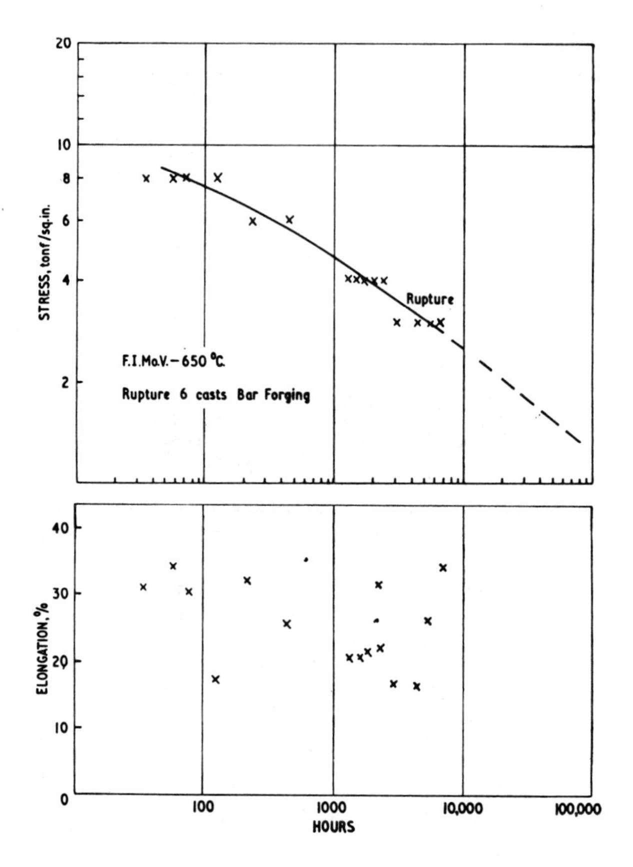

Rupture data at 650 °C for modified type 410 (12Cr-Mo-V, FIMoV) stainless steel.

Source: H.W. Kirkby and R.J. Truman, "12%Cr Steels: Creep-Resisting and High-Strength Variants," in High Temperature Properties of Steels (Proceedings of a Conference in Eastbourne, England), Iron and Steel Institute, London, 1967, p 369

Type 410 (Modified) Stainless Steel: 12Cr-Mo-V-Nb-Co (FV 535) at 450 to 550 °C

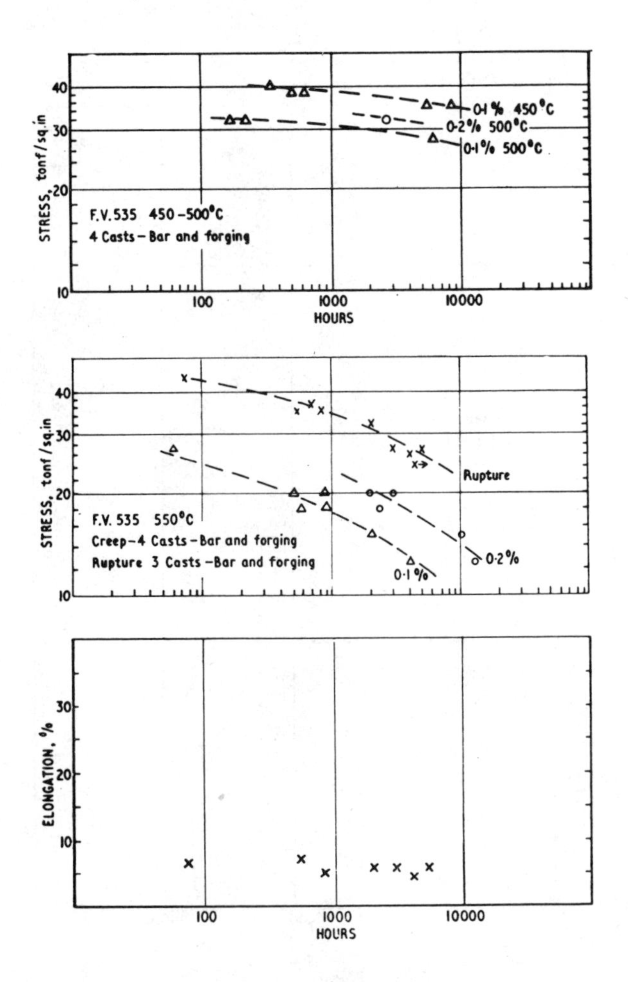

Creep data at 450 and 500 °C (top) and creep and rupture data at 550 °C (bottom) for modified type 410 (12Cr-Mo-V-Nb-Co, FV 535) stainless steel.

Source: H.W. Kirkby and R.J. Truman, "12%Cr Steels: Creep-Resisting and High-Strength Variants," in High Temperature Properties of Steels (Proceedings of a Conference in Eastbourne, England), Iron and Steel Institute, London, 1967, p 363

Type 410 (Modified) Stainless Steel: 12Cr-Mo-V-Nb (FV 448E)

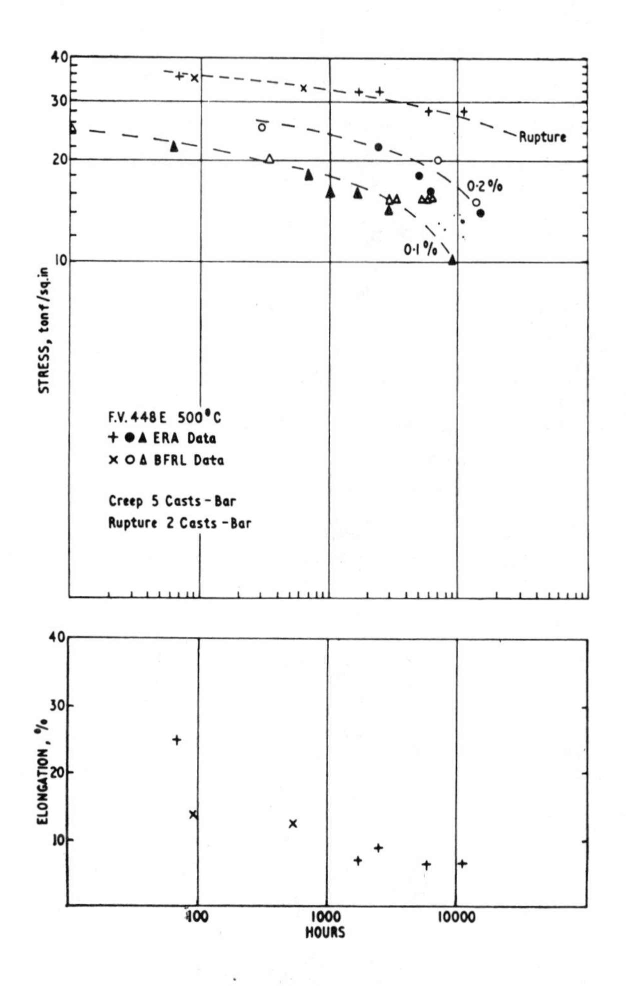

Creep and rupture data at 500 °C for modified type 410 (12Cr-Mo-V-Nb, FV 448E) stainless steel.

Type 410 (Modified) Stainless Steel: 12Cr-Mo-V-Nb (FV 448E) (Continued)

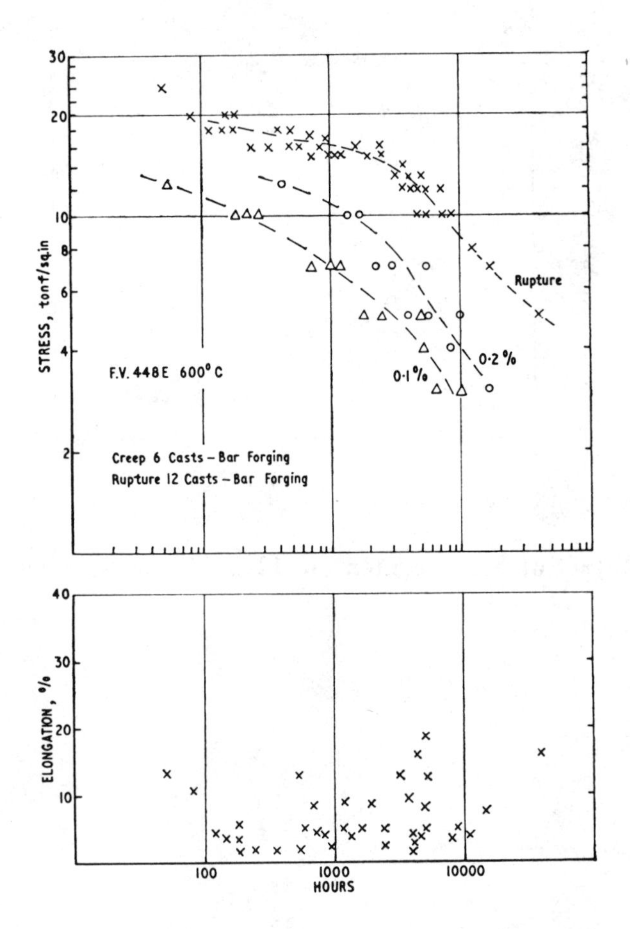

Creep and rupture data at 600 °C for modified type 410 (12Cr-Mo-V-Nb, FV 448E) stainless steel.

Source: H.W. Kirkby and R.J. Truman, "12%Cr Steels: Creep-Resisting and High-Strength Variants," in High Temperature Properties of Steels (Proceedings of a Conference in Eastbourne, England), Iron and Steel Institute, London, 1967, p 371

Type 410 (Modified) Stainless Steel: Effect of Boron Content

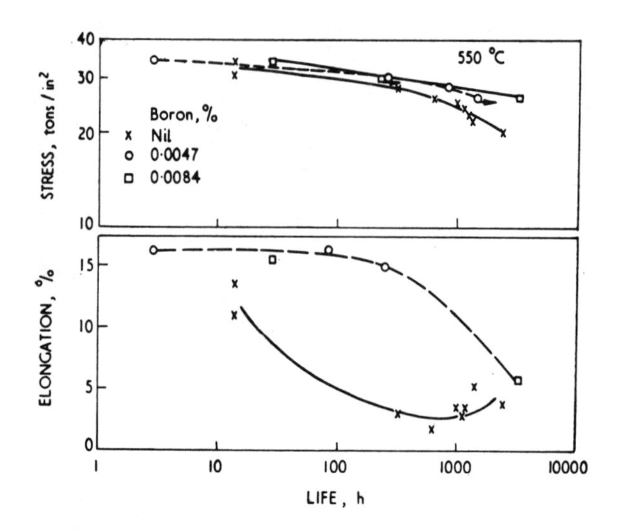

Effect of boron content on 12Cr-Mo-V-Nb steel.

Source: P.G. Stone, The Effect of Carbon, Nitrogen, and Boron on Elevated-Temperature Properties, in High-Temperature Properties of Steels (Proceedings of a Conference in Eastbourne, England), Iron and Steel Institute, London, 1967, p 507

Type 410 (Modified) Stainless Steel: Effect of Hardening Temperature

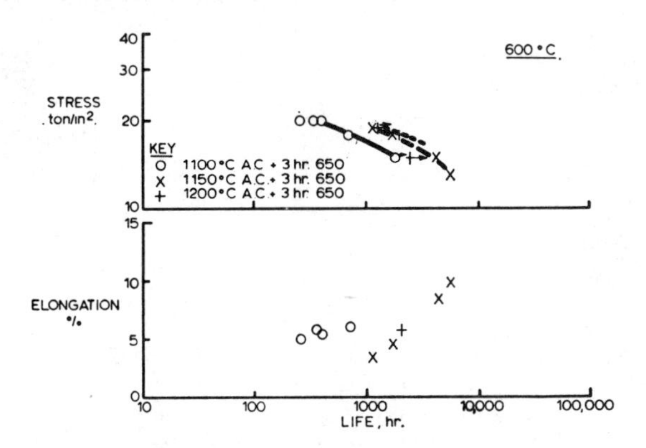

Effect of hardening temperature on stress-rupture properties of modified type 410 (12Cr-Mo-Nb) stainless steel.

It is well known that the rupture ductility of many ferritic steels frequently descends to a minimum value and then increases again at long testing times. This feature was apparently exhibited by the 12Cr-Mo-V-Nb steel used in constant-strain-rate work, as shown by the results obtained in stress-rupture testing carried out on this material.

Source: M.C. Murphy, "Discussion Four: Factors that Affect the Properties," in High Temperature Properties of Steels (Proceedings of a Conference in Eastbourne, England), Iron and Steel Institute, London, 1967, p 525

Type 410 (Modified) Stainless Steel: Effect of Irradiation

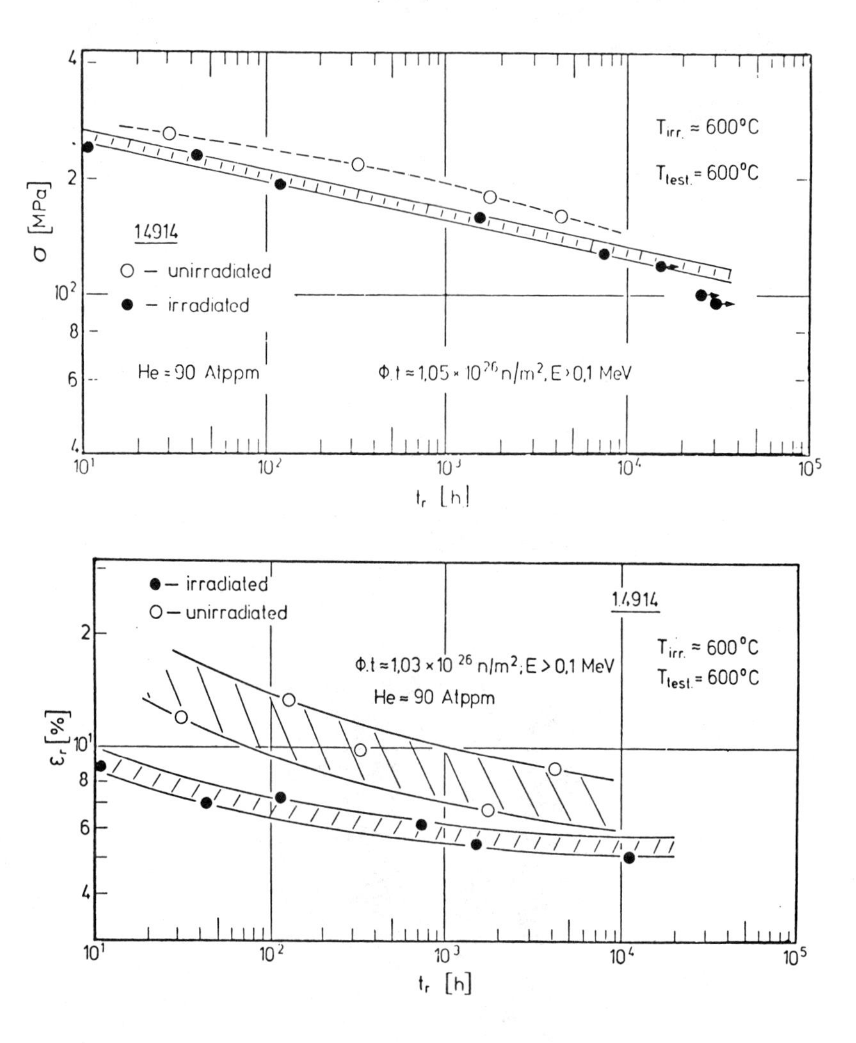

Effects of irradiation on the stress-rupture life (top) and the rupture ductility (bottom) of 1.4914 martensitic stainless steel.

Composition (wt%): 0.21 C, 0.37 Si, 0.50 Mn, 11.20 Cr, 0.42 Ni, 0.83 Mo, 0.21 V. The 600 °C stress-rupture lives and ductilities of 1.4914 steel irradiated at 600 °C in the Mol BR2 reactor to about 5 dpa and a helium concentration of 90 Atppm are shown in the figures above. The small irradiation-induced reductions in rupture lives may be atttributed to the combined effects of increases in minimum creep rates and reductions in rupture strains; nevertheless, the post-irradiation ductilities remained reasonably high at 5-10%.

Source: D.R. Harries, "Ferritic/Martensitic Steels for Use in Near-Term and Commercial Fusion Reactors," in Topical Conference on Ferritic Alloys for Use in Nuclear Energy Technologies (Proceedings of a Conference in Snowbird, UT), J.W. Davis and D.J. Michel, Eds., Metallurgical Society of AIME, Warrendale, PA, 1984, p 148

Type 410 (Modified) Stainless Steel: Effect of Irradiation and Pretreatment Conditions

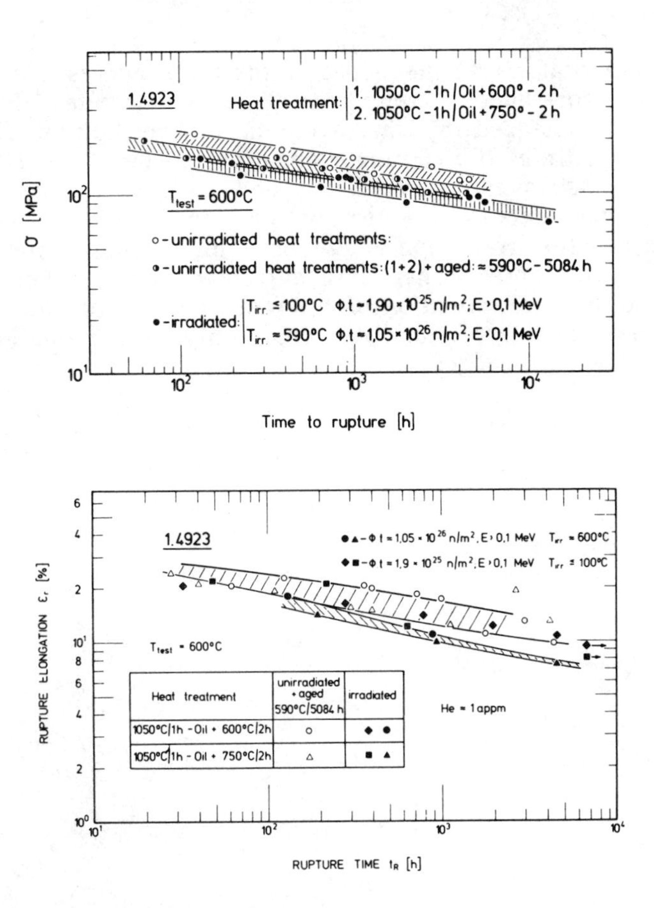

Time to rupture (top) and creep-rupture ductility (bottom) of unirradiated and irradiated martensitic stainless steel 1.4923 at different irradiation temperatures.

The top figure shows the time to rupture of steel 1.4923 for virgin material in two pretreatment conditions, as well as data for thermally aged or irradiated material. Several conclusions can be drawn:

* Changes in the tempering conditions do not affect the rupture time.
* Thermally aged samples, where the annealing time is identical to the irradiation time at 590 °C, are reduced in their rupture time.
* Irradiations at room-temperature or at 590 °C result in the same rupture strength, and t_r are reduced by about a factor of 5 when compared with the unirradiated or unaged specimens.

13.18 Type 410 (Modified) Stainless Steel: Effect of Irradiation and Pretreatment Conditions (Continued)

It is of great importance that the slopes of the three curves in the top figure, represented as error bands, are identical. This indicates that the creep mechanism is not changed by the irradiation. The bottom figure on the preceding page illustrates the rupture elongation of steel 1.4923 for both heat treatment conditions in the unirradiated and irradiated state. The rupture elongation is little affected by irradiation, and there is no significant difference between the irradiation at low or at high temperatures as well as at low or high neutron doses. This is in agreement with earlier experimental observations. A closer examination of the creep curves for equivalent rupture lives showed that the loss of ductility occurs only by a reduction in the last part of the tertiary creep stage.

Source: C. Wassilew, K. Herschbach, E. Materna-Morris, and K. Ehrlich, "Irradiation Behavior of 12% Cr Martensitic Steels," in Topical Conference on Ferritic Alloys for Use in Nuclear Energy Technologies (Proceedings of a Conference in Snowbird, UT), J.W. Davis and D.J. Michel, Eds., Metallurgical Society of AIME, Warrendale, PA, 1984, p 610

Type 410 (Modified) Stainless Steel: Effect of Nitrogen on 12Cr-2Mo Steel

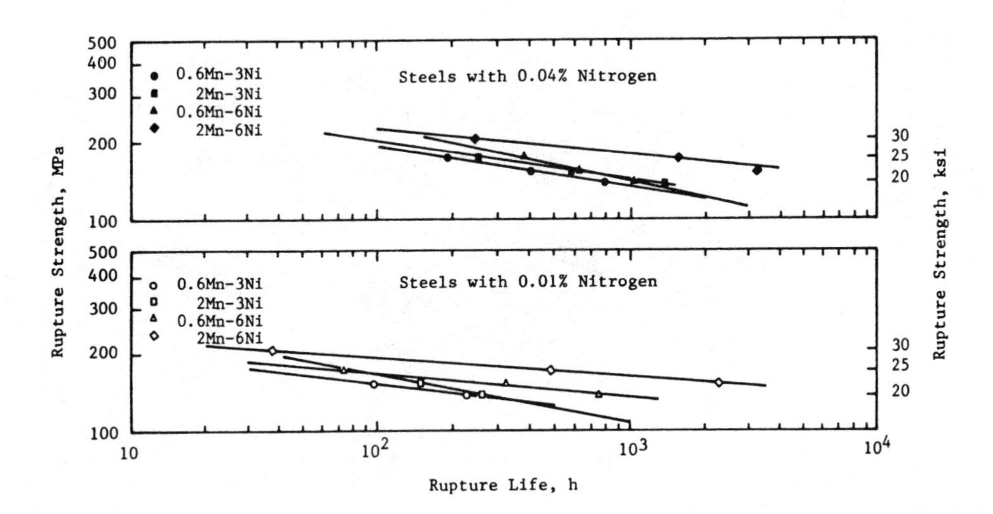

Rupture strength of modified type 410 (12Cr-2Mo) stainless steels at 649 °C (1200 °F).

This 12Cr-2Mo steel has excellent creep-rupture characteristics. Its outstanding weldability results from its low C content, nominally 0.075%. Its high creep resistance is due to its martensitic microstructure, which is strengthened with dispersed austenite. In addition to 12% Cr and 2% Mo, the steel contains 0.6% Mn, 6% Ni, 0.25% V, 0.1% Nb, and 0.04% N. The tempering response is essentially flat for a wide range of tempering conditions. When tempered for 1 h at 700 °C (1290 °F), the steel exhibits room-temperature yield and tensile strengths of 790 and 1080 MPa (115 and 156 ksi), respectively, with 15% elongation and 64% reduction in area. Elevated-temperature tensile properties at 649 °C (1200 °F) include yield and tensile strengths of 345 and 405 MPa (50 and 58 ksi), respectively, with 32% elongation and 89% reduction in area. The steel exhibits 100% ductile fracture in room-temperature Charpy V-notch impact tests, with a typical impact energy of 135 J (100 ft·lb). In creep-rupture tests at 649 °C (1200 °F), the steel exhibits rupture strengths and minimum creep rates at least comparable to those of type 316 stainless steel.

Source: E.J. Vineberg and T.B. Cox, "Development of a Ferritic Low-Carbon Steel for Elevated Temperature Service," in Topical Conference on Ferritic Alloys for Use in Nuclear Energy Technologies (Proceedings of a Conference in Snowbird, UT), J.W. Davis and D.J. Michel, Eds., Metallurgical Society of AIME, Warrendale, PA, 1984, p 297

13.20 Type 410 (Modified) Stainless Steel: Effect of Solution Heat-Treatment Temperatures

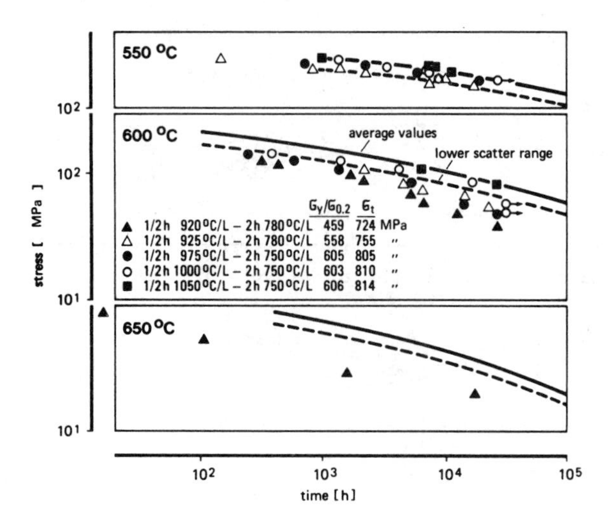

Effect of solution treatment on creep-rupture strength of modified type 410 (12Cr-1Mo-V) stainless steel.

This steel shows a fully martensitic structure with fine-dispersely precipitated Cr-Mo-mixed carbides as a requirement for optimum long-time properties at high temperatures. Accordingly, solution heat-treatment at relatively high temperatures of 1020 to 1050 °C (1870 to 1920 °F) is required. Lower solution heat-treatment temperatures will lead to undissolved coarser carbide particles and therefore to a smaller amount of superfine precipitations and to a non-fully martensitic structure. A distinct dependence of creep-rupture strength on the solution heat-treatment temperature is shown in the figure above.

Source: G.P. Kalwa, K. Haarmann, and K.J. Janssen, "Experience with Ferritic and Martensitic Steel Tubes and Piping in Nuclear and Non-Nuclear Applications," in Topical Conference on Ferritic Alloys for Use in Nuclear Energy Technologies (Proceedings of a Conference in Snowbird, UT), J.W. Davis and D.J. Michel, Eds., Metallurgical Society of AIME, Warrendale, PA, 1984, p 240

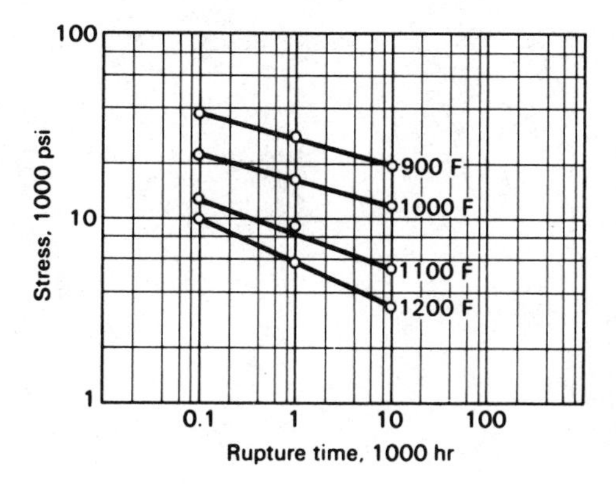

High-temperature data for quenched-and-tempered type 414 (12% chromium, 2% nickel) stainless steel (two heats).

Source: J.G. Parr and A. Hanson, An Introduction to Stainless Steel, American Society for Metals, Metals Park, OH, 1965, p 43

Type 420 Stainless Steel

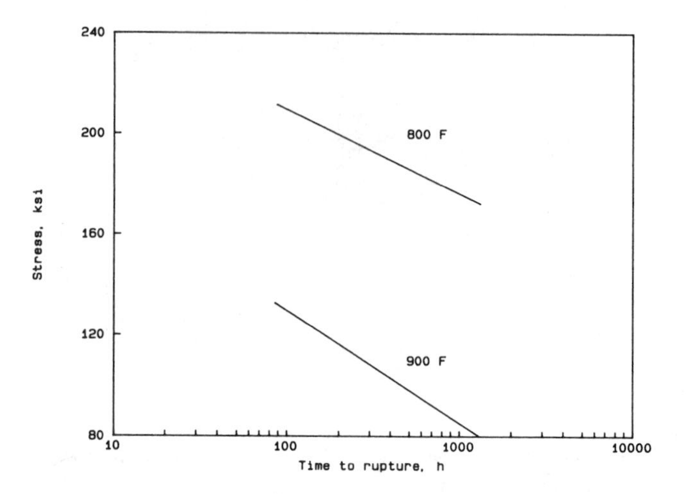

Type 420 (AISI 420) stress-rupture strength; specimens were heat treated at 1850 °F, oil quenched, and tempered at 50 °F above test temperature.

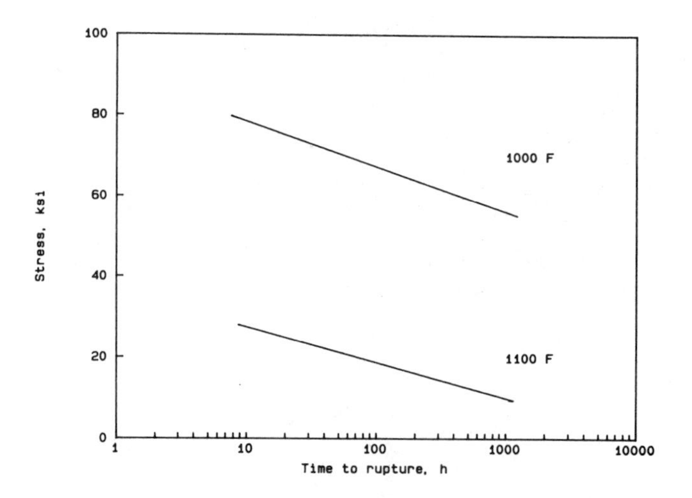

Type 420 (AISI 420) stress-rupture strength; specimens were heat treated at 1850 °F, oil quenched, and tempered at 50 °F above test temperature.

Source: High-Temperature Property Data: Ferrous Alloys, M.F. Rothman, Ed., ASM International, Metals Park, OH, 1988, p 9.86

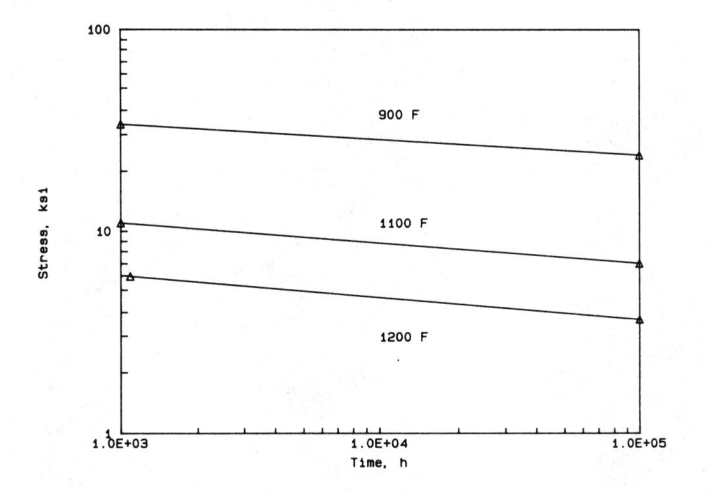

Type 431 (AISI 431) stress-rupture strength; specimens were oil quenched and tempered at 1200 °F.

Source: High-Temperature Property Data: Ferrous Alloys, M.F. Rothman, Ed., ASM International, Metals Park, OH, 1988, p 9.95

13.24 Class I Super 12% Cr (0.2C-14.5Cr-1.07Mo) vs 0.2C-13.8Cr Martensitic Stainless Steels: A Comparison

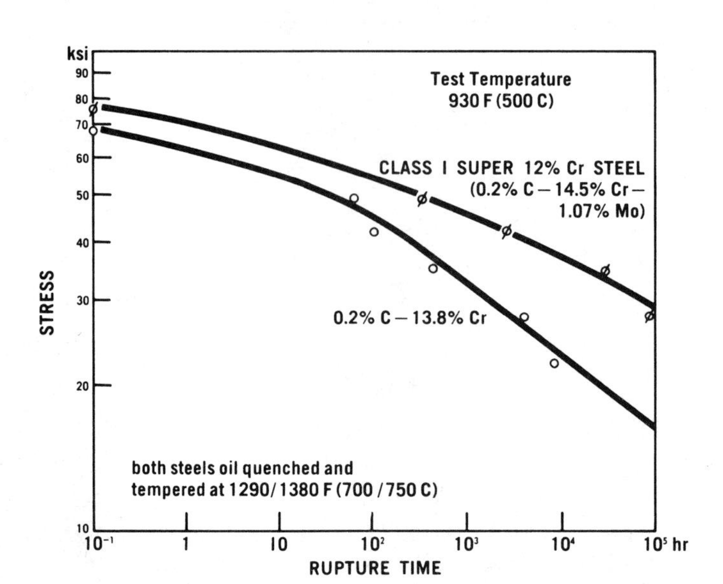

Stress vs rupture time for two nonstandard martensitic stainless steels.

Source: J.Z. Briggs and T.D. Parker, "The Super 12% Cr Steels," in Source Book on Materials for Elevated-Temperature Applications, E.F. Bradley, Ed., American Society for Metals, Metals Park, OH, 1979, p 147

Types 410 and Modified 410 Stainless Steel Weld Compositions: A Comparison

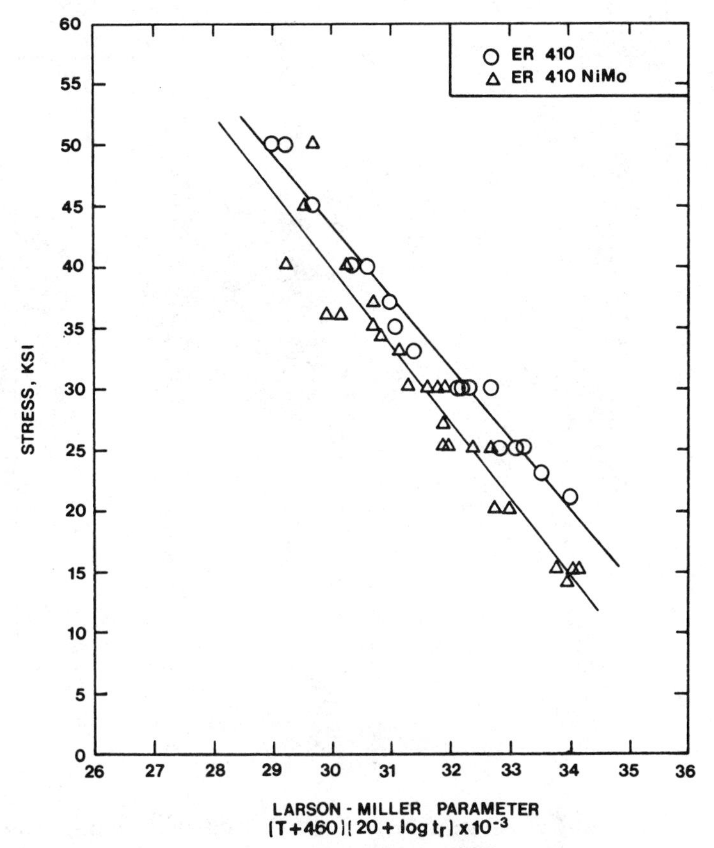

Rupture stress vs Larson-Miller parameter.

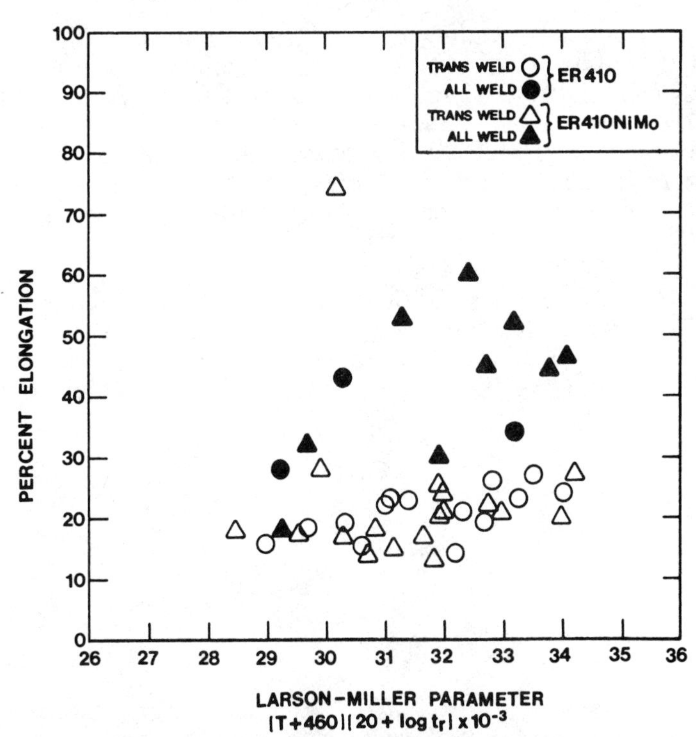

Percent elongation vs Larson-Miller parameter.

Types 410 and Modified 410 Stainless Steel Weld Compositions: A Comparison (Continued)

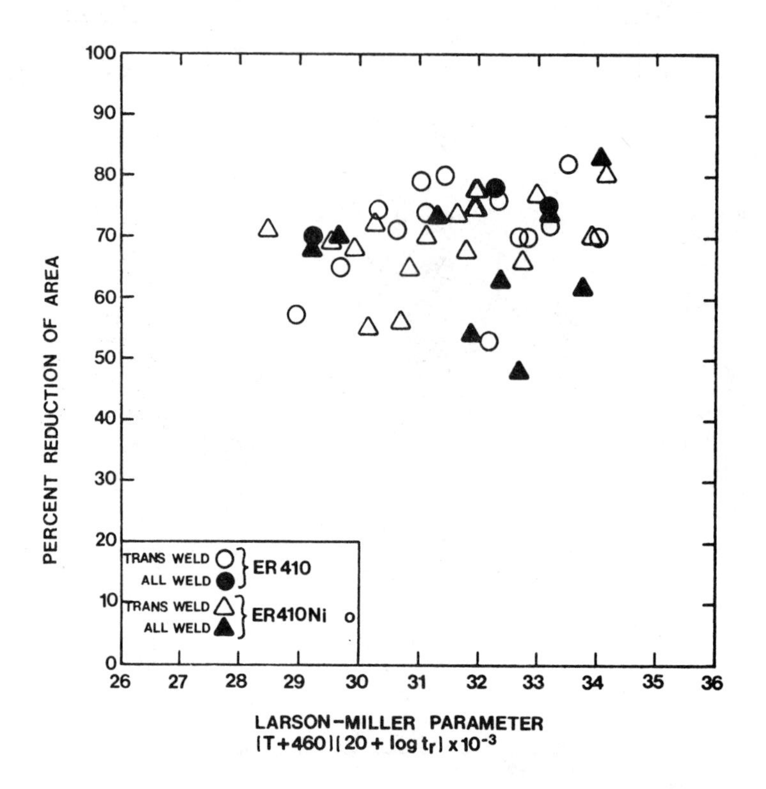

Percent reduction in area vs Larson-Miller parameter.

Stress-rupture tests were conducted at 900, 950, and 1000 °F on both ER 410 and ER 410 NiMo weld compositions. Specimens were both all-weld (0.357-in. diam by 1.400-in. gage length, plain bar, notch bar) and transverse weld (0.357-in. diam by 1.400-in. gage length, plain bar). K_T (stress concentration factor) for the notch bar was 3.9. The welds for the transverse specimens were located in the center of the specimen with at least one fusion zone also present in the gage length. Test results are shown in the table and first figure.

The ER 410 NiMo weld has considerable scatter and much lower properties than the ER 410 weld. All plain-bar notch-bar (all-weld) specimens failed in the plain bar first. No-notch specimens were tested following the initial plain-bar failure. All transverse specimens (both weld compositions) failed in the weld metal.

Source: R.E. Clark and L.E. Wagner, "Comparison of ER 410 and ER 410 Ni Mo Gas Metal Arc Weld Metal," in Cast Metals for Structural and Pressure Containment Applications, G.V. Smith, Ed., American Society of Mechanical Engineers, New York, 1979, p 155, 167-169

Type 410 (Modified) Stainless Steel: Comparison of Rupture Strengths of 12Cr (FI), 12Cr-Mo (FIMo), 12Cr-Mo-V (FIMoV), and 12Cr-Mo-V-Nb (FV 448E)

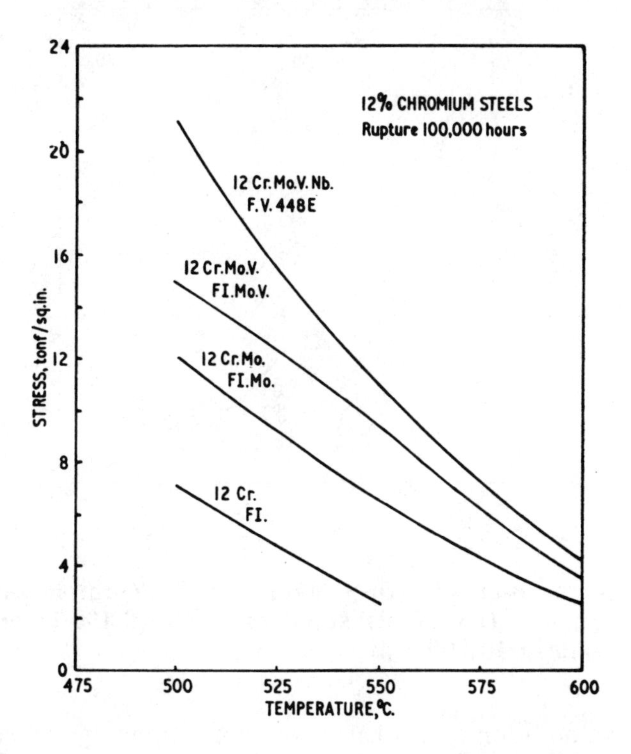

Summary of 100,000-h rupture strengths for modified type 410 (12Cr) stainless steels.

Source: H.W. Kirkby and R.J. Truman, "12%Cr Steels: Creep-Resisting and High-Strength Variants," in High Temperature Properties of Steels (Proceedings of a Conference in Eastbourne, England), Iron and Steel Institute, London, 1967, p 371

Type 410 (Modified) Stainless Steel: Creep Data Comparison for High-Chromium Steels

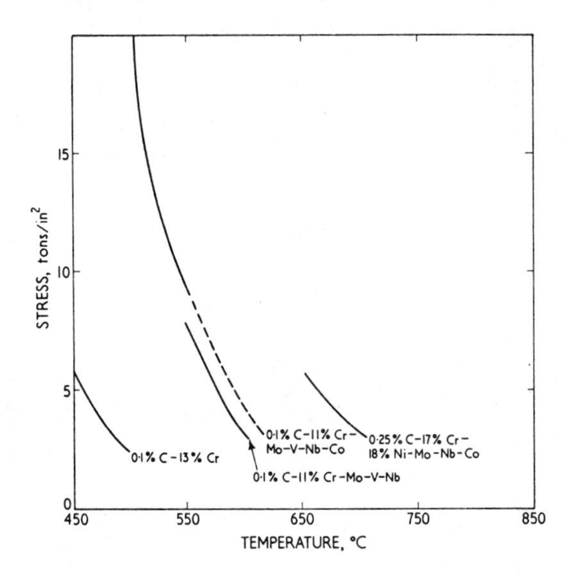

Comparison of creep data for high-chromium type 410 (modified) stainless steels (0.1% creep strain in 10,000 h).

An addition of Co has a solution-hardening effect on the matrix structure and provides improved creep strength up to 500-550 °C. The creep properties of this type of steel, with and without Co, are shown in the figure above. The creep properties of an austenitic alloy are included for comparison.

Source: R. Brook and J.E. Russell, "Highly-Alloyed Martensitic Steels, in Metallurgical Developments in High-Alloy Steels (Proceedings of a Conference in Scarborough, England), Iron and Steel Institute, London, 1964, p 25

14

Precipitation Hardening Stainless Steels

17-4 PH: Smooth vs Notched Bars

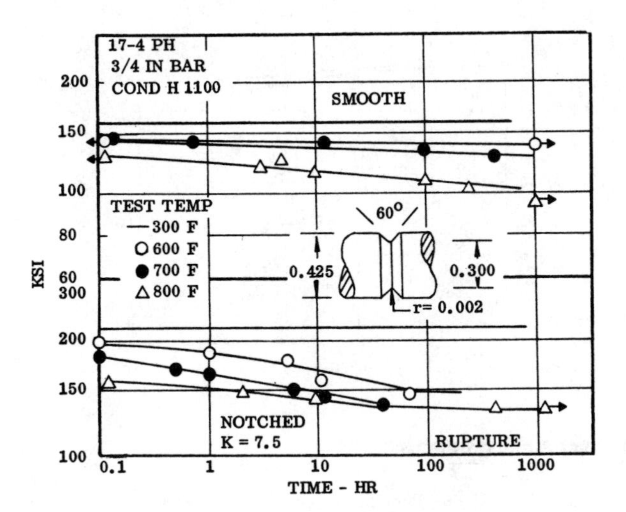

Creep-rupture curves at 300 to 800 °F for smooth and notched bars in condition H 1100.

Source: Aerospace Structural Materials Handbook, Volume 2, Mechanical Properties Data Center, Battelle Columbus Laboratories, 1980, p 1501-19

17-7 PH: TH1050 Condition

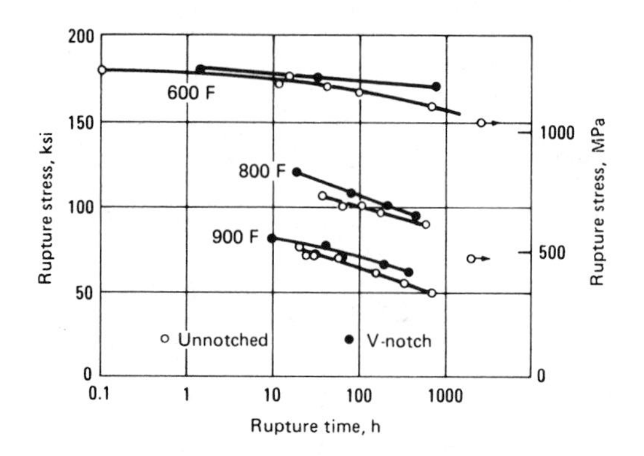

Average short-time properties

Temperature °C	°F	Tensile yield strength(a) (b) MPa	ksi	Compressive yield strength(a) MPa	ksi	Elongation in 50 mm or 2 in., %
24	75	1340	194	1525	221	6.7
315	600	1125	163	1280	186	4.8
425	800	950	138	1015	147	11.8
480	900	765	111	785	114	19.6

(a) At 0.2% offset. (b) Tensile strengths (low to high temperature) were: 1400, 1185, 1020 and 840 MPa (203, 172, 148 and 122 ksi). Hardness at 24 °C (75 °F) was 44 HRC.

Stress-rupture properties of 17-7 PH stainless steel, TH1050 condition.

Specimens were taken transverse to the rolling direction from sheet 915 by 3050 by 1.6 mm (36 by 120 by 0.063 in.) and were heat treated by air cooling from 760 to 260 °C (1400 to 500 °F), water quenching, holding 10 h at 15 °C (60 °F), aging 1½ h at 565 °C (1050 °F) and air cooling. V notches on both edges of specimens were 7.6 mm (0.300 in.) wide with 0.61-mm (0.024-in.) root radii; K_t = 3.1. Strain of 1 to 2% occurred within 10% of rupture life in smooth specimens.

Source: Metals Handbook, Ninth Edition, Volume 3, Properties and Selection: Stainless Steels, Tool Materials and Special-Purpose Metals, American Society for Metals, Metals Park, OH, 1980, p 204

17-7 PH: Isochronous Stress-Strain Curves for 1.3-mm (0.050-in.) Sheet, TH1050 Condition

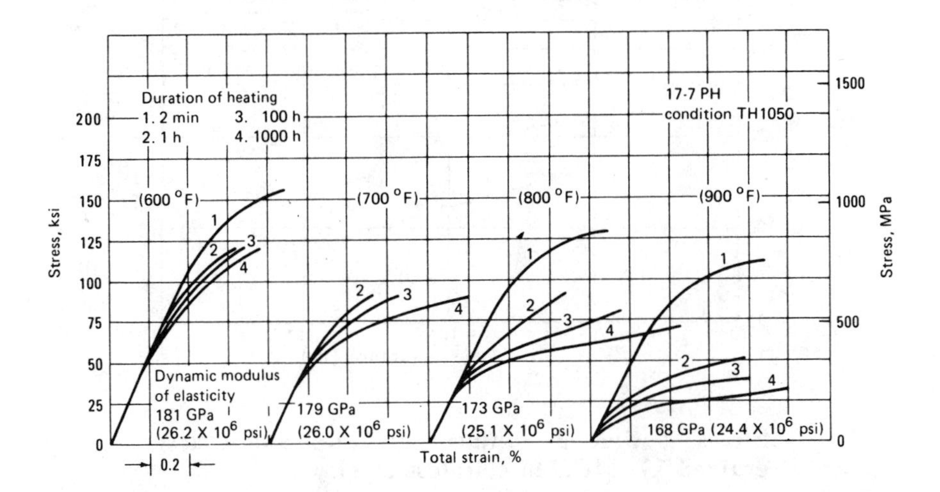

Room-temperature properties were: tensile strength, 1290 MPa (187 ksi); yield strength at 0.2% offset, 1225 MPa (178 ksi); dynamic modulus of elasticity, 200 GPa (29 × 10^6 psi). Total strain was adjusted to the indicated modulus values.

Isochronous stress-strain curves such as those shown in the figure above are useful in selection of design stresses for permissible total deformations during short and long periods of time. Because these data are taken from creep curves, extension due to thermal expansion is not included.

Source: Metals Handbook, Ninth Edition, Volume 3, Properties and Selection: Stainless Steels, Tool Materials and Special-Purpose Metals, American Society for Metals, Metals Park, OH, 1980, p 192

17-14CuMo Stainless Steel: Fine- vs Coarse-Grained

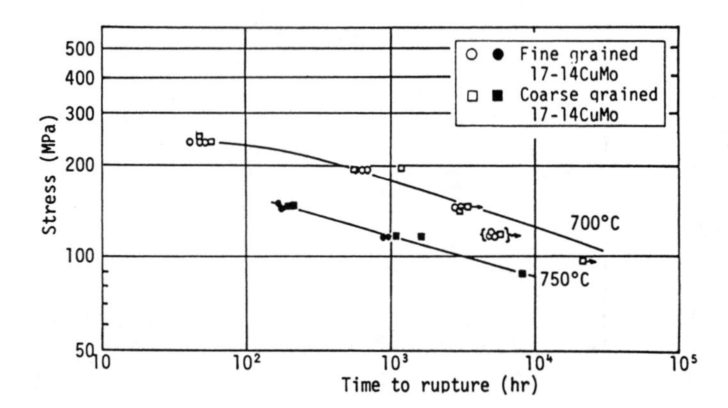

Stress vs time to rupture for fine- and coarse-grained 17-14CuMo stainless steel.

Source: H. Teranishi, K. Yoshikawa, and Y. Sawargi, "Application of Fine-Grained Austenitic Steel Tubes to Superheaters in Power Boiler," in Creep (Proceedings of the International Conference in Tokyo), Japan Society of Mechanical Engineers, Tokyo, 1986, p 235

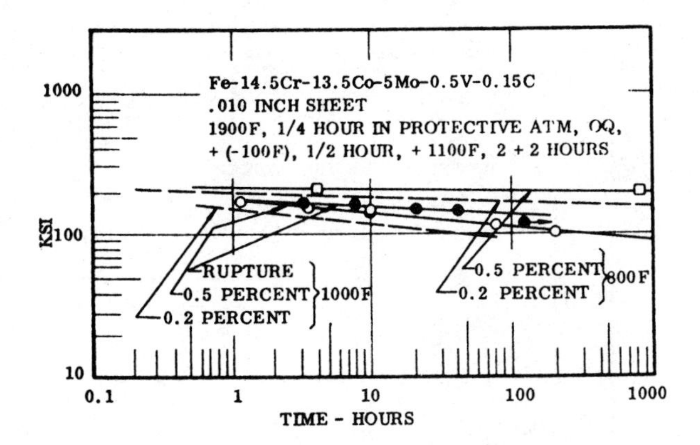

Curves for 0.2 and 0.5% deformation and for rupture of sheet tempered at 1100 °F.

Source: Aerospace Structural Materials Handbook, Volume 2, Mechanical Properties Data Center, Battelle Columbus Laboratories, 1980, p 1509-18

AFC-77: Wire

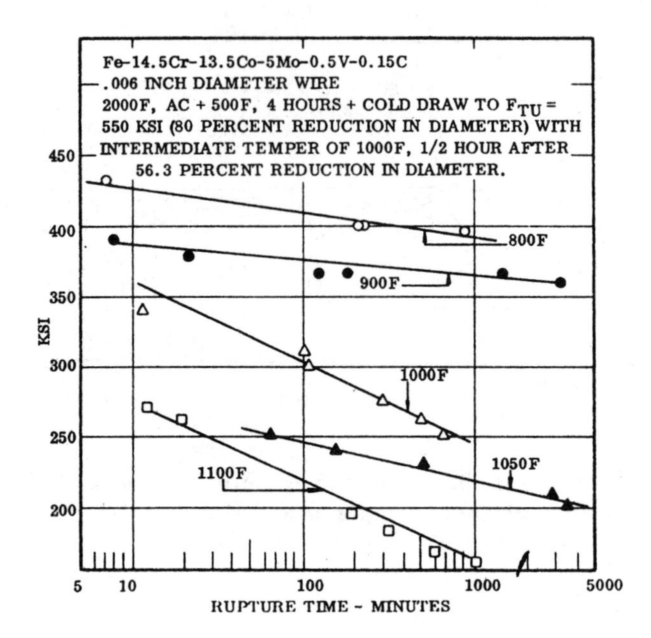

Creep-rupture properties of 0.006-in. diameter wire drawn to a tensile strength of 550 ksi and tested at 800 to 1100 °F.

Source: Aerospace Structural Materials Handbook, Volume 2, Mechanical Properties Data Center, Battelle Columbus Laboratories, 1980, p 1509-18

AISI 602: Notched Stress-Rupture Behavior

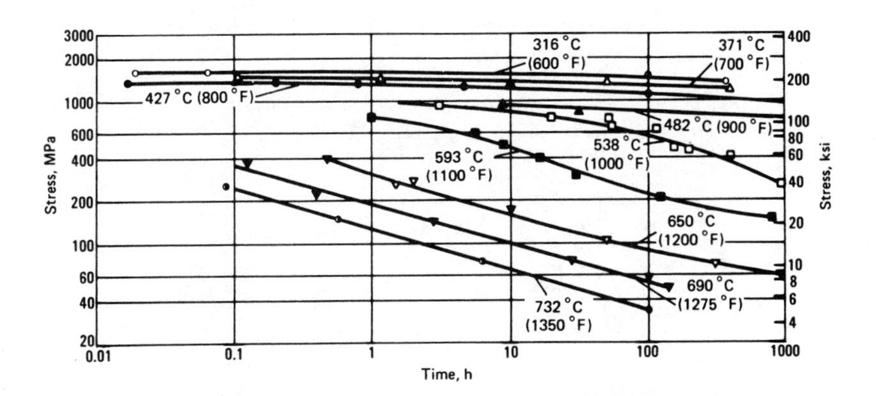

Stress-rupture behavior of AISI 602 steel at the indicated temperatures. Specimens were notched (K = 10). Steel was normalized at about 940 °C (1725 °F) and tempered 6 h at 650 °C (1200 °F).

Source: Metals Handbook, Ninth Edition, Volume 1, Properties and Selection: Irons and Steels, American Society for Metals, Metals Park, OH, 1978, p 651

AM-350: Sheet, Creep-Rupture Properties

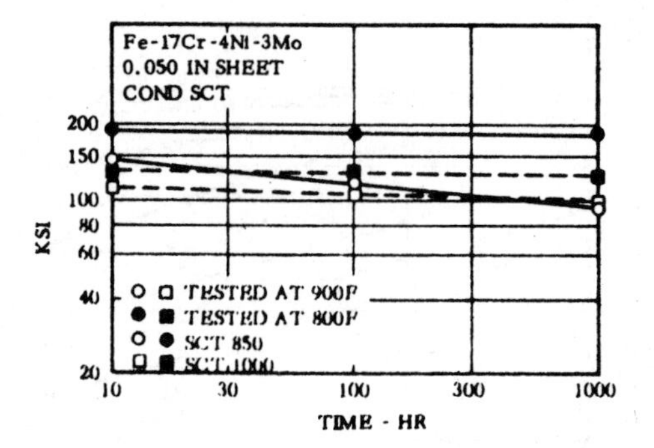

Creep-rupture properties at 800 and 900 °F of sheet in the subzero cooled and tempered (SCT) condition.

Source: Aerospace Structural Materials Handbook, Volume 2, Mechanical Properties Data Center, Battelle Columbus Laboratories, 1980, p 1504-19

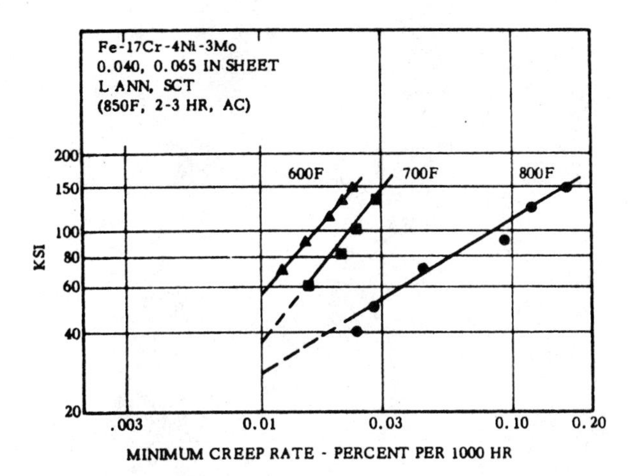

Minimum creep rate curves at 600, 700, and 800 °F for sheet in the subzero cooled and tempered (SCT) condition.

Source: Aerospace Structural Materials Handbook, Volume 2, Mechanical Properties Data Center, Battelle Columbus Laboratories, 1980, p 1504-19

AM-355: Sheet

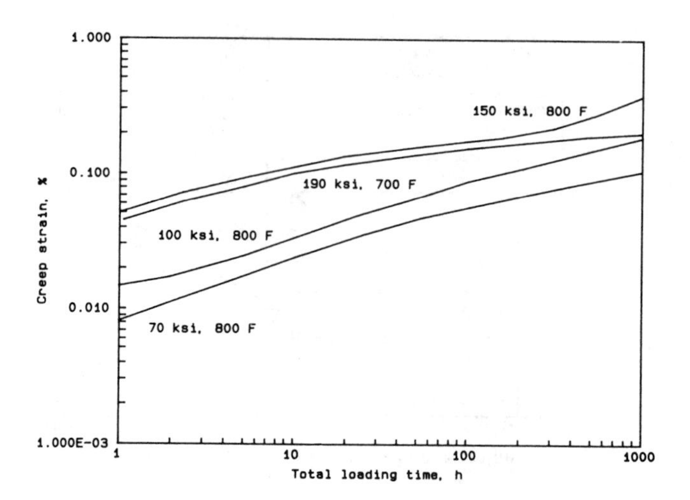

Total creep strain at 700 and 800 °F for four levels of stress for 0.056-in. AM-355 sheet. Specimens were cold rolled and tempered.

Source: High-Temperature Property Data: Ferrous Alloys, M.F. Rothman, Ed., ASM International, Metals Park, OH, 1988, p 9.108

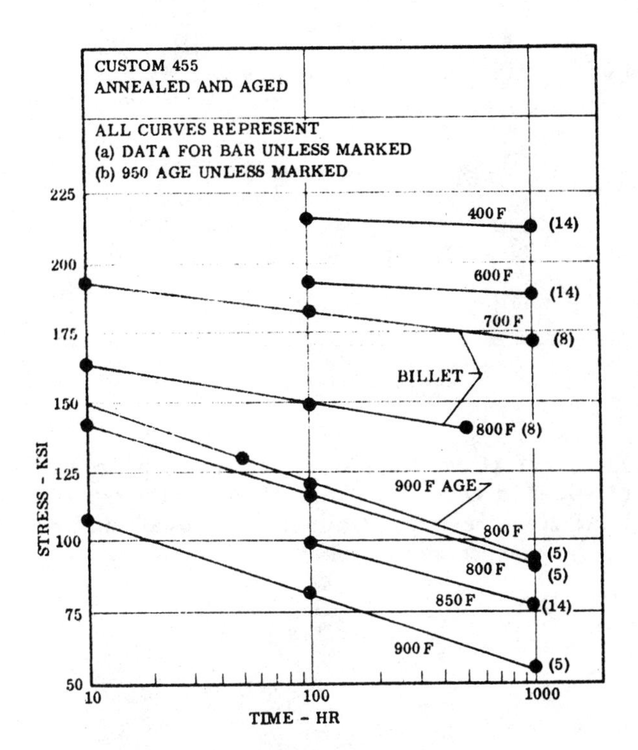

Creep-rupture curves at 400 to 900 °F for aged billet and bar.

Source: Aerospace Structural Materials Handbook, Volume 2, Mechanical Properties Data Center, Battelle Columbus Laboratories, 1980, p 1514-16

PH 14-8 Mo: Sheet

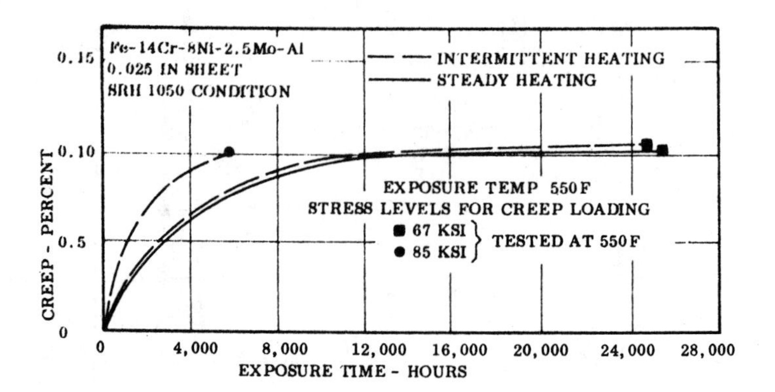

Effect of stress level on smooth creep properties of sheet in condition SRH 1050. Data represents total creep, elastic strain not included, and none of the creep is recoverable.

Source: Aerospace Structural Materials Handbook, Volume 2, Mechanical Properties Data Center, Battelle Columbus Laboratories, 1980, p 1507-16

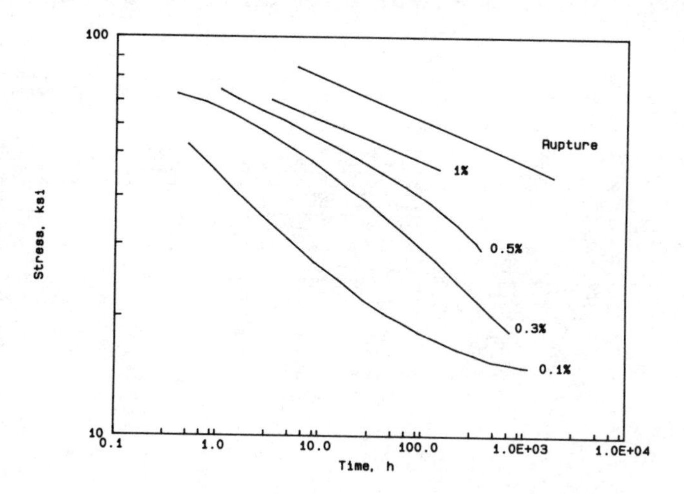

Creep-rupture curves at 1000 °F for PH 15-7 Mo sheet, refrigeration hardened at 950 °F. Creep: curve 1: 0.1%, curve 2: 0.3%, curve 3: 0.5%, curve 4: 1.0%, and curve 5: rupture.

Source: High-Temperature Property Data: Ferrous Alloys, M.F. Rothman, Ed., ASM International, Metals Park, OH, 1988, p 9.140

14.14 Precipitation-Hardening Stainless Steels: Comparison of Short-Time Tensile, Rupture, and Creep Properties

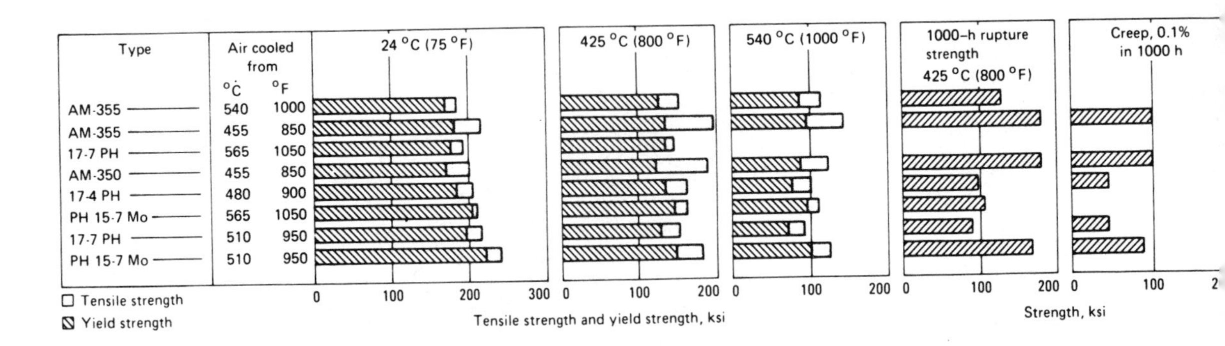

Short-time tensile, rupture, and creep properties of precipitation-hardening stainless steels.

AM-355 was finish hot worked from a maximum temperature of 980 °C (1800 °F), reheated to 930 to 955 °C (1710 to 1750 °F), water quenched, treated at –75 °C (–100 °F) and aged at 540 and 455 °C (1000 and 850 °F). 17-7 PH and PH 15-7 Mo were solution treated at 1040 to 1065 °C (1900 to 1950 °F). 17-7 PH (TH1050) and PH 15-7 Mo (TH1050) were reheated to 760 °C (1400 °F), air cooled to 15 °C (60 °F) within 1 h and aged 90 min at 565 °C (1050 °F). 17-7 PH (RH950 and PH 15-7 Mo (RH950) were reheated to 955 °C (1750 °F) after solution annealing, cold treated at –75 °C (–100 °F) and aged at 510 °C (950 °F). 17-4 PH was aged at 480 °C (900 °F) after solution annealing. AM-350 was solution annealed at 1040 to 1065 °C (1900 to 1950 °F) and reheated to 930 °C (1710 °F), air cooled, treated at –75 °C (–100 °F) and then aged at 455 °C (850 °F).

Source: Metals Handbook, Ninth Edition, Volume 3, Properties and Selection: Stainless Steels, Tool Materials and Special-Purpose Metals, American Society for Metals, Metals Park, OH, 1980, p 200

15

Higher-Nickel Austenitic Alloys

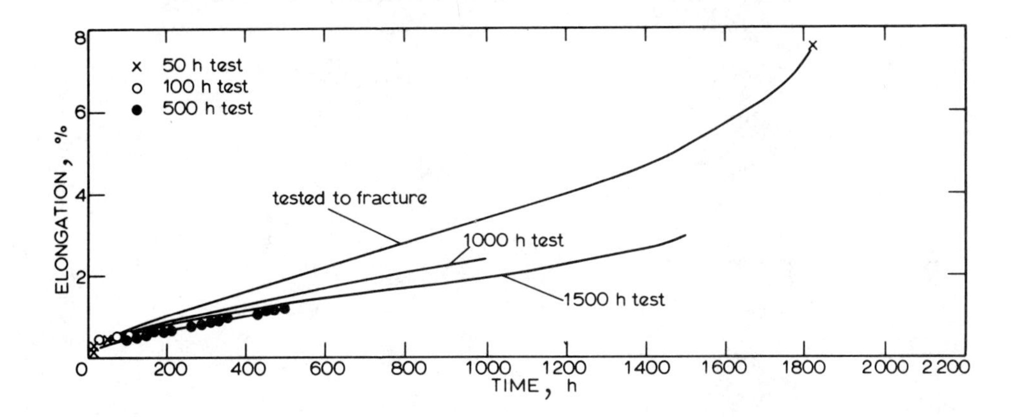

Creep curves for 17Cr-26Ni-0.03C austenitic stainless steel, 7.75 tons/in.^{2} at 600 °C.

The alloy used for this investigation was a commercial-purity, air-melted austenitic stainless steel with the composition 0.033% C, 0.010% N, 17.5% Cr, and 26.4% Ni. This alloy was one of a series originally planned to study the effect of stacking-fault energy on various mechanical properties. This particular alloy had a high stacking-fault energy owing to the high Ni content and at ambient temperature deformed by a process of cross-slip with a low work-hardening rate. The interstitial content of the material was at the low end of commercial austenitic steel specifications. The alloy was solution treated at 1100 °C for 30 min and water quenched to dissolve and prevent the reprecipitation of $Cr_{23}C_6$. Creep testing was carried out at 600 °C using a load of 7.75 tons/in.^{2}.

Source: S.R. Keown, "Microstructural Changes Occurring During the Creep Deformation of a Simple Austenitic Steel at 600 °C," in Creep and Strength in Steel and High-Temperature Alloys, The Metals Society, London, 1972, p 78-79

15.2

18Cr-37Ni Higher-Nickel Austenitic Alloy: Effect of Carbon Content on Rupture Strength of Spun-Cast Material

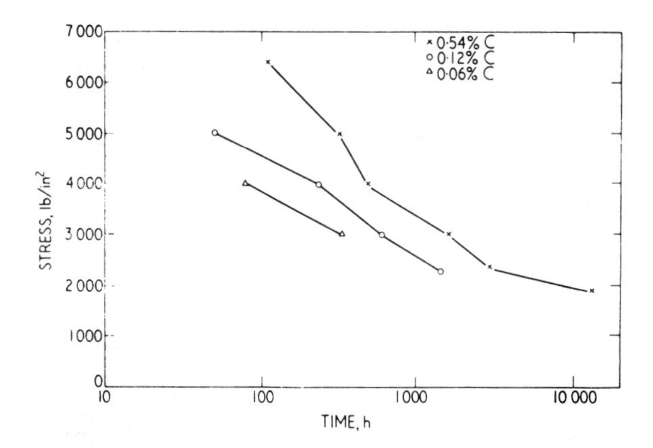

Effect of carbon content on the rupture strength of spun-cast 18Cr-37Ni alloy at 900 °C.

Source: C. Edeleanu and B. Estruch, "Alloy Steels in a Reforming Furnace of a Chemical Plant," in Metallurgical Developments in High-Alloy Steels (Proceedings of a Conference in Scarborough, England), Iron and Steel Institute, London, 1964, p 225

18Cr-37Ni Higher-Nickel Austenitic Alloy: Stress-Rupture Data

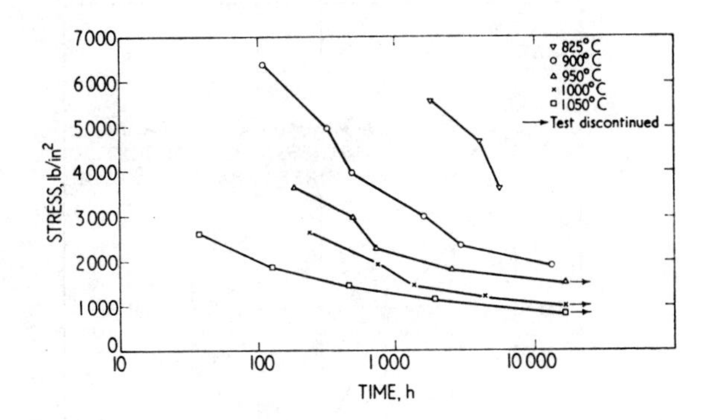

Rupture stresses for a cast of 18Cr-37Ni alloy.

Source: C. Edeleanu and B. Estruch, "Alloy Steels in a Reforming Furnace of a Chemical Plant," in Metallurgical Developments in High-Alloy Steels (Proceedings of a Conference in Scarborough, England), Iron and Steel Institute, London, 1964, p 224

RA-330: Creep Properties

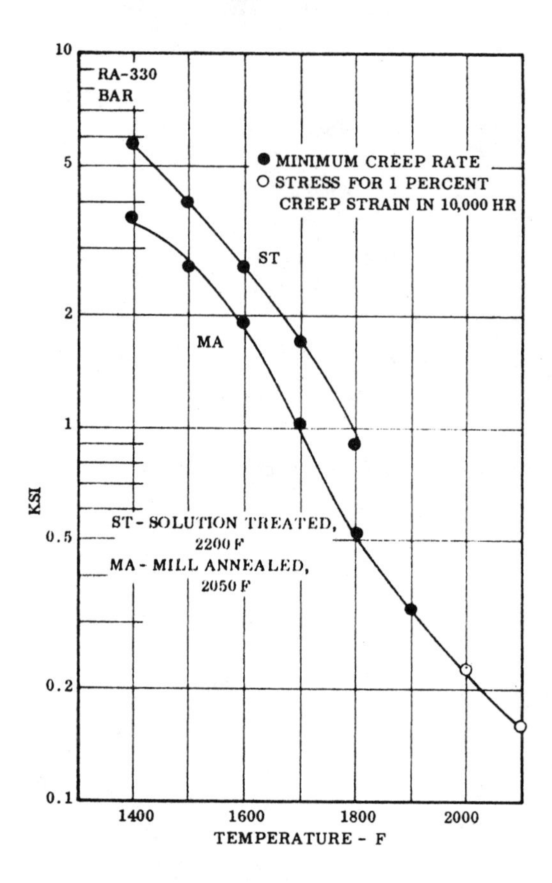

Minimum creep rate curve (0.0001%/h) for mill-annealed and solution-treated bar.

Source: Aerospace Structural Materials Handbook, Volume 2, Mechanical Properties Data Center, Battelle Columbus Laboratories, 1980, p 1611-4

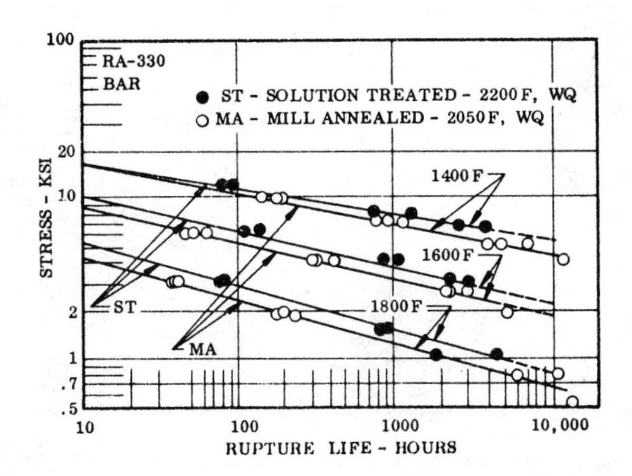

Creep-rupture curves for mill-annealed and solution-treated bar at 1400 to 1800 °F.

Source: Aerospace Structural Materials Handbook, Volume 2, Mechanical Properties Data Center, Battelle Columbus Laboratories, 1980, p 1611-4

RA-330 HC: Creep-Rate Data

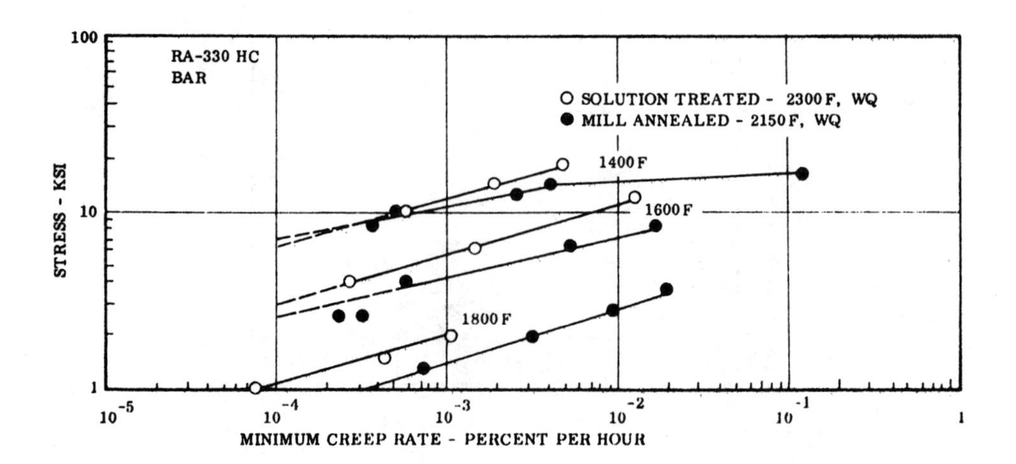

Minimum creep-rate curves for bar in the solution-treated and mill-annealed condition at 1400 to 1800 °F.

Source: Aerospace Structural Materials Handbook, Volume 2, Mechanical Properties Data Center, Battelle Columbus Laboratories, 1980, p 1612-5

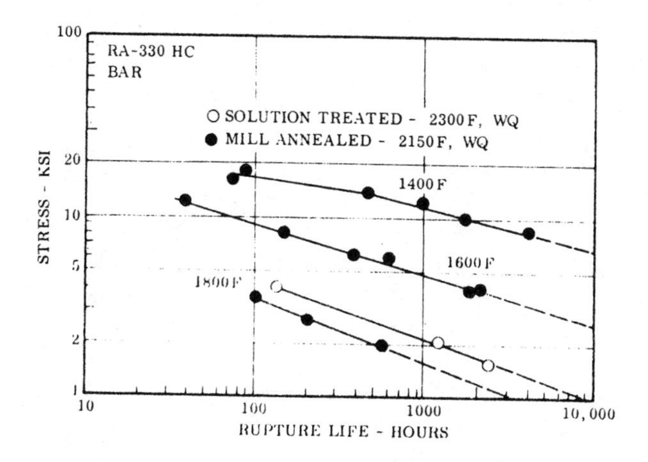

Creep-rupture curves for mill-annealed and solution-treated bar at 1400 to 1800 °F.

Source: Aerospace Structural Materials Handbook, Volume 2, Mechanical Properties Data Center, Battelle Columbus Laboratories, 1980, p 1612-4

16

Stainless Steels Comparisons

1.4914, 1.4962, and 1.4970 Stainless Steels: Comparison of Rupture Lives, as Influenced by Irradiation

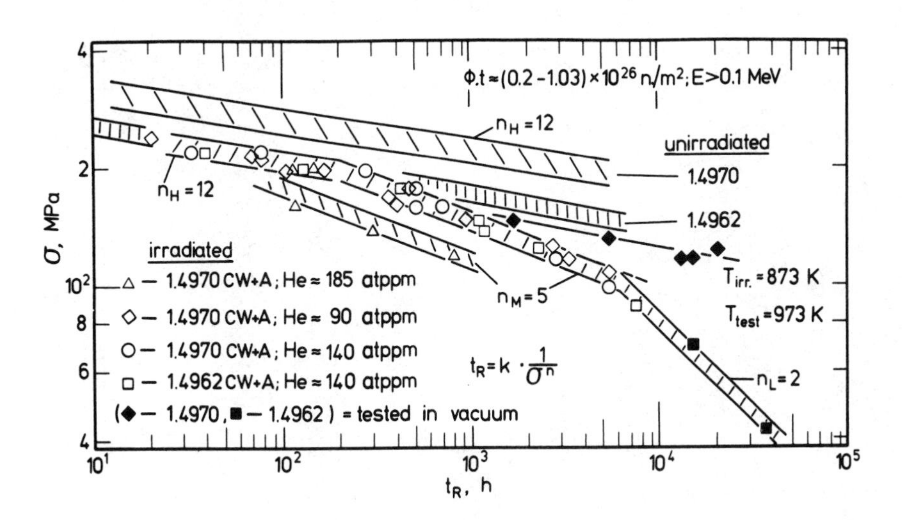

Rupture lives of unirradiated and irradiated steels 1.4970 and 1.4962.

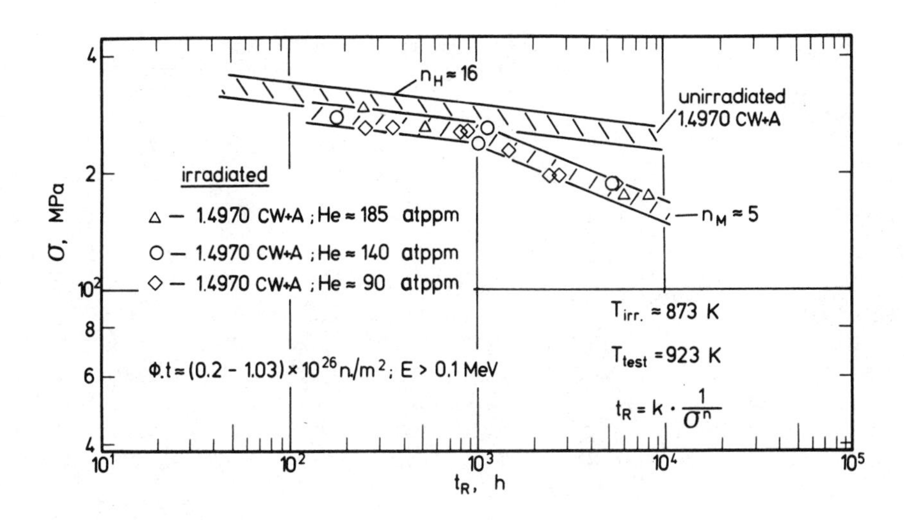

Rupture life of unirradiated and irradiated austenitic steel 1.4970.

1.4914, 1.4962, and 1.4970 Stainless Steels: Comparison of Rupture Lives, as Influenced by Irradiation (Continued)

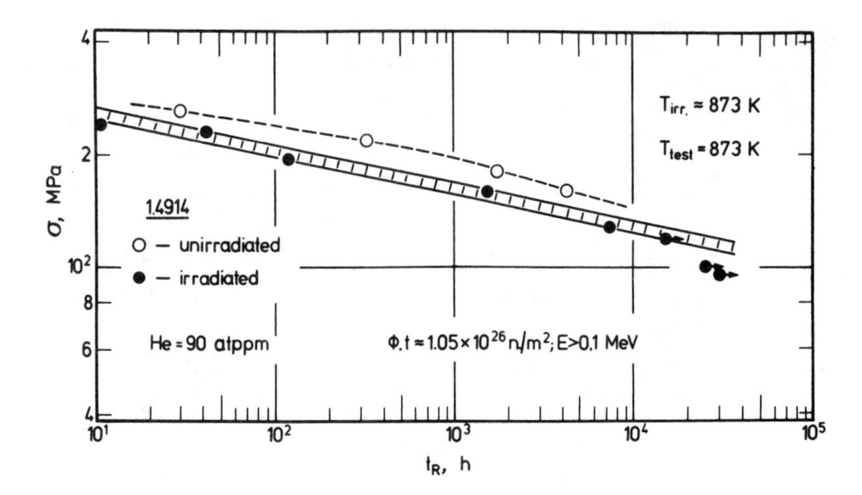

Rupture life of unirradiated and irradiated martensitic steel 1.4914.

Chemical composition (wt%) of steels

Steel grade	C	Si	Mn	Cr	Ni	Mo	W	V	Nb	Ti	B
1.4970 (aus.)	0.1	0.43	1.94	14.5	15.0	1.3	—	—	—	0.58	0.01
1.4962 (aus.)	0.1	0.64	1.0	17.3	13.5	—	3.3	0.3	—	0.49	0.01
1.4914 (mar.)	0.11	0.45	0.35	11.3	0.7	0.5	—	0.3	0.25	—	0.007

The top figure on the preceding page shows the rupture life of the austenitic steels 1.4970 and 1.4962 in the unirradiated and irradiated conditions at the test temperature of 973 K. It can be seen that 1.4970 has a higher rupture life in the unirradiated condition than 1.4962. The rupture lives of both materials after irradiation under equivalent irradiation conditions (neutron fluence, helium content, and irradiation temperature) are practically identical. Rupture life is evidently independent of variations in chemical composition and of the state of the material within the stress range covered by the experiment.

The creep-rupture strength of martensitic steel 1.4914 at 873 K is shown in the bottom figure on this page. It can be observed that the rupture life is decreased after irradiation.

Source: C. Wassilew, "Influence of Helium Embrittlement on Post-Irradiation Creep Rupture Behaviour of Austenitic and Martensitic Stainless Steels," in Mechanical Behaviour and Nuclear Applications of Stainless Steel at Elevated Temperatures (Proceedings of an International Conference in Varese, Italy), The Metals Society, London, 1982, p 173, 175

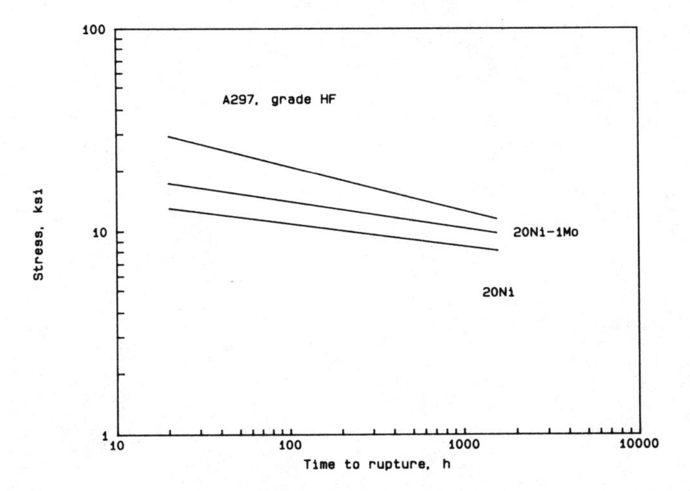

Stress-rupture curves for 19Cr-9Ni cast stainless steel (ASTM A297, grade HF) and high-nickel heat-resistant irons at 705 °C.

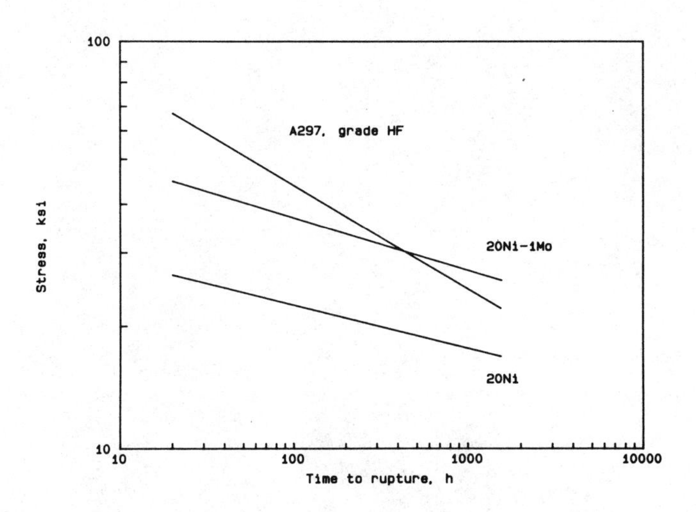

Stress-rupture curves for 19Cr-9Ni cast stainless steel (ASTM A297, grade HF) and high-nickel heat-resistant irons at 595 °C.

Source: High-Temperature Property Data: Ferrous Alloys, M.F. Rothman, Ed., ASM International, Metals Park, OH, 1988, p 1.22

19Cr-9Ni, 30Ni, and 30Ni-1Mo: A Comparison

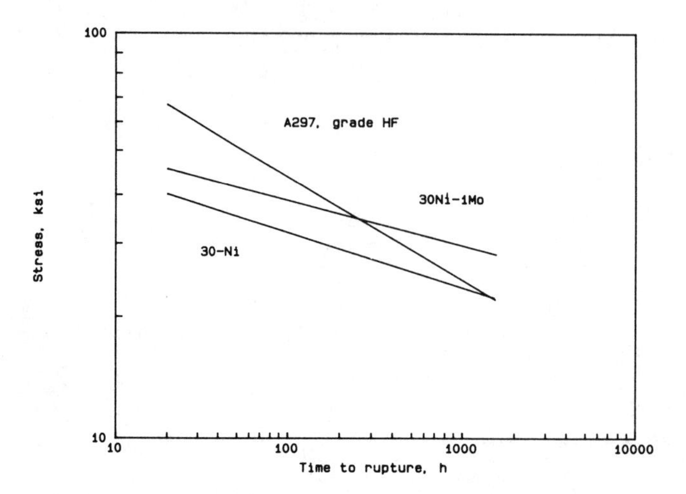

Stress-rupture curves for 19Cr-9Ni cast stainless steel (ASTM A297, grade HF) and high-nickel heat-resistant irons at 595 °C.

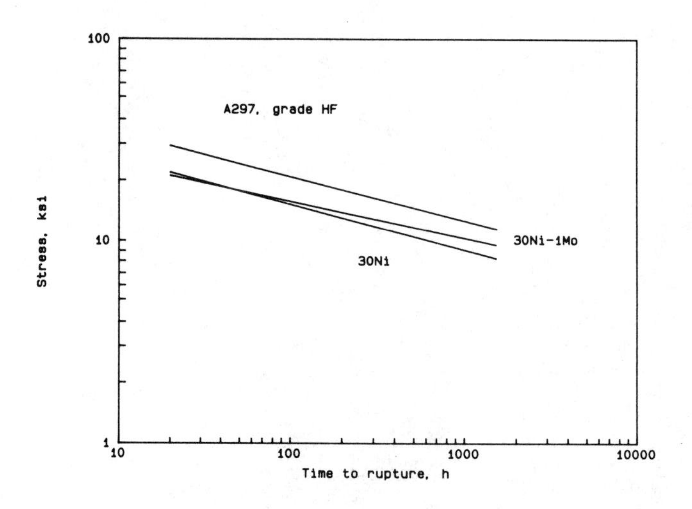

Stress-rupture curves for 19Cr-9Ni cast stainless steel (ASTM A297, grade HF) and high-nickel heat-resistant irons at 705 °C.

Source: High-Temperature Property Data: Ferrous Alloys, M.F. Rothman, Ed., ASM International, Metals Park, OH, 1988, p 1.22

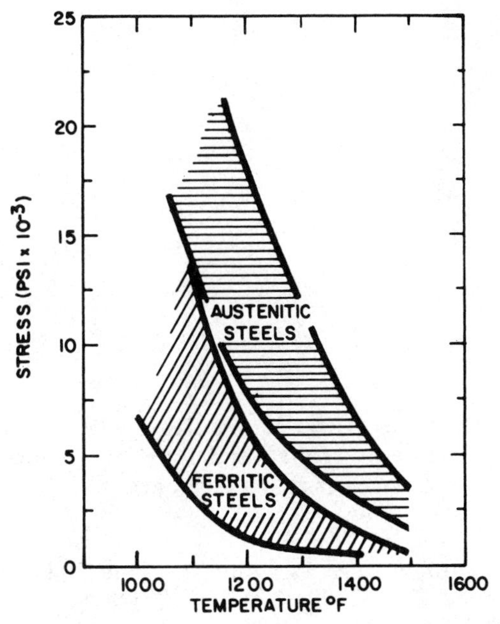

Rupture strength (rupture in 10,000 h).

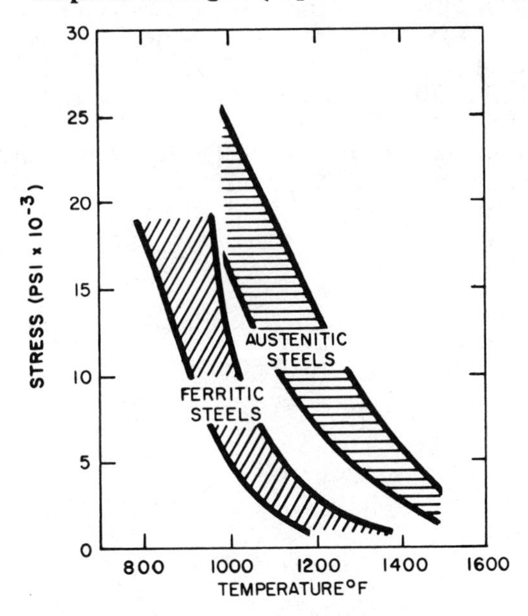

Creep strength (1% in 10,000 h).

A study of materials for anti-smog devices indicated that the austenitic steels exhibit high-temperature creep and rupture strengths that are superior to those of ferritic steels, as shown in the figures above. Where both strength and corrosion resistance are required, the austenitic materials are generally considered the superior choice.

Source: H.T. Michels, "Corrosion Performance of Heat-Resisting Alloys in Automobile Exhausts," in Source Book on Materials Selection, Volume II, American Society for Metals, Metals Park, OH, 1977, p 62-63

Austenitic Stainless Steels vs Iron-Base Superalloys: A Comparison

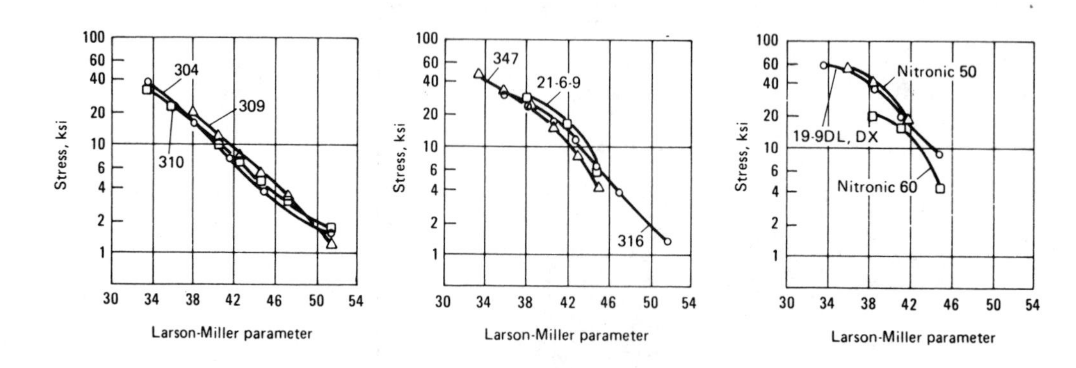

Stress-rupture plots for three austenitic stainless steels (left), two austenitic stainless steels and the iron-base superalloy 21-6-9 (center), and three iron-base superalloys (right).

Heat treating schedules were as follows: Type 304: 1065 °C (1950 °F), water quench. Type 309: 1090 °C (2000 °F), water quench. Type 310: 1120 °C (2050 °F), water quench. Type 316: 1090 °C (2000 °F), water quench. Type 347: 1065 °C (1950 °F), water quench. 21-6-9: 1065 °C (1950 °F), water quench. 19-9 DX, DL: for tests above 705 °C (1300 °F), 1065 °C (1950 °F) and water quench, plus 705 °C and air cool; for tests below 705 °C, 705 °C and air cool. Nitronic 50: 1090 °C (2000 °F), water quench. Nitronic 60: 1065 °C (1950 °F), water quench. Larson-Miller parameter = $T/1000\ (20 + \log t)$, where T is temperature in °R and t is time in h; all data taken from 1000-h tests.

Source: Metals Handbook, Ninth Edition, Volume 3, Properties and Selection: Stainless Steels, Tool Materials and Special-Purpose Metals, American Society for Metals, Metals Park, OH, 1980, p 205

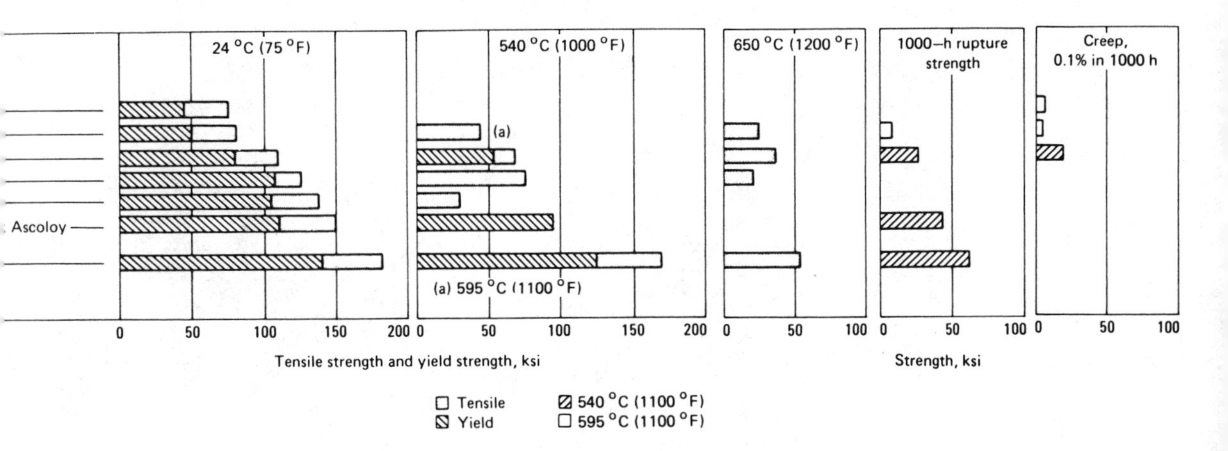

Tensile, yield, rupture, and creep strengths for seven ferritic and martensitic stainless steels

Types 430 and 446 were annealed. Type 403 was quenched from 870 °C (1600 °F) and tempered at 620 °C (1150 °F). Type 410, quenched from 955 °C (1750 °F) and tempered at 590 °C (1100 °F). Type 431, quenched from 1025 °C (1875 °F) and tempered at 590 °C. Greek Ascoloy, quenched from 955 °C and tempered at 590 °C. Type 422, quenched from 1040 °C (1900 °F) and tempered at 590 °C.

Source: Metals Handbook, Ninth Edition, Volume 3, Properties and Selection: Stainless Steels, Tool Materials and Special-Purpose Metals, American Society for Metals, Metals Park, OH, 1980, p 199

Heat-Resisting Metals and Alloys: A Comparison

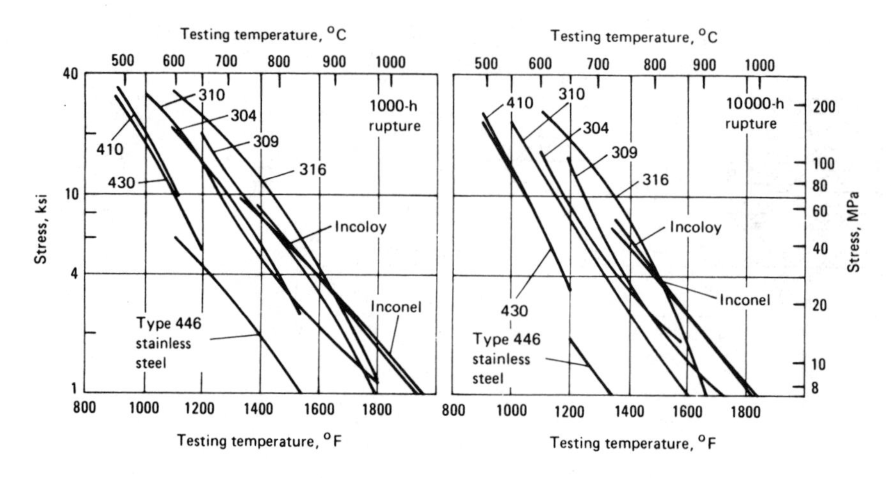

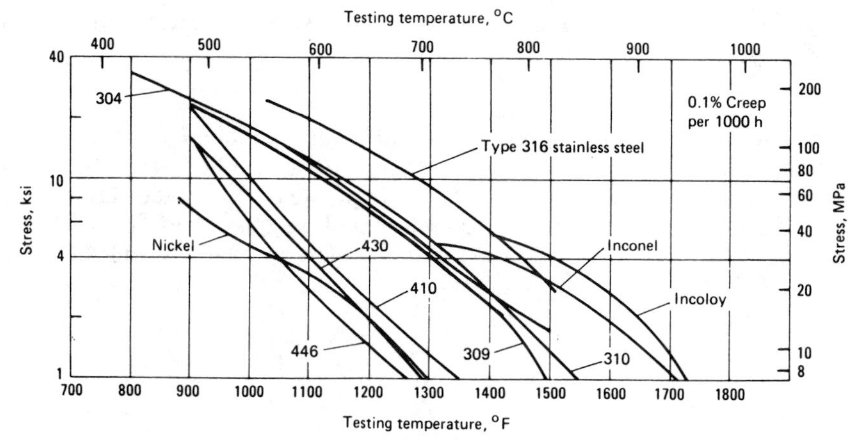

Creep and stress-rupture curves for heat-resisting metals and alloys, including stainless steels.

Source: Metals Handbook, Ninth Edition, Volume 3, Properties and Selection: Stainless Steels, Tool Materials and Special-Purpose Metals, American Society for Metals, Metals Park, OH, 1980, p 195

Low-Carbon, Heat-Resisting, and Stainless Steels: Comparison of 1000-h Rupture Strengths

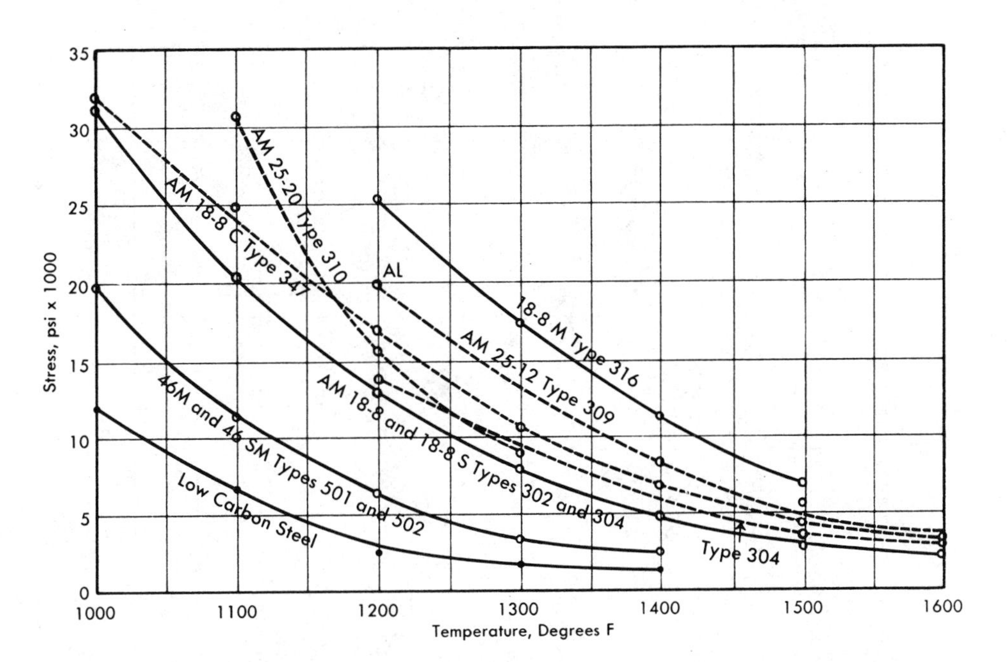

Stress to produce rupture in 1000 h.

Source: Stainless Steel Handbook, Allegheny Ludlum Steel Corp., Pittsburgh, 1959, p 58

Low-Carbon, Heat-Resisting, and Stainless Steels: Comparison of Creep Strengths

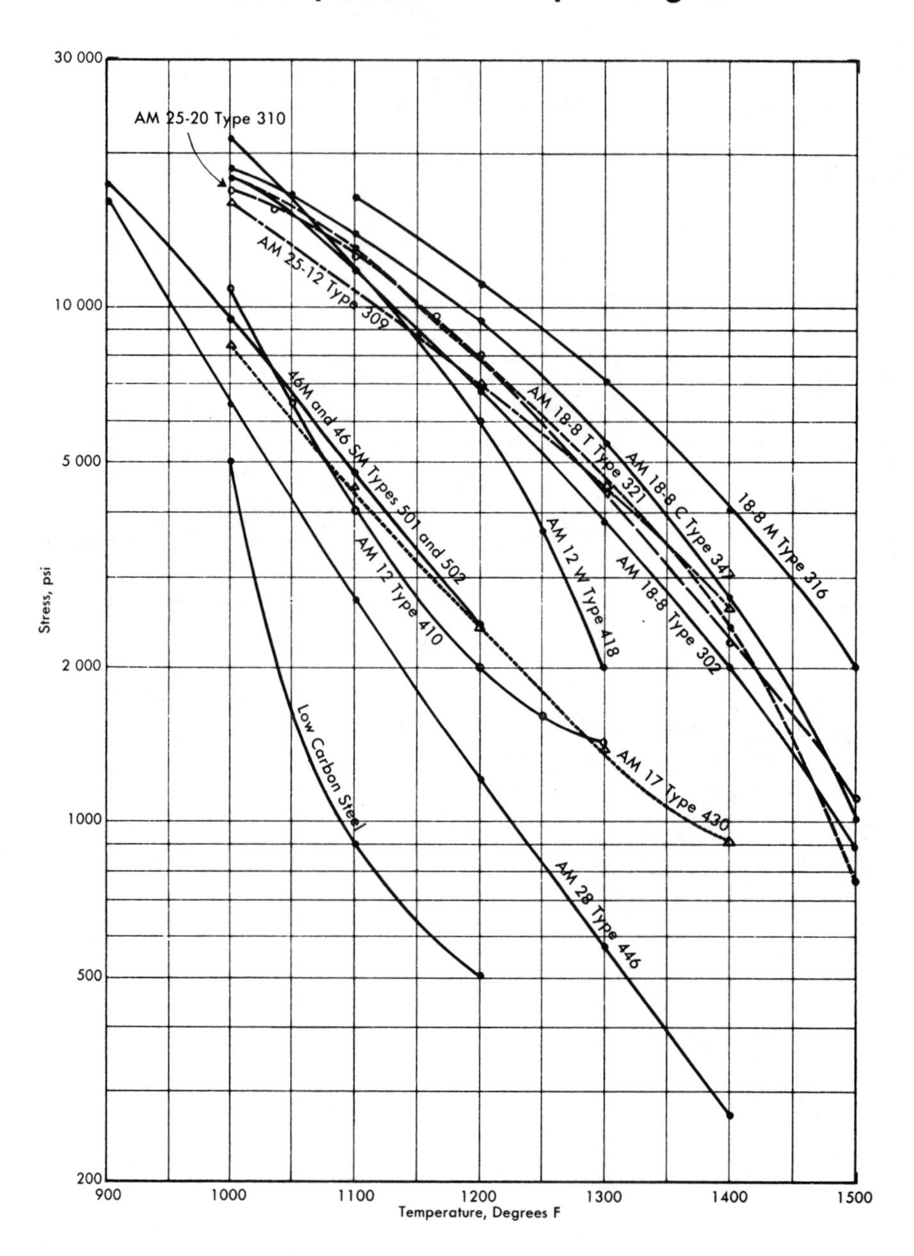

Creep strength of several stainless steels (1% creep in 10,000 h).

Source: Stainless Steel Handbook, Allegheny Ludlum Steel Corp., Pittsburgh, 1959, p 57

Stainless Steels: Comparison of the 10,000-h Rupture Strengths of Annealed Material

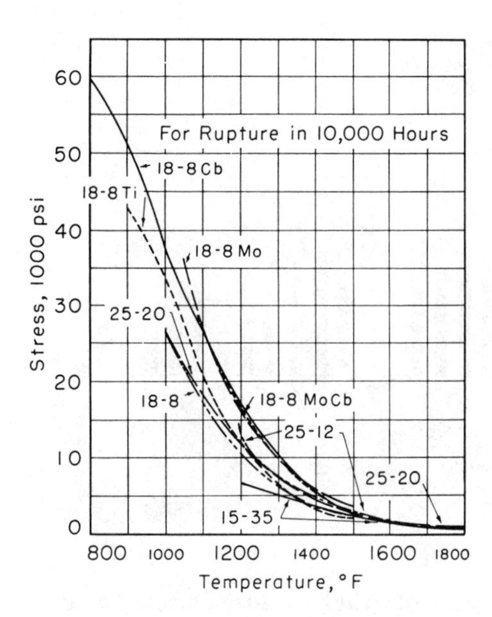

Rupture strength of several stainless steels and a higher-nickel austenitic alloy in the annealed condition.

Note: 25-20 is austenitic type 310; 25-12 is austenitic type 309; 15-35 is a higher-nickel austenitic alloy.

Source: H.C. Cross and Ward F. Simmons, "Alloys and Their Properties for Elevated Temperature Service," in Utilization of Heat Resistant Alloys, American Society for Metals, Cleveland, OH, 1954, p 75

16.12 Stainless Steel Weld Metals vs Wrought Material of Similar Composition: A Comparison

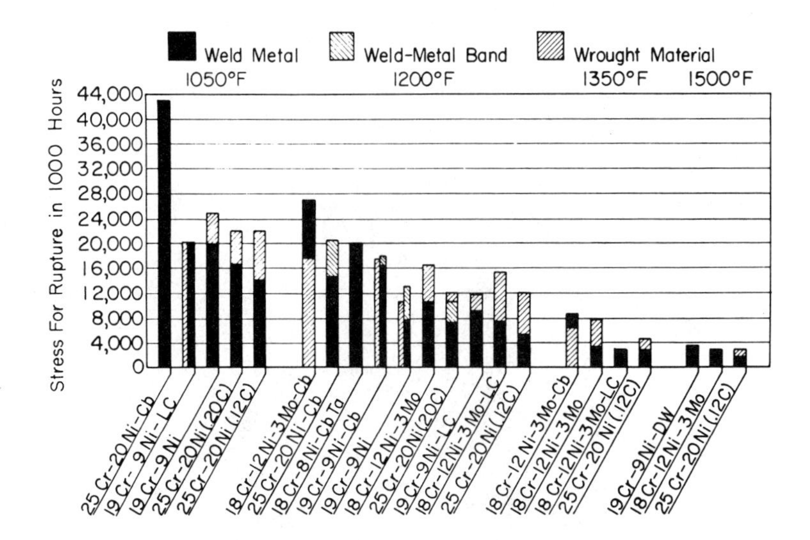

Comparison of the rupture strength of stainless-steel-weld metals with wrought products of similar composition.

Source: J.W. Freeman, C.L. Corey, and A.I. Rush, "Metallurgical Variables Influencing Properties of Heat Resisting Alloys," in Utilization of Heat Resistant Alloys, American Society for Metals, Cleveland, OH, 1954, p 263

Types 317 (Modified) and 410 Stainless Steel vs Low-Alloy Steels: A Comparison

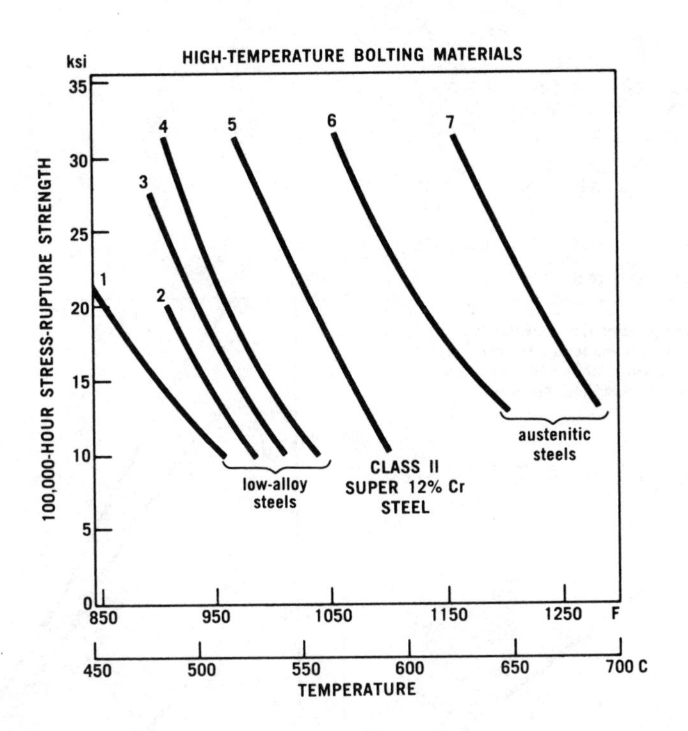

	% C	% Cr	% Mo	% Ni	% V
1	0.12	0.7	1.0	1.5	—
2	0.31	2.7	0.3	—	0.3
3	0.24	1.3	0.6	—	0.2
4	0.21	1.3	1.1	—	0.3
5	0.22	12.0	1.0	0.5	0.3
6	0.09	16.0	1.8	15.0	— 1% Cb/Ta + 0.1% N
7	0.07	16.5	1.8	16.5	— 0.7% Cb/Ta

Summary of 100,000-h stress-rupture strength vs temperature.

Source: J.Z. Briggs and T.D. Parker, "The Super 12% Cr Steels," in Source Book on Materials for Elevated-Temperature Applications, E.F. Bradley, Ed., American Society for Metals, Metals Park, OH, 1979, p 153

16.14 Type 410 Stainless Steel, Precipitation-Hardened Stainless Steel, and Low-Alloy Steels: A Comparison

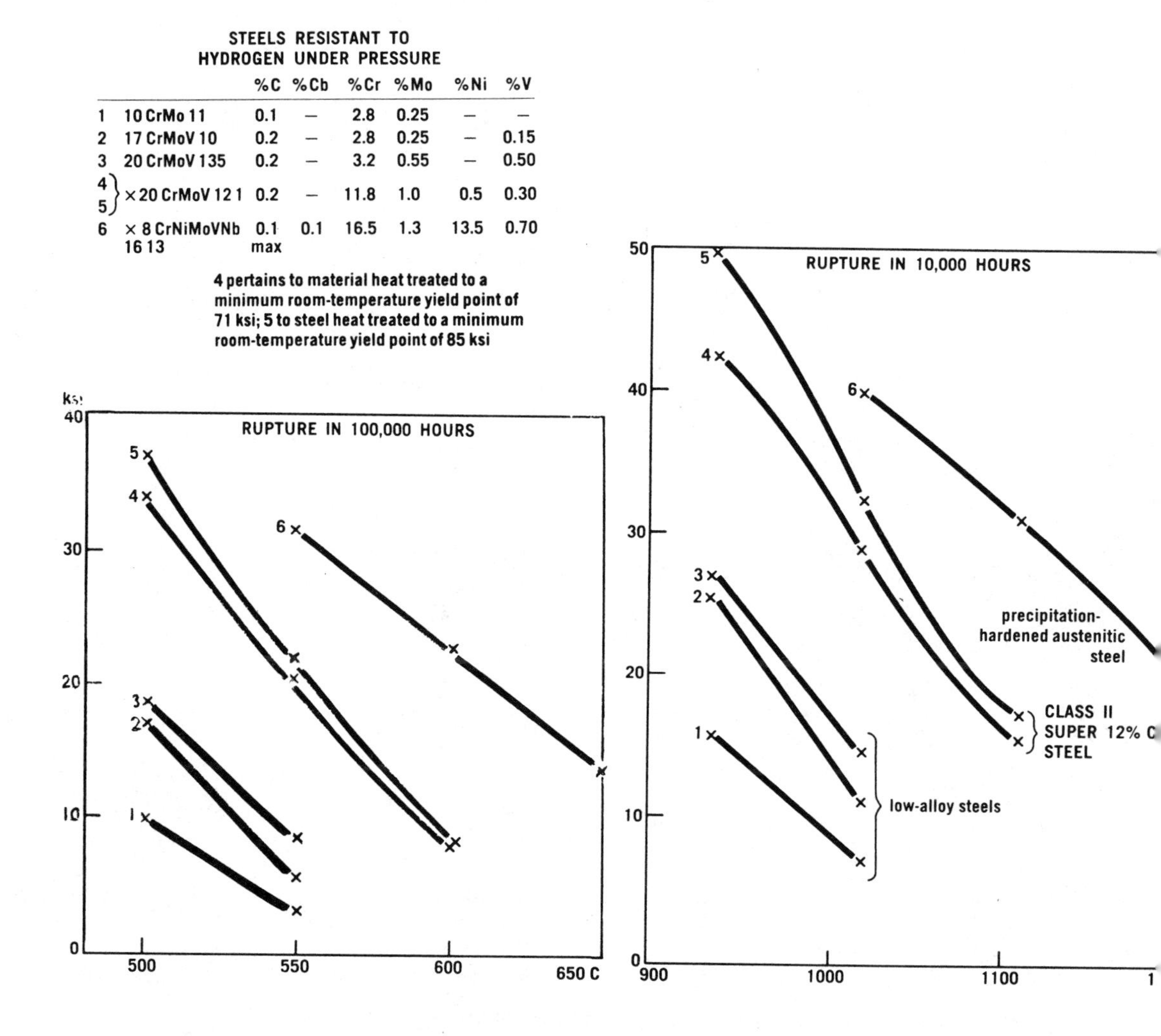

STEELS RESISTANT TO HYDROGEN UNDER PRESSURE

		%C	%Cb	%Cr	%Mo	%Ni	%V
1	10 CrMo 11	0.1	–	2.8	0.25	–	–
2	17 CrMoV 10	0.2	–	2.8	0.25	–	0.15
3	20 CrMoV 135	0.2	–	3.2	0.55	–	0.50
4, 5	× 20 CrMoV 12 1	0.2	–	11.8	1.0	0.5	0.30
6	× 8 CrNiMoVNb 16 13	0.1 max	0.1	16.5	1.3	13.5	0.70

4 pertains to material heat treated to a minimum room-temperature yield point of 71 ksi; 5 to steel heat treated to a minimum room-temperature yield point of 85 ksi

Mean stress vs temperature curves for rupture in 100,000 and 10,000 h.

Source: J.Z. Briggs and T.D. Parker, "The Super 12% Cr Steels," in Source Book on Materials for Elevated-Temperature Applications, E.F. Bradley, Ed., American Society for Metals, Metals Park, OH, 1979, p 153

Type 410 (Modified) vs Types 304 and 316 Stainless Steel: A Comparison

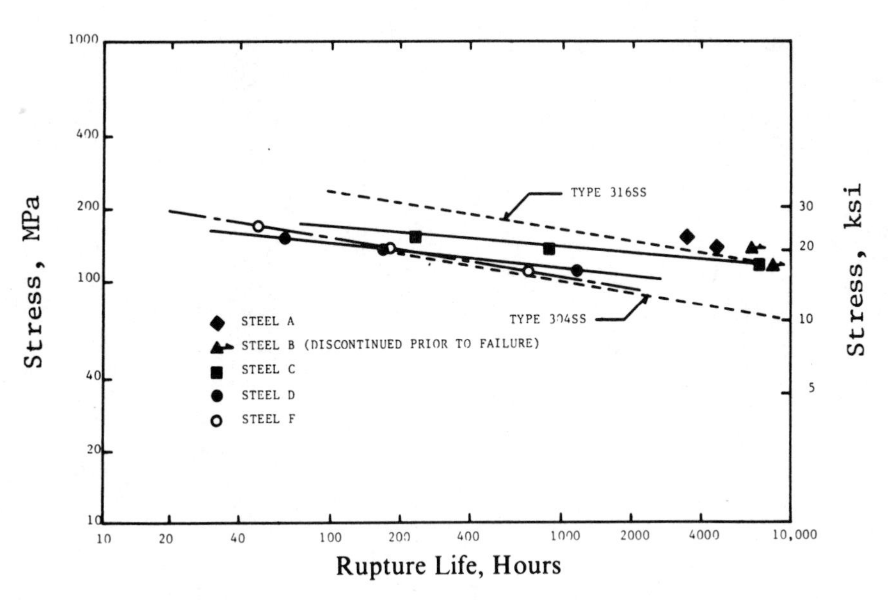

Rupture life curves for steels tested at 649 °C (1200 °F) in the normalized and tempered condition.

Creep-rupture results for steels tested at 649 °C (1200 °F)

Steel	Type of Steel	Stress MPa (ksi)	Minimum Creep Rate % Hr	Rupture Life Hr	El., %	R.A., %
A	0.05C-12Cr-2Mo-6.5Ni-V-N	138 (20) 152 (22)	0.0009 0.00095	4625.7 3340.6	11.5 11.3	[illegible]1.6 33.9
B	0.05C-12Cr-2Mo-8Ni-V	117 (17) 138 (20)	0.00017 0.0002	Discontinued at 8200 h Discontinued at 6700 h		
C	0.09C-12Cr-2Mo-6.5Ni-V	117 (17) 138 (20) 152 (22)	0.0007 0.009 0.045	8039.1 873.3 231.7	 30.0 32.5	 64.0 73.7
D	0.09C-12Cr-1.5Mo-5.5Ni-W	117 (17) 138 (20) 152 (22)	0.0063 0.052 0.14	1130.4 164.0 62.5	25.0 36.0 34.5	63.4 76.5 77.0
D	Aged 1000 h at 650 C (1200 F)	124 (18) 138 (20) 152 (22)	0.045 0.14 0.45	220.5 67.5 42.1	45.0 51.5 42.0	78.9 80.3 79.1

Results of creep-rupture tests are summarized in the table and are presented graphically in the figure above. Also shown are the stress vs rupture life curves for types 304 and 316 stainless steel. All four low-carbon 12Cr-Mo steels tested in the present investigation have significantly better rupture properties than either higher carbon 12Cr-Mo or type 304 stainless steels.

Source: E.J. Vineberg, P.J. Grobner, and V.A. Biss, "12Cr-Mo Steels with Improved Rupture Strength and Weldability," in Ferritic Steels for High-Temperature Applications (Proceedings of an International Conference in Warren, PA), A.K. Khare, Ed., American Society for Metals, Metals Park, OH, 1983, p 309-311

17

Nickel-Base Alloys

CP Nickel: Alloy 201

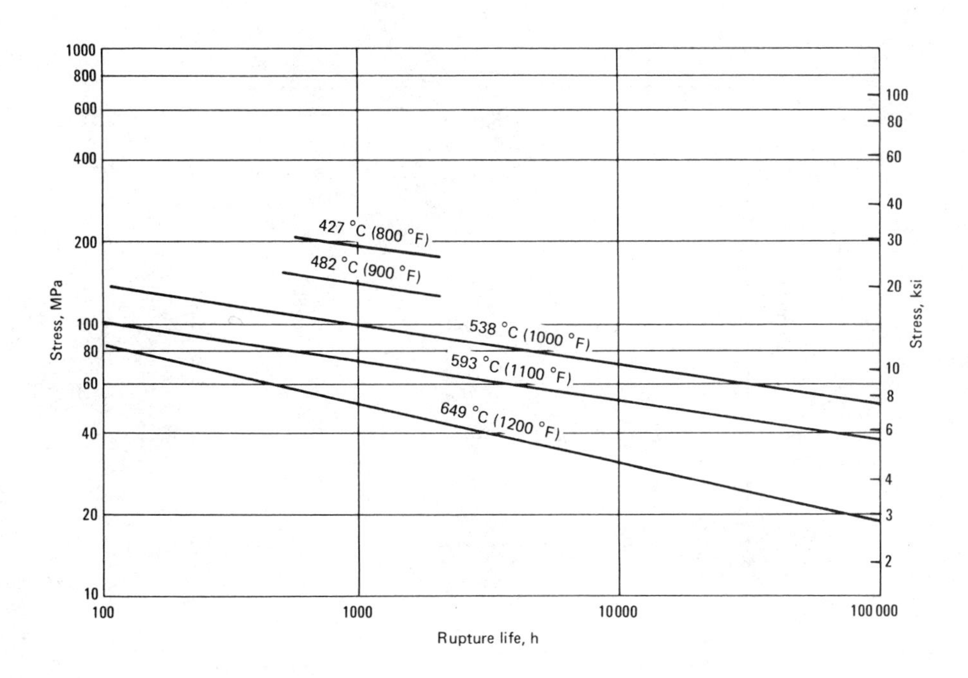

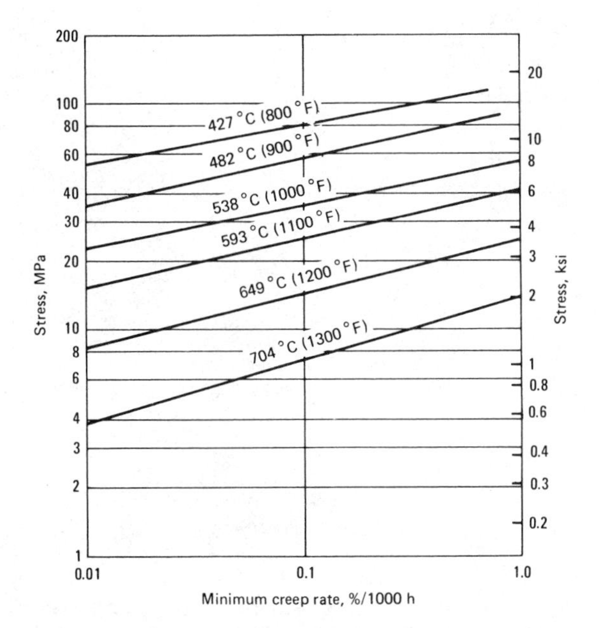

Typical stress-rupture strength (top) and creep strength (bottom) of annealed CP nickel alloy 201 at 800-1200 °F (top) or 1300 °F (bottom).

Source: Metals Handbook, Ninth Edition, Volume 3, Properties and Selection: Stainless Steels, Tool Materials and Special-Purpose Metals, American Society for Metals, Metals Park, OH, 1980, p 130

CP Nickel: Effect of Air and Vacuum Environments on Creep Properties

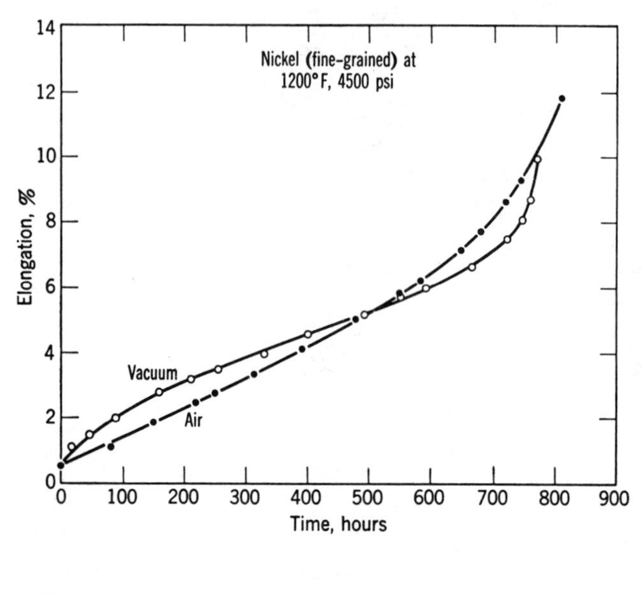

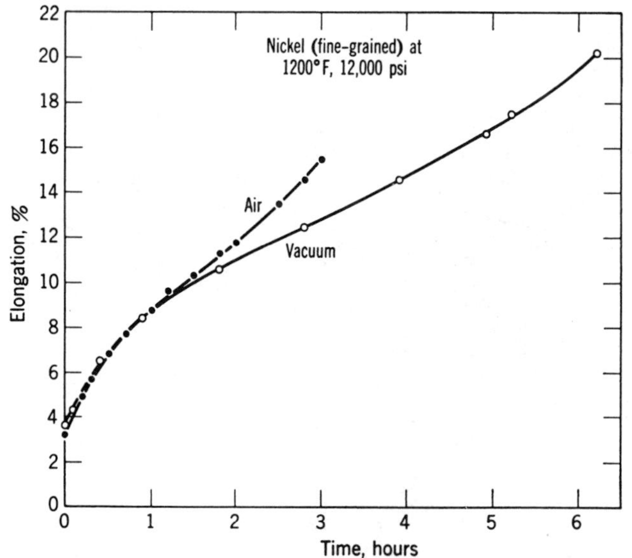

Creep curves in air and vacuum for commercially pure nickel at 1200 °F and 4.5 ksi (top) and 12.0 ksi (bottom).

Source: P. Shahinian and M.R. Achter, "A Comparison of the Creep-Rupture Properties of Nickel in Air and in Vacuum," in High Temperature Materials (Proceedings of a Conference in Cleveland), R.F. Hehemann and G.M. Ault, Eds., John Wiley & Sons, New York, 1959, p 455

CP Nickel: Effect of Air and Vacuum Environments on Minimum Creep Rate

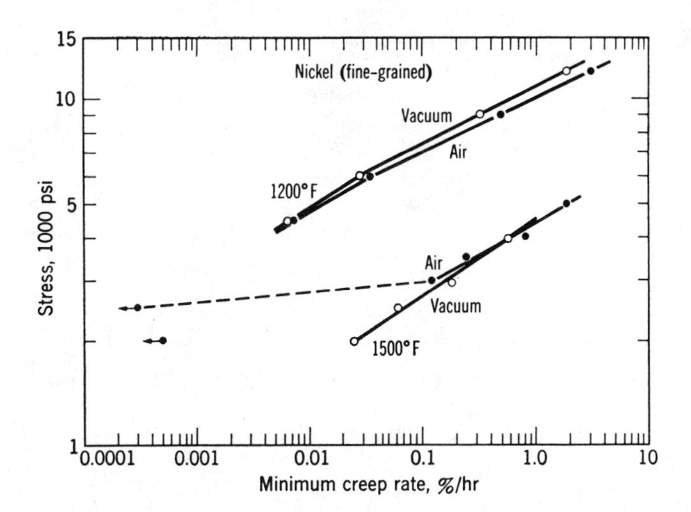

Comparison of vacuum (10^{-5} torr) vs air effects on creep capability of commercially pure nickel at 1200 and 1500 °F.

Source: P. Shahinian and M.R. Achter, "A Comparison of the Creep-Rupture Properties of Nickel in Air and in Vacuum," in High Temperature Materials (Proceedings of a Conference in Cleveland), R.F. Hehemann and G.M. Ault, Eds., John Wiley & Sons, New York, 1959, p 452

17.4 CP Nickel: Effect of Air and Vacuum Environments on Stress-Rupture Properties

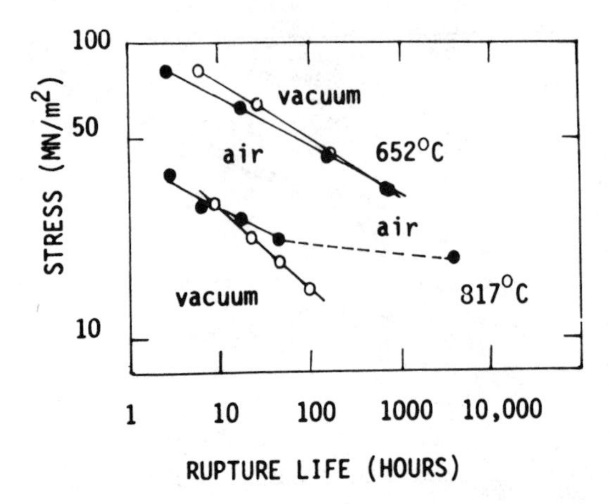

Comparison of vacuum (10^{-5} torr) vs air effects on stress-rupture capability of commercially pure nickel at 652 °C (1200 °F) and 817 °C (1500 °F).

Source: J.M. Davidson, K. Aning, and J.K. Tien, "Hot Environment Effects on Alloy Mechanical Properties," in Properties of High Temperature Alloys, Z.A. Foroulis and F.S. Pettit, Eds., Electrochemical Society, Princeton, NJ, 1976, p 197

CP Nickel: Effect of Grain Size on Rupture Properties in Air and Vacuum

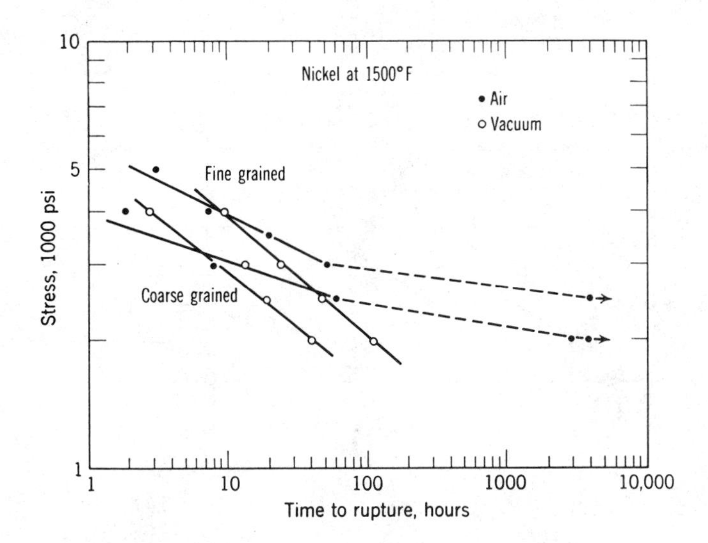

Effect of grain size on stress-rupture behavior of commercially pure nickel in vacuum (10^{-5} torr) and air at 1500 °F.

Source: P. Shahinian and M.R. Achter, "A Comparison of the Creep-Rupture Properties of Nickel in Air and in Vacuum," in High Temperature Materials (Proceedings of a Conference in Cleveland), R.F. Hehemann and G.M. Ault, Eds., John Wiley & Sons, New York, 1959, p 452

CP Nickel: Effect of Prestrain and Recovery Treatment

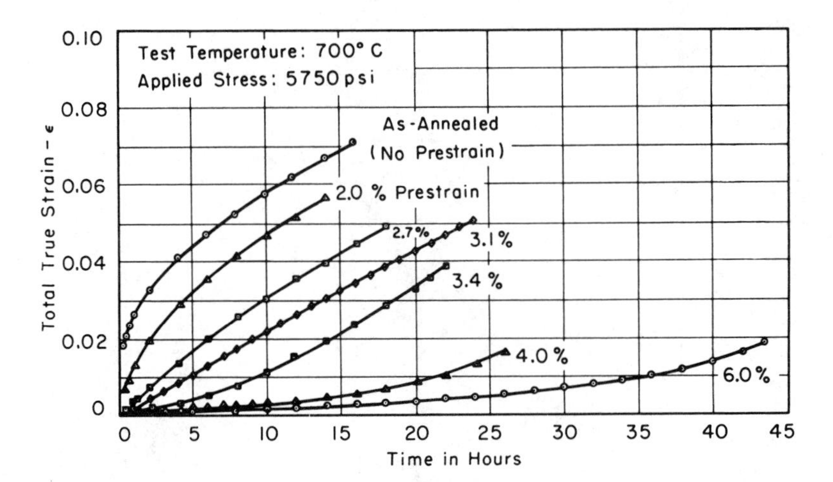

Effect of prestrain followed by 1-h recovery treatment at 800 °C (1470 °F) on the shape of the creep curve for commercially pure nickel at 700 °C (1292 °F) and 5.75 ksi.

Source: E.R. Parker and J. Washburn, "The Role of the Boundary in Creep Phenomena," in Creep and Recovery, American Society for Metals, Cleveland, OH, 1957, p 244

CP Nickel: Effect of Stress on Secondary Creep Rate

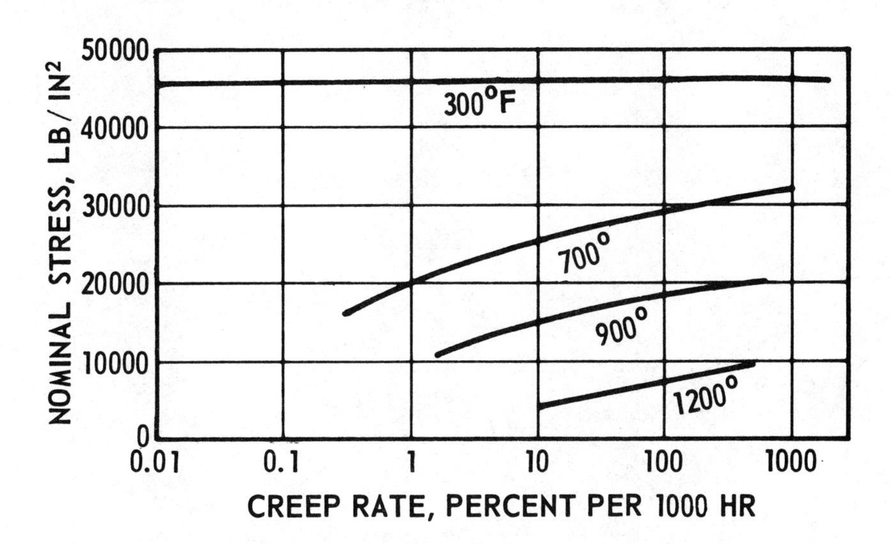

Influence of stress on the secondary creep rate of commercially pure nickel at four temperatures.

Source: S.J. Rosenberg, Nickel and Its Alloys, National Bureau of Standards Monograph 106, Washington, D.C., May 1968, p 46

Monel 400: Creep Properties of Cold-Drawn, Stress-Relieved Material

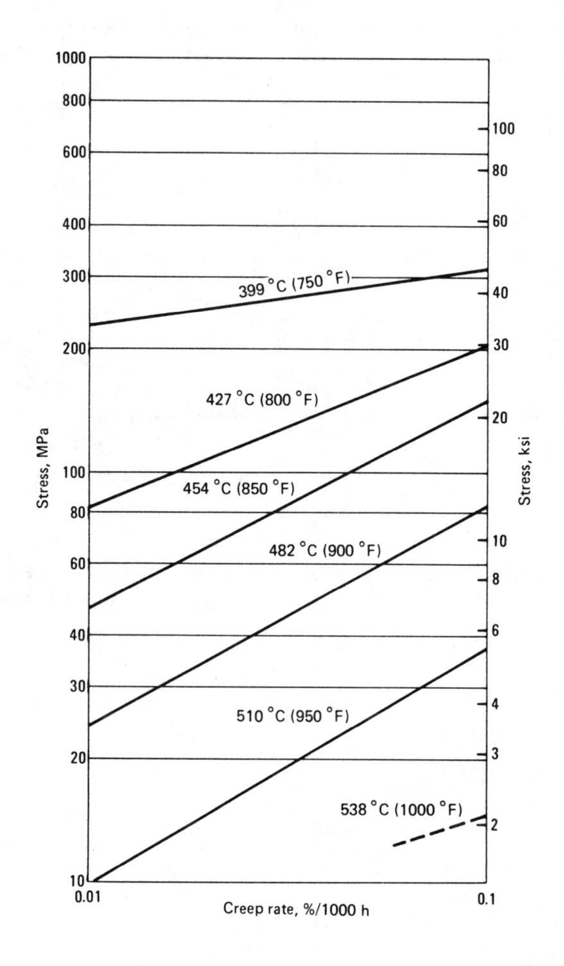

Creep properties of Monel 400 cold drawn 20% and stress relieved at 538 °C (1000 °F) for 8 h.

Source: Metals Handbook, Ninth Edition, Volume 3, Properties and Selection: Stainless Steels, Tool Materials and Special-Purpose Metals, American Society for Metals, Metals Park, OH, 1980, p 135

Monel 400: Log-Log Plot of Rupture Life vs Minimum Creep Rate

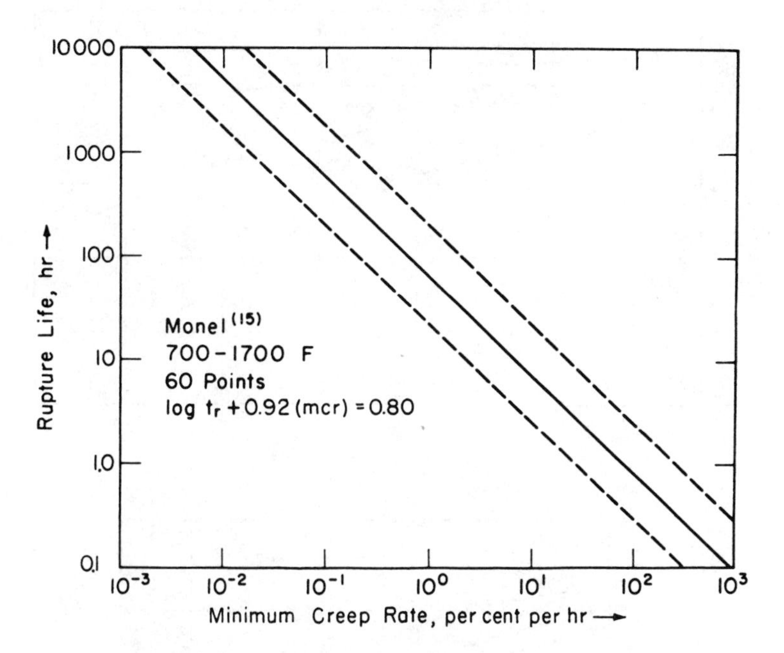

Grant-Monkman correlation of stress-rupture life and minimum creep rate for Monel 400.

Source: F.C. Monkman and N.J. Grant, "An Empirical Relationship between Rupture Life and Minimum Creep Rate in Creep-Rupture Tests," in Deformation and Fracture at Elevated Temperatures, N.J. Grant and A.W. Mullendore, Eds., M.I.T. Press, Cambridge, 1965, p 93

Monel 400: Log-Log Plot of Stress vs Rupture Life

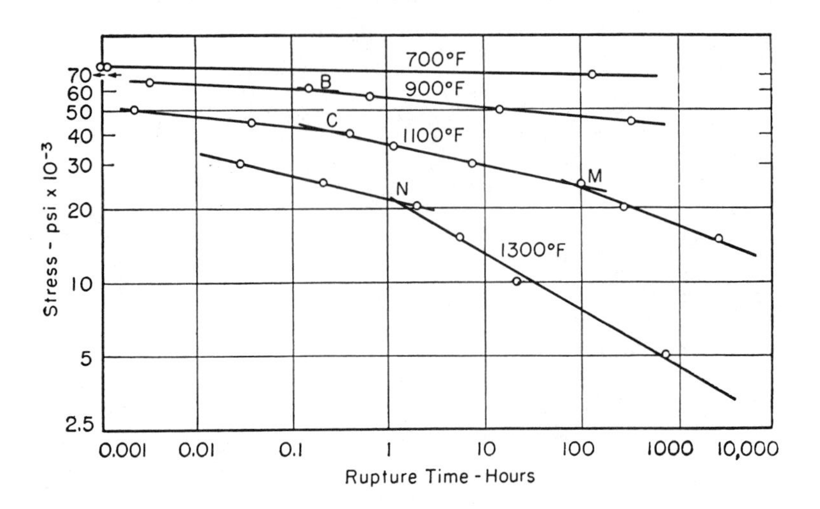

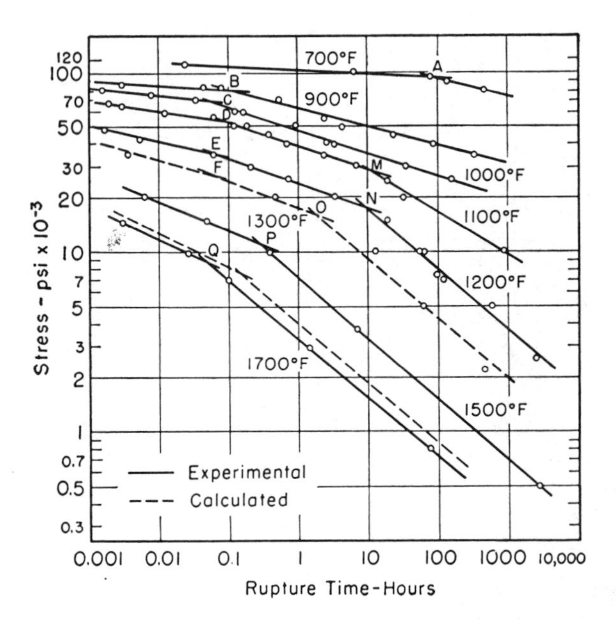

Log-stress vs log-rupture time plot for Monel 400 in two conditions - annealed (top) and cold worked 30% (bottom) - when tested at temperatures from 700 to 1300 °F (annealed) and to 1700 °F (cold worked).

Breaks in curve at A-F mark the transition from transcrystalline fracture (left of break) to intercrystalline (also called intergranular) fracture (right of break). Breaks in curve at M-Q represent effects of recrystallization and oxidation (right of break).

Source: N.J. Grant and A.G. Bucklin, "Creep-Rupture and Recrystallization of Monel from 700 to 1700 °F," Preprint No. 5, American Society for Metals, Cleveland, OH, 1952, p 6

Monel 400: Rupture Properties of Cold-Drawn, Stress-Relieved Material

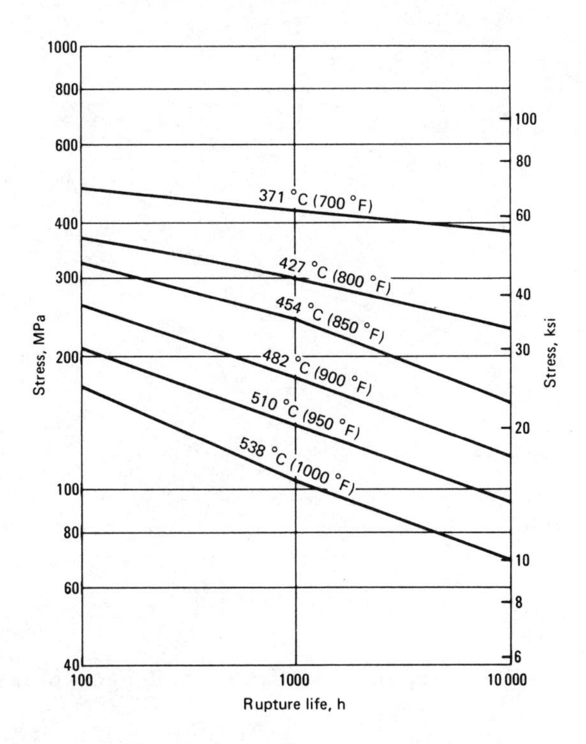

Rupture properties of cold-drawn Monel 400 stress relieved at 538 °C (1000 °F) for 8 h.

Source: Metals Handbook, Ninth Edition, Volume 3, Properties and Selection: Stainless Steels, Tool Materials and Special-Purpose Metals, American Society for Metals, Metals Park, OH, 1980, p 135

Monel K-500: Creep Properties of Cold-Drawn and Aged Material

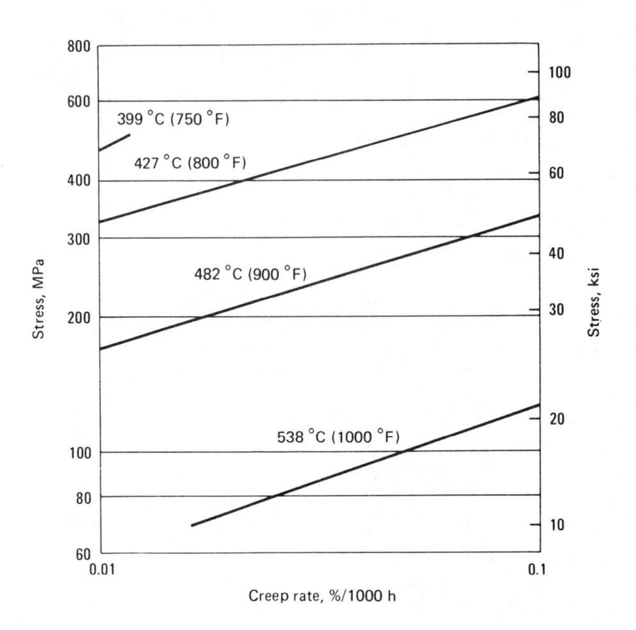

Creep properties of cold-drawn and aged Monel K-500.

Source: Metals Handbook, Ninth Edition, Volume 3, Properties and Selection: Stainless Steels, Tool Materials and Special-Purpose Metals, American Society for Metals, Metals Park, OH, 1980, p 139

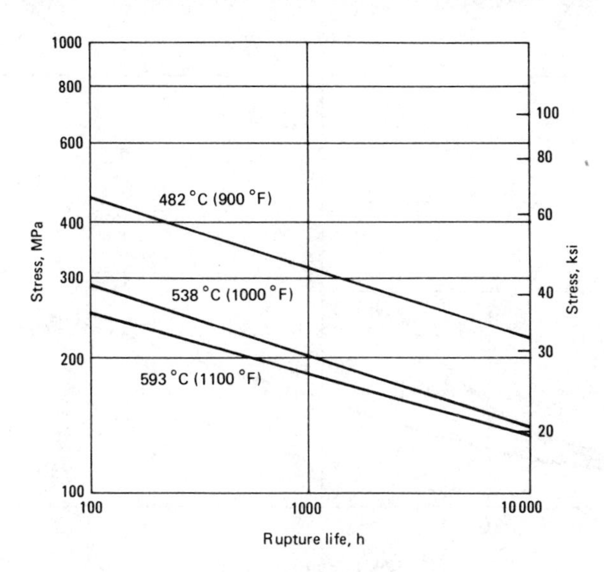

Rupture life of hot-finished aged Monel K-500.

Source: Metals Handbook, Ninth Edition, Volume 3, Properties and Selection: Stainless Steels, Tool Materials and Special-Purpose Metals, American Society for Metals, Metals Park, OH, 1980, p 139

17.14 Ni-Mo, Ni-W, and Ni-Cr: Comparison of 100-h Rupture Lives as Affected by Binary Alloy Additions

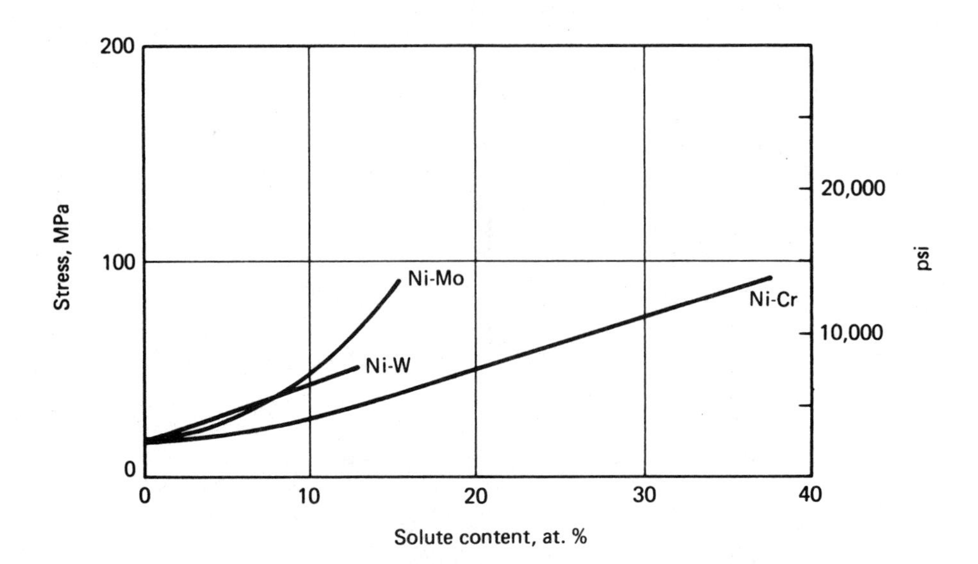

Effect of binary alloy additions of tungsten, molybdenum, or chromium on the 100-h stress-rupture life of nickel-base solid-solution alloys.

Source: C.R. Brooks, Heat Treatment, Structure, and Properties of Nonferrous Alloys, American Society for Metals, Metals Park, OH, 1982, p 147

TD Nickel: 1000-h Rupture Strength

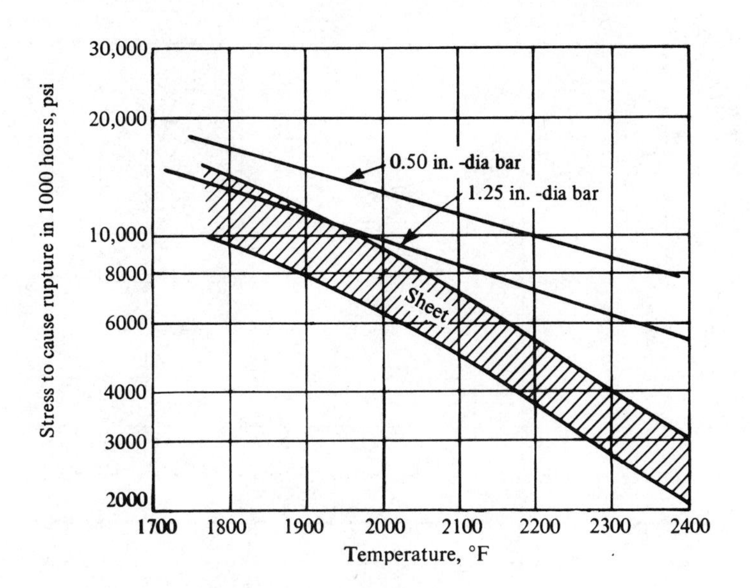

Comparison of bar and sheet stock; stress for 1000-h rupture life vs temperature for oxide-dispersion strengthened nickel (thoria-dispersed nickel).

Thoria was the oxide used to harden nickel.

Source: F.J. Clauss, Engineer's Guide to High-Temperature Materials, Addison-Wesley, Reading, MA, 1969, p 174

TD Nickel-Chromium: Influence of Temperature Gradient

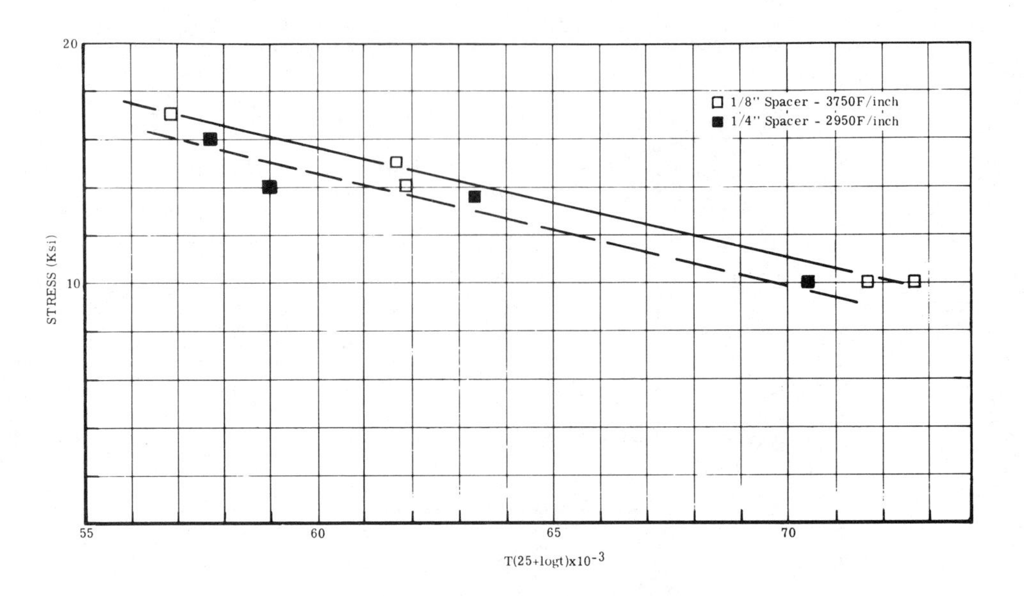

Influence of temperature gradient on stress-rupture properties of TD nickel-chromium bar TC2760-1B.

Specimens were prepared at 24 in./h using 1/8- and 1/4-in. separations between the induction coil and the water-cooled Cu chill block and were stress-rupture tested at 1800, 2000, and 2200 °F. Note that the material prepared with the 1/8-in. spacer (2450 °F/in.) appears to be at least two parameters better than material prepared with the 1/4-in. spacer (3750 °F/in.).

Source: R.E. Allen, "Directionally Recrystallized TD NiCr," in Superalloys – Processing (Proceedings of the Second International Conference, Champion, PA), American Institute of Mining, Metallurgical, and Petroleum Engineers, Warrendale, PA, 1972, p X-7, X-22

18

Cast Irons

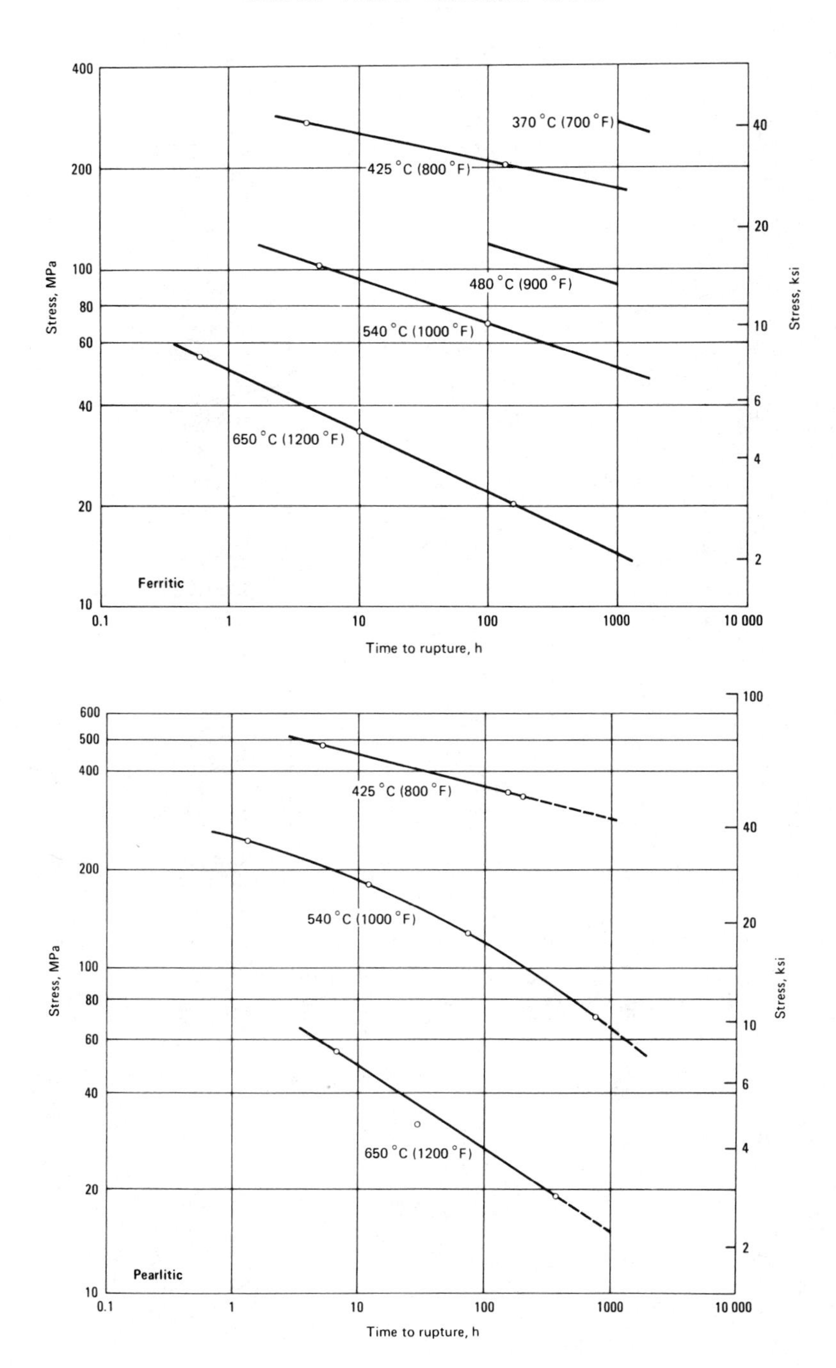

Stress-rupture properties of 2.5Si-1.0Ni ductile iron.

Source: Metals Handbook, Ninth Edition, Volume 1, Properties and Selection: Irons and Steels, American Society for Metals, Metals Park, OH, 1978, p 50

4Si Ductile Iron: Effect of 2% Molybdenum

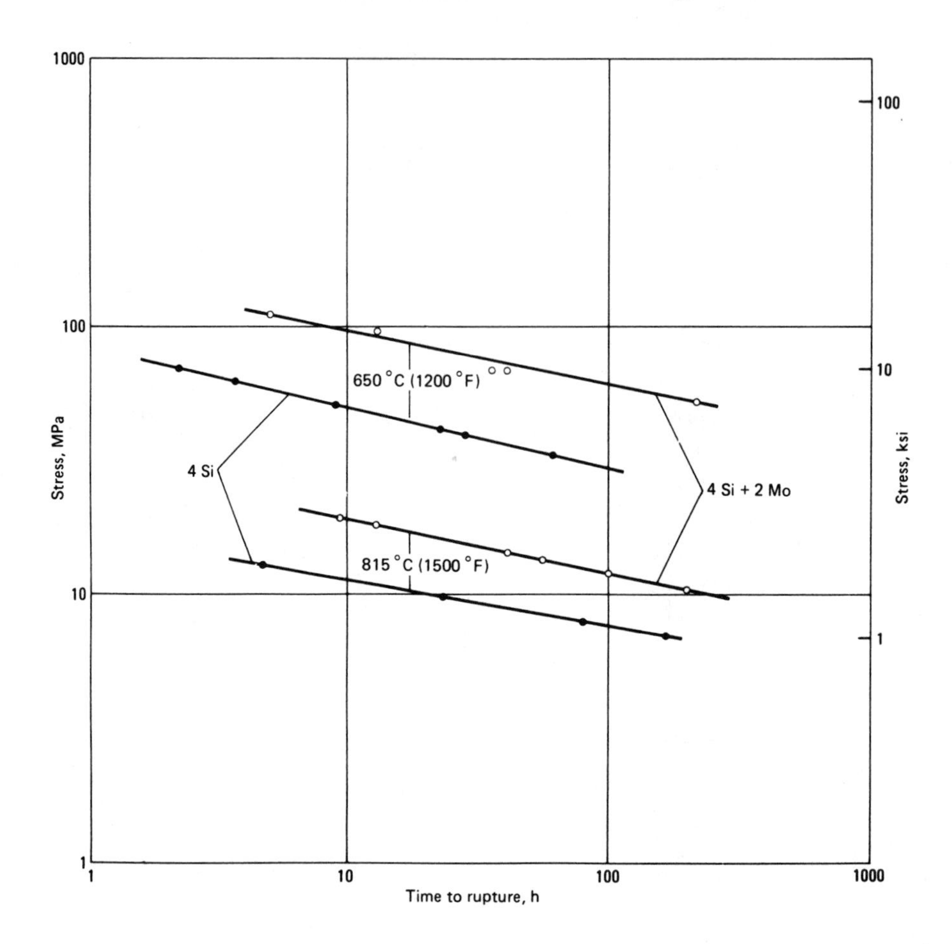

Effect of 2% molybdenum on the stress rupture of 4Si ductile iron.

Source: Metals Handbook, Ninth Edition, Volume 1, Properties and Selection: Irons and Steels, American Society for Metals, Metals Park, OH, 1978, p 51

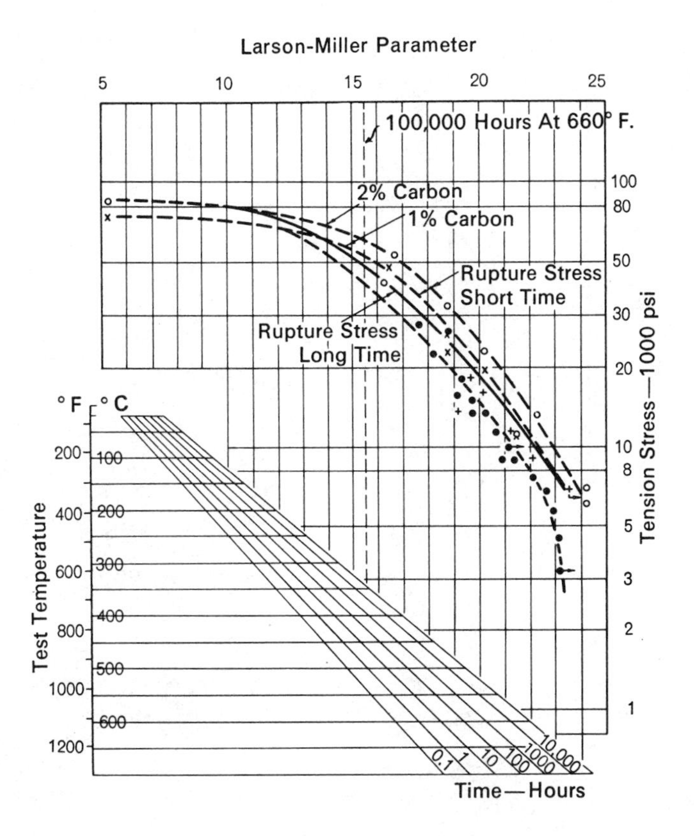

Stress-rupture characteristics of 24-34% chromium irons, with an example of 100,000 h at 660 °F. The Larson-Miller parameter (top scale) is equal to $T(20 + \log t)$, with T in Kelvin and t in hours.

Comparison of these data to those for gray and ductile irons clearly indicates that the higher Cr irons are markedly superior in creep properties. Data are available that indicate that the hot hardness values of these irons may be used to assess the creep-strength properties.

Source: Gray and Ductile Iron Castings Handbook, C.F. Walton, Ed., Gray and Ductile Iron Founders' Society, Cleveland, 1971, p 285

Class 40 Gray Iron: Growth and Creep at 500 °C

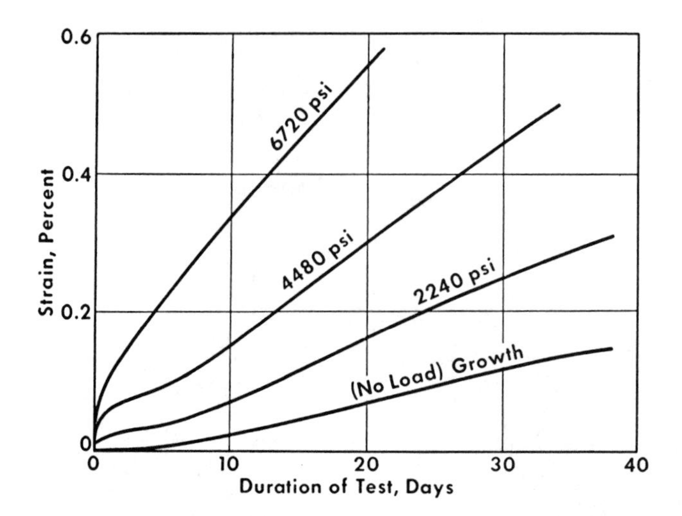

Growth and creep of a typical class 40 unalloyed gray iron at 500 °C.

Analysis: 3.19% TC, 0.85% CC, 1.66% Si, 0.91% Mn, 0.09% S, 0.08% P. Tensile strength: 42,000 psi.

Source: The Gray Iron Castings Handbook, C.F. Walton, Ed., Gray Iron Founders' Society, Cleveland, 1958, p 177

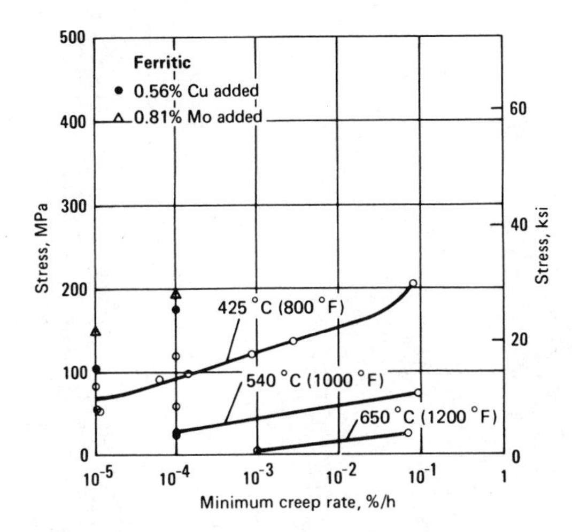

Effect of molybdenum and copper additions on the creep characteristics of ductile iron.

Source: Metals Handbook, Ninth Edition, Volume 1, Properties and Selection: Irons and Steels, American Society for Metals, Metals Park, OH, 1978, p 49

Ductile Iron: Creep Characteristics

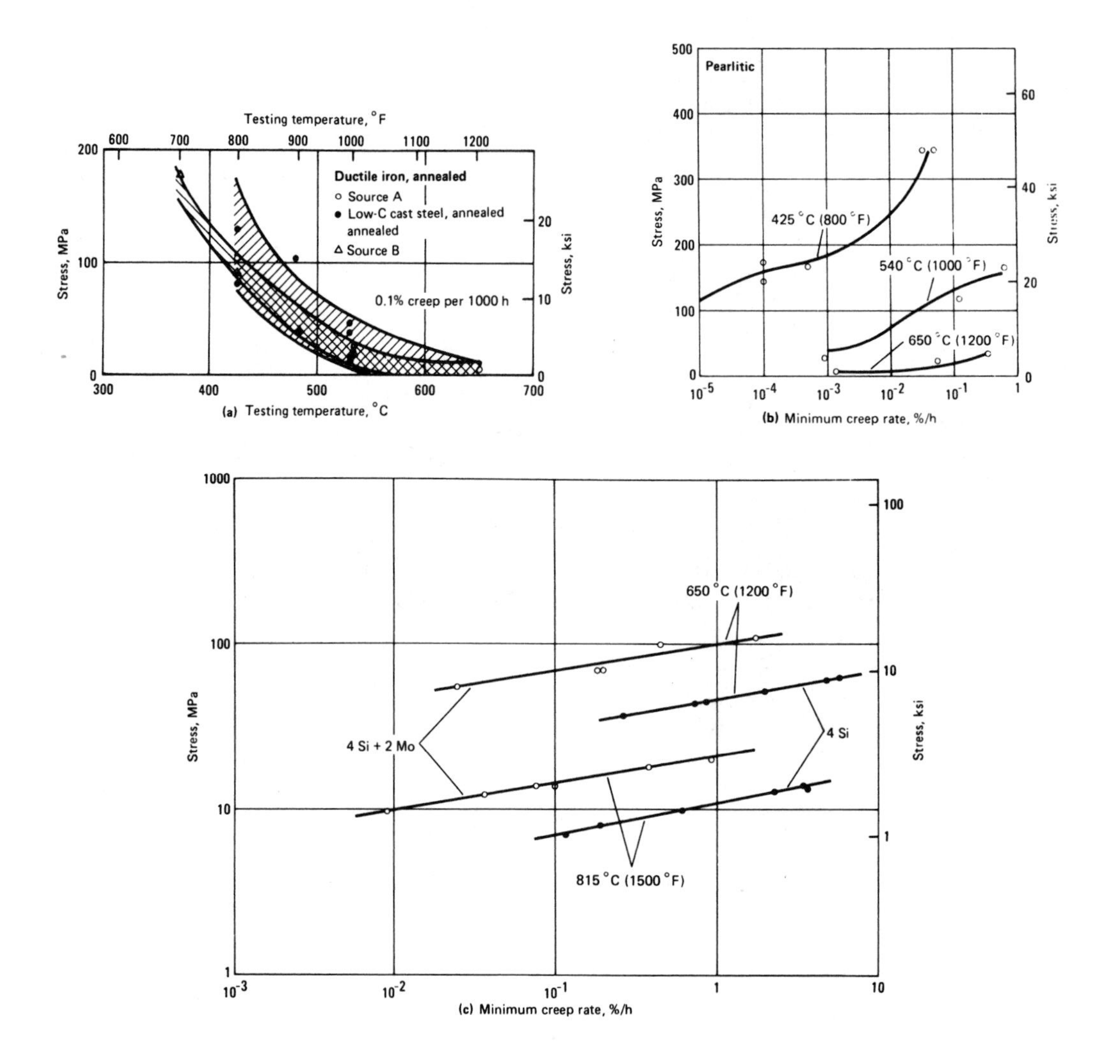

Creep characteristics of ductile iron.

Source: Metals Handbook, Ninth Edition, Volume 1, Properties and Selection: Irons and Steels, American Society for Metals, Metals Park, OH, 1978, p 48

Ductile Iron: Larson-Miller Curve of Stress-Rupture Properties

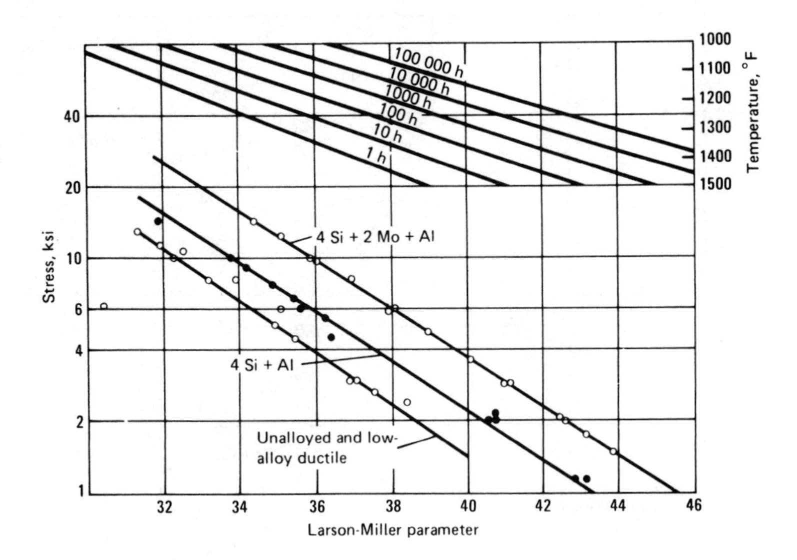

Master curves for stress rupture of ductile iron.

Larson-Miller parameter is $10^{-3}\ T(20 + \log t)$, where T is temperature in °R and t is time to rupture in hours.

Source: Metals Handbook, Ninth Edition, Volume 1, Properties and Selection: Irons and Steels, American Society for Metals, Metals Park, OH, 1978, p 51

Ductile Iron: Molybdenum-Free Material

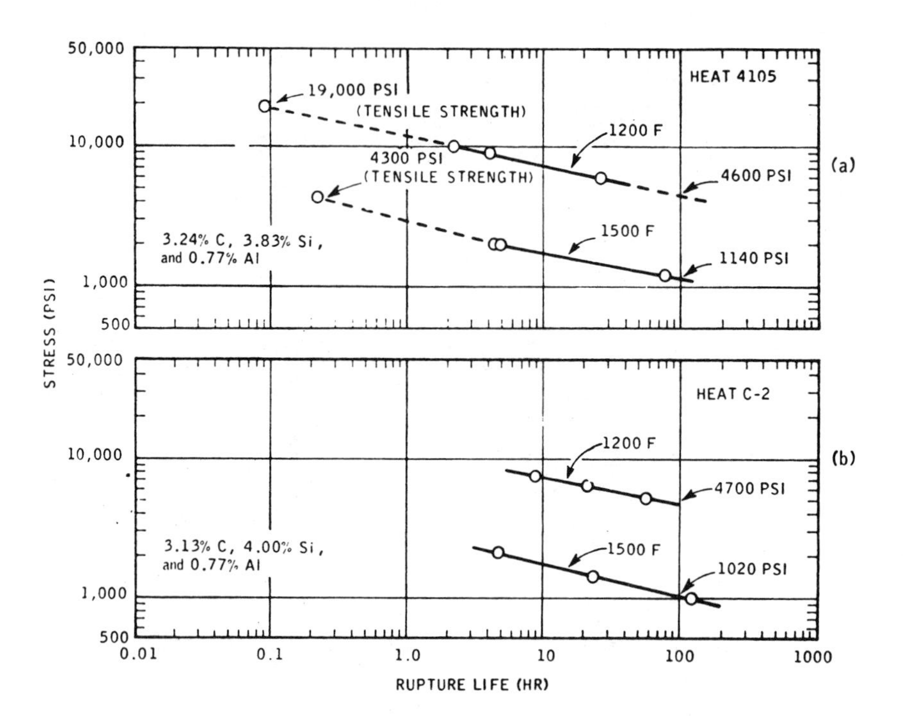

Stress-rupture curves for molybdenum-free ductile iron alloys tested at 1200 and 1500 °F (649 and 816 °C). These alloys were cast as 1-in. Y-blocks and normalized at 1750 °F (954 °C).

Source: D.L. Sponseller, W.G. Scholz, and D.F. Rundle, "Development of Low-Alloy Ductile Irons for Service at 1200-1500 F," in Source Book on Ductile Iron, A.H. Rauch, Ed., American Society for Metals, Metals Park, OH, 1977, p 382

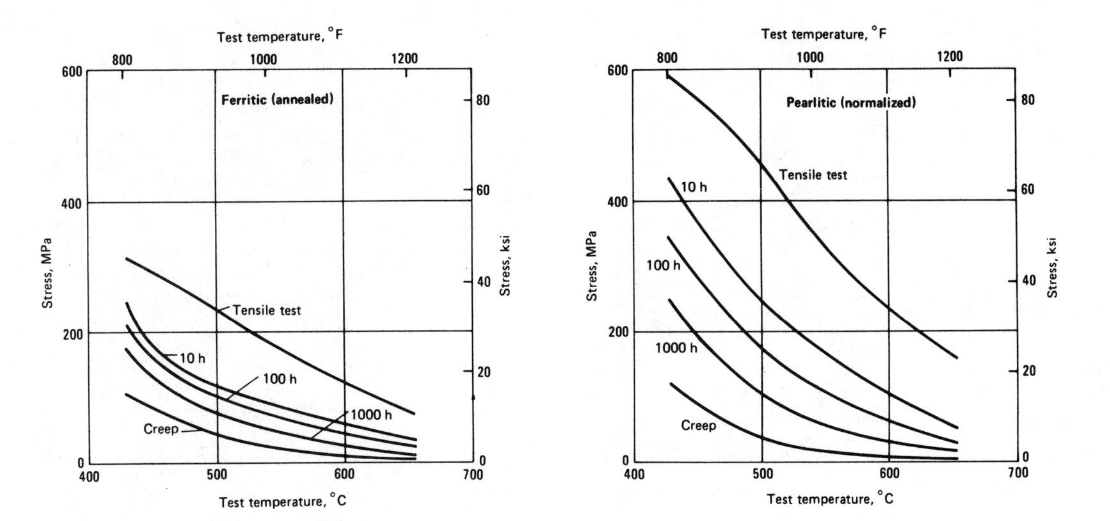

Stress-rupture properties of ductile iron.

The curve labeled "creep" shows the stress-temperature combination that will result in a creep rate of 0.0001%/h.

Source: Metals Handbook, Ninth Edition, Volume 1, Properties and Selection: Irons and Steels, American Society for Metals, Metals Park, OH, 1978, p 49

Ferritic Malleable Iron

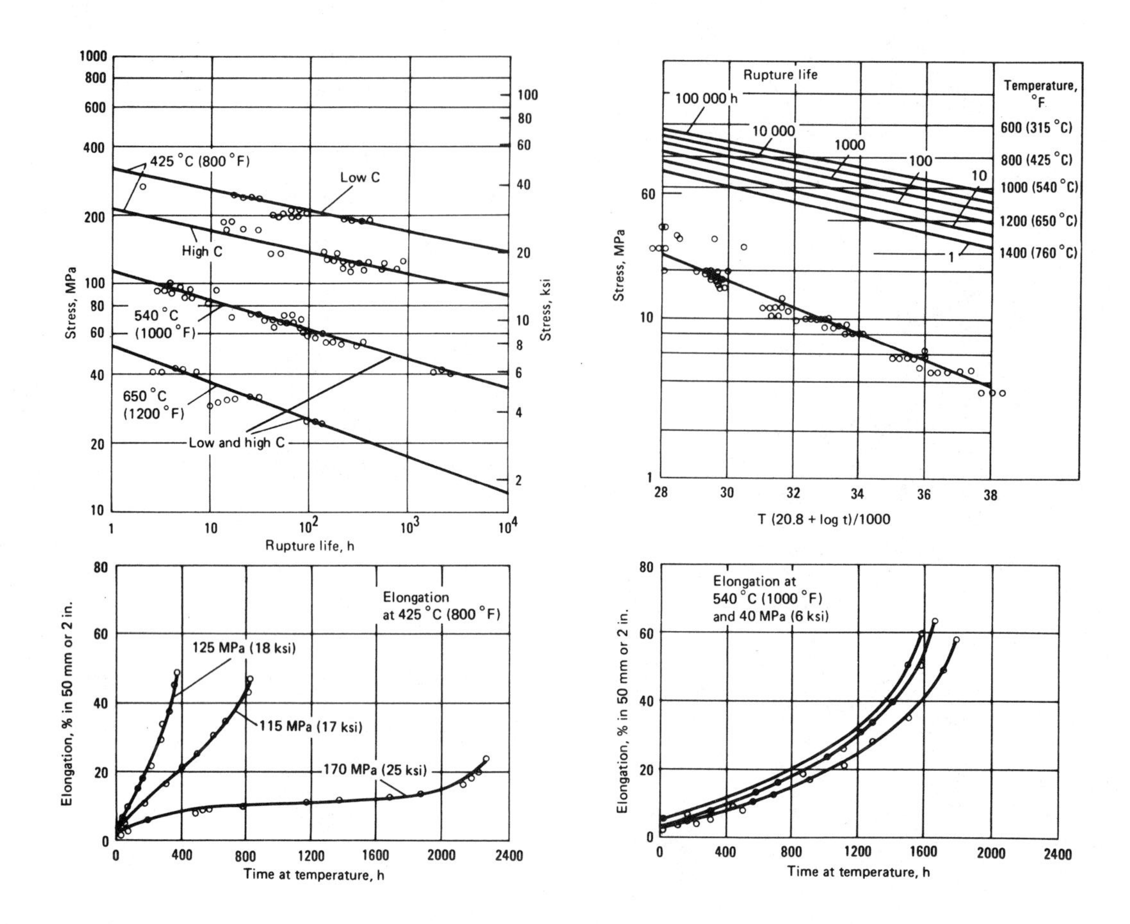

Stress-rupture and creep data for ferritic malleable iron.

The two lower charts show typical time-elongation curves, clearly indicating the three stages of creep; the normally ductile behavior of ferritic malleable iron at all times and temperatures is emphasized.

Source: Metals Handbook, Ninth Edition, Volume 1, Properties and Selection: Irons and Steels, American Society for Metals, Metals Park, OH, 1978, p 66

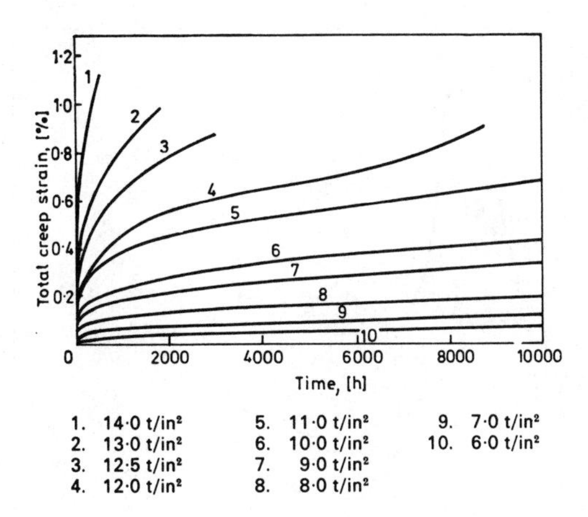

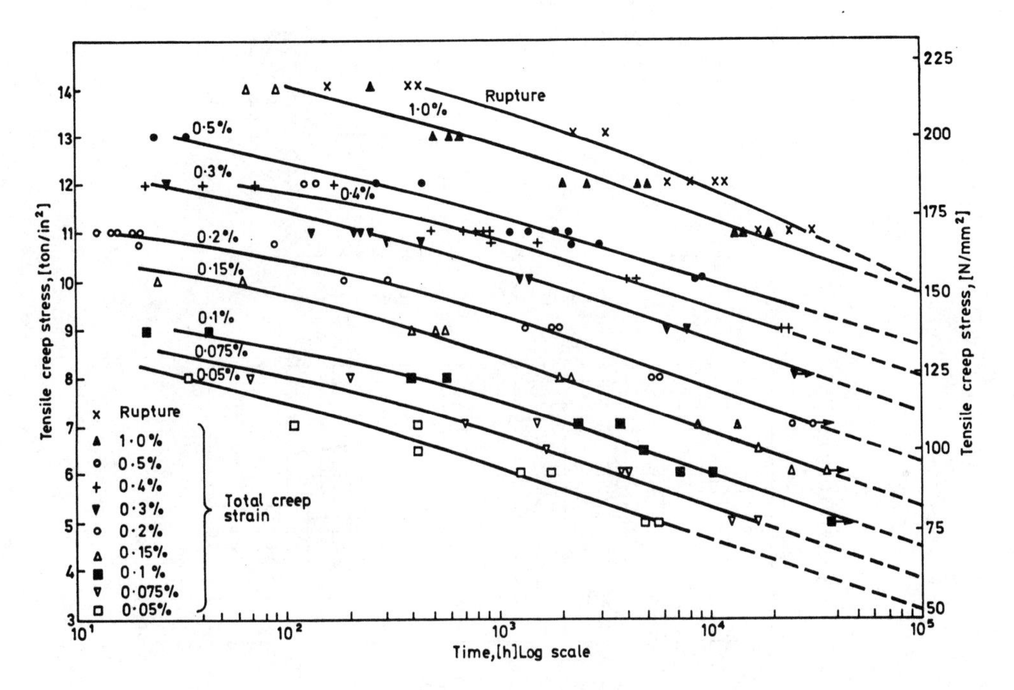

Creep curves of strain against time at 350 °C for 1.2-in. cast section size (top) and stress/log time curves for given strains and rupture of 3-in. cast section size at 350 °C.

Typical total creep strain curves for tests at stresses ranging from 6 tons/in.^{2} (90 N/mm^2) to 14 tons/in.^{2} (215 N/mm^2) for times up to 10,000 h are shown for a 1.2-in. (30-mm) section size in the top figure. A similar family of curves was obtained on a 3-in. (76-mm) section size, and from these curves, the stresses to produce various creep strains and rupture are shown plotted against the logarithm of time in the bottom figure. For a grade 17 gray iron, variations in cast section size from 1.2-in. (30-mm) to 3-in. (76-mm) diameter have little effect on stress-rupture properties at 350 °C. The rupture specimens had elongations after failure of 1.2-2.3%.

Source: K.B. Palmer, "High Temperature Properties of Cast Irons," Iron and Steel, Vol. 44, Feb. 1971, p 43

Gray and Ductile Iron: Effect of Aluminum Content

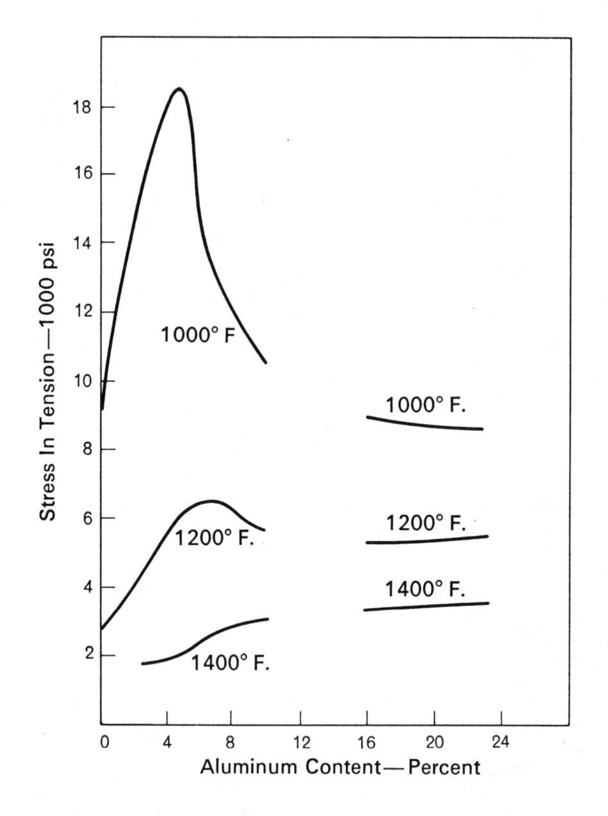

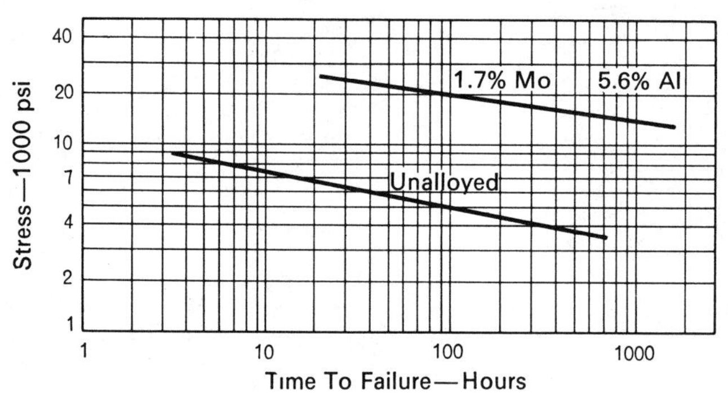

Effect of aluminum content on the stress to fracture in 100 h at three temperatures (top) and stress-rupture characteristics of alloyed ductile iron (bottom).

The test temperature was not stated, but the unalloyed iron data appear to be for 1100 °F, and the alloyed iron tests may have been at 1000 °F.

Source: Gray and Ductile Iron Castings Handbook, C.F. Walton, Ed., Gray and Ductile Iron Founders' Society, Cleveland, 1971, p 283

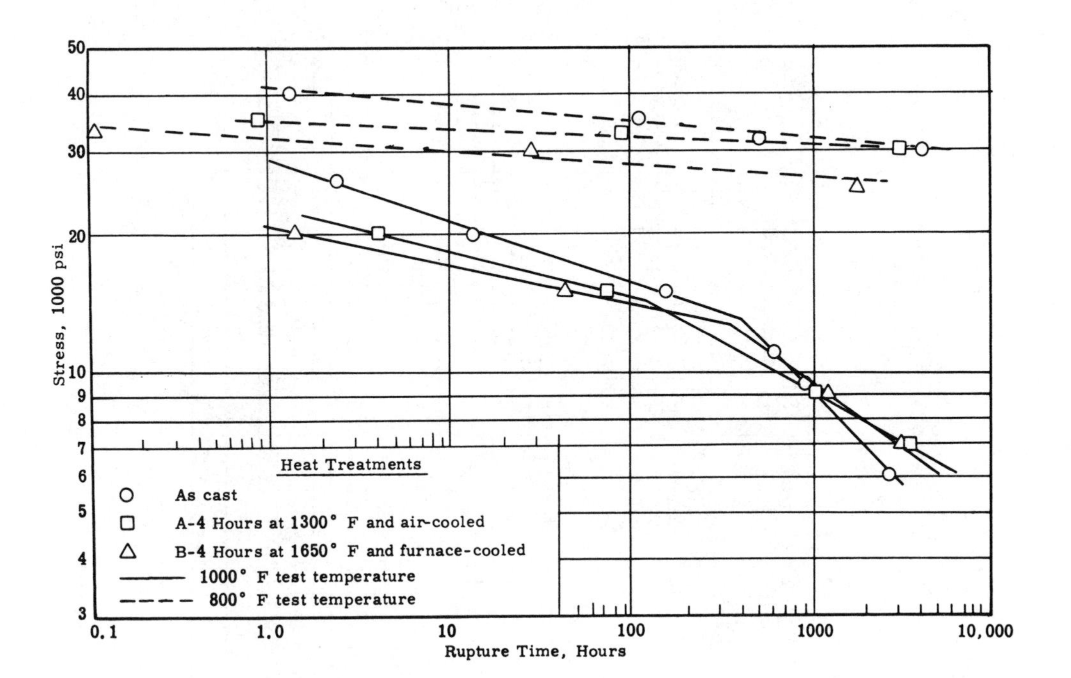

Rupture times of molybdenum-alloyed gray iron at 1000 and 800 °F after various heat treatments.

Source: Properties of Cast Iron at Elevated Temperatures, ASTM Special Technical Publication No. 248, American Society for Testing Materials, Philadelphia, 1959, p 53

Gray Iron: Unalloyed

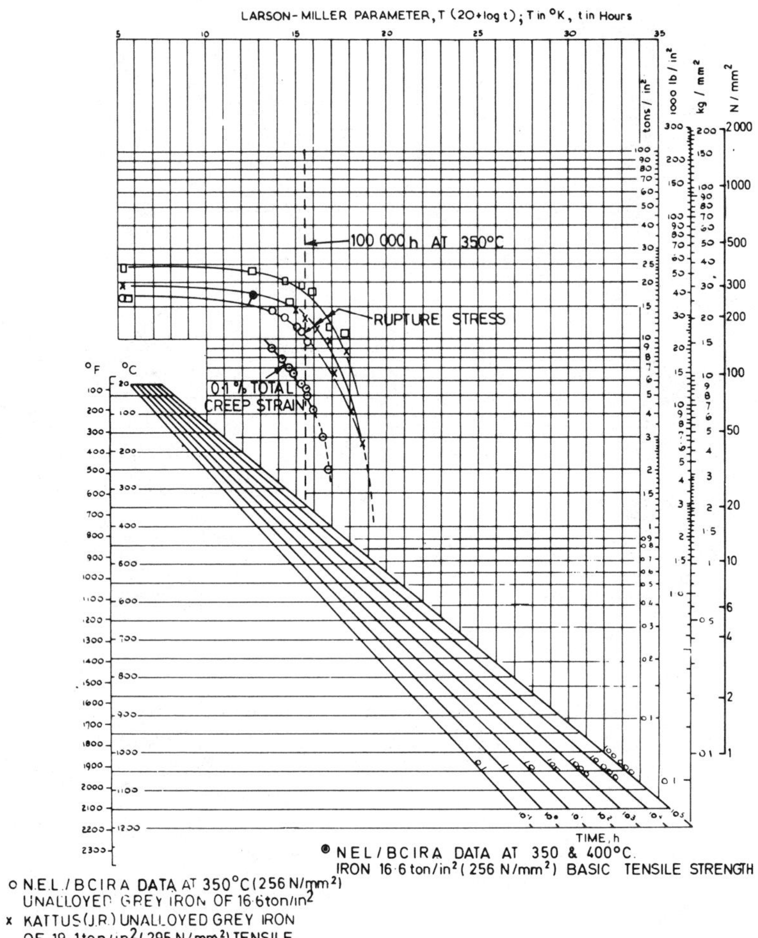

Rupture strength of gray iron, unalloyed; 16-19 ton/in.² (247-293 N/mm²) tensile strength. Rupture strength of gray iron alloyed 0.51% molybdenum 24.8 ton/in.² (383 N/mm²) tensile strength. NEL/BCIRA data from 1-, 2-, and 3-in. diam. bars (25-, 51-, 76-mm). Tottle data from 0.875-in. (22-mm) diam. bars.

If the applied stress is sufficiently high at the test temperature for the tertiary stage of creep to develop, rupture will ultimately occur if the stress is applied for a sufficient time. Using the Larson-Miller parameter, it is possible to make use of short-time tests to determine long-time properties, and a series of curves is given for various cast irons, which makes it possible to predict the stress to rupture in times up to 100,000 h for a range of temperatures. These curves are based on tests of duration up to 30,000 h and include short-time tests. The results are remarkably consistent from various experimental data where the materials under test are comparable. This method of plotting appears to be valid also for total creep strain where information is available.

The curves for gray irons and alloyed gray irons in the figure above are plotted by the method devised by the National Engineering Laboratory, in which time, temperature, and stress are included on one curve of the Larson-Miller parameter. Short-time tensile tests up to 600 °C, in which the duration of the test is assumed to be 0.1 h, fall consistently with little scatter on the curves for rupture tests up to 400 °C and with rupture times up to 30,000 h. Therefore, short-time tensile tests at varying temperatures can be used to give a preliminary assessment of the suitability of materials for long-term stressed conditions at various temperatures.

Source: H.T. Angus, Cast Iron: Physical and Engineering Properties, Butterworths, London, 1976, p 271-272

Pearlitic Malleable Iron: Creep Properties

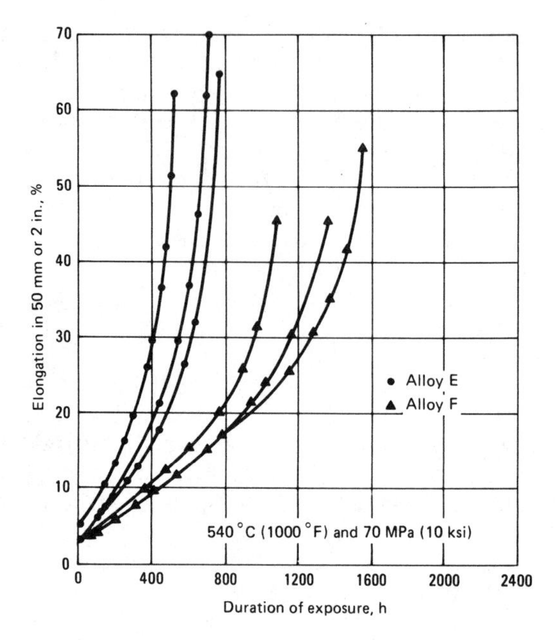

Creep properties of pearlitic malleable iron.

Source: Metals Handbook, Ninth Edition, Volume 1, Properties and Selection: Irons and Steels, American Society for Metals, Metals Park, OH, 1978, p 51

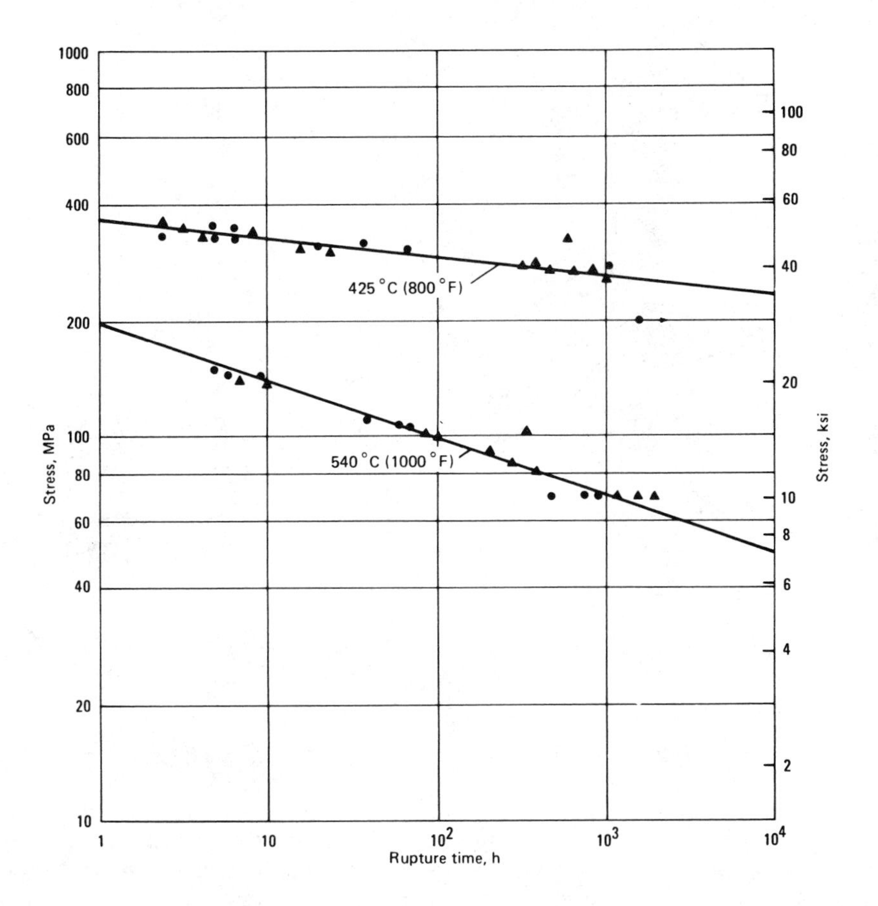

Stress-rupture properties of pearlitic malleable iron.

Source: Metals Handbook, Ninth Edition, Volume 1, Properties and Selection: Irons and Steels, American Society for Metals, Metals Park, OH, 1978, p 51

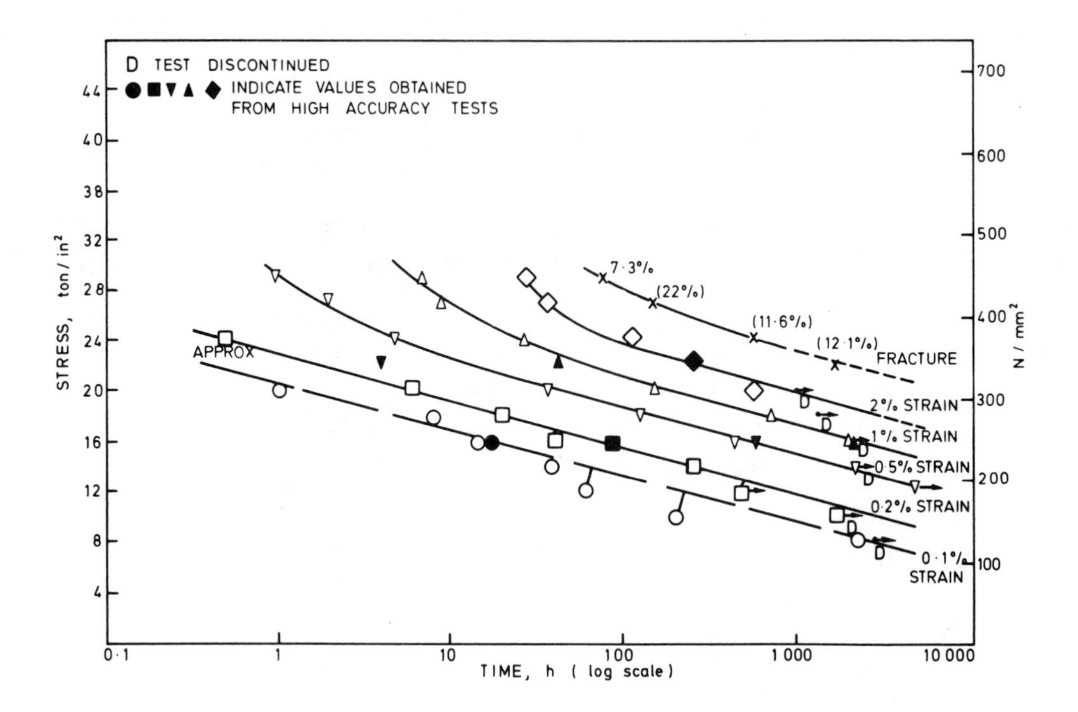

Creep tests at 400 °C for pearlitic nodular cast iron.

Although the pearlitic iron had the highest creep resistance at 426 °C, its structural instability at the higher temperature of 538 °C gave poor long-term load-bearing properties, being inferior to the ferritic matrix.

The creep resistance of an as-cast pearlitic nodular iron at 400 °C has been evaluated. The results indicate that the stresses to produce 0.1, 0.2 and 0.5% strain in 1000 h are, respectively, 9.5, 11.7, and 15.0 ton/in.^{2} (147, 181, and 232 N/mm^2) and compare with 5.5 and 6.8 ton/in.^{2} (85 and 105 N/mm^2) for 0.1 and 0.2% strain, respectively, when using gray cast iron.

Source: H.T. Angus, Cast Iron: Physical and Engineering Properties, Butterworths, London, 1976, p 212, 214

4Si-2Mo-Al Alloy vs Unalloyed and Low-Alloy Ductile Irons: A Comparison

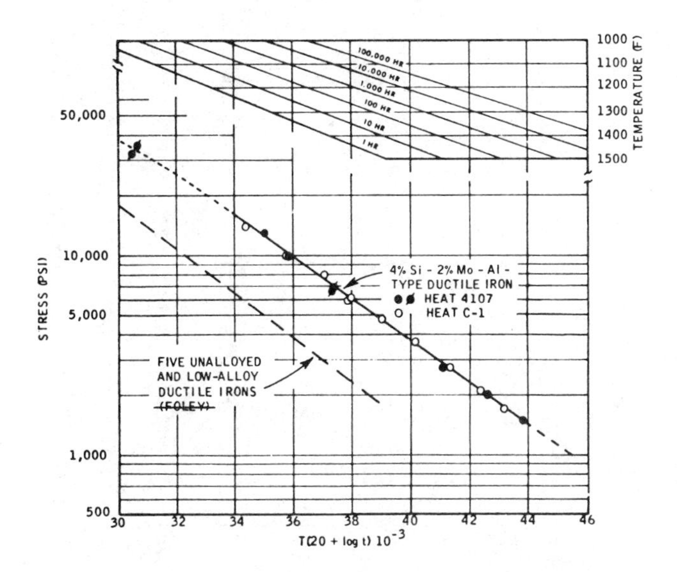

Variation of stress with Larson-Miller parameter for various normalized ductile irons of the 4Si-2Mo-Al type (cast as 1-in. Y-blocks) and for five unalloyed and low-alloy ductile irons.

Because of the promising elevated-temperature strength of the normalized high-Si alloy containing Al and 2% Mo, it seems appropriate to provide an indication of the rupture strength of this alloy at temperatures and times that are different from those employed in this investigation. This is done by plotting the results of elevated-temperature tests for heats 4107 and C-1 in the stress-parameter diagram of the figure above. All rupture tests for these alloys are represented by plotting stress as a function of the Larson-Miller parameter, $T(20 + \log t)\ 10^{-3}$, where T is the test temperature in degrees Rankine (degrees fahrenheit + 460) and t is rupture life in hours.

Source: D.L. Sponseller, W.G. Scholz, and D.F. Rundle, "Development of Low-Alloy Ductile Irons for Service at 1200-1500 F," in Source Book on Ductile Iron, A.H. Rauch, Ed., American Society for Metals, Metals Park, OH, 1977, p 386

Austenitic Flake, Nodular, and 30Cr Irons: A Comparison

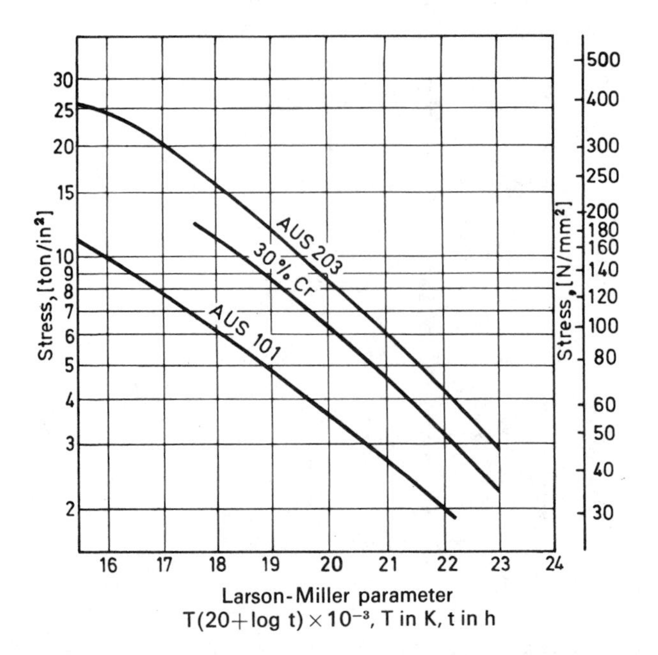

Stress-to-rupture properties of austenitic nodular iron AUS 203, austenitic flake graphite iron AUS 101, and a 1C-30Cr iron.

The highly alloyed austenitic and 30% Cr irons have excellent growth and scaling resistance at high temperautres. Larson-Miller plots of some of the available rupture data on austenitic flake, nodular (SG) irons, and 30% Cr irons are shown in the figure above. The data on the Ni-Resist flake graphite iron, AUS 101, is extremely limited and subject to some scatter due to variations in composition, but has been included in the figure above for comparison with the other irons. The low-Cr, high-Mn nodular (SG) iron, AUS 203, has the highest rupture properties, and rupture elongations between 7 and 17% are reported. This iron has good structural stability and scaling resistance at temperatures up to 650 ° C. It has been shown that the best rupture properties in austenitic nodular (SG) irons are obtained with a high-Ni, high-Cr grade containing Mo.

Source: K.B. Palmer, "High Temperature Properties of Cast Irons," Iron and Steel, Vol. 44, Feb. 1971, p 45

Cast Iron: Comparison of the Creep Strengths of Four Cast Irons at 1000 °F

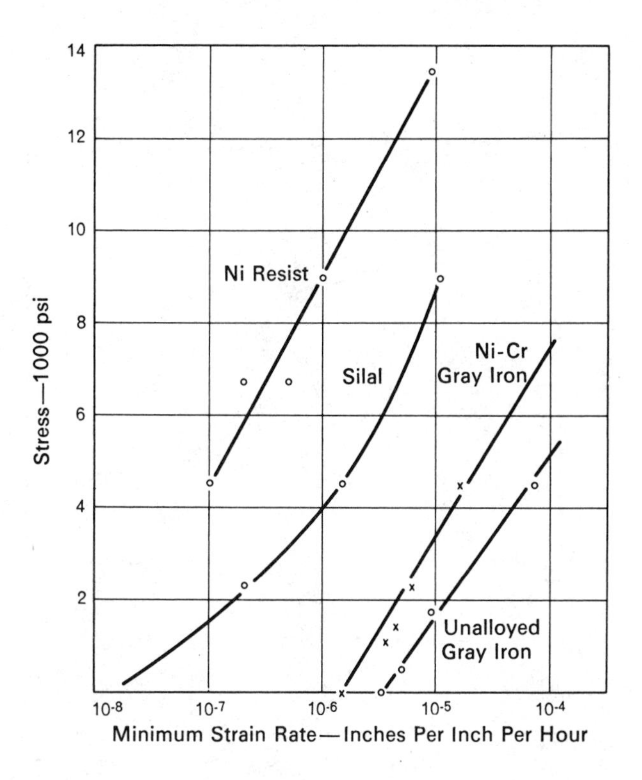

Comparative creep strength of four cast irons at 1000 °F.

Creep data for high-alloy cast irons compared to plain and low-alloy gray irons are shown in the above figure. The creep rate at 1000 °F for even the best iron is fairly high, and the irons are all inferior to cast 18-8 stainless steel.

Source: Gray and Ductile Iron Castings Handbook, C.F. Walton, Ed., Gray and Ductile Iron Founders' Society, Cleveland, 1971, p 283-284

Cast Iron: Comparison of Stress-Rupture Properties of High-Chromium Ferritic and Austenitic Material

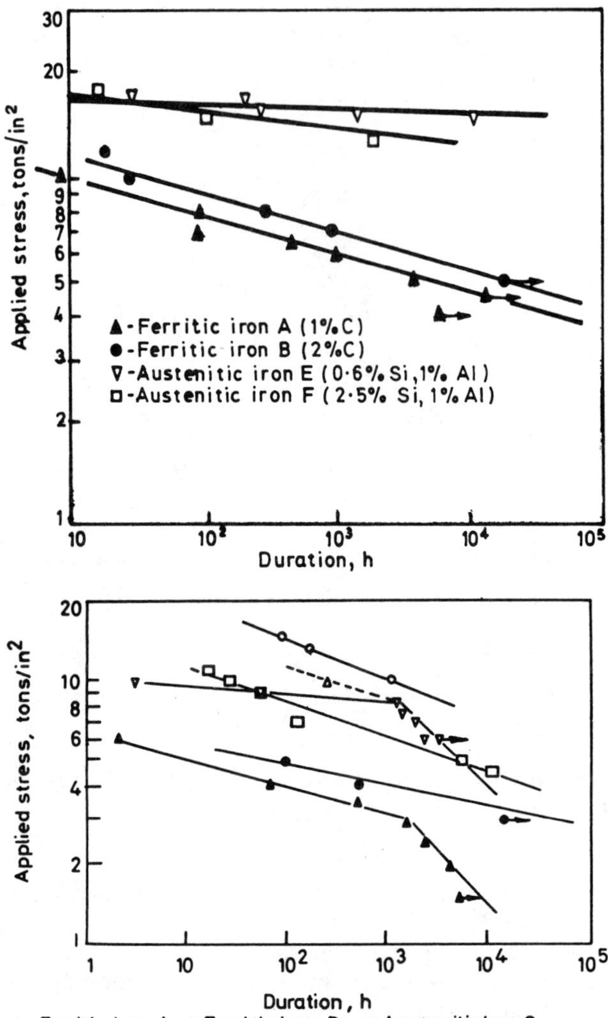

Stress-rupture data at 600 °C (top) and at 700 °C (bottom).

Cast Iron: Comparison of Stress-Rupture Properties of High-Chromium Ferritic and Austenitic Material (Continued)

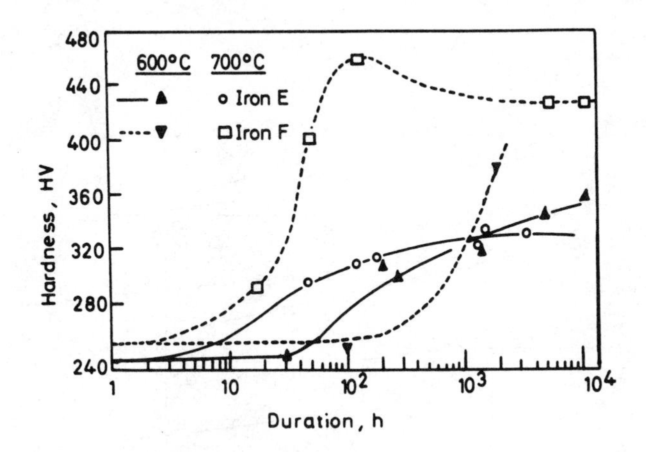

Age hardening of the austenitic irons E and F.

The strengths of all the irons exceed the requirement for their proposed use (i.e., stress to rupture in 10^5 h, >0.7 ton/in.^{2}) and are comparable to those of wrought alloy steels currently used, for example, in modern power stations. This demonstrates that these irons could be useful and economic alternatives to high alloy steels for many high-temperature applications, especially in view of their excellent corrosion resistance. Where reasonsable ductility is required, the 1% C, ~34% Cr ferritic iron composition is to be preferred; greater strength can be obtained, at greater cost, by the use of an austenitic iron. The rupture strengths of these materials are superior to those of the low-alloy cast irons and most Ni-rich austenitic cast irons. Extrapolation of the data on cast 25% Cr, 20% Ni, 0.4% C steels commonly used in reformer furnaces shows them to have similar strengths at 700 °C to austenitic iron F.

Source: C.W. Corti, J.W. Boyes, and C.W. King, "The Stress-Rupture Properties and Structural Stability of High Chromium Cast Irons at Elevated Temperature," Iron and Steel, Vol. 42, Oct. 1969, p 338, 341

18.24 Ductile Iron: Comparison of Stress-Rupture Properties of High-Nickel Heat-Resistant Cast Irons

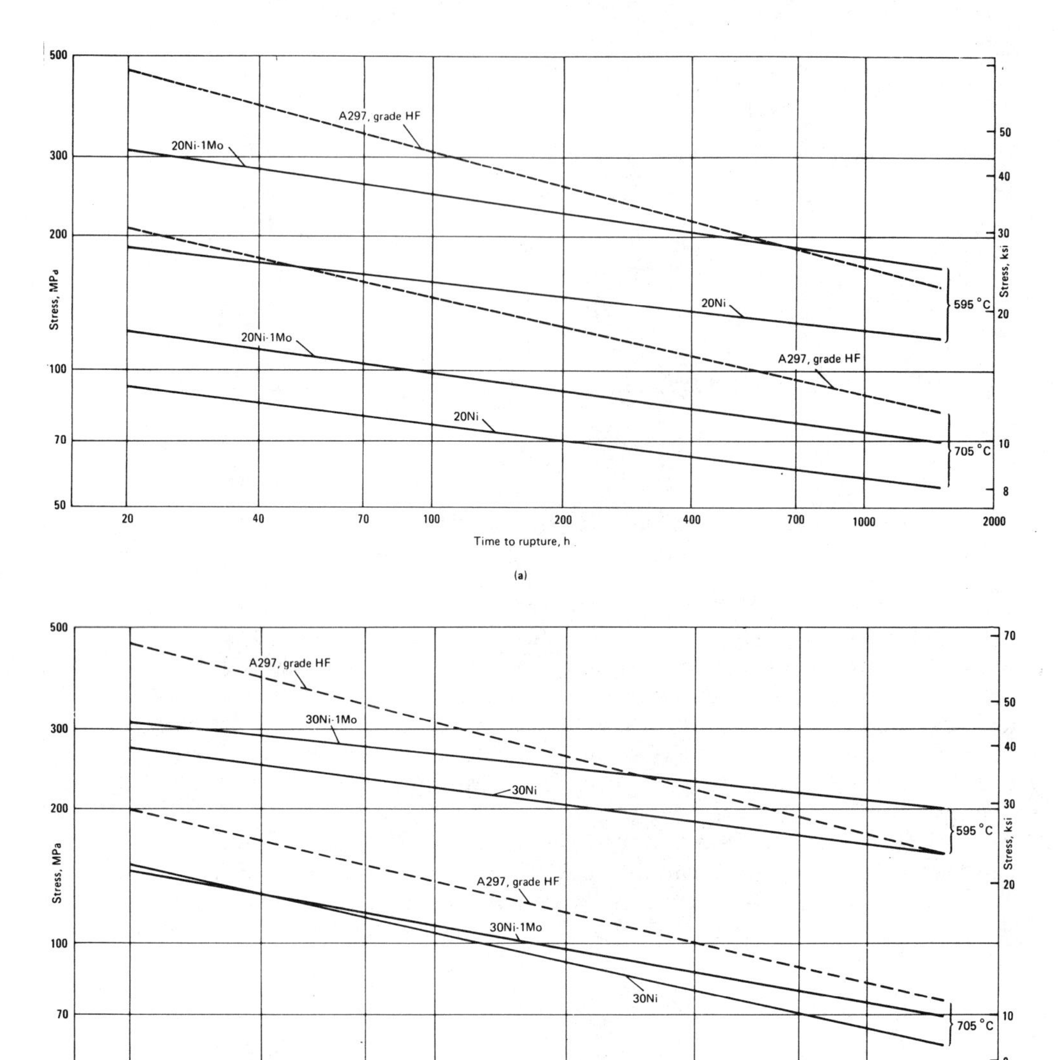

Typical stress-rupture properties of high-nickel heat-resistant ductile irons.

(a) Stress vs time to rupture for 20Ni cast iron and 20Ni-1Mo cast iron at 595 °C (1100 °F) and 705 °C (1300 °F). (b) Similar data for 30Ni and 30Ni-1Mo cast irons. 19Cr-9Ni cast stainless steel (ASTM A297 grade HF) shown for reference.

Source: Metals Handbook, Ninth Edition, Volume 1, Properties and Selection: Irons and Steels, American Society for Metals, Metals Park, OH, 1978, p 95

Grade 17 Gray Cast Iron vs Pearlitic and Ferritic Nodular Irons: A Comparison

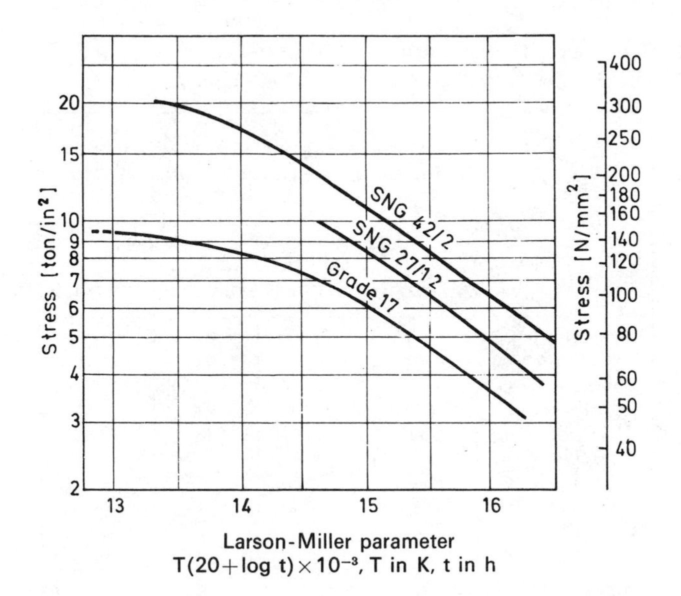

Creep strain properties (0.1%) of pearlitic nodular iron SNG 42/2, ferritic nodular iron SNG 27/12, and flake graphite iron grade 17.

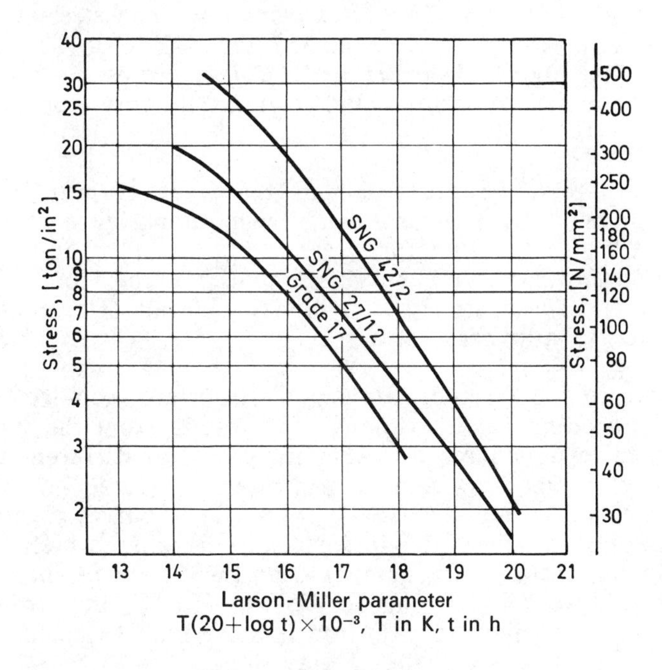

Stress-to-rupture properties of pearlitic nodular iron SNG 42/2, ferritic nodular iron SNG 27/12, and flake graphite iron grade 17.

Grade 17 Gray Cast Iron vs Pearlitic and Ferritic Nodular Irons: A Comparison (Continued)

Stresses to produce 0.1% creep strain for unalloyed flake and nodular (SG) irons

Material	Stresses to give 0·1% creep strain for temperatures and hours stated ton/in²*			
	350°C		400°C	
	10 000 h	100 000 h	10 000 h	100 000 h
Flake Graphite iron Grade 17 3 in Section	6·1 (94)	4·5 (70)	3·3 (51)	2·2 (34)
Ferritic Nodular (SG) iron SNG 27/12	8·6 (133)	6·1 (94)	4·4 (68)	3·0 (46)
Pearlitic Nodular (SG) iron SNG 42/2	10·9 (168)	7·9 (121)	5·9 (91)	— —

*Figures in brackets are N/mm²

All available 0.1% creep strain data are shown plotted to the Larson-Miller parameter in the top figure on the preceding page. The curves for pearlitic and ferritic nodular (SG) irons were obtained from 5,000-h tests at temperatures of 400 °C and 427 °C, respectively. The stresses that produce 0.1% creep strain for the three irons at 350 and 400 °C in 10^4 and 10^5 h are shown in the table above. Pearlitic nodular (SG) irons clearly have higher creep properties than ferritic irons and a grade 17 gray iron has the lowest properties.

All the available stress-to-rupture data on the grade 17 gray iron are shown plotted to the Larson-Miller parameter in the bottom figure on the preceding page. Results are also reported for stress-to-rupture tests at 427 and 538 °C on six flake graphite irons with tensile strengths ranging from 19-28 tons/in.^{2} (290-430 N/mm^2). These clearly show the beneficial effects of Mo on the ruputre properties of gray cast irons.

Curves for pearlitic and ferritic nodular (SG) irons are included in the bottom figure on the preceding page. Although the results from the two sources on each iron were determined for different times and at different temperatures, they were in very good agreement and are considered to be typical of commercially available nodular (SG) irons. The curves again show the superior properties of pearlitic nodular (SG) irons, which had rupture elongations ranging from 7-22% compared to 13-30% in ferritic nodular (SG) irons. At temperatures in the range of 350 to 450 °C, the stress to produce rupture in the pearlitic nodular (SG) iron is more than twice that required, at the same temperatures, in the grade 17 gray iron.

Source: K.B. Palmer, "High Temperature Properties of Cast Irons, Iron and Steel, Vol. 44, Feb. 1971, p 44-45

Gray Iron Alloys and Ferritic Nodular Iron: Comparison of Rupture Times

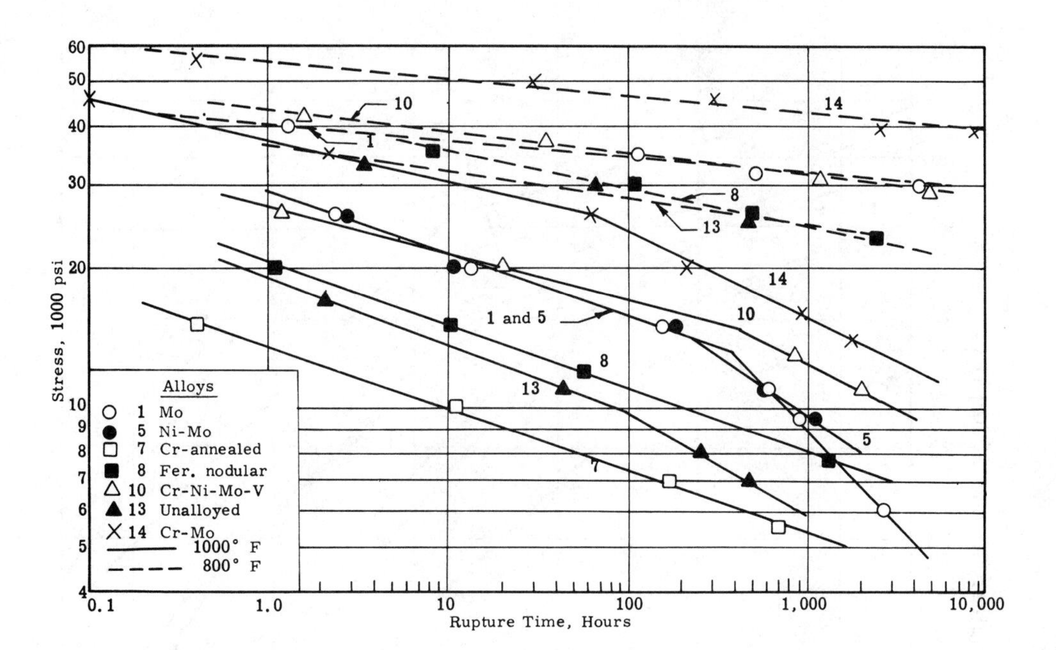

Rupture times of six gray cast iron alloys and one ferritic nodular iron at various stresses at 1000 and 800 °F.

Source: Properties of Cast Iron at Elevated Temperatures, ASTM Special Technical Publication No. 248, American Society for Testing Materials, Philadelphia, 1959, p 37

18.28 Gray Iron Alloys, Ferritic Nodular Iron, and Wrought Steels: Comparison of Master Rupture Curves

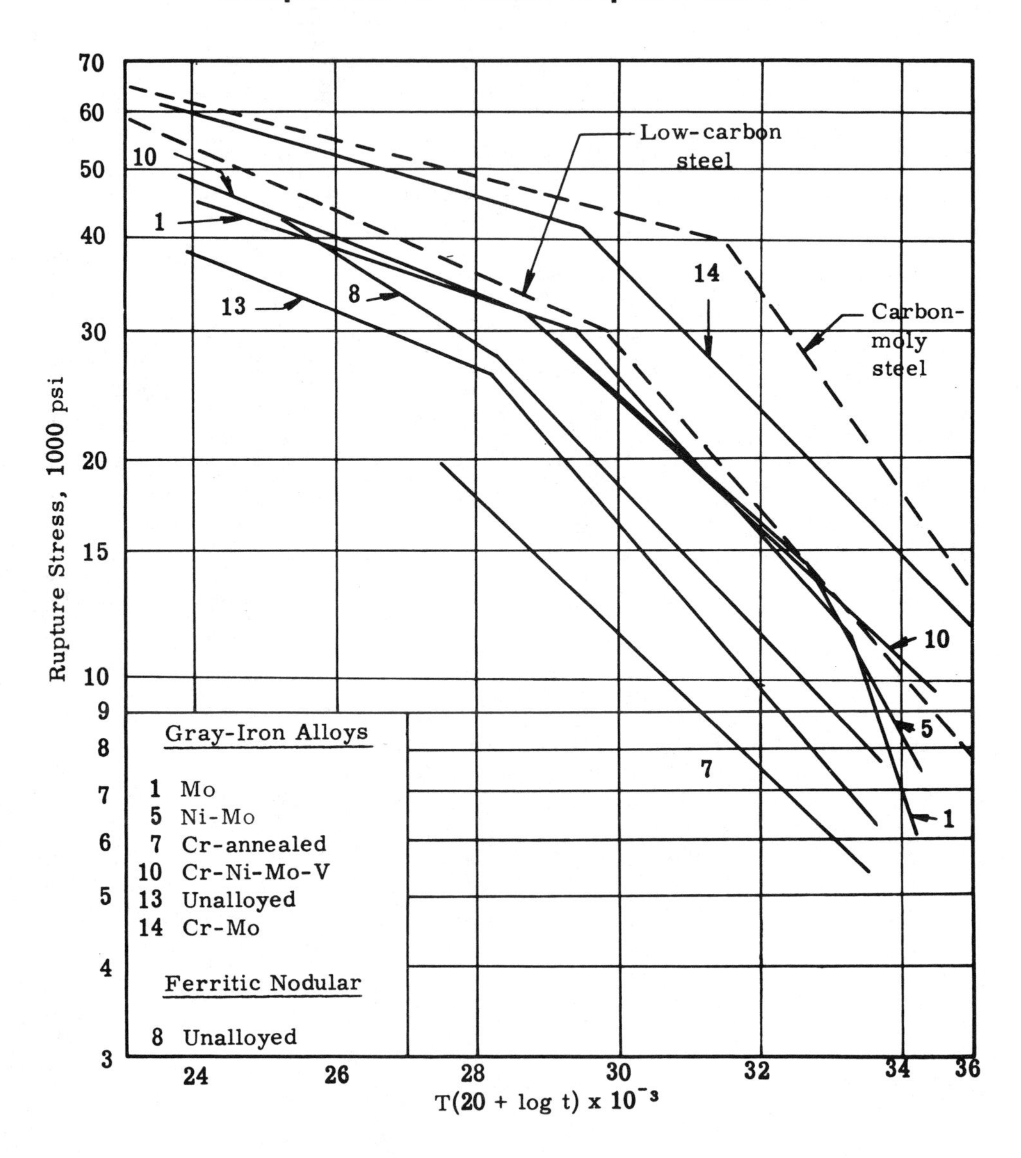

Comparison of master rupture curves for six gray cast iron alloys, one ferritic nodular iron, and two wrought steels.

Source: Properties of Cast Iron at Elevated Temperatures, ASTM Special Technical Publication No. 248, American Society for Testing Materials, Philadelphia, 1959, p 45

Pearlitic, Whiteheart, and Ferritic Malleable Irons: A Comparison

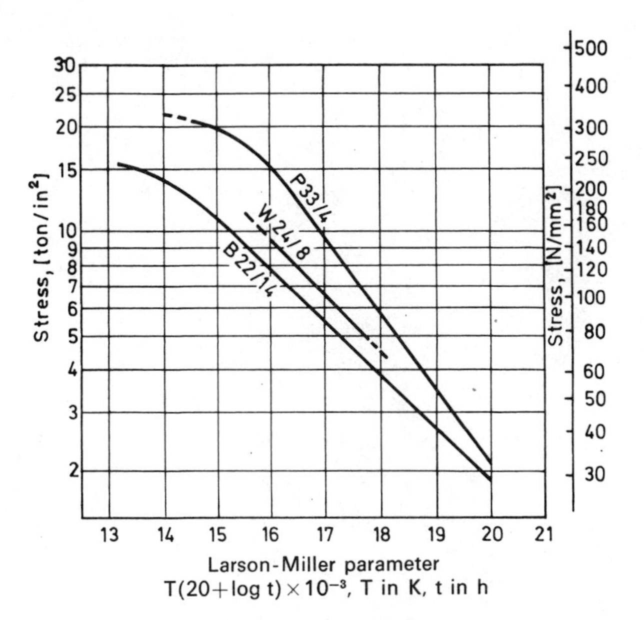

Stress-to-rupture properties of pearlitic malleable iron P33/4, Whiteheart malleable iron W24/8, and ferritic malleable iron B22/14.

The stress-to-rupture properties of ferritic and pearlitic malleable irons have been investigated in detail, and the combined results on each iron are shown plotted to the Larson-Miller parameter in the figure above. The results again reveal the superior properties of pearlitic irons. Rupture elongations in ferritic malleable irons were above 15% and were generally in the range of 10-20% in pearlitic malleable irons.

Source: K.B. Palmer, "High Temperature Properties of Cast Irons," Iron and Steel, Vol. 44, Feb. 1971, p 44-45

19

Carbon and Alloy Steels

Carbon Steels

1010 Steel: Short-Time Creep Rate

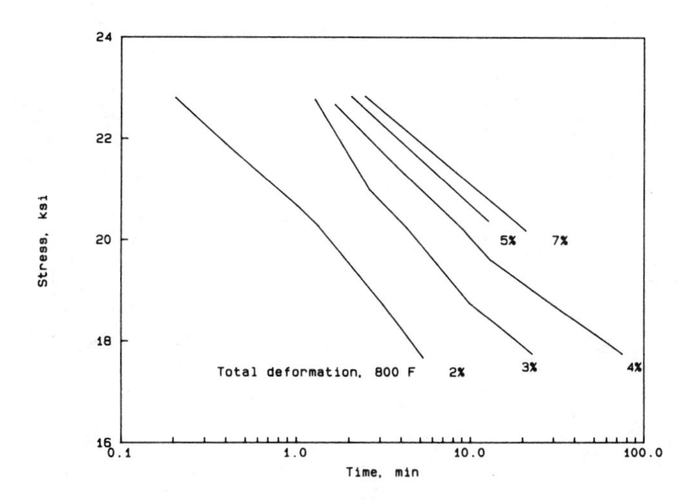

Short-time creep data for AISI 1010.

Deformation includes thermal expansion of 0.61%; heating rate, 115 °F/s. Hot rolled commercial quality (0.058-in. thick) specimens were 17 in. by 2 in.; gage section 8 in. by 0.5 in. Tensile strength: 43 ksi. Yield strength: 28.2 ksi. Elongation: 64.4%.

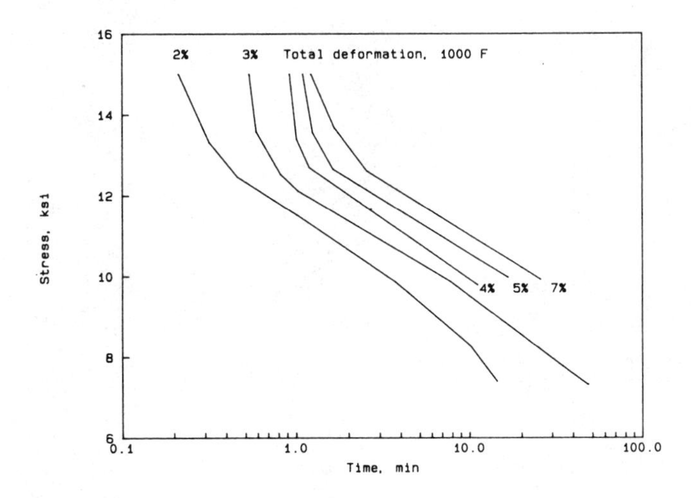

Short-time creep data for AISI 1010.

Deformation includes thermal expansion of 0.80%; heating rate, 125 °F/s. Hot rolled commercial quality (0.058-in. thick) specimens were 17 in. by 2 in.; gage section 8 in. by 0.5 in. Tensile strength: 43 ksi. Yield strength: 28.2 ksi. Elongation: 64.4%.

Source: High-Temperature Property Data: Ferrous Alloys, M.F. Rothman, Ed., ASM International, Metals Park, OH, 1988, p 2.8

1012 Steel: Short-Time Creep Rate

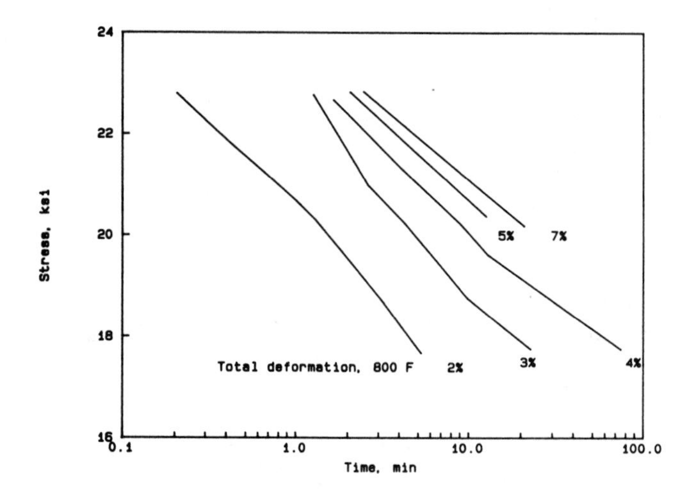

Short-time creep data for AISI 1012.

Deformation includes thermal expansion of 0.61%; heating rate, 115 °F/s. Hot rolled commercial quality (0.058-in. thick) specimens were 17 by 2 in.; gage section 8 by 0.5 in. Tensile strength: 43 ksi. Yield strength: 28.2 ksi. Elongation: 64.4%.

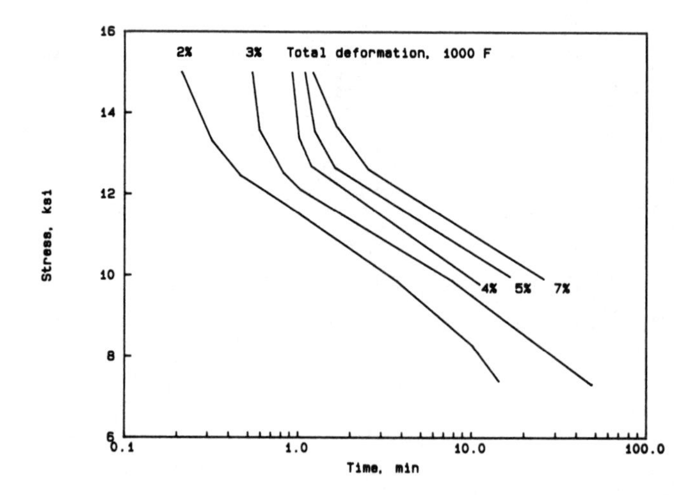

Short-time creep data for AISI 1012.

Deformation includes thermal expansion of 0.80%; heating rate, 125 °F/s. Hot rolled commercial quality (0.058-in. thick) specimens were 17 by 2 in.; gage section 8 in. by 0.5 in. Tensile strength: 43 ksi. Yield strength: 28.2 ksi. Elongation: 64.4%.

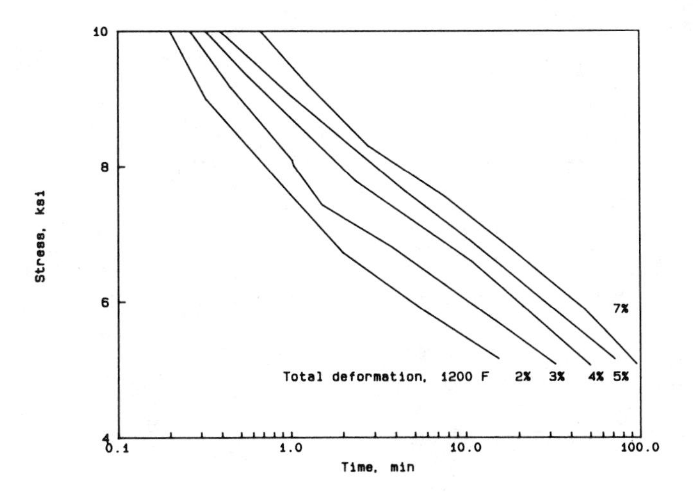

Short-time creep data for AISI 1012.

Deformation includes thermal expansion of 0.90%; heating rate, 115 °F/s. Hot rolled commercial quality (0.058-in. thick) specimens were 17 in. by 2 in.; gage section 8 in. by 0.5 in. Tensile strength: 43 ksi. Yield strength: 28.2 ksi. Elongation: 64.4%.

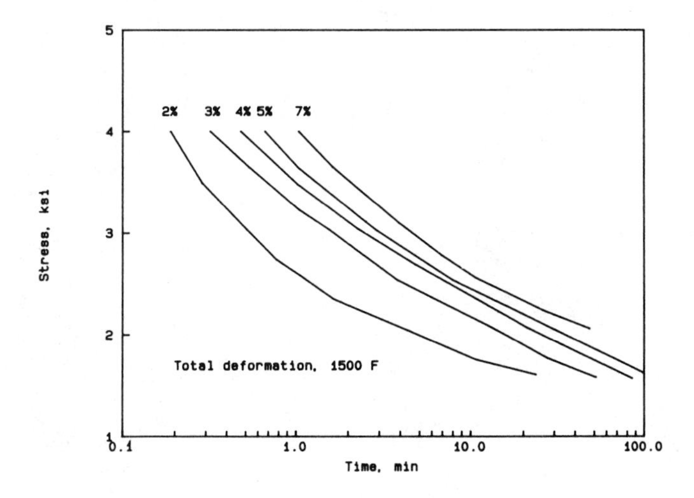

Short-time creep data for AISI 1012.

Deformation includes thermal expansion of 1.14%; heating rate, 125 °F/s. Hot rolled commercial quality (0.058-in. thick) specimens were 17 in. by 2 in.; gage section 8 in. by 0.5 in. Tensile strength: 43 ksi. Yield strength: 28.2 ksi. Elongation: 64.4%.

Source: High-Temperature Property Data: Ferrous Alloys, M.F. Rothman, Ed., ASM International, Metals Park, OH, 1988, p 2.10

1015 Steel: 100-h Rupture Strength

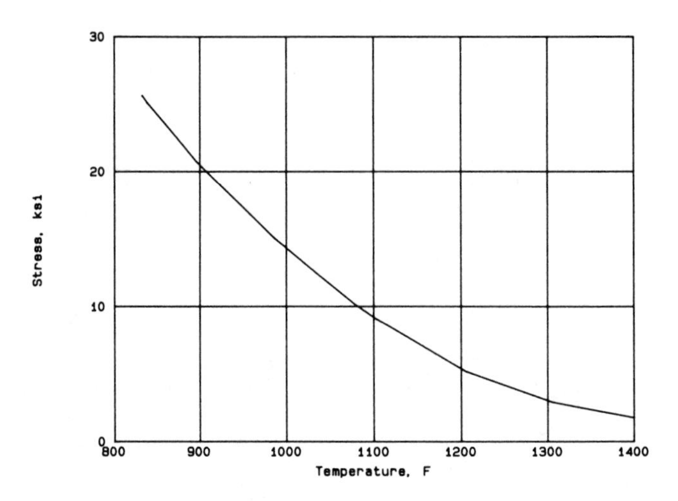

Stress to cause rupture in 100 h for AISI 1015.

Specimens were annealed, normalized, hot rolled, stress relieved, oil quenched and tempered, and cold worked.

Source: High-Temperature Property Data: Ferrous Alloys, M.F. Rothman, Ed., ASM International, Metals Park, OH, 1988, p 2.13

1015 Steel: Creep Strength

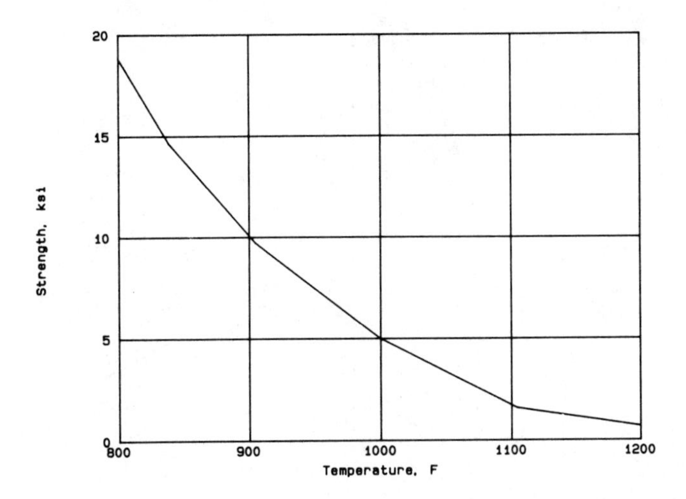

Creep strength of AISI 1015.

Creep rate of 0.0001%/h (1% in 10,000 h). Specimens were annealed, normalized, hot rolled, stress relieved, oil quenched and tempered, and cold worked.

Source: High-Temperature Property Data: Ferrous Alloys, M.F. Rothman, Ed., ASM International, Metals Park, OH, 1988, p 2.12

1018 Steel: 1000- and 10,000-h Rupture Strength

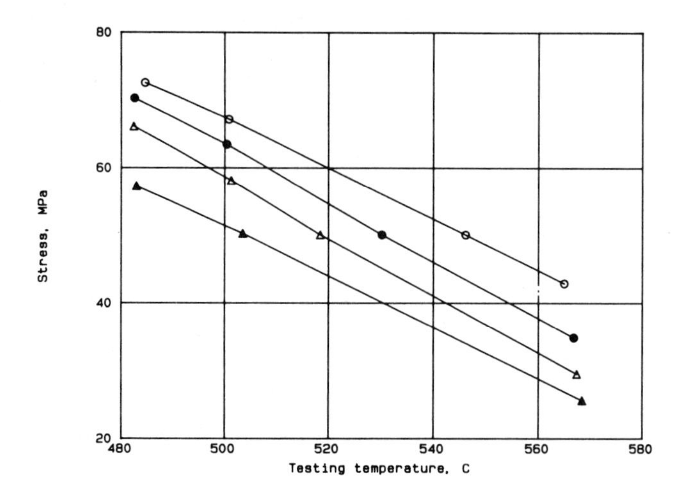

Stress-to-rupture for AISI 1018.

Circles: 1000 h. Triangles: 10,000 h. Solid symbols indicate specimens that were held at test temperature for 83,000 h prior to testing.

Source: High-Temperature Property Data: Ferrous Alloys, M.F. Rothman, Ed., ASM International, Metals Park, OH, 1988, p 2.14

1020 Steel: Creep Strength

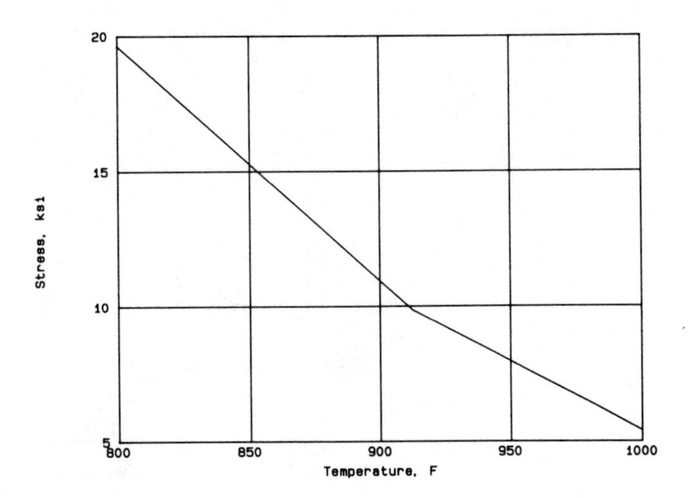

Stress required to produce 1% creep in 10,000 h for AISI 1020.

Source: High-Temperature Property Data: Ferrous Alloys, M.F. Rothman, Ed., ASM International, Metals Park, OH, 1988, p 2.16

1020 Steel: Rupture Strength

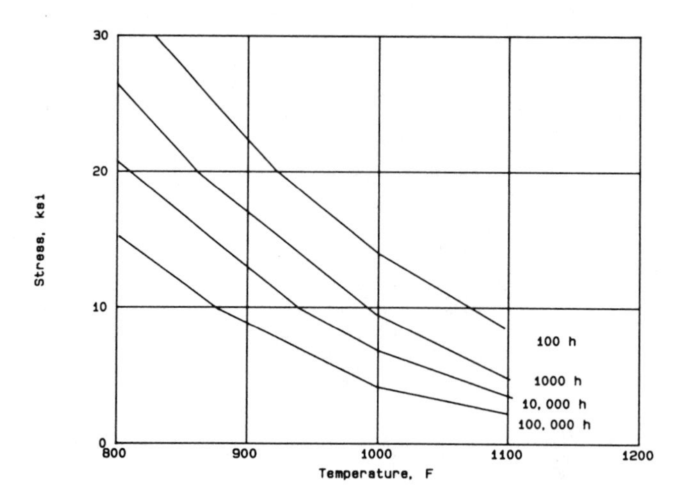

Effect of temperature and exposure time on stress rupture for AISI 1020.

Source: High-Temperature Property Data: Ferrous Alloys, M.F. Rothman, Ed., ASM International, Metals Park, OH, 1988, p 2.17

1038, 1039, and 1040 Steels: Tensile, Yield, and Rupture Strength

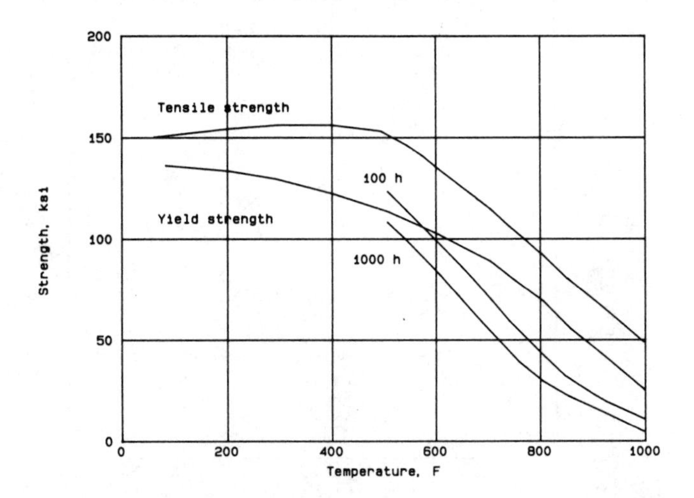

Tensile, yield (0.20% offset), and rupture strength of quenched and tempered ANSI 1038, 1039, and 1040 specimens.

Source: High-Temperature Property Data: Ferrous Alloys, M.F. Rothman, Ed., ASM International, Metals Park, OH, 1988, p 2.22

Carbon Steel: Aluminum-Treated vs Aluminum-Free

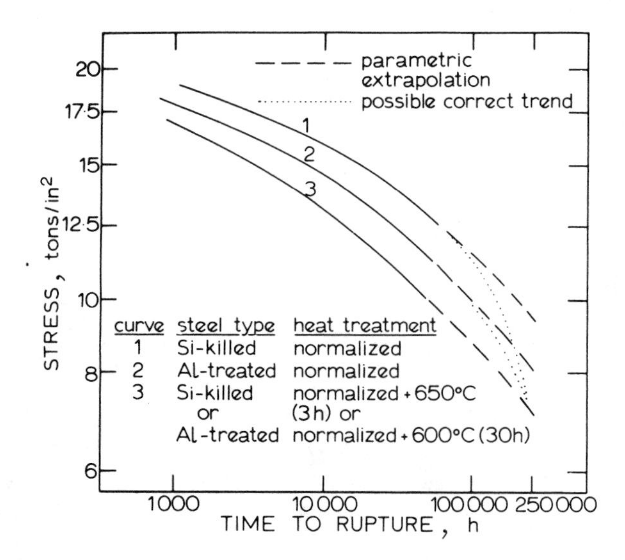

Log stress vs log rupture life: typical behavior for three types of steel. Melting (deoxidation) practice, at 400 °C (752 °F).

Dashed lines show life extrapolations predicted by parametric extrapolation at the time of publication of the data, whereas dotted lines were estimated to represent more accurately the future trends of the data.

Source: R.F. Johnson and J. Glen, "Some Problems in the Assessment of High-Temperature Properties for Engineering Purposes," in Creep Strength in Steel and High-Temperature Alloys, The Metals Society, London, 1972, p 39

Carbon Steel: Effect of Deoxidation Practice and Ultimate Tensile Strength Level on 100,000-h Stress-Rupture Strength

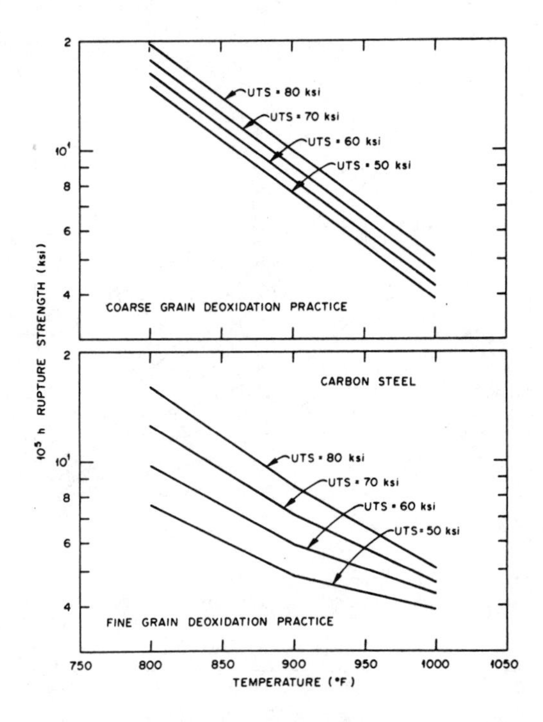

Estimated 100,000-h stress-rupture strength vs temperature capability for carbon steels of various ultimate strength capabilities, showing the additional effects of deoxidation practice.

Source: M.K. Booker and B.L.P. Booker, "Analysis of Elevated Temperature Tensile and Creep Properties of Wrought Carbon Steels," in Factors Influencing the Time-Dependent Properties of Carbon Steels for Elevated Temperature Pressure Vessels (Proceedings of the 4th National Congress on Pressure Vessel and Piping Technology in Portland, OR), M. Prager, Ed., American Society of Mechanical Engineers, New York, 1983, p 13

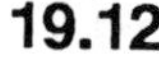

Carbon Steel: Effect of Deoxidation Practice and Ultimate Tensile Strength Level on Creep Strength

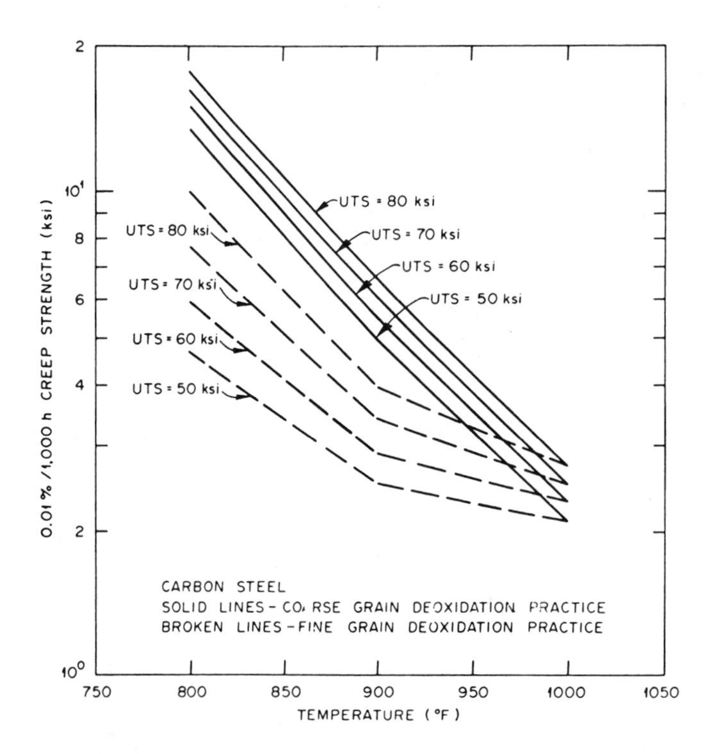

Estimated 0.01%/1000-h creep strength vs temperature capability for carbon steels of various ultimate strength capabilities, showing the additional effects of deoxidation practice.

Source: M.K. Booker and B.L.P. Booker, "Analysis of Elevated Temperature Tensile and Creep Properties of Wrought Carbon Steels," in Factors Influencing the Time-Dependent Properties of Carbon Steels for Elevated Temperature Pressure Vessels (Proceedings of the 4th National Congress on Pressure Vessel and Piping Technology in Portland, OR), M. Prager, Ed., American Society of Mechanical Engineers, New York, 1983, p 15

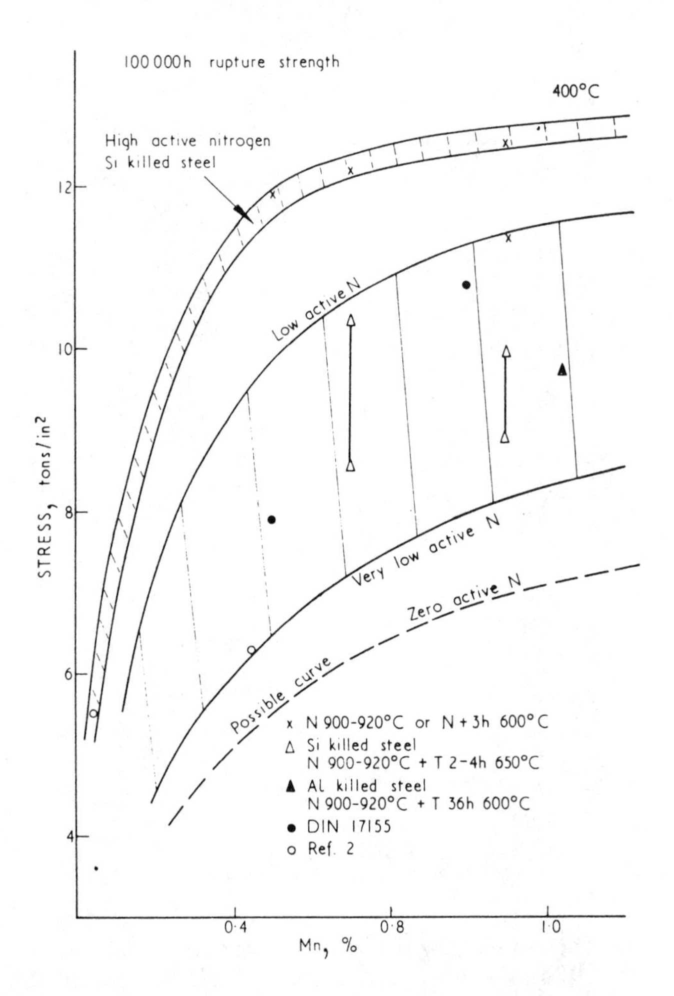

Effect of manganese and active nitrogen on 100,000-h rupture strength of carbon steel at 400 °C (752 °F).

Carbon Steel: Effect of Manganese and Active Nitrogen (Continued)

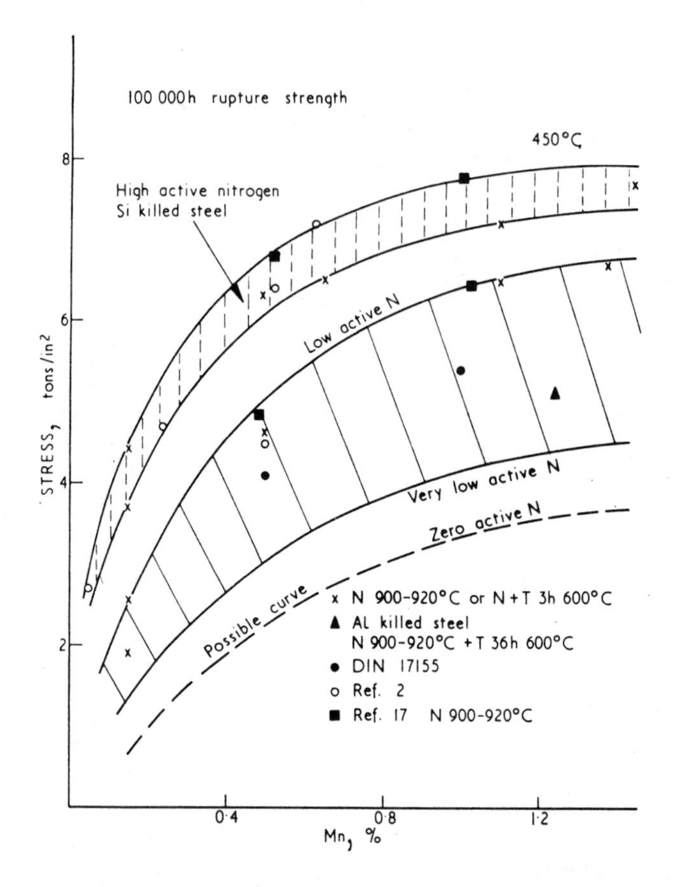

Effect of manganese and active nitrogen on 100,000-h rupture strength of carbon steel at 450 °C (842 °F).

Source: J. Glen, "Factors Controlling the Rupture Strength of Carbon Steel," in Metallurgical Developments in Carbon Steels, Iron and Steel Institute, London, 1963, p 50

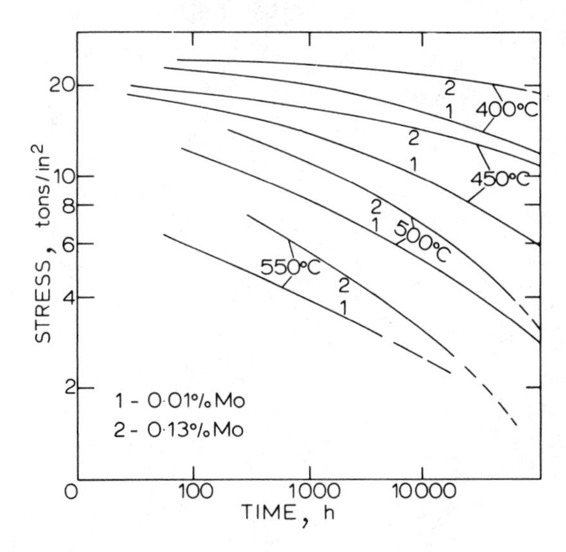

Effects of residual molybdenum on the stress-rupture properties of carbon steel at 400, 450, 500, and 550 °C (752, 842, 932, and 1022 °F).

Source: R.F. Johnson and J. Glen, "Some Problems in the Assessment of High-Temperature Properties for Engineering Purposes," in Creep Strength in Steel and High-Temperature Alloys, The Metals Society, London, 1972, p 40

Alloy Steels

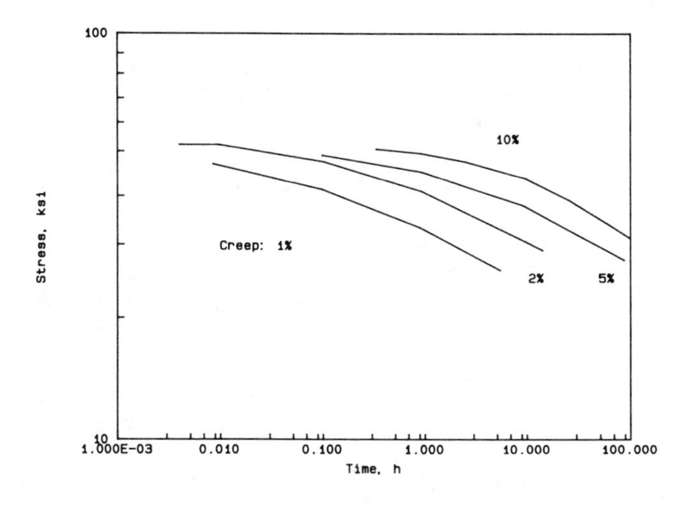

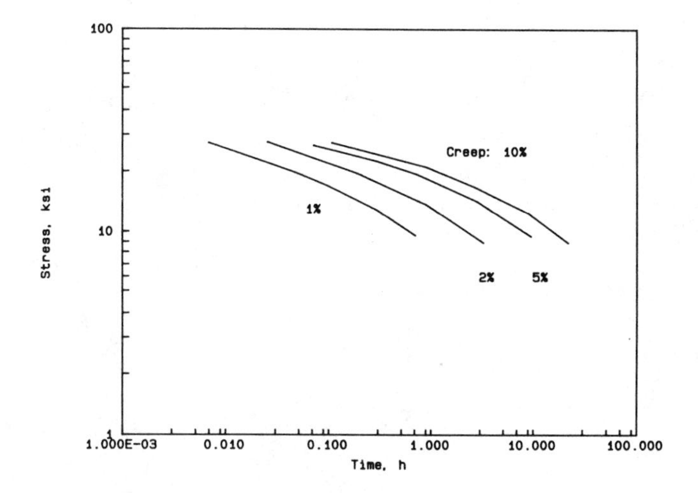

Creep properties at 1000 °F (top) and 1200 °F (bottom) of 0.050-in. AISI 4140 sheet annealed at 1550 °F.

Source: High-Temperature Property Data: Ferrous Alloys, M.F. Rothman, Ed., ASM International, Metals Park, OH, 1988, p 3.13

19.18 4140 Steel: Creep Strength of Austenitized, Oil Quenched, and Tempered Sheet

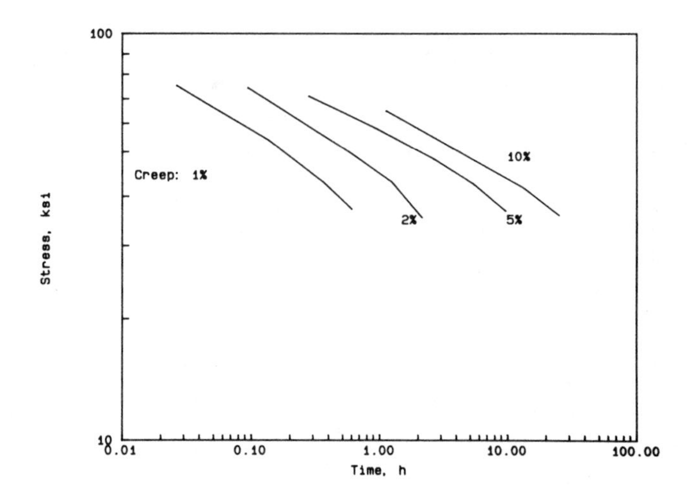

Creep properties at 1000 °F of 0.050-in. AISI 4140 sheet, austenitized at 1575 °F, oil quenched, and tempered at 1000 °F.

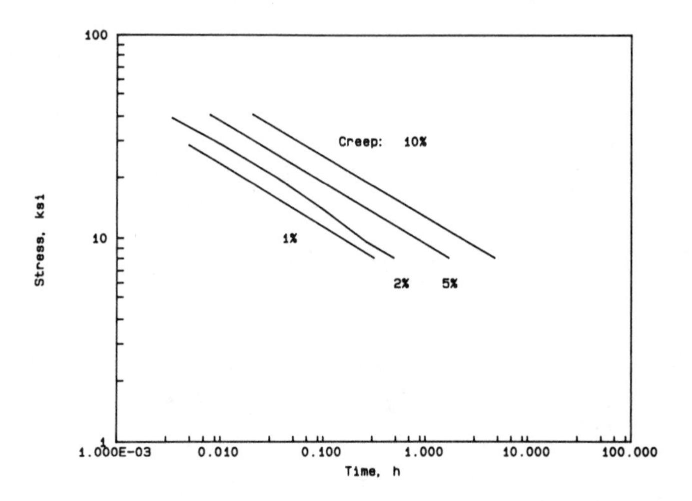

Creep properties at 1200 °F of 0.050-in. AISI 4140 sheet, austenitized at 1575 °F, oil quenched, and tempered at 1200 °F.

Source: High-Temperature Property Data: Ferrous Alloys, M.F. Rothman, Ed., ASM International, Metals Park, OH, 1988, p 3.14

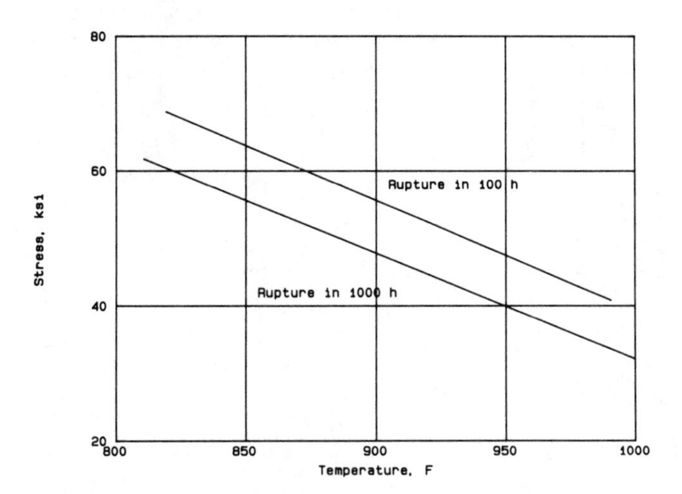

Stress-rupture of AISI 4140 1-in. diam specimens heated to 1650 °F for 1 h, air cooled, heated to 1200 °F for 1 h, and furnace cooled. Hardness was 233 HB.

Source: High-Temperature Property Data: Ferrous Alloys, M.F. Rothman, Ed., ASM International, Metals Park, OH, 1988, p 3.14

4340 Steel: Creep Strength of Sheet

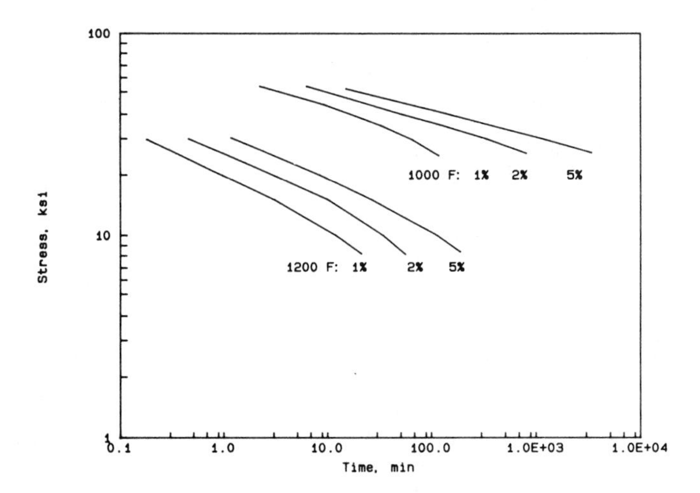

Creep strength of 0.050-in. AISI 4340 sheet. Specimens were heated to 1600 °F, air cooled, and tempered.

Source: High-Temperature Property Data: Ferrous Alloys, M.F. Rothman, Ed., ASM International, Metals Park, OH, 1988, p 3.19

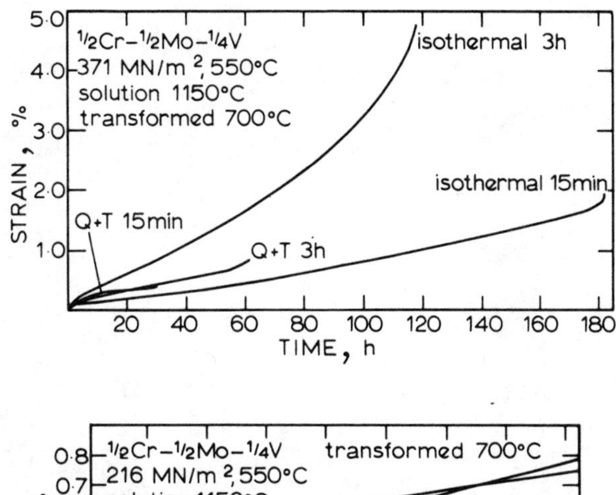

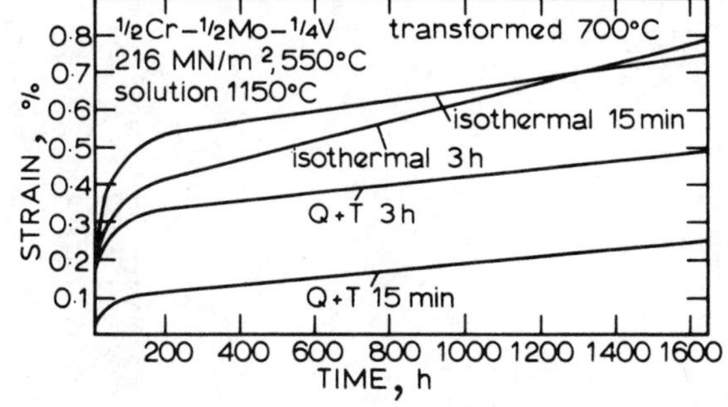

Creep strain vs time curves at 350 °C and 371 MN/m² (1020 °F and 53.8 ksi) (top) and at 550 °C and 216 MN/m² (1020 °F and 31.3 ksi) (bottom) for 0.5Cr-0.5Mo-0.25V steel following four different heat treatments.

The heat treatments are:

* Isothermal 3 h = Isothermally transformed at 1292 °F for 3 h
* Isothermal 15 min = Isothermally transformed at 1292 °F for 15 min
* Q + T 15 min = Oil quenched and tempered at 1292 °F for 15 min
* Q + T 3 h = Oil quenched and tempered at 1292 °F for 3 h

Source: G.L. Dunlop, D.V. Edmonds, and R.W.K. Honeycombe, "Some Effects of Microstructure on the Creep Properties of Low-Alloy Steels Containing Vanadium Carbide," in Creep Strength in Steel and High-Temperature Alloys, The Metals Society, London, 1972, p 225

19.22 0.5Cr-0.5Mo-0.25V Steel: Long-Time Creep and Rupture Strength

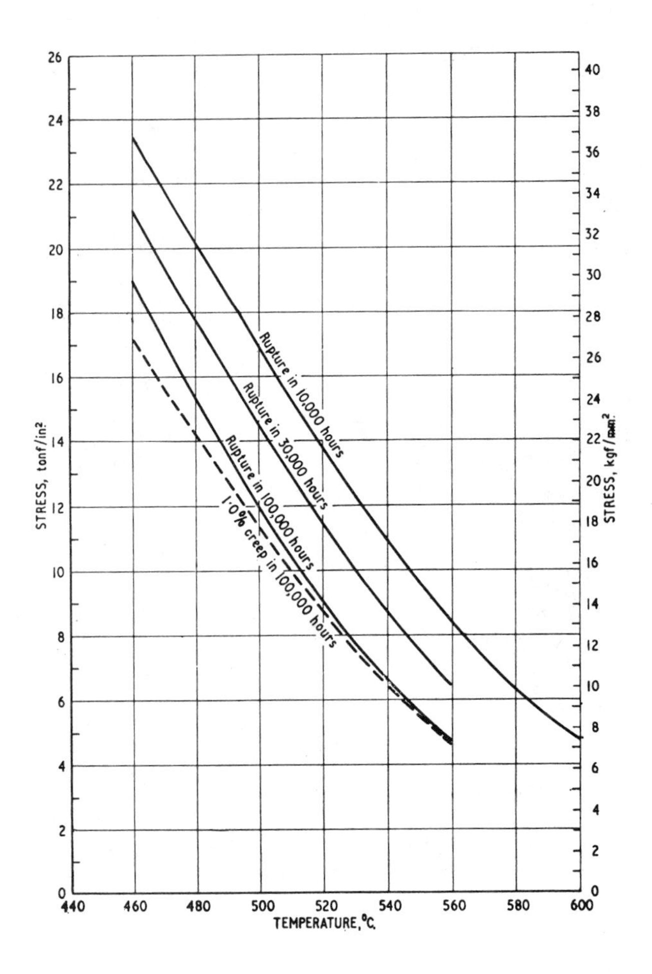

Estimated long-time creep and rupture strength vs temperature curves for 0.5Cr-0.5Mo-0.25V steel.

Source: R.F. Johnson and M.J. May, "Elevated-Temperature Tensile, Creep, and Rupture Properties," in High-Temperature Properties of Steels (Proceedings of a Conference in Eastbourne, England), Iron and Steel Institute, London, 1967, p 263

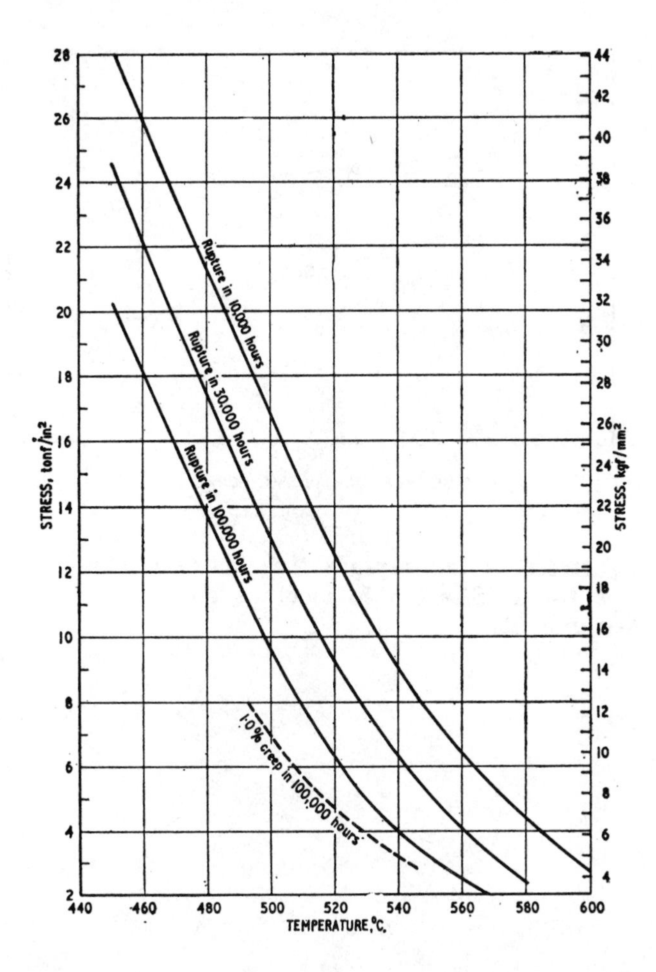

Estimated long-time creep and rupture strength vs temperature curves for 1.0Cr-0.5Mo steel.

Source: R.F. Johnson, M.J. May, et al., "Elevated-Temperature Tensile, Creep, and Rupture Properties," in High-Temperature Properties of Steels (Proceedings of a Conference in Eastbourne, England), Iron and Steel Institute, London, 1967, p 257

1.25Cr-0.5Mo Steel: Air vs Helium Environment

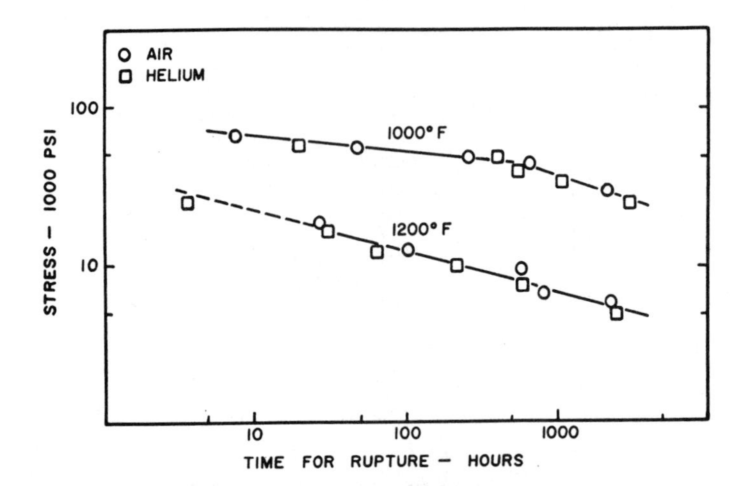

Log stress vs log rupture time at two temperatures for 1.25Cr-0.5Mo steel in air and helium environments.

Source: R.G. Shepheard and M.J. Donachie, Jr., "Elevated Temperature Behavior of a Low Alloy and a Stainless Steel in Helium," Transactions Quarterly, Vol. 55, March 1962, p 48

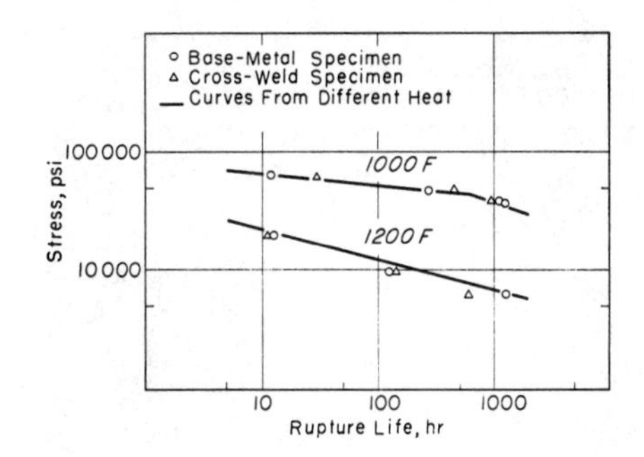

Log stress vs log rupture life at two temperatures for 1.25Cr-0.5Mo steel showing effects of welding on strength.

Source: R.G. Shepheard and M.J. Donachie, Jr., "Creep-Rupture Behavior of 1.25Cr-0.5Mo Steel Weldments," Materials Research & Standards, Vol. 2, April 1962, p 276

2.25Cr-1Mo Steel: Air vs TGR Helium Environment

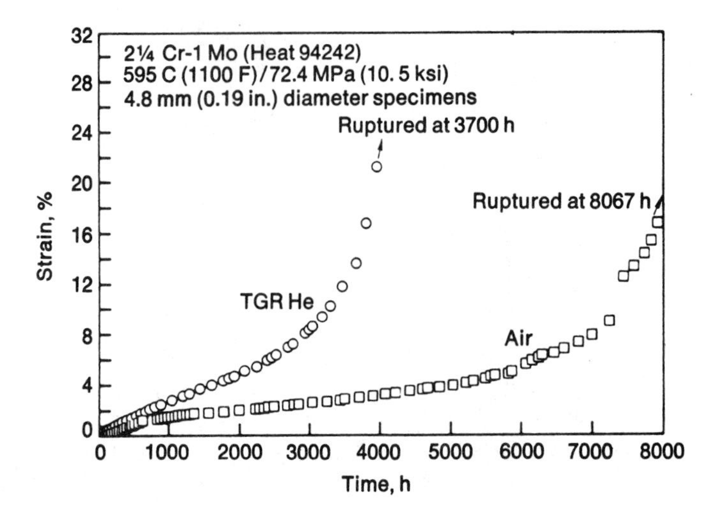

Creep strain vs time curves for annealed 2.25Cr-1.0Mo steel at 1100 °F and 10.5 ksi in thermal gas-cooled reactor (TGR) helium and in air.

Source: D.I. Roberts, S.N. Rosenwasser, and J.F. Watson, "Materials Selection for Gas Cooled and Fusion Reactor Applications," in Alloys for the Eighties, R.Q. Barr, Ed., Climax Molybdenum Co., Greenwich, CT, 1981, p 123

2.25Cr-1Mo Steel: Annealed vs Hardened and Tempered Material

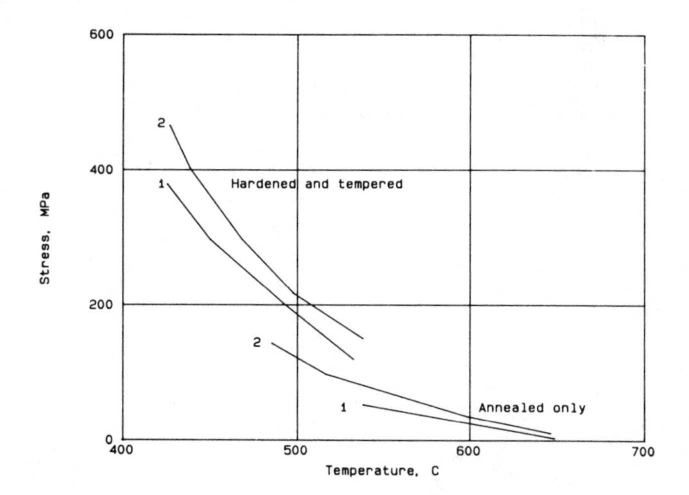

Stress required to produce a minimum creep rate of 0.001%/h (1) and rupture (2) in 100,000 h for specimens of 2.25Cr-1Mo steel with two different heat treatments.

Source: High-Temperature Property Data: Ferrous Alloys, M.F. Rothman, Ed., ASM International, Metals Park, OH, 1988, p 5.5

2.25Cr-1Mo Steel: Annealed vs Normalized and Tempered

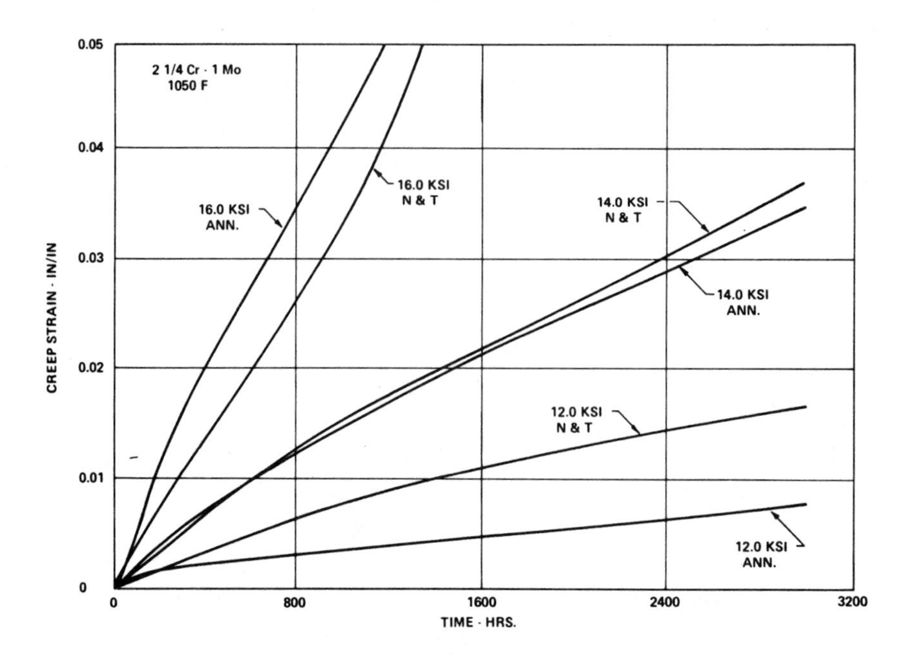

Creep strain vs time curves for 2.25Cr-1.0Mo steel at 1050 °F and the stress conditions shown, comparing annealed (Ann.) vs normalized and tempered (N & T) material.

Source: J.E. Bynum, F.V. Ellis, and B.W. Roberts, "Tensile and Creep Properties for an Annealed Versus Normalized and Tempered 2.25Cr-1Mo Steel Plate," in Chrome Moly Steel in 1976, G.V. Smith, Ed., American Society of Mechanical Engineers, New York, 1976, p 7

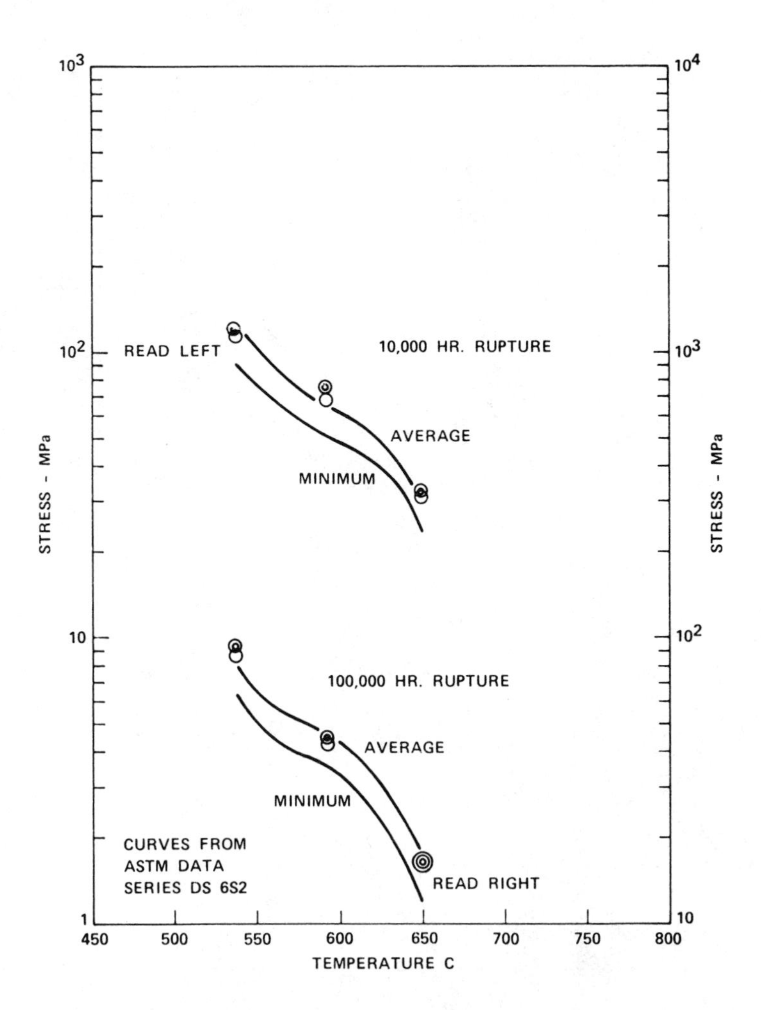

Comparison of long-time rupture strengths of cast (symbols) vs wrought (lines) 2.25Cr-1.0Mo steel as a function of temperature.

Source: W.E. Leyda, "Tensile and Creep-Rupture Behavior of Centrifugally Cast Carbon Steel, 1.25Cr-0.5Mo and 2.25Cr-1Mo Pipe," in Cast Metals for Structural and Pressure Containment Applications, G.V. Smith, Ed., American Society of Mechanical Engineers, New York, 1979, p 372

2.25Cr-1Mo Steel: Creep Data

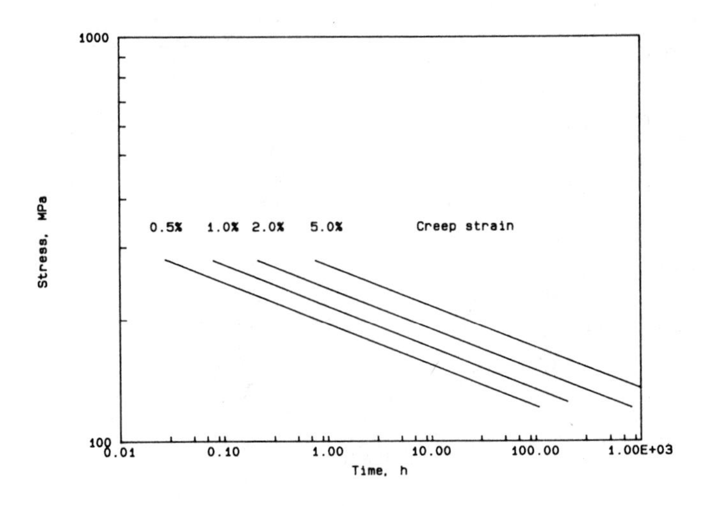

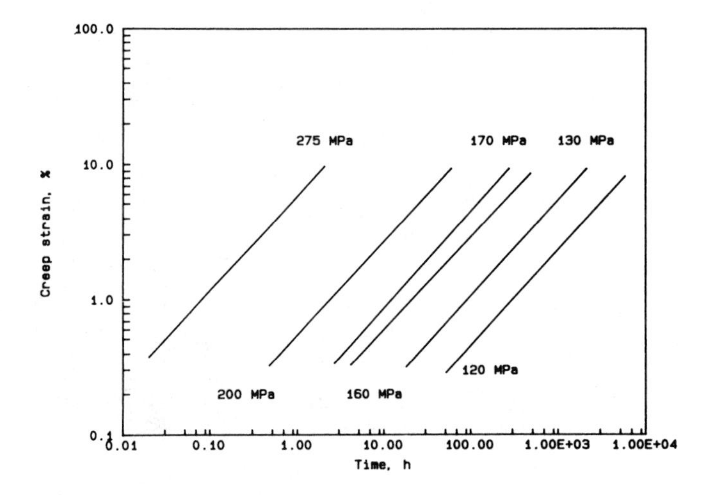

Stress-time plot (top) and creep strain-time plot (bottom) at 540 °C. Constant-stress lines have been drawn parallel.

Source: High-Temperature Property Data: Ferrous Alloys, M.F. Rothman, Ed., ASM International, Metals Park, OH, 1988, p 5.5

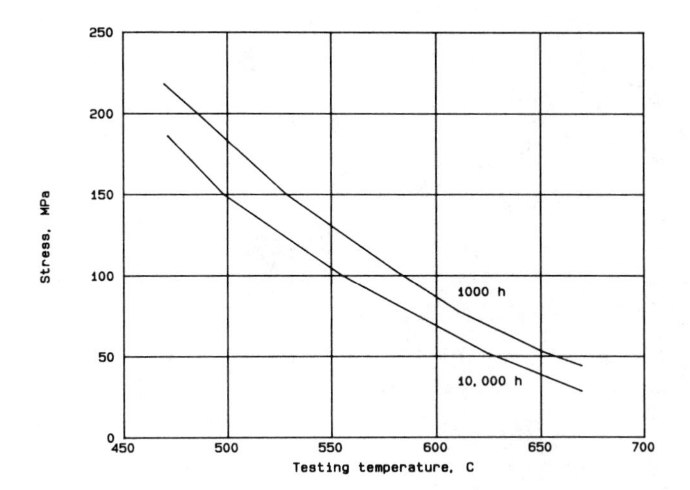

Effect of prior exposure at test temperature for 10,000 h (without stress) on stress required to cause rupture in 1000 and 10,000 h for normalized and tempered specimens.

Allowable design stress at elevated temperatures

TEMP C	TEMP F	Stress MPa	Stress 10E3 ksi
480	900	90	13.1
510	950	76	11.0
540	1000	54	7.8
565	1050	40	5.8
595	1100	29	4.2
620	1150	21	3.0
650	1200	14	2.0

Source: High-Temperature Property Data: Ferrous Alloys, M.F. Rothman, Ed., ASM International, Metals Park, OH, 1988, p 5.5

19.32

2.25Cr-1Mo Steel: Effect of Exposure to Elevated Temperature on Annealed Material

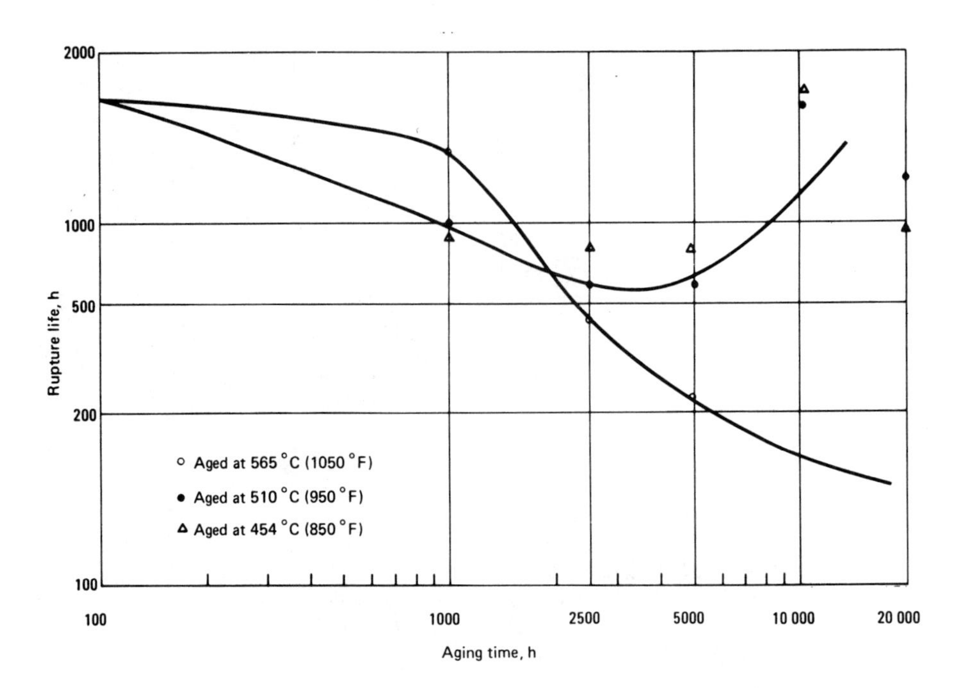

Variation in rupture life for specimens of annealed 2.25Cr-1Mo steel exposed to various elevated temperatures for the durations indicated; after aging, all specimens were stressed to 140 MPa (20 ksi) and tested at 565 °C (1050 °F).

Source: Metals Handbook, Ninth Edition, Volume 1, Properties and Selection: Irons and Steels, American Society for Metals, Metals Park, OH, 1978, p 658

2.25Cr-1Mo Steel: Larson-Miller Plot of Stress-Rupture Behavior of Annealed Material

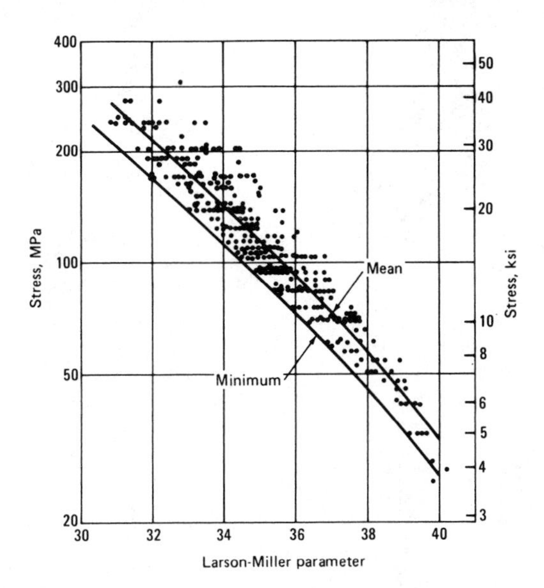

Variation in Larson-Miller parameter with rupture stress for annealed 2.25Cr-1Mo steel tested between 425 and 650 °C (800 and 1200 °F) for rupture life to 10,000 h.

Source: Metals Handbook, Ninth Edition, Volume 1, Properties and Selection: Irons and Steels, American Society for Metals, Metals Park, OH, 1978, p 646

19.34

2.25Cr-1Mo Steel: Larson-Miller Plot of Stress-Rupture Behavior of Normalized and Tempered, and Hardened and Tempered Material

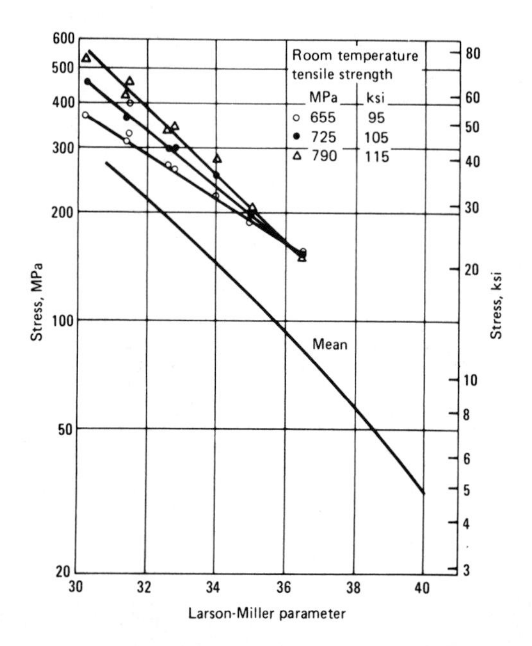

Variation in Larson-Miller parameter with stress to rupture for normalized and tempered and hardened and tempered specimens of 2.25Cr-1Mo steel tested between 425 and 650 °C (800 and 1200 °F) for rupture life to 10,000 h.

Data are grouped according to room-temperature tensile strength of the steel.

Source: Metals Handbook, Ninth Edition, Volume 1, Properties and Selection: Irons and Steels, American Society for Metals, Metals Park, OH, 1978, p 656

2.25Cr-1Mo Steel: Long-Time Creep and Stress-Rupture Properties

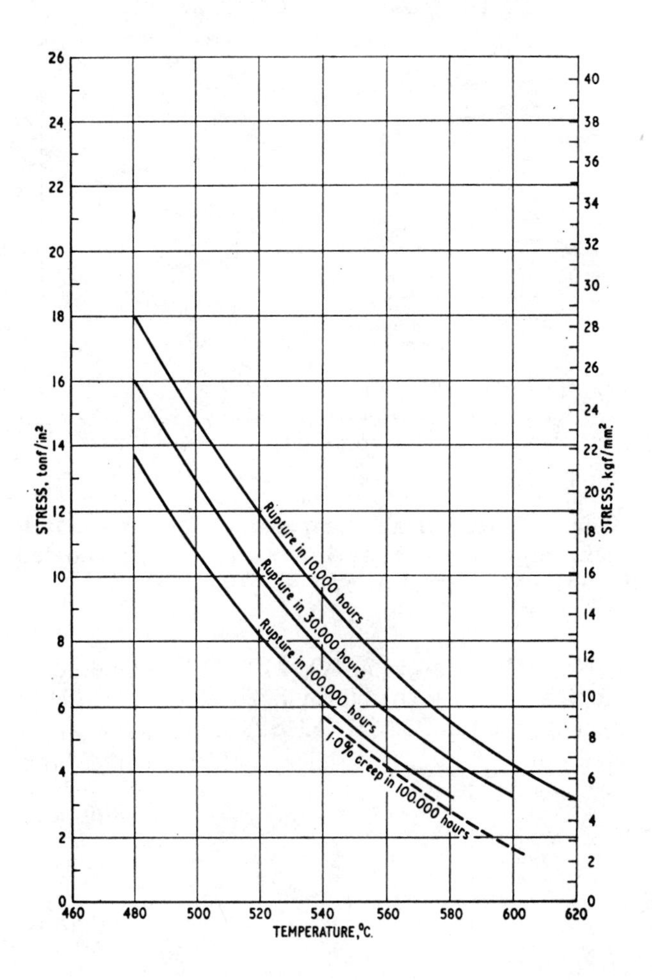

Estimated long-time creep and rupture strength vs temperature curves for 2.25Cr-1.0Mo steel.

Source: R.F. Johnson, M.J. May, et al., "Elevated-Temperature Tensile, Creep, and Rupture Properties," in High-Temperature Properties of Steels (Proceedings of a Conference in Eastbourne, England), Iron and Steel Institute, London, 1967, p 260

3Cr-Mo-W-V Steel: Effect of Tempering Temperature

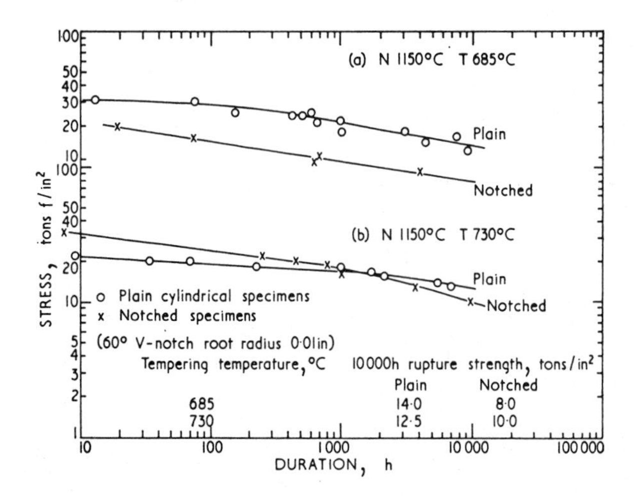

Effect of tempering temperature on the notch stress-rupture behavior of normalized 3Cr-Mo-W-V steel at 550 °C (1022 °F).

(a) Normalized at 1150 °C (2102 °F), tempered at 685 °C (1265 °F). (b) Normalized at 1150 °C (2102 °F), tempered at 730 °C (1346 °F).

Source: J.S. Blair, "Factors That Affect the Properties," in High-Temperature Properties of Steels (Proceedings of a Conference in Eastbourne, England), Iron and Steel Institute, London, 1967, p 531

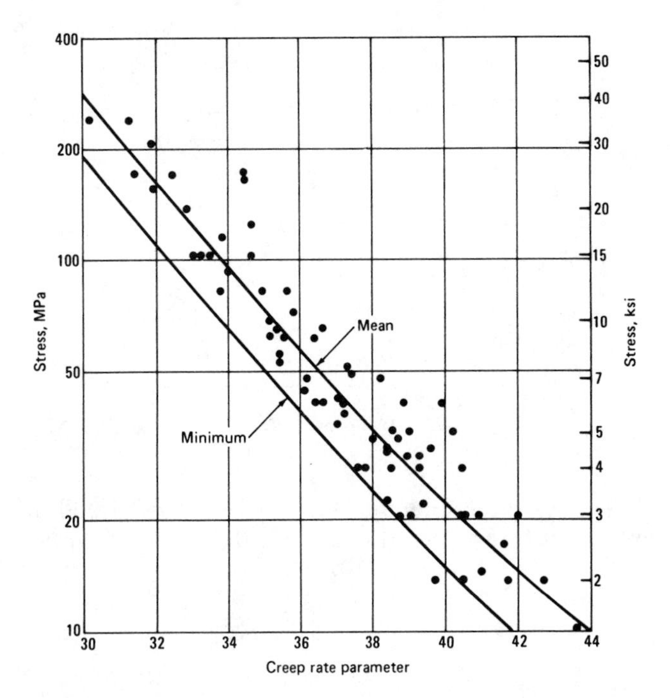

Variation in creep rate parameter with creep stress to give strain rates of 0.1 and 1.0 μm/m·h at test temperatures of 480 to 705 °C (900 to 1300 °F) for 9Cr-1Mo steel in either the normalized and tempered or the annealed condition.

Source: Metals Handbook, Ninth Edition, Volume 1, Properties and Selection: Irons and Steels, American Society for Metals, Metals Park, OH, 1978, p 646

10Cr-0.1V-0.05Nb Steel: Effect of Molybdenum

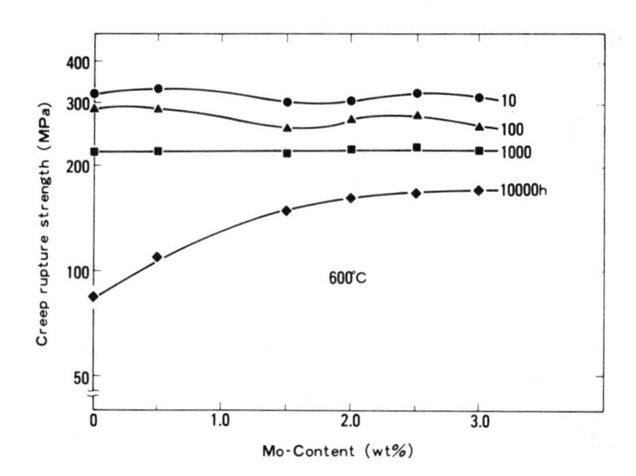

Effect of molybdenum on the 10 to 10,000-h creep-rupture strength of a 10Cr-0.1V-0.05Nb steel at 600 °C (1110 °F).

Source: K. Hashitomo, M. Yamanaka, et al., "Newly Developed 9Cr-2Mo-Nb-V (NSCR9) Steel," in Topical Conference on Ferritic Alloys for Use in Nuclear Energy Technologies (Proceedings of a Conference in Snowbird, UT) J.W. Davis and D.J. Michael, Eds., Metallurgical Society of AIME, Warrendale, PA, 1984, p 308

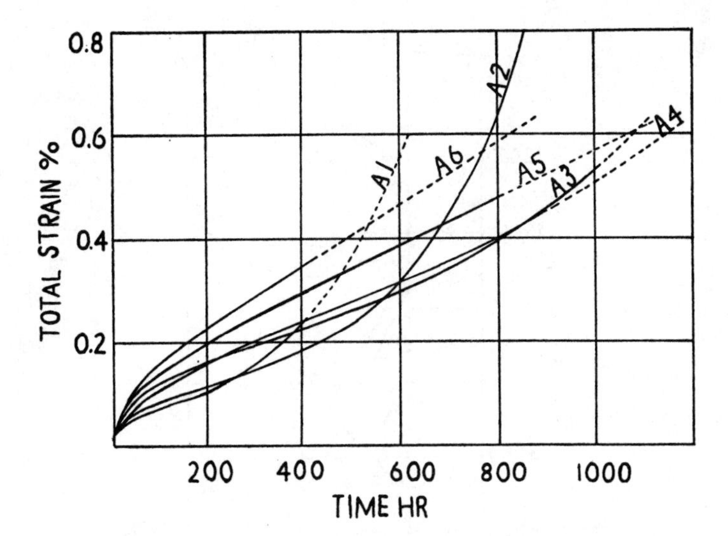

Effect of carbon content on the creep curves of 0.5Mo steels tested at 600 °C and 4 tons/in.^{2} (1112 °F and 8 ksi).

Carbon content: A1 0.055%, A2 0.09%, A3 0.145%, A4 0.21%, A5 0.31%, and A6 0.395%.

Source: F.H. Clark, Metals at High Temperatures, Reinhold, New York, 1950, p 127

4 to 6% Cr Steels: Effect of Molybdenum and Niobium

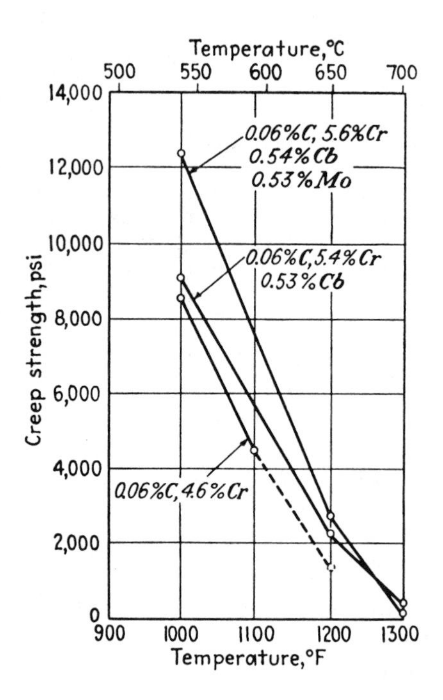

Creep strength of 4 to 6% chromium steels with or without molybdenum and/or niobium.

The primary function of Nb in the hardenable Cr steels is to eliminate air hardening. However, Nb plays a dual role, as it also produces some improvement in creep strength.

Source: R.A. Grange, F.J. Shortsleeve, D.C. Hilty, W.O. Binder, et al., Boron, Calcium, Columbium and Zirconium in Iron and Steel, John Wiley & Sons, New York, 1957, p 185, 194

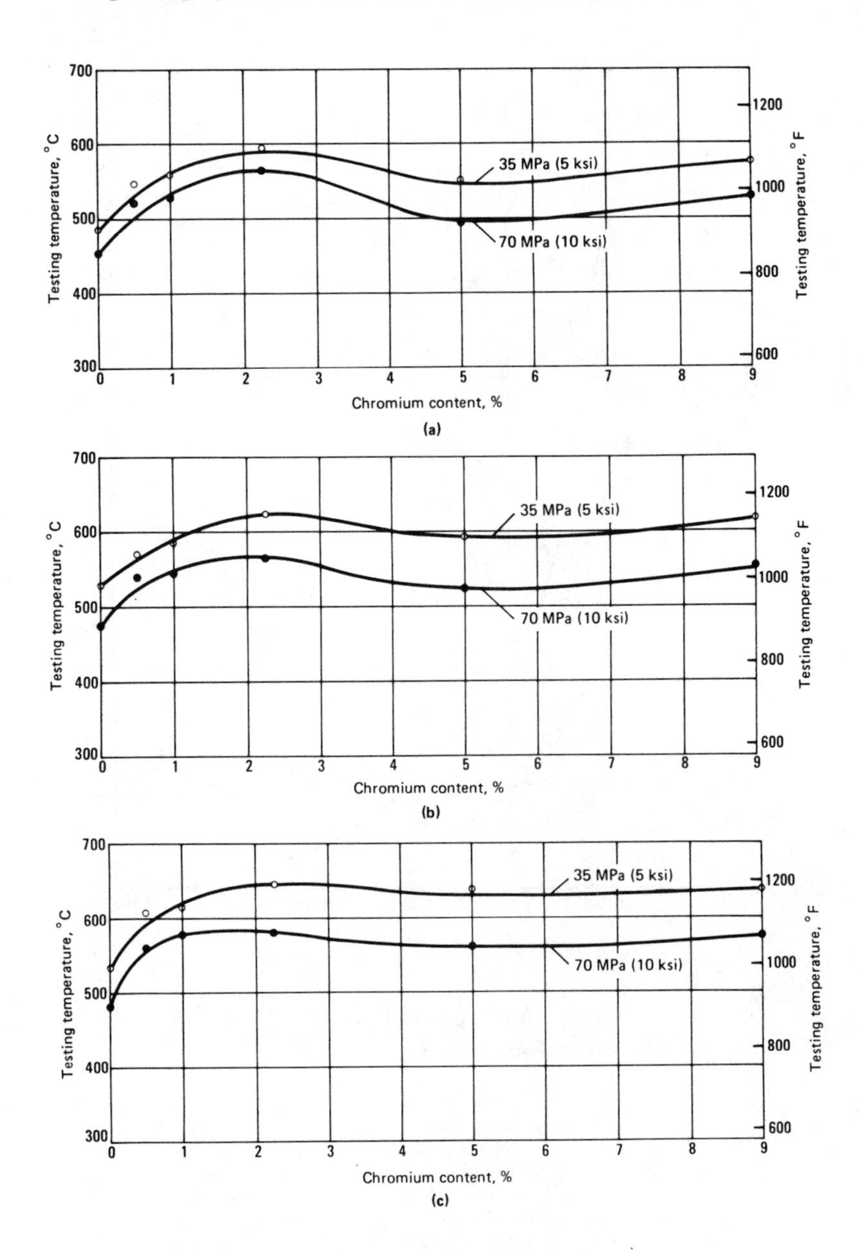

Effect of chromium content on the minimum creep rate and rupture life of Cr-Mo steels containing 0.5 to 1.0% molybdenum.

Test temperature required for (a) minimum creep rate of 0.1 μm/m·h, (b) minimum creep rate of 1 μm/m·h and (c) rupture in 10,000 h. Data are given for stresses of 35 and 70 MPa (5 and 10 ksi).

Source: Metals Handbook, Ninth Edition, Volume 1, Properties and Selection: Irons and Steels, American Society for Metals, Metals Park, OH, 1978, p 650

0.15C, 0.5Mo, 4-6CrMo, and 1.25Cr-Mo Steels: A Comparison

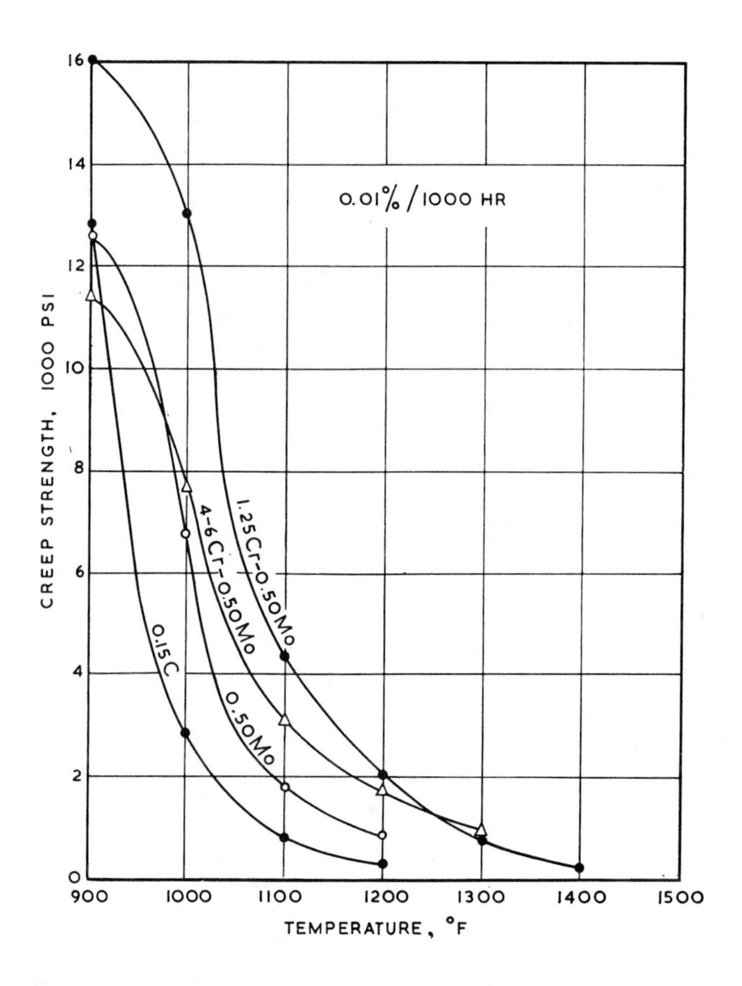

Creep strength for rate of 0.01%/1000 h vs temperature for four steels.

Steel	Chemical composition (%)				
	C	*Mn*	*Si*	*Cr*	*Mo*
0.15C	0.15	0.46	0.28		
0.50Mo	0.13	0.49	0.25		0.52
1.25Cr-0.50Mo	0.07	0.42	0.72	1.24	0.54
4–6Cr-0.50Mo	0.10	0.45	0.18	5.09	0.55

Source: C.L. Clark, High-Temperature Alloys, Pitman, New York, 1953, p 89

0.5Mo and 1Mo Steels: Comparison of Creep Strengths, as Influenced by Chromium Content

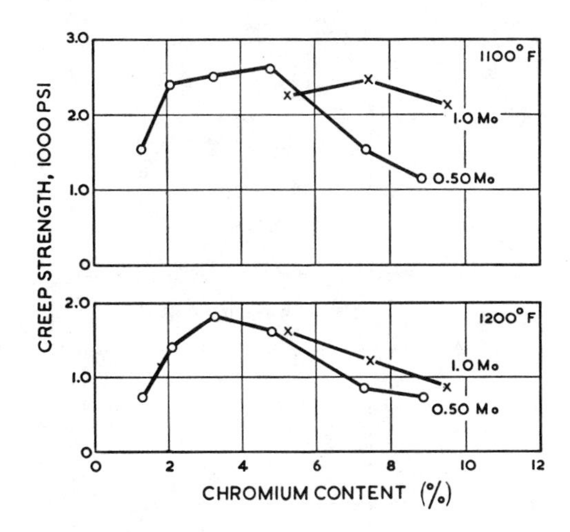

Effect of varying chromium content on the creep strength of 0.5Mo and 1.0Mo steels at 1100 and 1200 °F.

Source: C.L. Clark, High-Temperature Alloys, Pitman, New York, 1953, p 95

19.44 **0.5Mo and 1Mo Steels: Comparison of Rupture Strengths, as Influenced by Chromium Content**

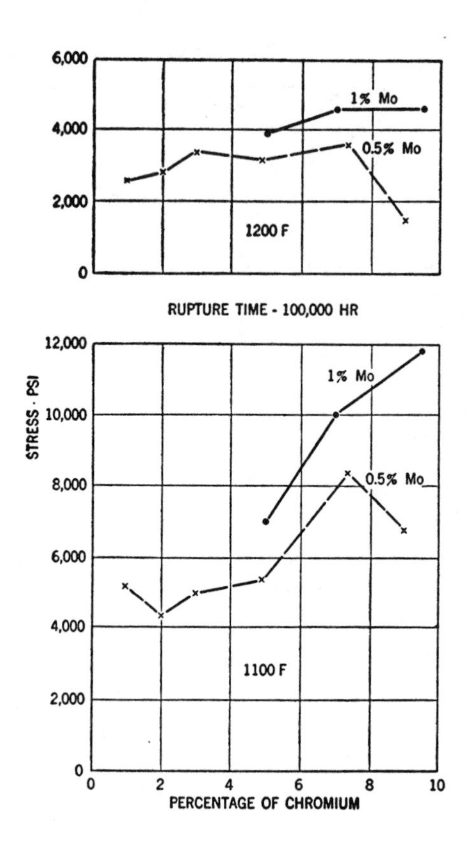

Effect of varying chromium content on the 100,000-h rupture strength of 0.5Mo and 1.0Mo steels at 1100 and 1200 °F.

Source: R.S. Archer, J.Z. Briggs, and C.M. Loeb, Jr., Molybdenum Steels, Irons, Alloys, Climax Molybdenum Co., New York, 1948, p 158

2.25Cr-1Mo, 9Cr-1Mo, and Type 304 Stainless Steel: A Comparison

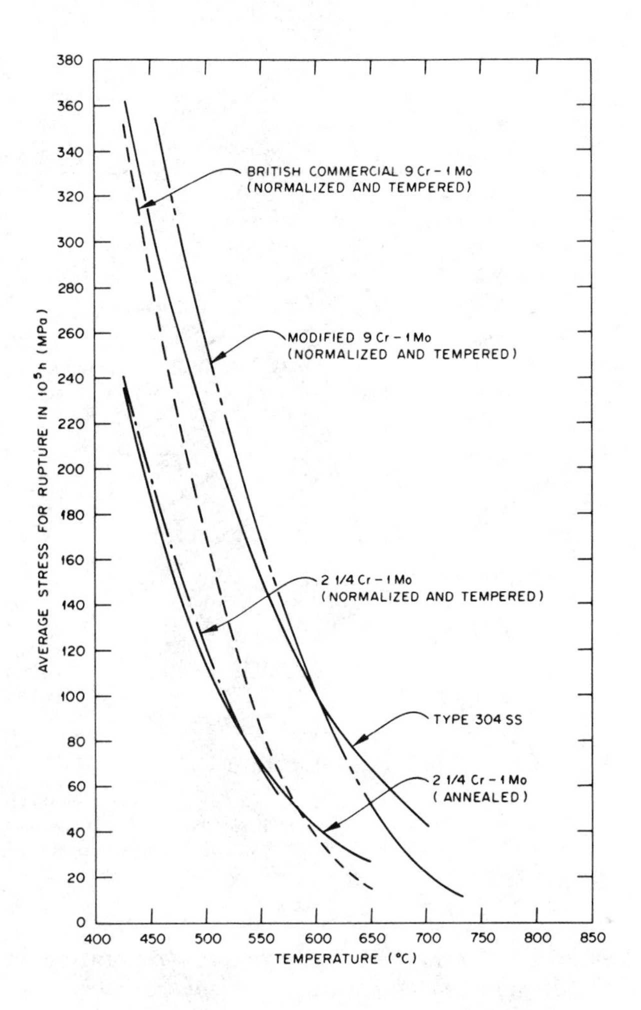

Stress for 100,000-h rupture life vs temperature for two variants of 9Cr-1Mo steel, two variants of 2.25Cr-1Mo steel, and a type 304 stainless steel.

Note: 10 MPa = 1.45 ksi.

Source: V.K. Sikka, "Development of Modified 9Cr-1Mo Steel for Elevated-Temperature Service," in Topical Conference on Ferritic Alloys for Use in Nuclear Energy Technologies (Proceedings of a Conference in Snowbird, UT), Metallurgical Society of AIME, Warrendale, PA, 1984, p 323

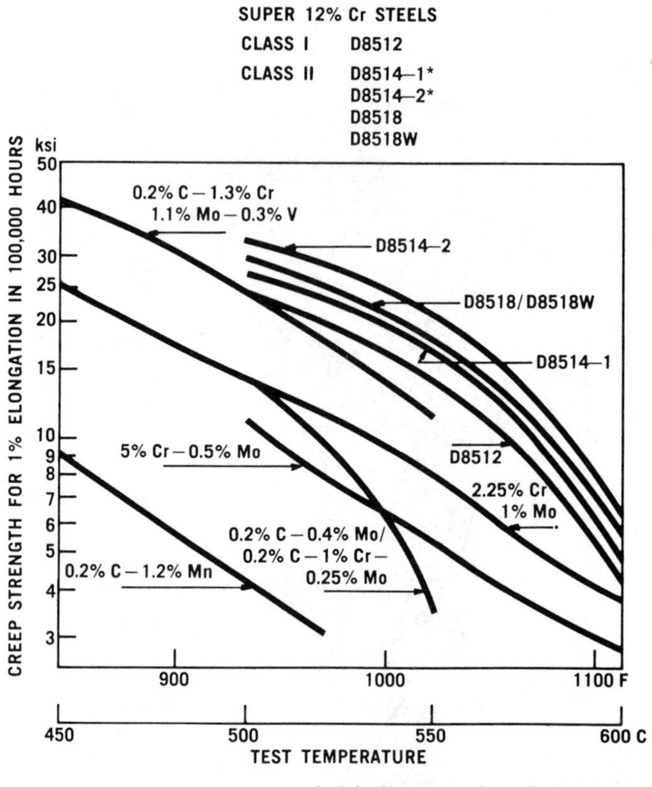

Variation of creep strength for 1% elongation in 100,000 h with temperature for several low-alloy steels (up to 5% Cr).

Source: J.Z. Briggs and T.D. Parker, "The Super 12% Cr Steels," in Source Book on Materials for Elevated-Temperature Applications, E.F. Bradley, Ed., American Society for Metals, Metals Park, OH, 1979, p 166

Mo, Cr-Mo, and Mo-V Steels: Comparison of 100,000-h Rupture Times

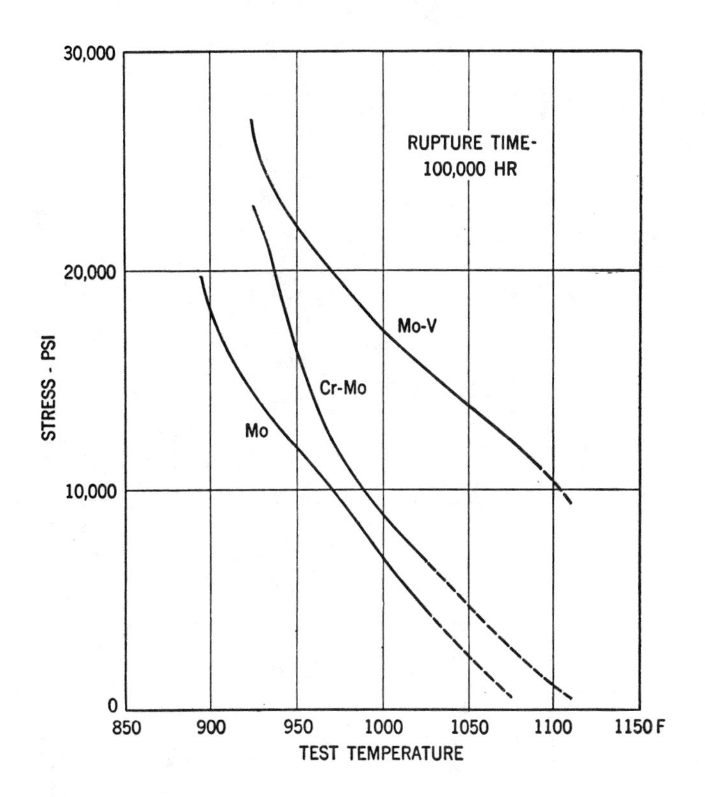

Stress for 100,000-h rupture time vs temperature for three types of steels.

	Type Compositions						Condition
	%C	%Mn	%Si	%Cr	%Mo	%V	
Mo	0.15*	0.60	0.3*	0.25*	0.55	....	1720 F air cooled
Cr–Mo	0.15*	0.50	0.3*	0.80	0.55	....	1720 F air cooled
Mo–V	0.13*	0.60	0.3*	....	0.55	0.28	1760 F air cooled; tempered five hours at 1275 F

*max

Source: R.S. Archer, J.Z. Briggs, and C.M. Loeb, Jr., Molybdenum Steels, Irons, Alloys, Climax Molybdenum Co., New York, 1948, p 153

Mo, Cr-Mo, and Mo-V Steels: Comparison of Creep Strengths

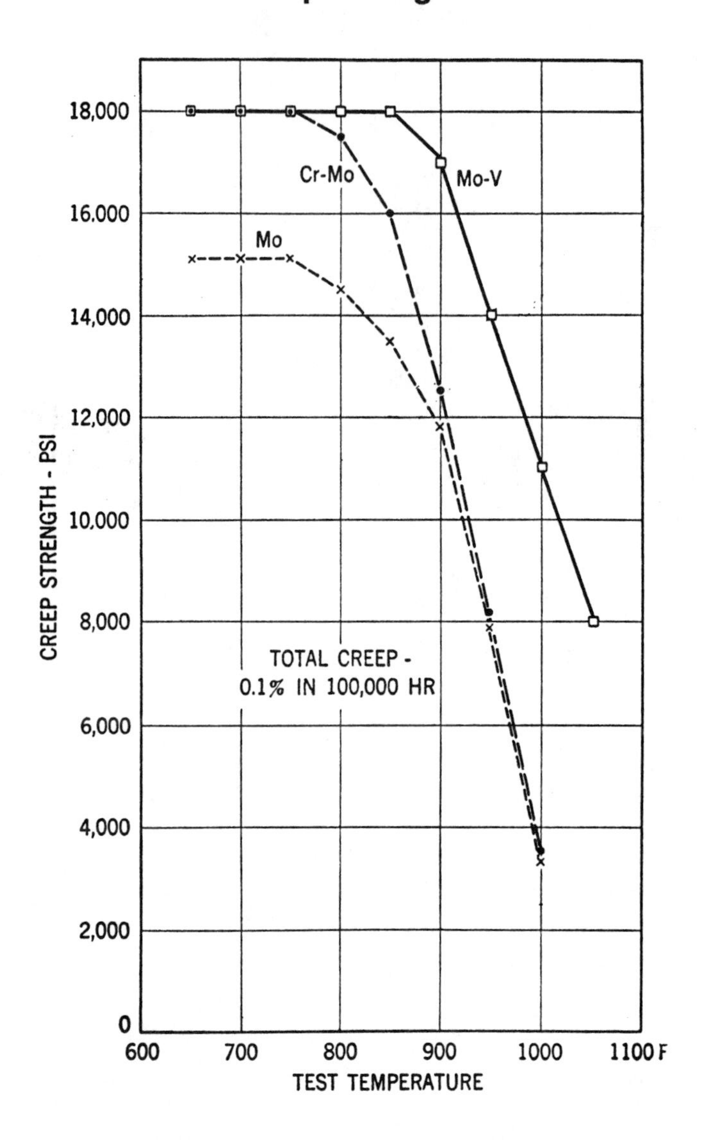

Creep strength for 0.1% total creep in 100,000 h vs temperature for three types of steels.

	Type Compositions						Condition
	%C	%Mn	%Si	%Cr	%Mo	%V	
Mo	0.15*	0.60	0.3*	0.25*	0.55	....	1720 F air cooled
Cr–Mo	0.15*	0.50	0.3*	0.80	0.55	....	1720 F air cooled
Mo–V	0.13*	0.60	0.3*	....	0.55	0.28	1760 F air cooled; tempered five hours at 1275 F

* max

Source: R.S. Archer, J.Z. Briggs, and C.M. Loeb, Jr., Molybdenum Steels, Irons, Alloys, Climax Molybdenum Co., New York, 1948, p 154

20

Copper and Copper Alloys

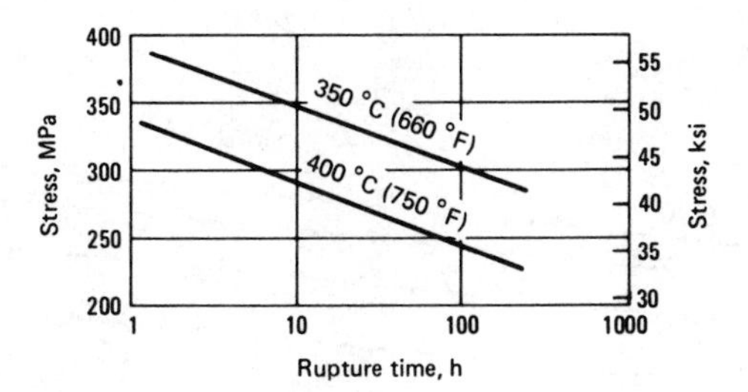

Stress-rupture properties of C15000, zirconium copper, TH08 temper.

Material was solution treated 1 h at a temperature of 950 °C (1740 °F), quenched, cold worked 85%, and aged 1 h at 425 °C (795 °F).

Source: Metals Handbook, Ninth Edition, Volume 2, Properties and Selection: Nonferrous Alloys and Pure Metals, American Society for Metals, Metals Park, OH, 1979, p 298

C19500: High-Copper Alloy; Creep Rate for Strip

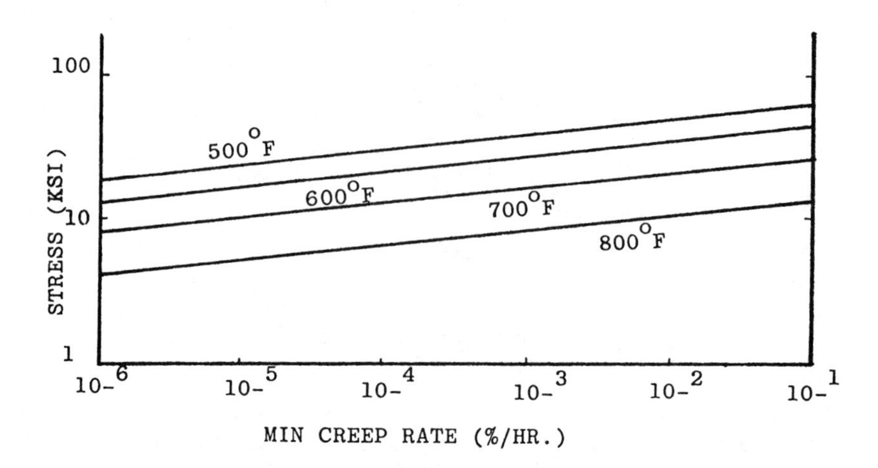

Stress vs minimum creep rate for high-copper alloy C19500 at 500, 600, 700, and 800 °F.

Source: J.J. Cronin, "Selecting High Conductivity Copper Alloys for Elevated Temperature Use," in Source Book on Materials Selection, Volume II, R.B. Gunia, Ed., American Society for Metals, Metals Park, OH, 1977, p 222

C19500: High-Copper Alloy; Stress-Rupture Strength of Strip

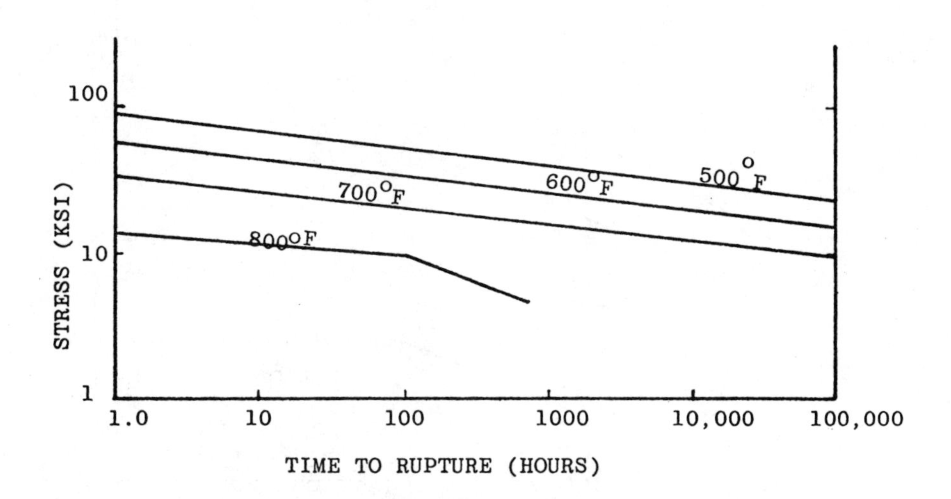

Stress-rupture curves for high-copper alloy C19500 at 500, 600, 700, and 800 °F.

Source: J.J. Cronin, "Selecting High Conductivity Copper Alloys for Elevated Temperature Use," in Source Book on Materials Selection, Volume II, R.B. Gunia, Ed., American Society for Metals, Metals Park, OH, 1977, p 222

C23000: Red Brass Wire

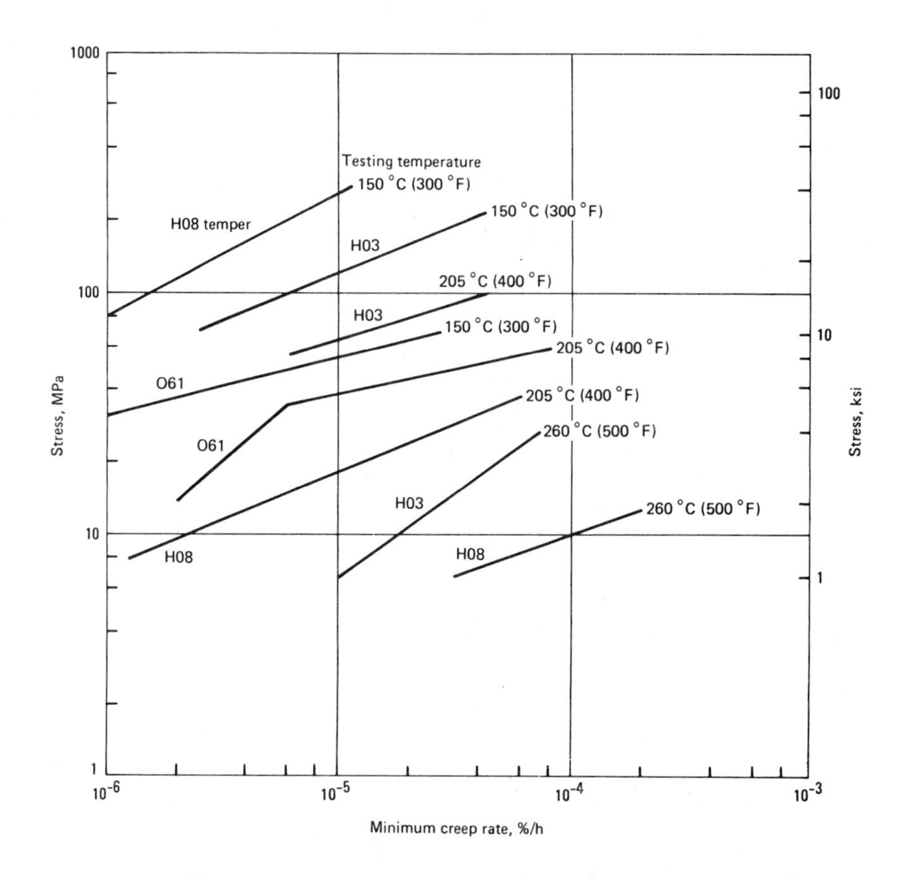

Minimum creep rates for red brass (C23000) wire.

Data are for red brass wire, 3.2-mm (0.125-in.) diam, cold drawn to size, then tested as drawn or annealed prior to testing.

Source: Metals Handbook, Ninth Edition, Volume 2, Properties and Selection: Nonferrous Alloys and Pure Metals, American Society for Metals, Metals Park, OH, 1979, p 322

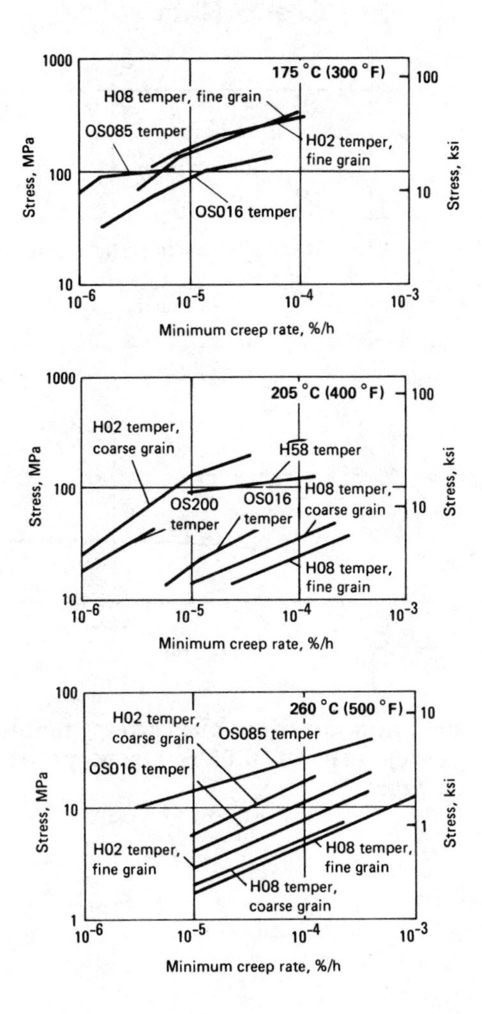

Minimum creep rates for C26000, cartridge brass, 70%.

Source: Metals Handbook, Ninth Edition, Volume 2, Properties and Selection: Nonferrous Alloys and Pure Metals, American Society for Metals, Metals Park, OH, 1979, p 326

20.6 C26000: Cartridge Brass, 70%; Effect of Grain Size on Creep Rate

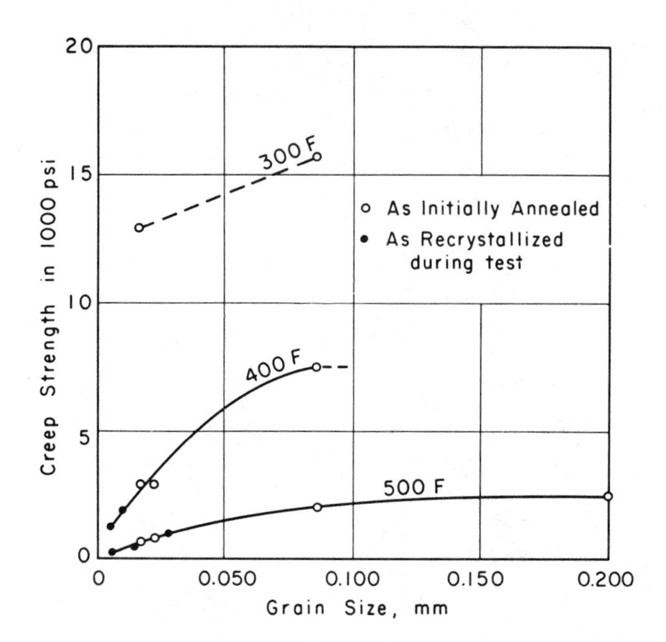

Effect of grain size on the stress required to produce creep rate of 0.01% strain per 1000 h for 70-30 brass.

Source: E.R. Parker and J. Washburn, "The Role of the Boundary in Creep Phenomena," in Creep and Recovery, American Society for Metals, Cleveland, OH, 1957, p 239

C27400: Yellow Brass; Effect of Titanium and Zirconium Additions

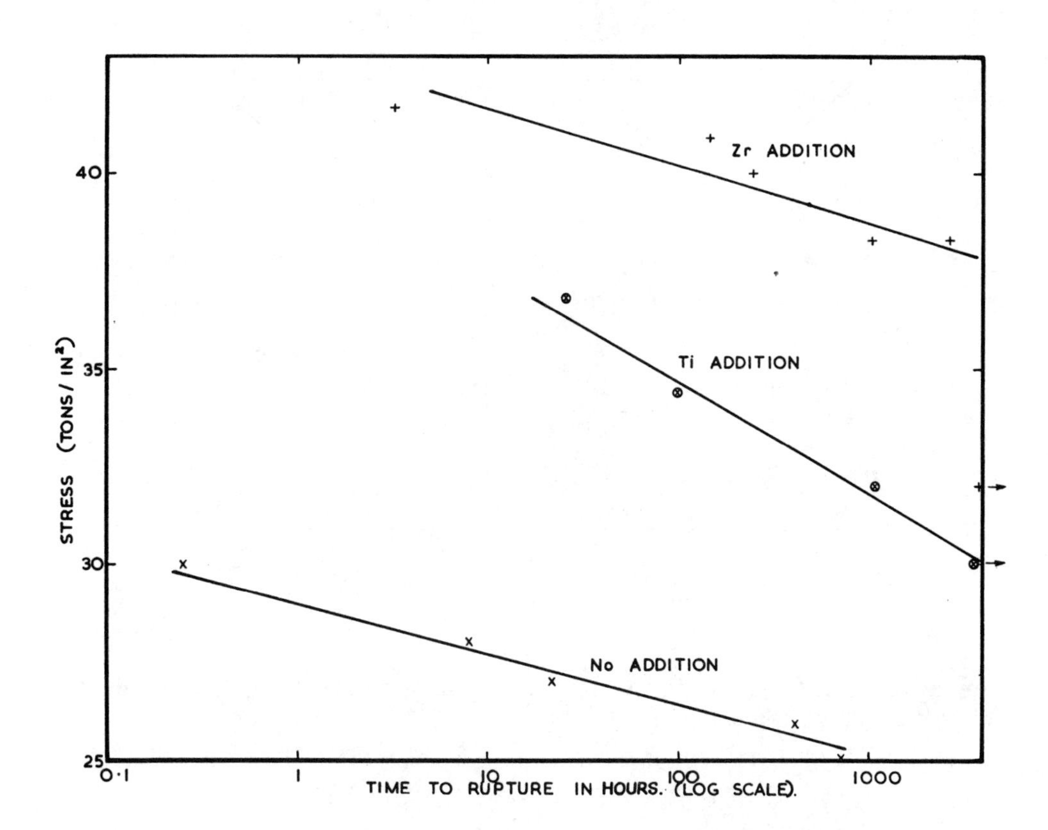

Effect of adding titanium (0.09%) and zirconium (0.09%) on time to fracture at room temperature of a wrought β-brass. Composition: 62.3 to 62.9% Cu, 33.1 to 33.4% Zn, 3.9 to 4.0% Al (high-purity basis).

Source: R. Eborall, "An Approach to the Problem of Intercrystalline Fracture," in Creep and Fracture of Metals at High Temperatures, Philosophical Library, New York, 1957, p 241

C68700: Aluminum Brass, Wire

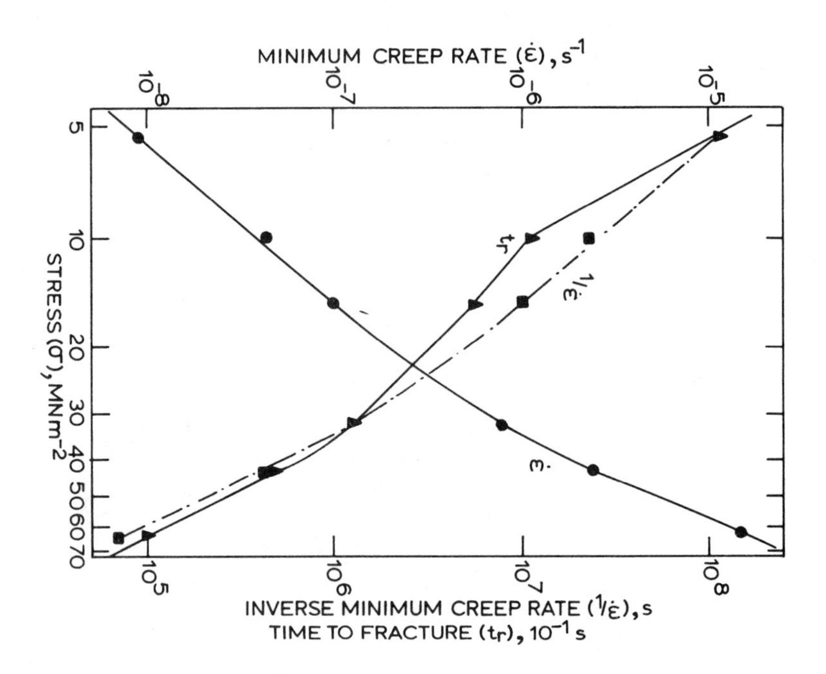

Minimum creep rate, $\dot{\varepsilon}$, its inverse, $1/\dot{\varepsilon}$, and time to fracture, t, plotted as a function of applied stress, σ.

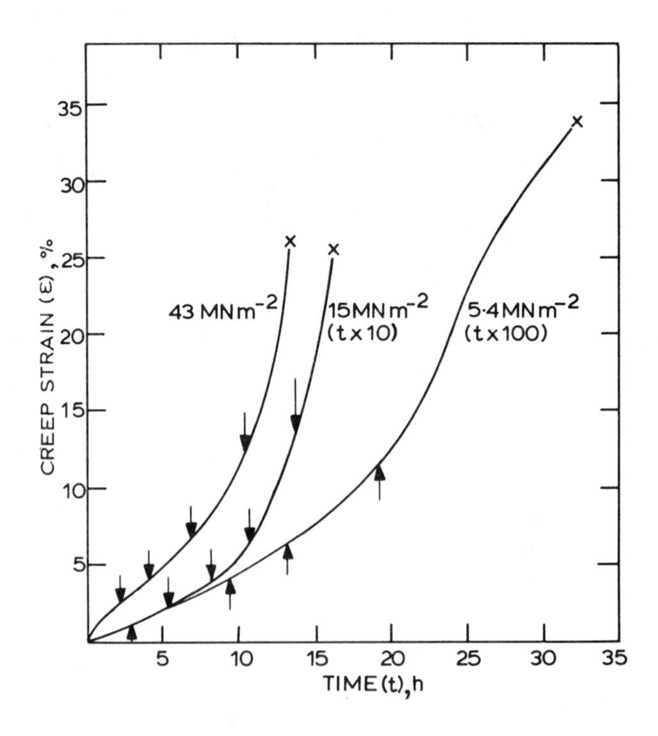

Creep curves for three applied stresses at which detailed cavity size measurements were carried out. Arrows mark strains and times at which creep tests were interrupted. Note that a reduced time scale is used for creep curves corresponding to σ = 5.4 MN/m^2 and σ = 15 MN/m^2.

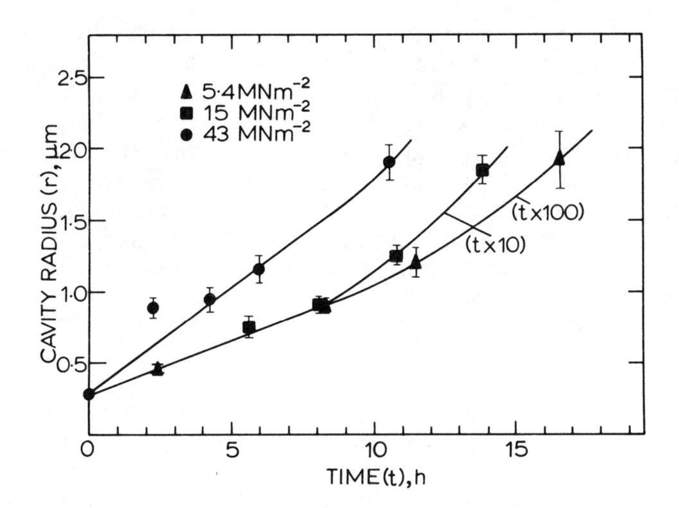

Average values of cavity radii, r, plotted as a function of time under creep. Note reduced time scale for $\sigma =$ 15 MN/m^2 and $\sigma =$ 5.4 MN/m^2.

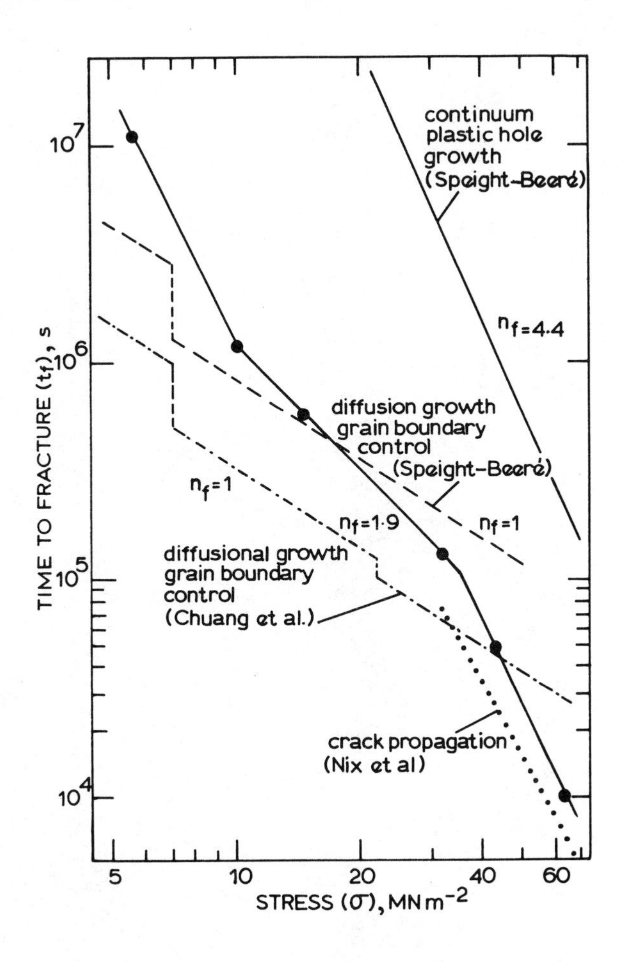

Time to fracture plotted as a function of applied stress. Also shown are predicted times to fracture for various cavity growth theories.

Intergranular cavities were introduced into α-brass by gas bubble precipitation during simple stress-free annealing. The growth rates of these cavities during creep at a temperature of 700 K and stresses of 5-43 MN/m^2 were measured by scanning electron microscopy. The results were consistent with cavity growth being controlled by grain-boundary diffusion. Time-to-fracture measurements indicated that grain-boundary diffusion-controlled cavity growth was the time-dominant step in the overall creep-fracture process for stresses of 10-30 MN/m^2. At higher stresses, the time to fracture was reduced by the formation and propagation of multifacet intergranular cracks during the final stages of the creep life.

Source: L.E. Svensson and G.L. Dunlop, "Growth Mechanism of Intergranular Creep Cavities in α-Brass," Metal Science, Vol. 16, Jan. 1982, p 58-59, 61, 63

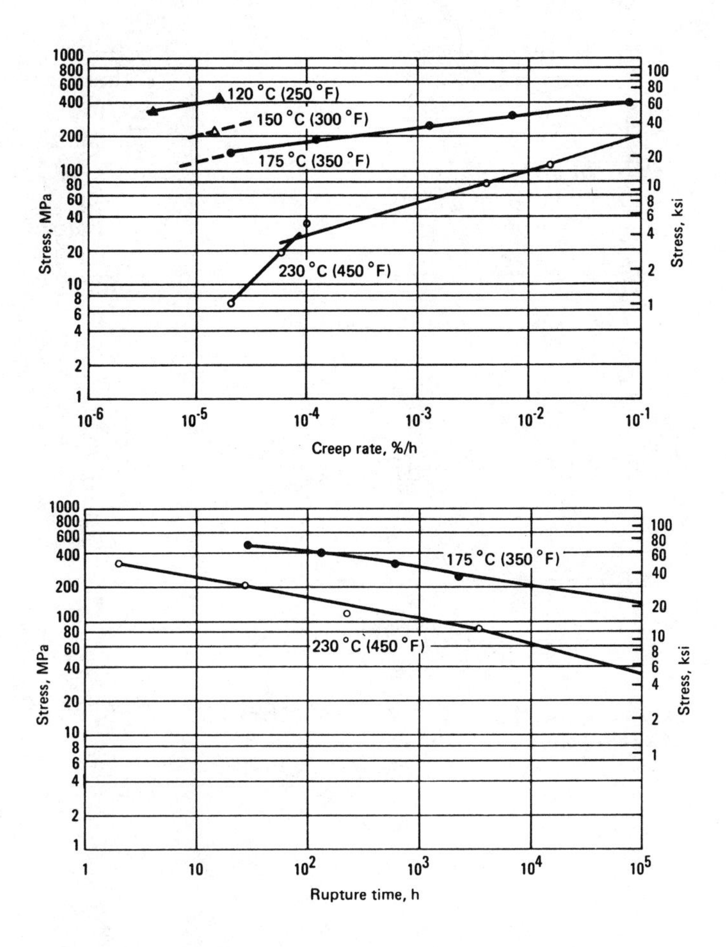

Creep-rupture properties of C86300, leaded manganese bronze.

Source: Metals Handbook, Ninth Edition, Volume 2, Properties and Selection: Nonferrous Alloys and Pure Metals, American Society for Metals, Metals Park, OH, 1979, p 410

C86500: Leaded Manganese Bronze

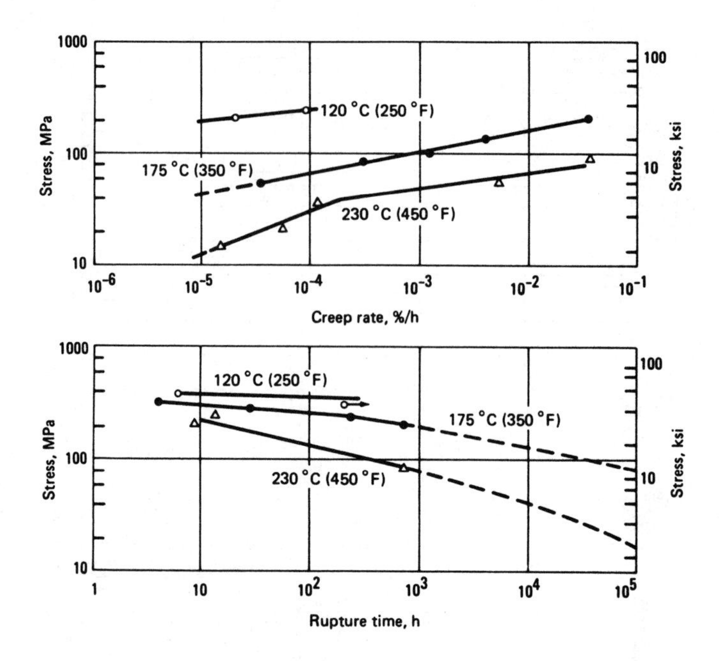

Typical creep-rupture properties of C86500, leaded manganese bronze.

Source: Metals Handbook, Ninth Edition, Volume 2, Properties and Selection: Nonferrous Alloys and Pure Metals, American Society for Metals, Metals Park, OH, 1979, p 414

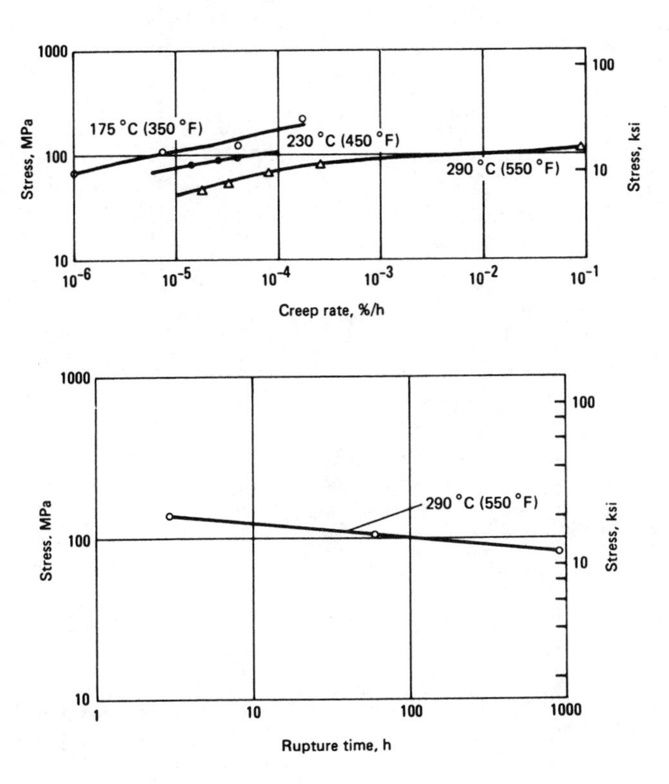

Creep-rupture properties of C92200, leaded tin bronze.

Source: Metals Handbook, Ninth Edition, Volume 2, Properties and Selection: Nonferrous Alloys and Pure Metals, American Society for Metals, Metals Park, OH, 1979, p 421

C93700: High-Leaded Tin Bronze

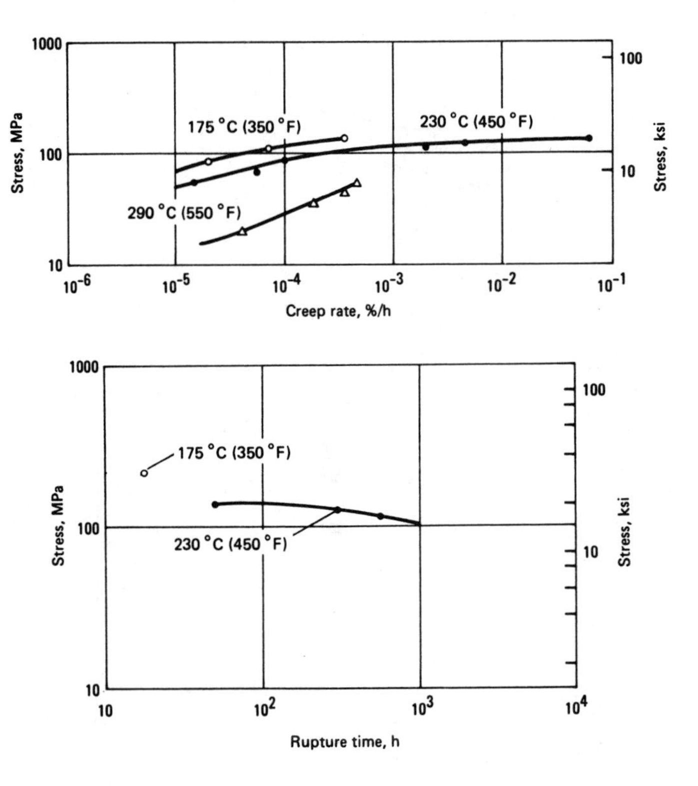

Typical creep-rupture properties for C93700, high-leaded tin bronze.

Source: Metals Handbook, Ninth Edition, Volume 2, Properties and Selection: Nonferrous Alloys and Pure Metals, American Society for Metals, Metals Park, OH, 1979, p 427

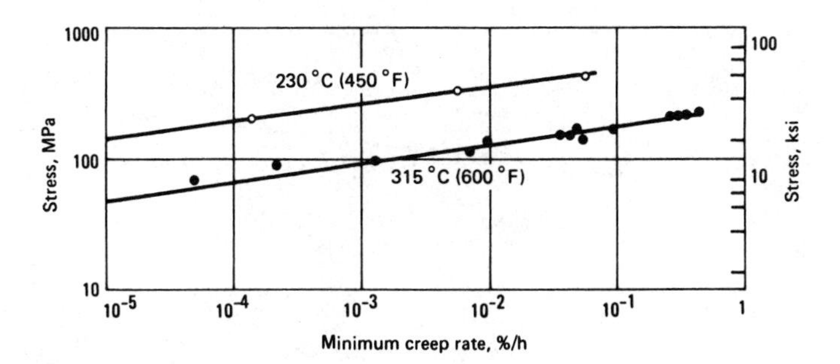

Typical creep properties of C95200, as-cast aluminum bronze.

Source: Metals Handbook, Ninth Edition, Volume 2, Properties and Selection: Nonferrous Alloys and Pure Metals, American Society for Metals, Metals Park, OH, 1979, p 431

C95400: Aluminum Bronze, As-Cast

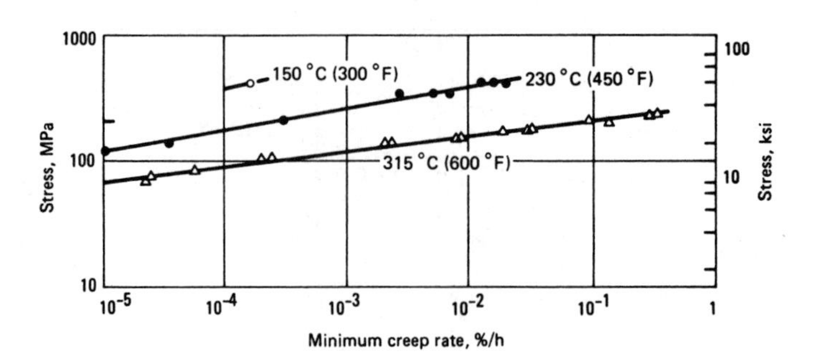

Typical creep properties for C95400, as-cast aluminum bronze.

Source: Metals Handbook, Ninth Edition, Volume 2, Properties and Selection: Nonferrous Alloys and Pure Metals, American Society for Metals, Metals Park, OH, 1979, p 432

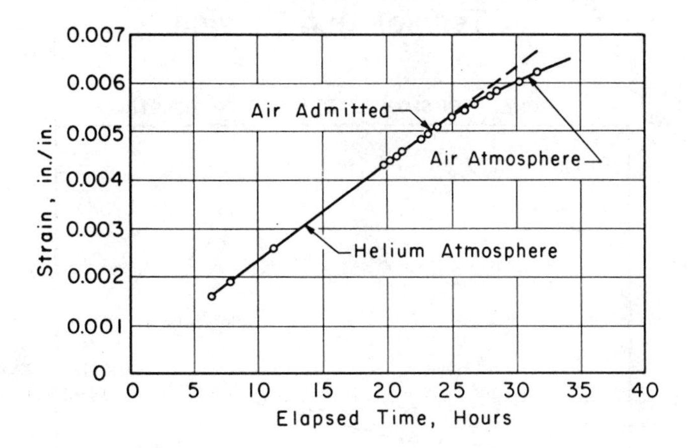

Creep curve for commercially pure copper obtained at 200 °C showing how an oxide surface layer reduces the creep rate.

Source: E.R. Parker and J. Washburn, "The Role of the Boundary in Creep Phenomena," in Creep and Recovery, American Society for Metals, Cleveland, OH, 1957, p 232

20.18 Commercially Pure Copper: Creep Properties for Tube Tested in Tension, Torsion, and Combined Tension and Torsion

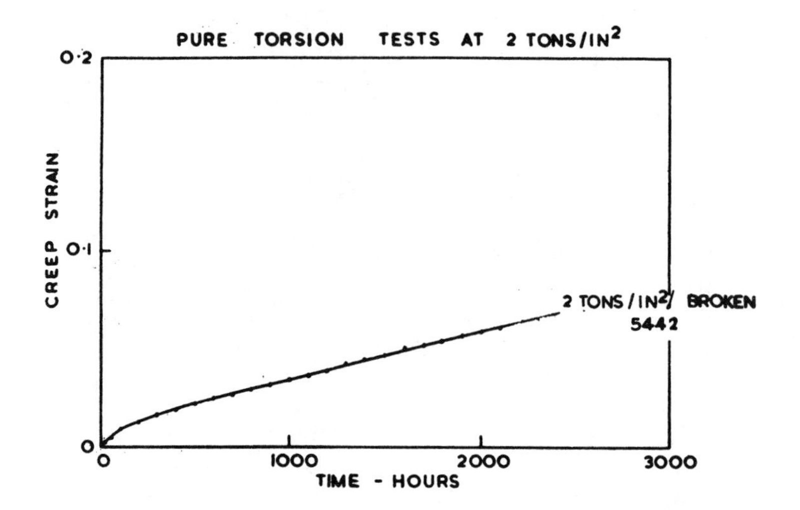

Creep strain-time curve for combined stress-creep tests on thin walled tubular specimens of commercially pure copper at 250 °C.

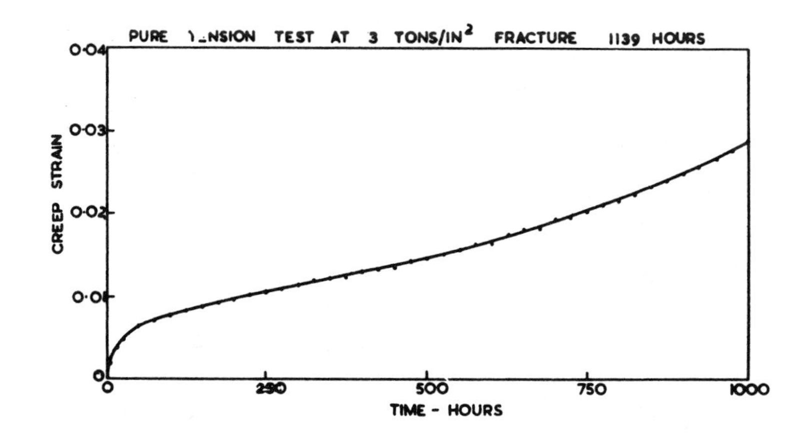

Creep strain-time curve for combined stress-creep tests on thin walled tubular specimens of commercially pure copper at 250 °C.

Commercially Pure Copper: Creep Properties for Tube Tested in Tension, Torsion, and Combined Tension and Torsion (Continued)

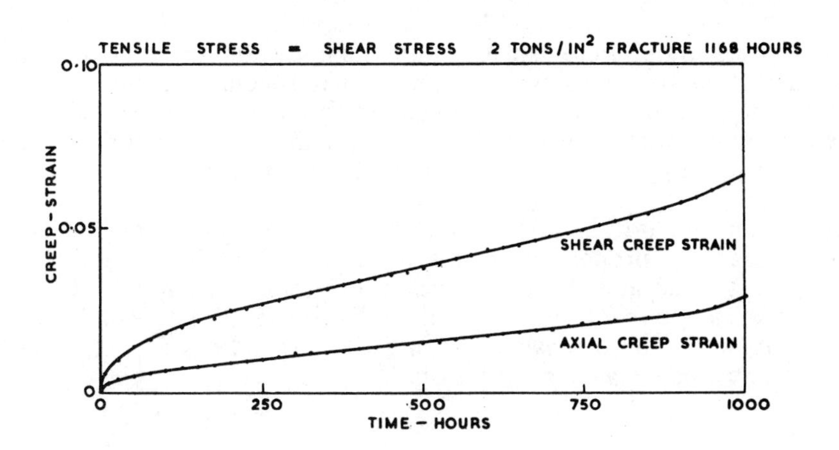

Creep strain-time curves for combined stress-creep tests on thin walled tubular specimens of commercially pure copper at 250 °C.

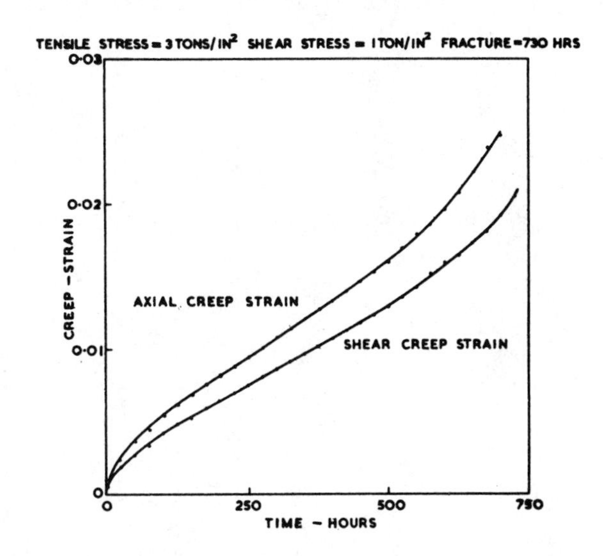

Creep strain-time curves for combined stress-creep tests on thin walled tubular specimens of commercially pure copper at 250 °C.

20.20 Commercially Pure Copper: Creep Properties for Tube Tested in Tension, Torsion, and Combined Tension and Torsion (Continued)

A series of tests under pure tension, pure torsion, and three combinations of tensile and torsion stresses was made on tubular specimens, the stresses being so chosen that the range of the "equivalent" stress (i.e., the octahedral stress) remained of much the same order as the equivalent stresses of the range of tensile stresses giving virtually intercrystalline fracture.

Tests consist of the initial tests of a comprehensive survey of the combined stress-creep-fracture properties of this material at 250 °C. Actually the two combined stress tests, and the pure tension and torsion tests, may be regarded as members of a group of tests of similar "equivalent," i.e., octahedral stress level. Of course the pure tension test and the test at tension 3 tons/in.^{2}, and torsion 1 ton/in.^{2} have the same hydrostatic stress.

Source: A.E. Johnson and N.E. Frost, "Note on the Fracture under Complex Stress Creep Conditions of a 0.5% Molybdenum Steel at 550 °C, and a Commercially Pure Copper at 250 °C," in Creep and Fracture of Metals at High Temperatures, Philosophical Library, New York, 1957, p 373

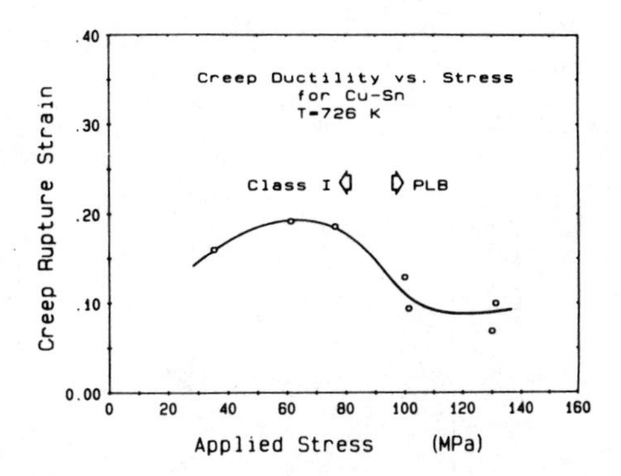

Stress dependence of creep ductility for a copper-tin alloy in the class I and power law breakdown regimes (tested in vacuum).

Source: J.S. Wang and W.D. Nix, "High Temperature Creep and Fracture Properties of a Class I Solid Solution Alloy: Cu-2.7 at.%Sn," Acta Metallurgica, Vol. 34, March 1986, p 551

Copper-Tin Alloy: Creep Ductility vs Temperature

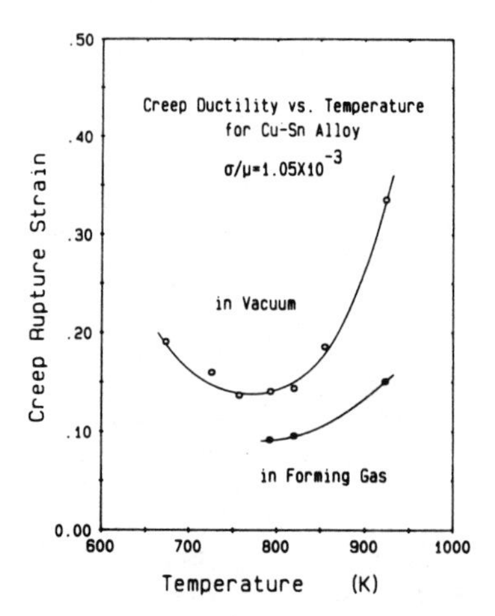

Creep-rupture strain of a copper-tin alloy as a function of temperature.

Source: J.S. Wang and W.D. Nix, "High Temperature Creep and Fracture Properties of a Class I Solid Solution Alloy: Cu-2.7 at.%Sn," Acta Metallurgica, Vol. 34, March 1986, p 551

Copper-Tin Alloy: Typical Creep Curve

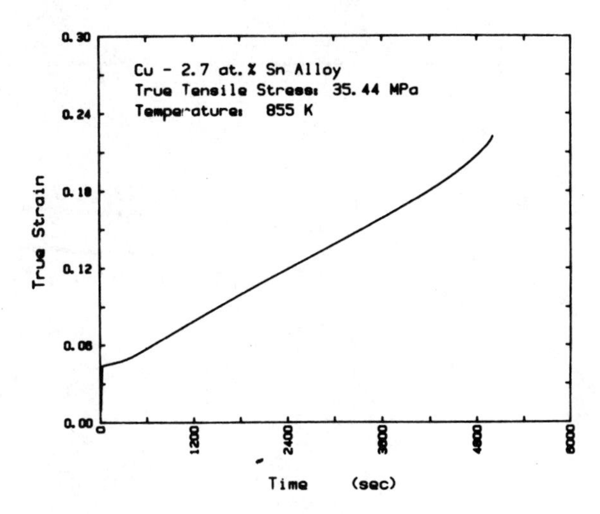

Typical creep curve for a copper-tin alloy in the class I regime.

Source: J.S. Wang and W.D. Nix, "High Temperature Creep and Fracture Properties of a Class I Solid Solution Alloy: Cu-2.7 at.%Sn," Acta Metallurgica, Vol. 34, March 1986, p 547

Copper-Zirconium Alloys: Air Melted and Cast vs Vacuum Melted and Cast

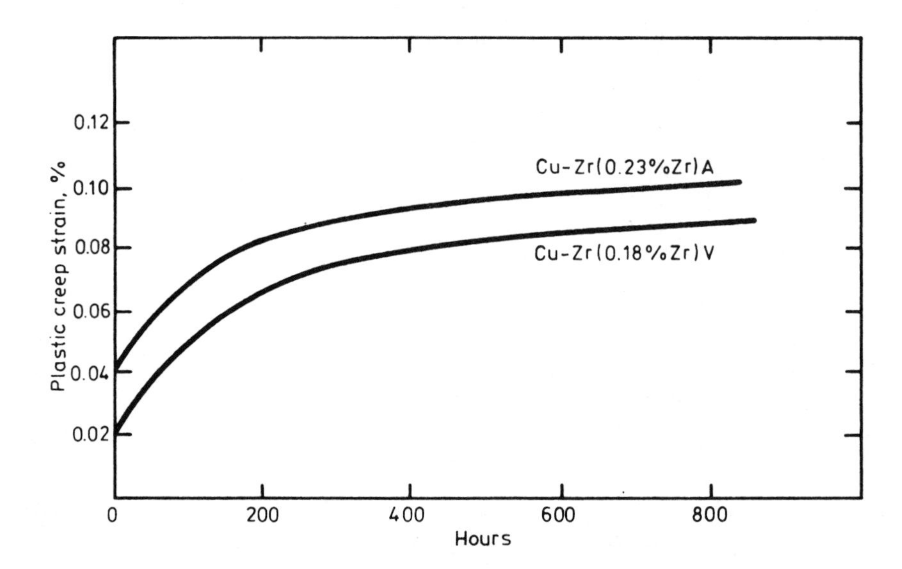

Creep resistance of copper-zirconium alloys at 230 °C and 308 N/mm². A = air melted and cast; V = vacuum melted and cast.

Source: E.G. West, Copper and Its Alloys, John Wiley & Sons, New York, 1982, p 147

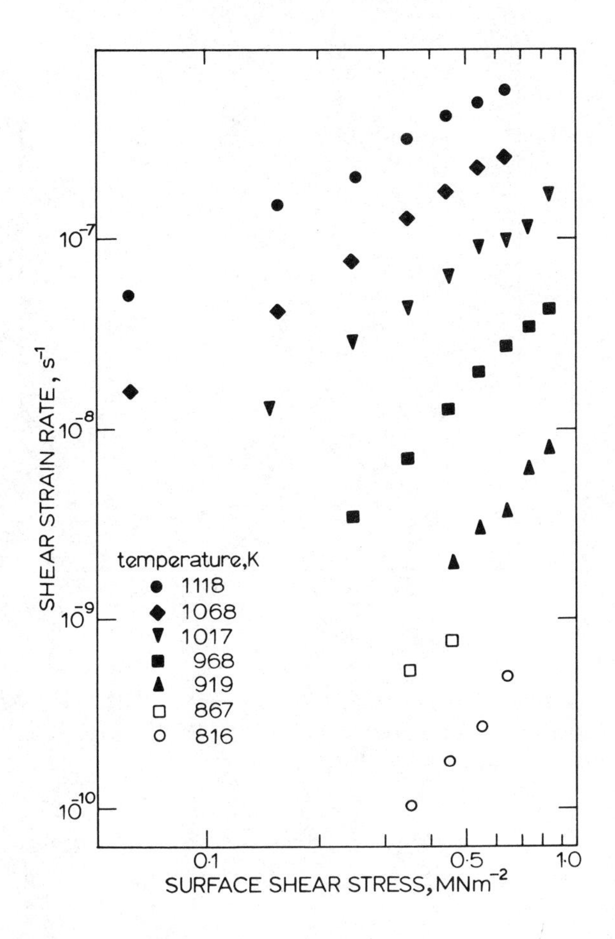

Stress-strain rate data obtained for $CuSiO_2$ plotted on logarithmic axes.

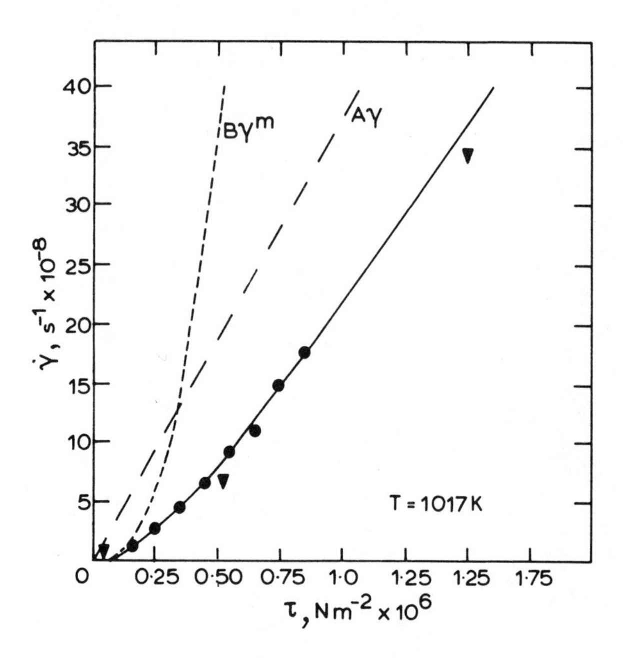

Strain rate vs shear stress for $CuSiO_2$ spring at 1017 K. Also included are data from springs of different geometries. Results are shown to be reproducible even if geometry is altered.

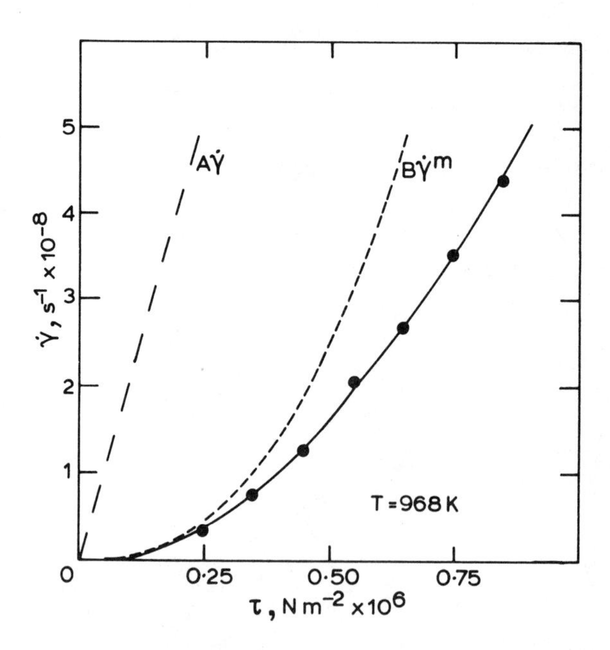

Strain rate vs shear stress for $CuSiO_2$ spring at 968 K. Broken lines represent diffusional process and accommodating process; full line represents sum of the two processes.

The creep of polycrystalline Cu containing a dispersion of silica particles has been investigated at stresses below 1 MN/m^2 and at temperatures from 820-1120 K. Previous work has shown that pure Cu of similar grain size deforms by Coble creep in this regime. The introduction of silica particles is shown to inhibit creep. It is found that the deformation characteristics can be represented by the operation of two coupled processes, one of which can be identified as Coble creep. It is suggested that the other process is the motion of grain-boundary dislocations controlled by the rate at which they can bypass particles. Using this approach, it is shown that this can lead to an apparent threshold stress for creep with characteristics similar to those observed by other workers.

Source: W.J. Clegg and J.W. Martin, "Diffusion Creep Threshold in Two-Phase Alloys," Metal Science, Vol. 16, Jan. 1982, p 65-67, 69

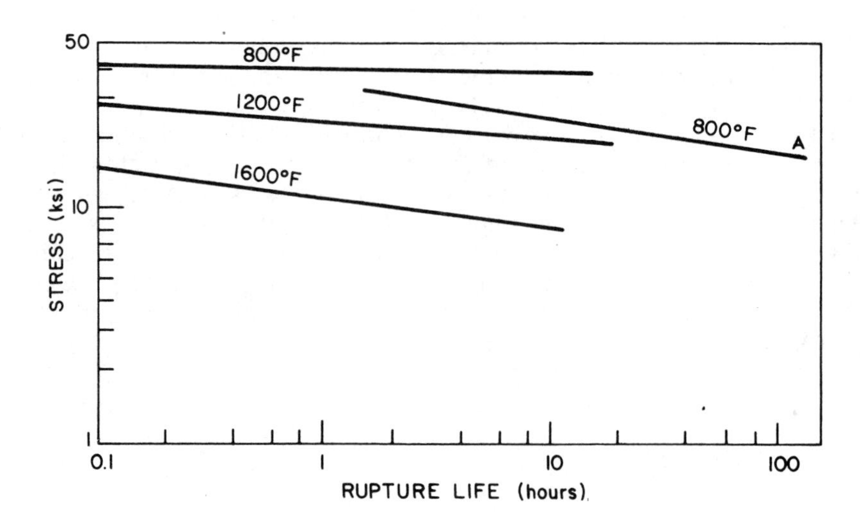

Stress-rupture response of dispersion-strengthened copper wire containing 0.7 vol% Al_2O_3 at various temperatures. Curve A is for Cu-0.6Be-2.5Co age-hardening alloy.

Source: A. Lawley, "Copper and Copper-Base P/M Specialty Alloys," in New Perspectives in Powder Metallurgy, Volume 7, Copper Base Powder Metallurgy, Metal Powder Industries Federation, Princeton, NJ, 1980, p 158

High-Purity Copper: Effect of Grain Size on Strain Rate during Creep

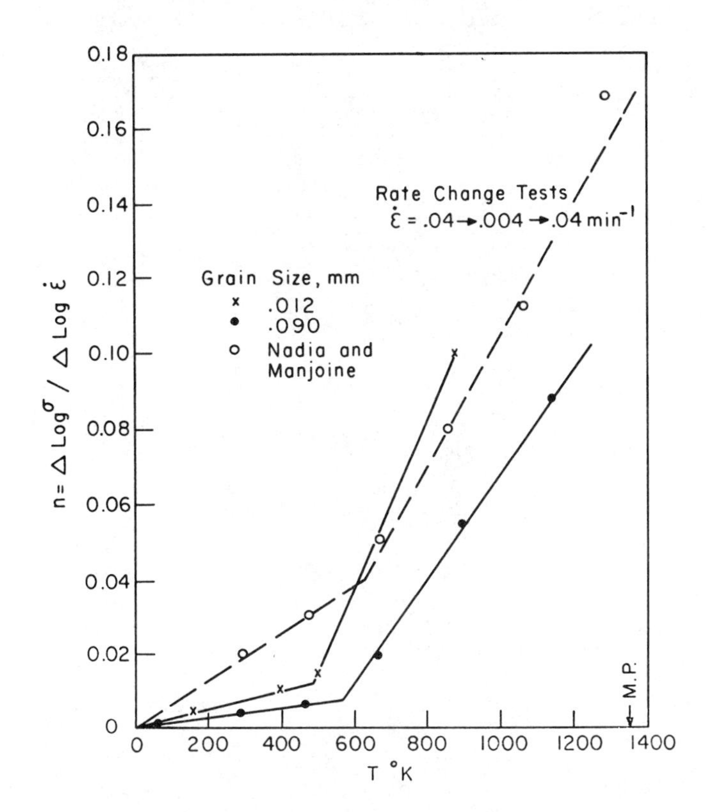

Logarithmic ratio of the change in flow stress produced by a change in strain rate and plotted against testing temperature.

Source: E.R. Parker and J. Washburn, "The Role of the Boundary in Creep Phenomena," in Creep and Recovery, American Society for Metals, Cleveland, OH, 1957, p 238

20.30 Oxygen-Free, Silver-Bearing Oxygen-Free, Tough-Pitch, and Silver-Bearing Tough-Pitch Copper: Effect of Cooling Rate on Stress-Rupture Properties of Wire Bar

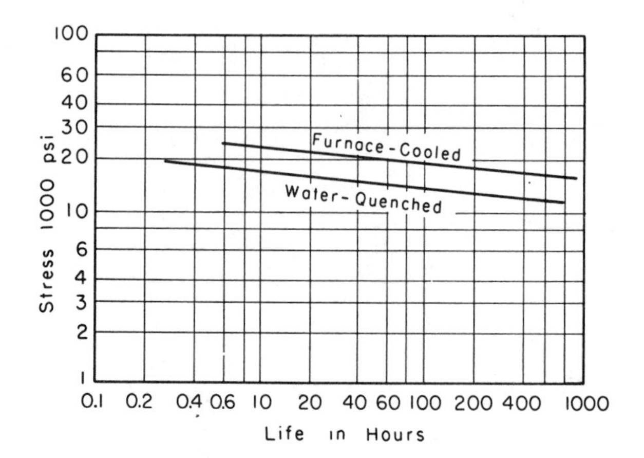

Stress-rupture curve for copper wire bar showing effect of cooling rate from 850 °C (1560 °F) on life at 200 °C (390 °F).

Treatment	*Life in Hours* At 20,000 psi	At 18,000 psi	*Material*
Furnace-cooled from 850 °C (1560 °F)	28	100	Commercial oxygen free
Water-quenched from 850 °C (1560 °F)	5	18	Commercial oxygen free
Furnace-cooled from 850 °C (1560 °F)	380	>1000	Commercial silver-bearing oxygen free
Water-quenched from 850 °C (1560 °F)	50	600	Commercial silver-bearing oxygen free
Furnace-cooled from 850 °C (1560 °F)	10	40	Tough pitch
Water-quenched from 850 °C (1560 °F)	3	15	Tough pitch
Furnace-cooled from 850 °C (1560 °F)	>1000	>>1000	Silver-bearing tough pitch
Water-quenched from 850 °C (1560 °F)	300	>1000	Silver-bearing tough pitch

Comparison of rupture properties of copper at 200 °C

Source: E.R. Parker, "Fundamental Considerations," in Effect of Residual Elements on the Properties of Metals, American Society for Metals, Cleveland, OH, 1957, p 19-20

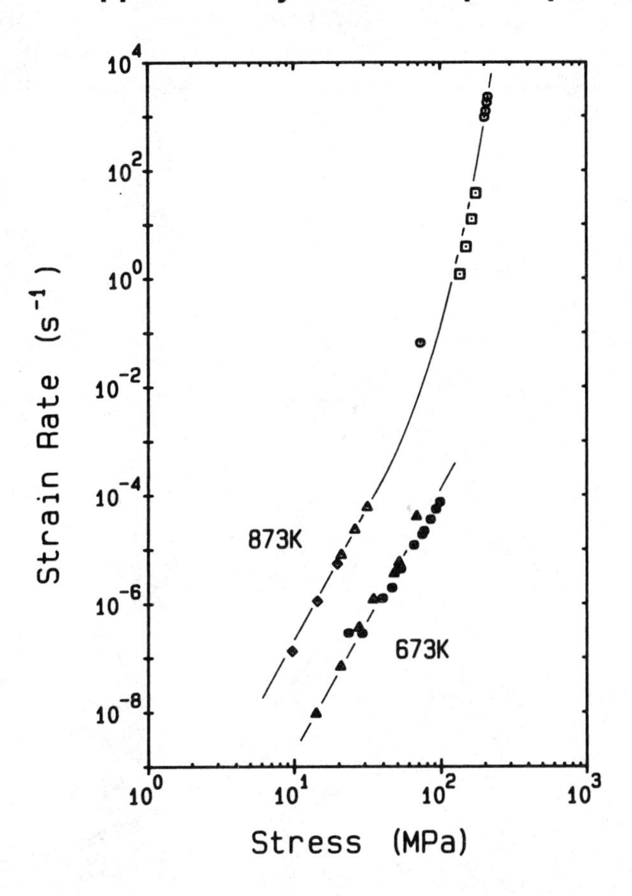

Steady-state creep properties of pure copper at two temperatures. Over a wide range of strain rates at 873 K, the stress exponent is not constant.

Source: W.D. Nix and J.C. Gibeling, "Mechanisms of Time-Dependent Flow and Fracture of Metals," in Flow and Fracture at Elevated Temperatures (paper presented at the ASM Materials Science Seminar, Philadelphia), R. Raj, Ed., American Society for Metals, Metals Park, OH, 1985, p 50

C12200, C14200, C26000, C44300, and C70600: A Comparison

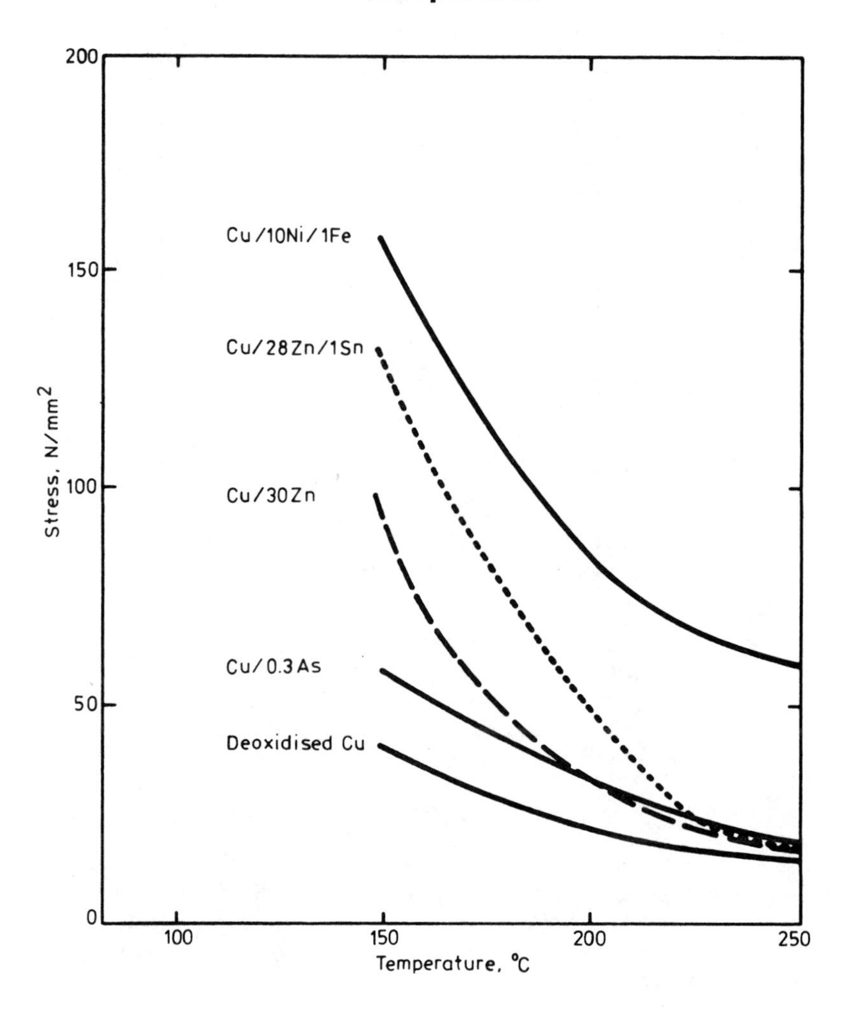

Creep-strength curves for two types of copper and three wrought copper alloys at a stress for 0.00001%/h.

Source: E.G. West, Copper and Its Alloys, John Wiley & Sons, New York, 1982, p 34

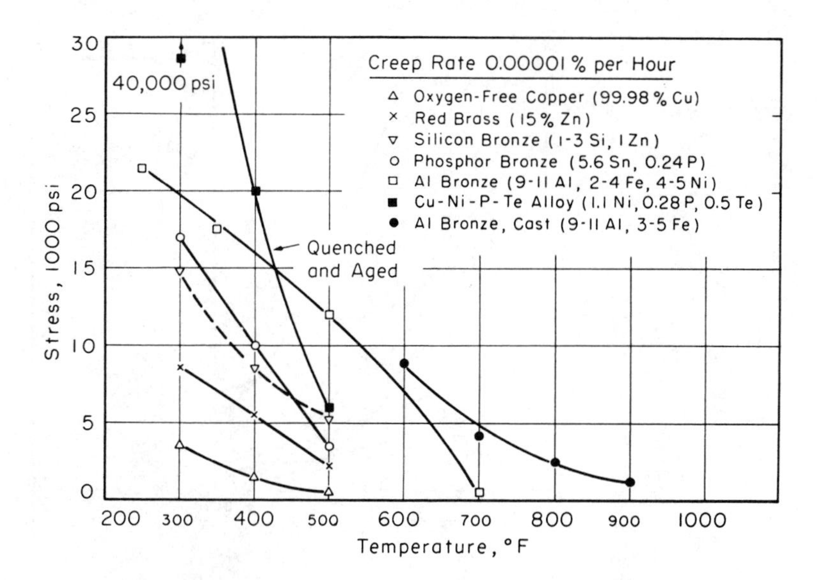

Creep strength curves for copper-base alloys in the annealed condition (except as noted).

Alloy	Chemical Composition, %							
	Zn	Sn	Si	Ni	Al	Fe	P	Te
Oxygen-free copper	99.98% Cu							
Arsenical copper	99.60% Cu, 0.32% As							
Red brass	15							
Cartridge brass	30							
Silicon bronze	1		1–3					
Phosphor bronze		5					0.24	
Aluminum bronze, wrought				4–5	9–11	2–4		
Aluminum bronze, cast					9–11	3–5		
Cu-Ni-P-Te				1.1			0.28	0.51

Copper alloys

Source: H.C. Cross and W.F. Simmons, "Alloys and Their Properties for Elevated Temperature Service," in Utilization of Heat Resistant Alloys, American Society for Metals, Cleveland, OH, 1954, p 62-63

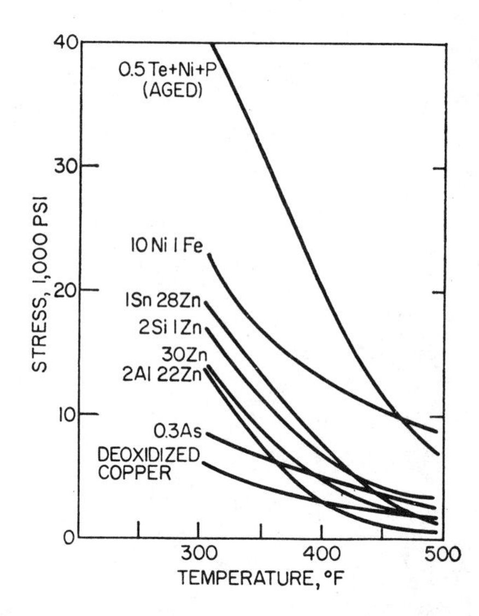

Stress for 1% creep in 100,000 h for copper and several copper alloys.

An alloy containing about 0.3% added Ni coprecipitates with P and Te to produce pronounced hardening. This material, when water quenched from 1450 °F, drawn 30%, and tempered at 770 °F for 1½ h, attains a creep strength of 8,300 psi at 500 °F for a creep rate of 0.01% in 1000 h.

Source: I. Finnie and W.R. Heller, Creep of Engineering Materials, McGraw-Hill, New York, 1959, p 108

Copper and High-Copper Alloys: Comparison of Creep Strengths of Strip

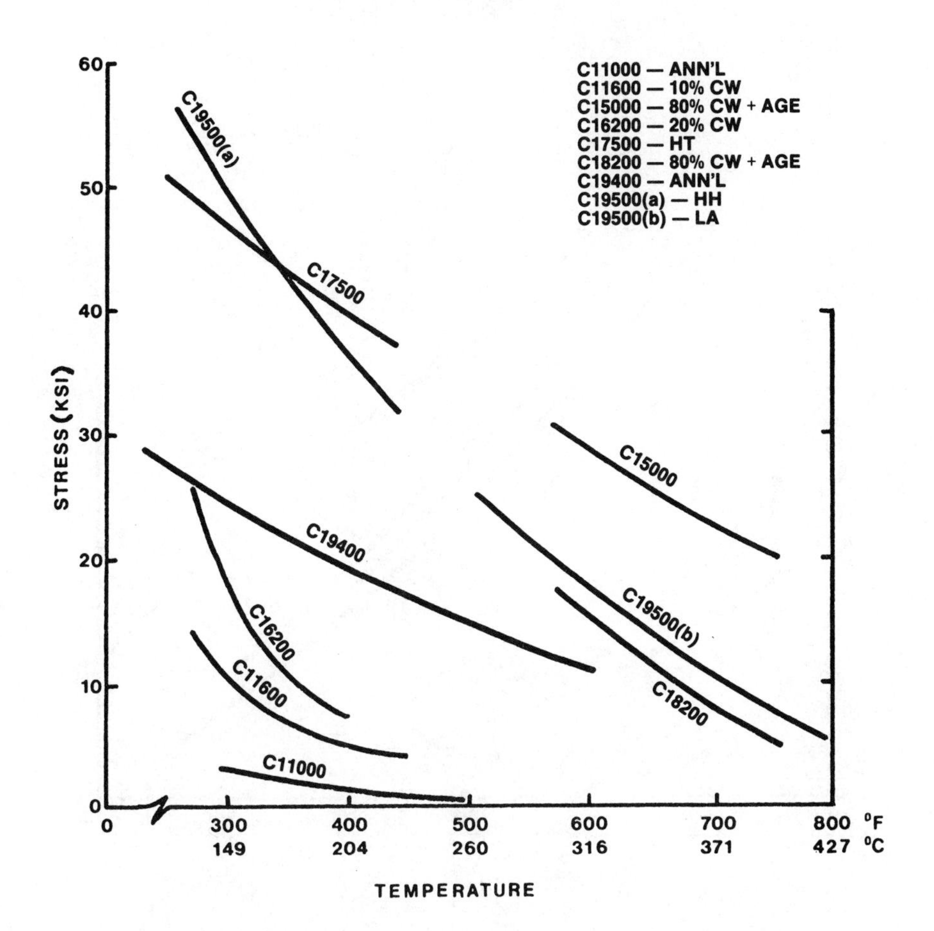

Stress vs temperature for copper and high-copper alloys.

Creep strengths of copper and copper alloys generally decrease as temperature increases.

Source: J.J. Cronin, "Selecting High Conductivity Copper Alloys for Elevated Temperature Use," in Source Book on Materials Selection, Volume II, R.B. Gunia, Ed., American Society for Metals, Metals Park, OH, 1977, p 226

20.36 Copper and High-Copper Alloys: Comparison of the 1000-h Rupture Strengths of Strip

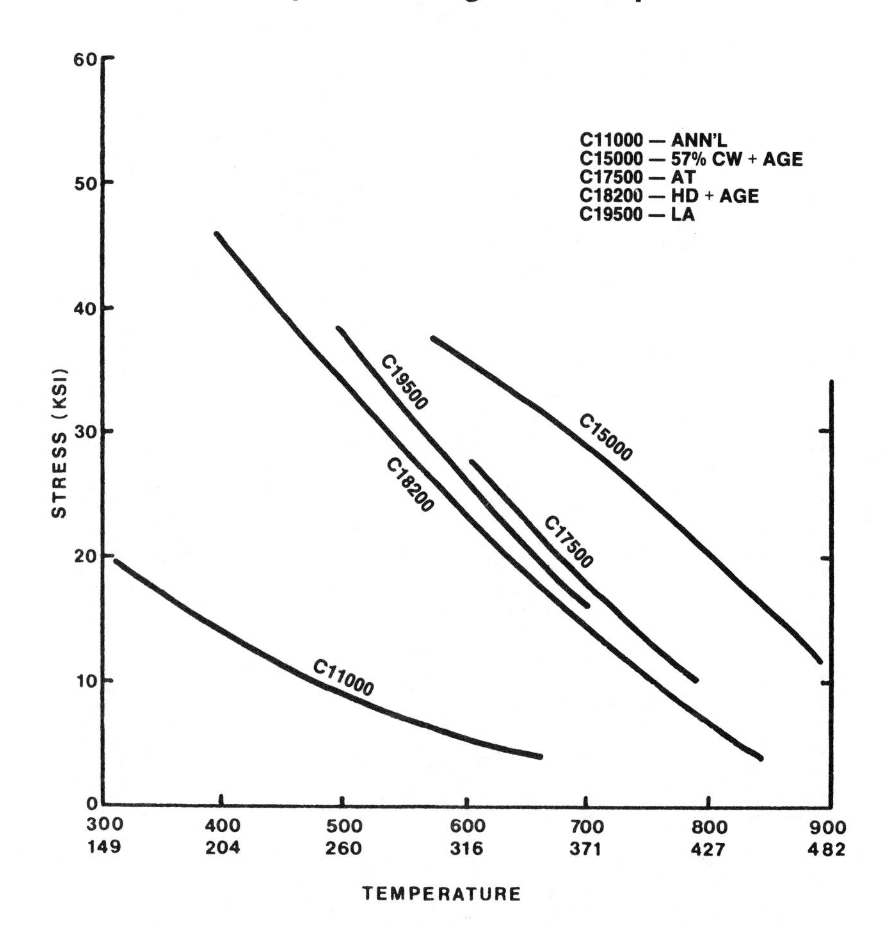

Rupture strengths of copper (C11000), zirconium copper (C15000), beryllium copper (C17500), chromium copper (C18200), and a high-copper alloy. Time = 1000 h.

Source: J.J. Cronin, "Selecting High Conductivity Copper Alloys for Elevated Temperature Use," in Source Book on Materials Selection, Volume II, R.B. Gunia, Ed., American Society for Metals, Metals Park, OH, 1977, p 227

Copper and High-Copper Alloys: Comparison of the 100,000-h Rupture Strengths of Strip

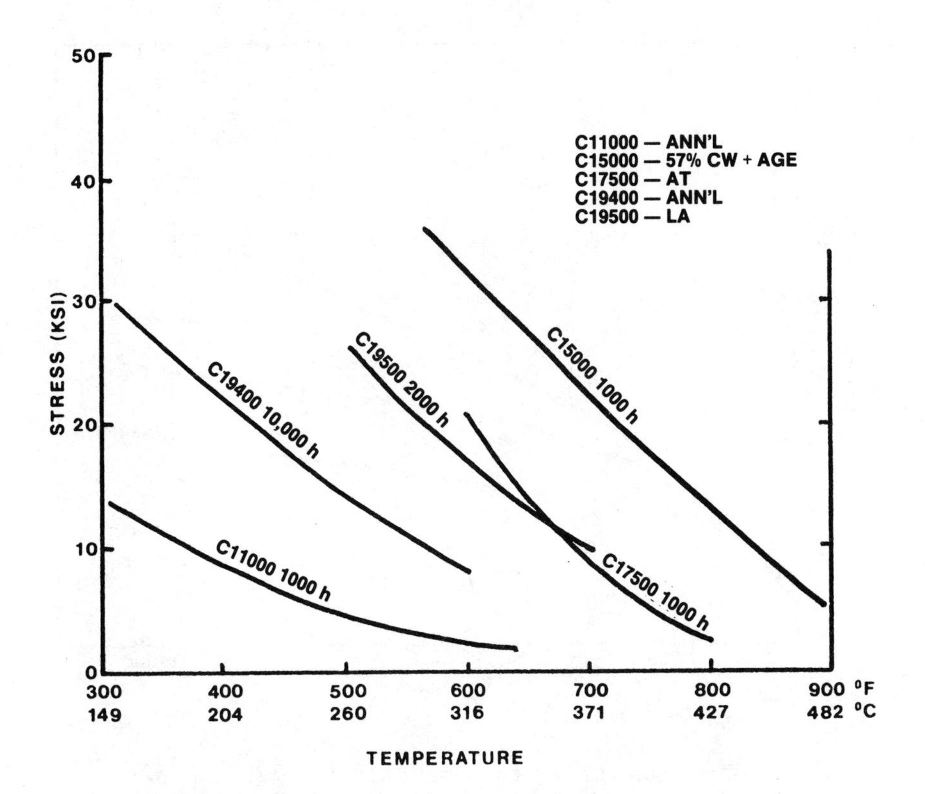

Rupture strengths of copper (C11000), zirconium copper (C15000), beryllium copper (C17500), and two high-copper alloys (C19400 and C19500). Time = 100,000 h.

Source: J.J. Cronin, "Selecting High Conductivity Copper Alloys for Elevated Temperature Use," in Source Book on Materials Selection, Volume II, R.B. Gunia, Ed., American Society for Metals, Metals Park, OH, 1977, p 228

20.38 Copper-Chromium Alloy vs High-Conductivity Coppers: A Comparison

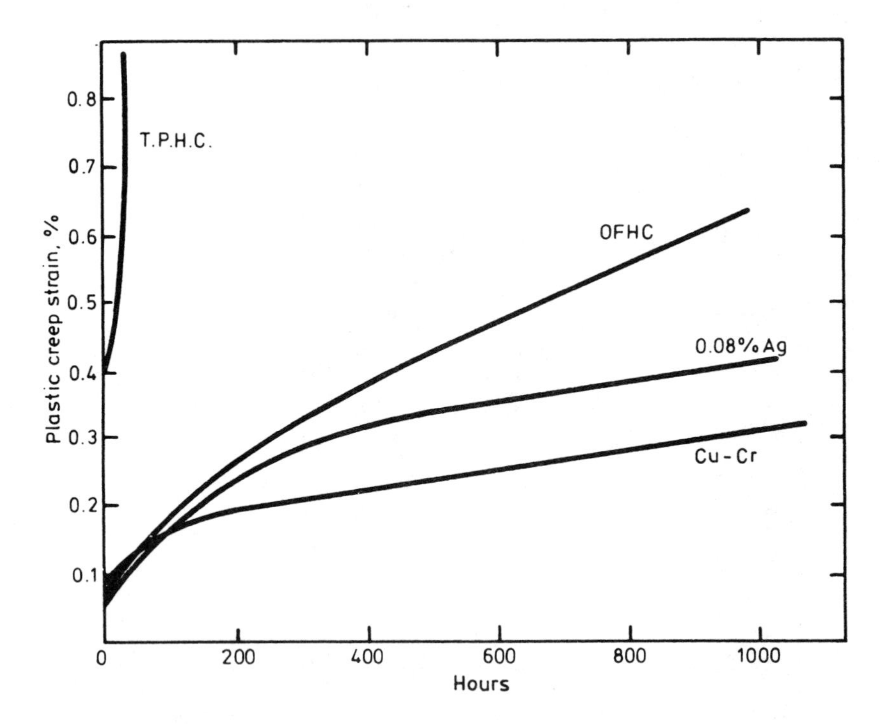

Comparison of creep resistance at 225 °C of copper-chromium alloy (cold worked 33%) at 231 N/mm² with three high-conductivity coppers (cold worked 25%).

Source: E.G. West, Copper and Its Alloys, John Wiley & Sons, New York, 1982, p 148

Copper-Magnesium-Zirconium-Chromium and Copper-Nickel-Titanium Alloys: Comparison of Stress-Rupture Properties of Wire

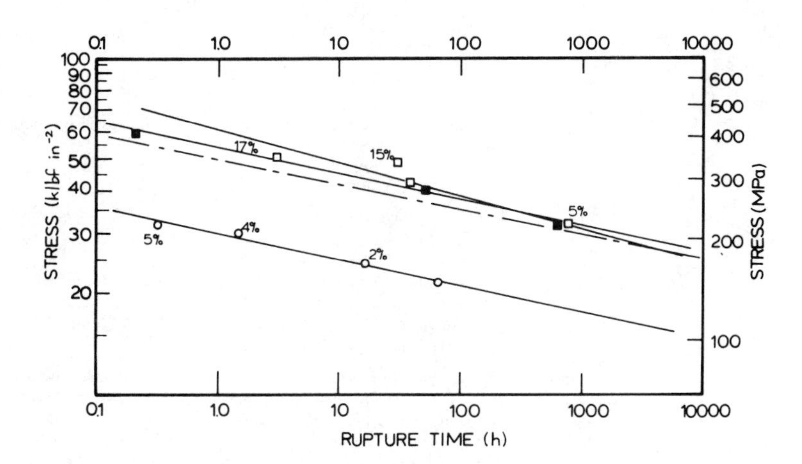

Stress-rupture plot at 673 K for various alloys. The open squares represent MZC ingot alloy, MI condition. Solid squares represent ZAC-1, P/M alloy with oxide, thermomechanically treated. Dashed line represents Cu-0.15Zr, ingot alloy, heat treated. Open circles represent Cu-Ni-Ti, ingot alloy, CI condition.

	MZC	*Cu-Ni-Ti*
Composition (wt.%)	0.06 Mg, 0.63 Cr, 0.13 Zr and balance Cu	4.94 Ni, 2.40 Ti and balance Cu
Fabrication history	(1) Semicontinuously cast into an ingot 55.88 cm in diameter (2) Cut into 2.54-3.81 cm slices (3) Then cut into strips 5.08 cm (maximum) thick (4) Hot rolled at 1173 K down to 1.90 cm	(1) Semicontinuously cast into an ingot 20.32 cm in diameter (2) Cut into 2.54-3.81 cm slices (3) Then cut into strips 5.08 cm (maximum) thick (4) Hot rolled at 1173 K down to 1.30 cm
Heat treatment (fully heat-treated state)	(1) Solution annealed at 1173 K for 1 h and water quenched (2) Cold drawn to 75% reduction in area (3) Aged at 748 K for 0.5 h	(1) Solution annealed at 1173 K for 1 h and water quenched (2) Cold drawn to 76% reduction in area (3) Aged at 773 K for 1 h
Final bar size	0.952 cm round	0.635 cm round

Data on Cu-Mg-Zr-Cr (MZC) and Cu-Ni-Ti alloys

20.40

Copper-Magnesium-Zirconium-Chromium and Copper-Nickel-Titanium Alloys: Comparison of Stress-Rupture Properties of Wire (Continued)

Alloy	*Condition*	*Designation*
MZC (ingot metallurgy)	MI	Fully heat treated
	MII	MI condition + annealed at 1223 K for 1h and water quenched + aged at 748 K for 1h
	MIII	MI condition + annealed at 1273 K for 1h and water quenched + aged at 748 K for 1h
Cu-Ni-Ti (ingot metallurgy)	CI	Fully heat treated
	CII	CI condition + solution annealed at 1138 K for 1h and water quenched + aged at 848 K for 1h
	CIII	CI condition + solution annealed at 1223 K for 1h and water quenched + aged at 848 K for 1h
	CIV	CI condition + solution annealed at 1273 K for 1h and water quenched + aged at 848 K for 1h
ZAC-1 (powder metallurgy) (0.1 wt.% Zr, 0.32 wt.% Cr, ZrO_2 and balance Cu)	Thermo-mechanically treated	Powder metallurgy product, −100 mesh (−149 μm) powder; hot extruded at 923 K with an extrusion ratio of 25 to 1 + solution annealed at 1253 K for 0.5 h and water quenched + cold worked at 293 K to a 50% reduction in area + aged at 723 K for 1 h and water quenched + further cold worked to a 25% reduction in area (total cold working to a 75% reduction in area)
Cu-0.15 wt.% Zr (ingot metallurgy)	Heat treated	Solution annealed at 1223 K + cold rolled to an 84% reduction in area + aged at 648 K for 1 h

Data on copper alloys

Source: A.K. Lee and N.J. Grant, "Properties of Two High Strength, High Temperature, High Conductivity Copper-Base Alloys," Materials Science and Engineering, Vol. 60, Sept. 1983, p 214-216

Copper-Tin Alloys: Comparison of Creep Rates, as Influenced by Grain Size, for Plate

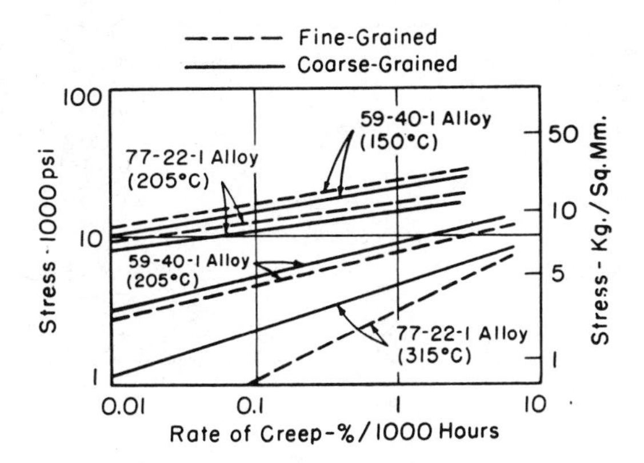

Influence of grain size on rate of creep of hot-rolled copper-zinc-tin alloys.

The grain size of the metal is of prime importance in determining its creep characteristics. At low temperatures, fine-grained materials have higher yield and ultimate strengths, but at elevated temperatures, the reverse is true. The reason for this anomaly is not clear, but it is known that grain boundaries at low temperatures inhibit plastic flow; at elevated temperatures, the grain boundaries can apparently act as centers for generation of dislocations that cause creep. Studies of the grain size effect have long been neglected. However, there are a few reports in the literature of experiments where this phenomenon was studied.

Source: E.R. Parker, "Creep of Metals," in High Temperature Properties of Metals, American Society for Metals, Cleveland, OH, 1951, p 22-23

20.42 Dispersion-Strengthened Copper vs Precipitation-Hardened Copper Alloys (C15000 and C18200): A Comparison

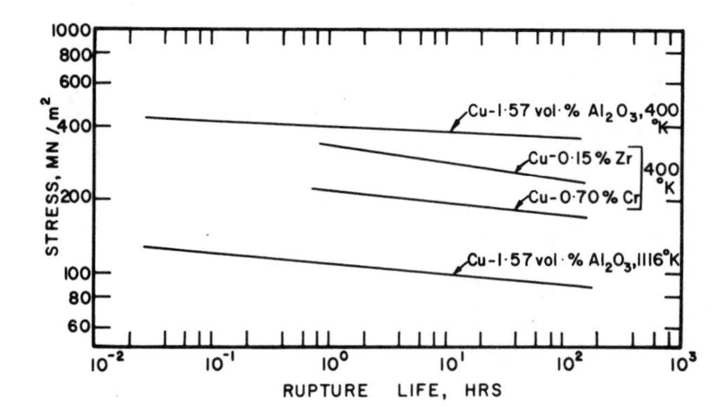

Comparison of stress-rupture data for dispersion-strengthened copper (1.57 vol% Al_2O_3) with zirconium copper (C15000) and chromium copper (C18200) rod.

Source: A.V. Nadkarni, E. Klar, and W.M. Shafer, "A Dispersion-Strengthened Copper," in Source Book on Materials Selection, Volume II, R.B. Gunia, Ed., American Society for Metals, Metals Park, OH, 1977, p 319

Silver-Bearing Copper vs a Copper-Magnesium-Zirconium Alloy: Comparison of Creep Data for Rod and Wire

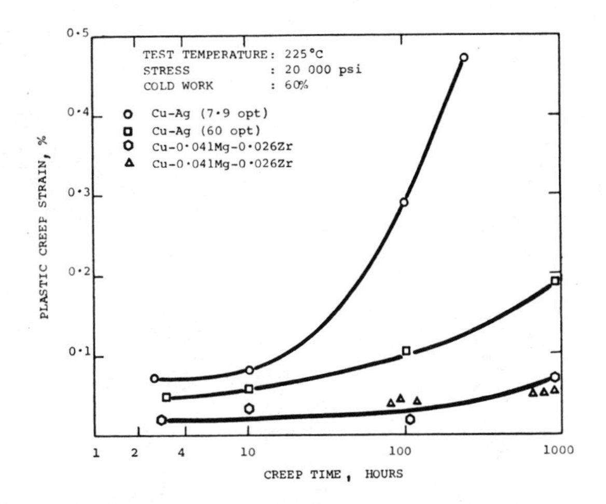

Comparative creep effects in 60% cold-worked copper alloys at 225 °C and 20,000 psi. Alloys containing zirconium were made from an oxygen-free copper base. Triangles represent specimens that were not solution annealed.

Alloy Composition	Test Temp. °C	Test Temp. K	100 000 h psi	100 000 h MN/m²	10 000 h psi	10 000 h MN/m²	1000 h psi	1000 h MN/m²
			Stress Required to Cause 1% Creep Strain					
					After 40% Cold Work			
1	300	575	35 000	241	45 000	310	58 000	400
2			16 000	110	21 000	145	26 500	183
3			28 000	193	32 000	221	36 000	248
4			32 000	221	39 000	269	47 000	324
					After 80% Cold Work			
1	300	575	30 000	207	38 000	262	46 000	317
2			17 000	117	22 000	152	29 000	200
3			32 000	221	36 000	248	39 000	269
4			40 000	276	45 000	310	50 000	345
					After 40% Cold Work			
1	400	675	15 000	103	19 000	131	24 000	165
2			5 000	34	8 500	59	14 000	97
3			12 000	83	15 000	103	20 000	138
4			13 000	90	18 500	128	27 000	186
					After 80% Cold Work			
1	400	675	20 000	138	23 000	159	27 000	186
2			5 000	34	8 000	55	12 000	83
3			12 000	83	17 000	117	22 000	152
4			21 000	145	26 500	183	33 000	228

Alloy No.	Composition	Condition at Test
1	Cu-0·16Zr	C.W. and aged, 425° C, 1 h
2	Cu-0·86Cr	C.W. and aged, 450° C, 2 h
3	Cu-0·05Mg-0·02Zr	C.W. and aged, 425° C, 1 h
4	Cu-0·28Cr-0·14Zr-0·055Mg	C.W. and aged, 450° C, 1 h

Comparative creep strength of cold-worked copper alloys (rod and wire)

Silver-Bearing Copper vs a Copper-Magnesium-Zirconium Alloy: Comparison of Creep Data for Rod and Wire (Continued)

Material	Temp.		% Cold Worked	Stress for 1% Creep in 100 000 h	
	°C	K		psi	MN/m²
Copper–magnesium–zirconium alloy	300	575	20	25 000	172
			40	26 000	179
			80	33 000	228
	400	675	20	12 000	83
			40	12 500	86
			80	13 000	90
Tough-pitch copper	150	425	84	9 400	65
	205	480	84	2 100	14
Silver-bearing copper, 10 opt (0·034%) Ag	150	425	84	36 000	248
Silver-bearing copper, 25 opt (0·085%) Ag	205	480	25	9 700	67
			50	7 200	50

Minimum creep stresses of copper-magnesium-zirconium alloy in comparison with tough-pitch and silver-bearing coppers (rod and wire)

Creep data on Cu alloys from different laboratories are difficult to compare, because there is little standardization of temperature, degree of cold work, and stress levels. The tables above, therefore, make a comparison in terms of the stress required, at temperature, for minimum creep to occur in Cu-Mg-Zr as well as in unalloyed tough-pitch Cu and in Ag-bearing Cu.

Source: W.R. Opie, Y.T. Hsu, and R.J. Smith, "Properties of Cu-Mg-Zr and Cu-Cr-Zr-Mg Alloys with Improved Conductivity/Strength Characteristics," in Copper and Its Alloys (Proceedings of an International Conference in Amsterdam), Monograph and Report Series No. 34, Institute of Metals, London, 1970, p 264-266

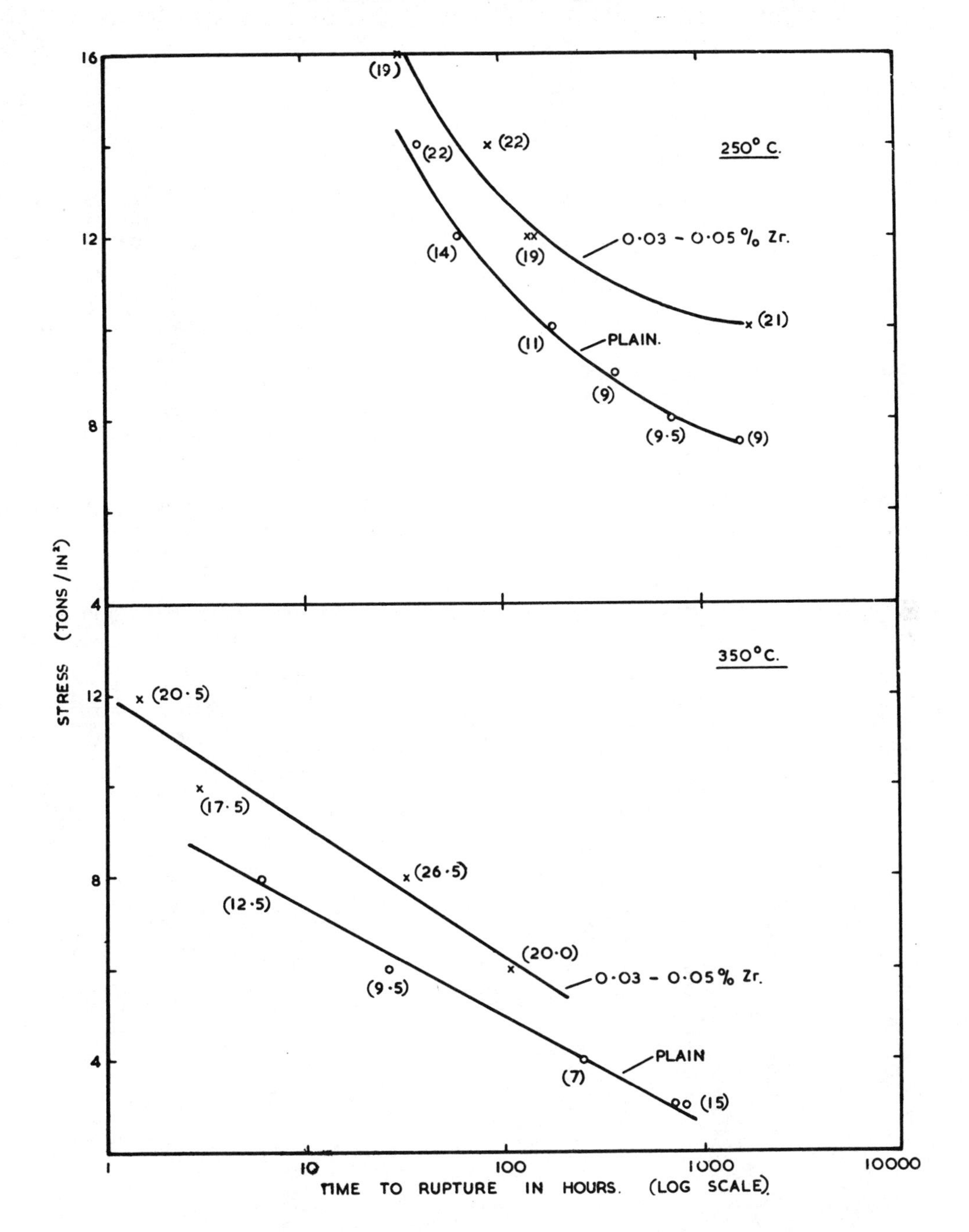

Stress-rupture time tests on wrought 5% tin bronzes at 250 and 350 °C. Grain size, obtained by the intercept method, was 0.032 ± 0.005 mm. Bracketed figures indicate general elongation at fracture.

Source: R. Eborall, "An Approach to the Problem of Intercrystalline Fracture," in Creep and Fracture of Metals at High Temperatures, Philosophical Library, New York, 1957, p 240

21

Magnesium and Magnesium Alloys

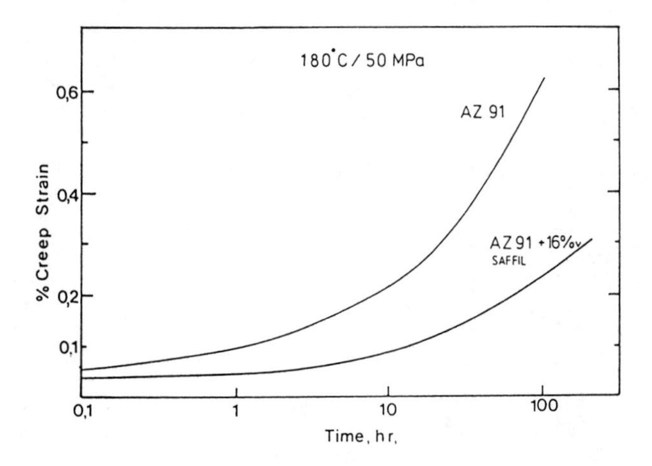

Creep curves of AZ91 and MMC of AZ91 containing 16 vol% Saffil®, a glass/graphite mixture.

Source: G.A. Chadwick, "Squeeze Casting of Magnesium Alloys and Magnesium Based Metal Matrix Composites," in Magnesium Technology (Proceedings of a Conference in London), The Institute of Metals, London, 1987, p 82

21.2 Magnesium: Minimum Creep Rate as a Function of Stress and Temperature

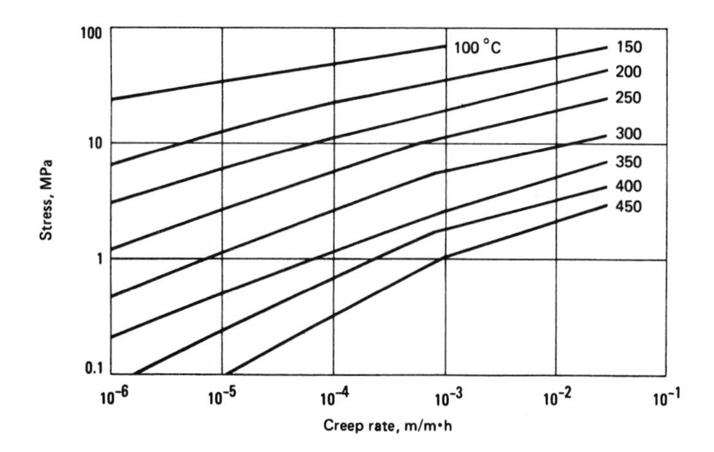

Minimum creep rate of magnesium as a function of stress and temperature.

Source: Metals Handbook, Ninth Edition, Volume 2, Properties and Selection: Nonferrous Alloys and Pure Metals, American Society for Metals, Metals Park, OH, 1979, p 765

Magnesium: Stress-Rupture Life as a Function of Stress and Temperature

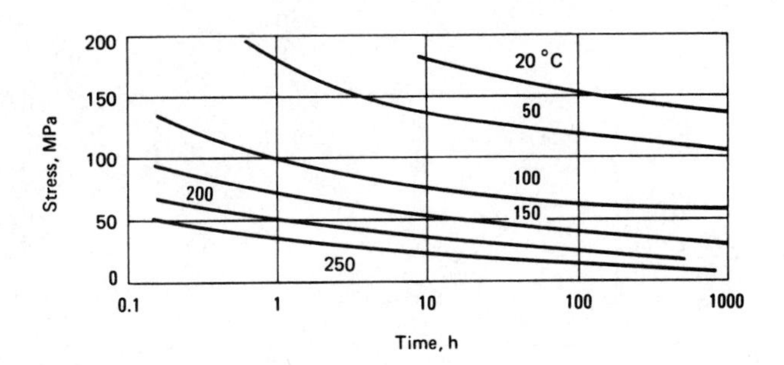

Stress-rupture life of magnesium as a function of stress and temperature.

Source: Metals Handbook, Ninth Edition, Volume 2, Properties and Selection: Nonferrous Alloys and Pure Metals, American Society for Metals, Metals Park, OH, 1979, p 766

21.4 Magnox AL80: Relationship Between Ductility and Time to Rupture for Coarse-Grained Material

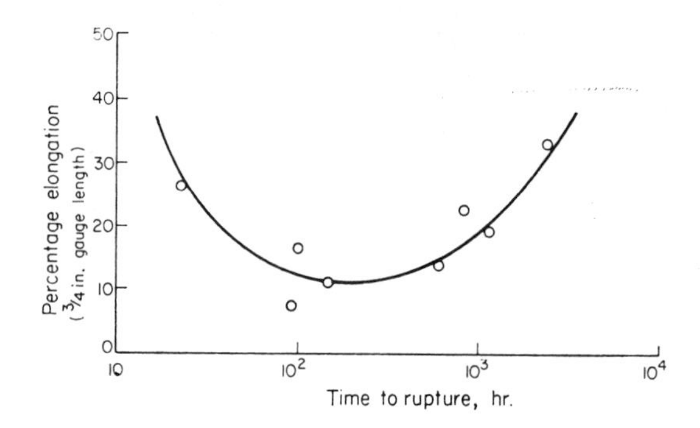

Relationship between ductility and time to rupture at 250 °C for coarse-grained Magnox AL80 alloy. Grain size: 0.016 in.

Source: E.F. Emley, Principles of Magnesium Technology, Pergamon, Oxford, 1966, p 883

QE22A-T6: Sand Castings, Short-Time Creep-Rupture Properties

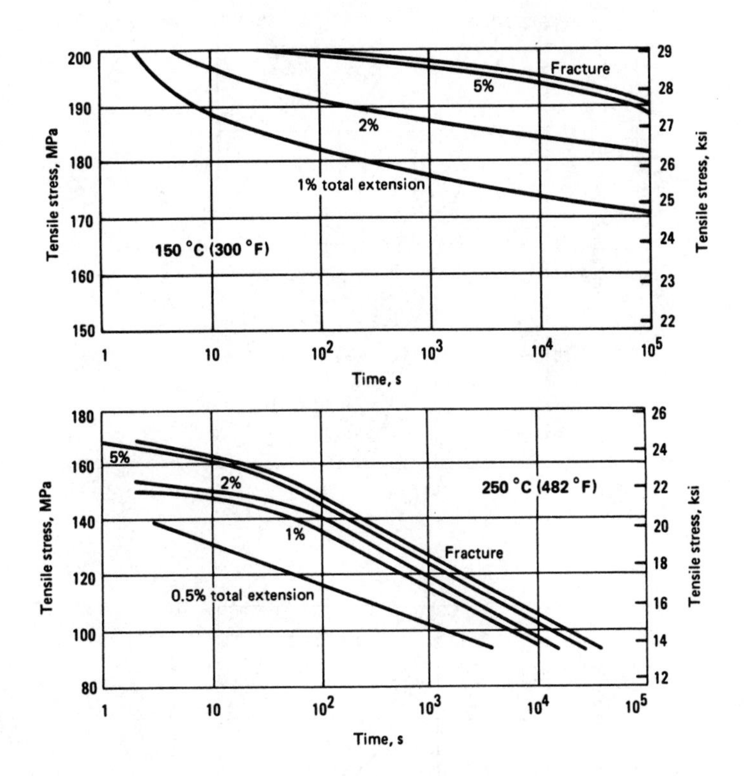

Short-time creep-rupture properties of QE22A-T6 sand castings.

Source: Metals Handbook, Ninth Edition, Volume 2, Properties and Selection: Nonferrous Alloys and Pure Metals, American Society for Metals, Metals Park, OH, 1979, p 589

QH21A-T6: Sand Castings

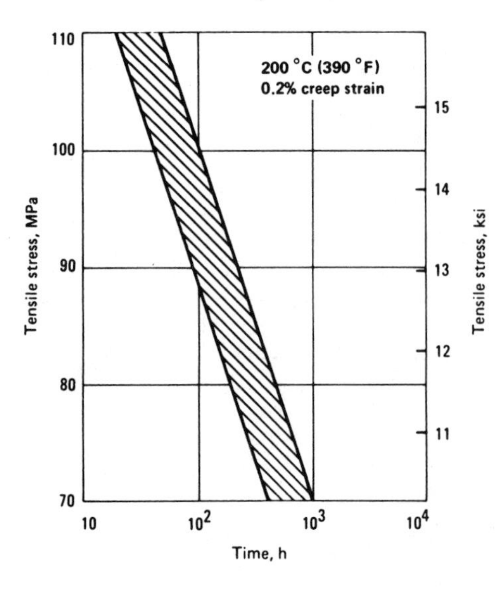

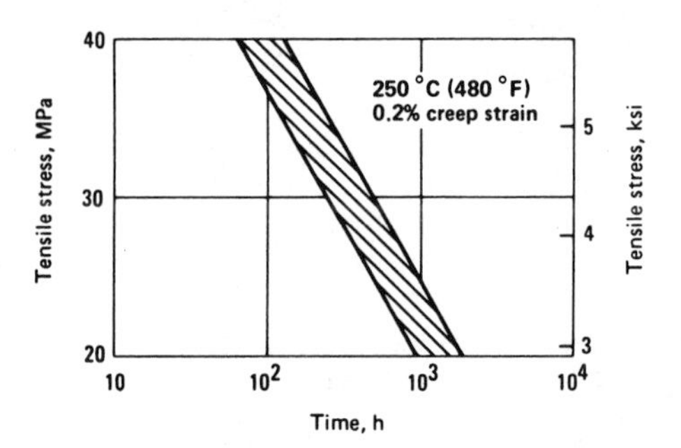

Creep properties of QH21A-T6 sand castings.

Source: Metals Handbook, Ninth Edition, Volume 2, Properties and Selection: Nonferrous Alloys and Pure Metals, American Society for Metals, Metals Park, OH, 1979, p 591

Mg-Al-Zn Alloys and EM-62: Comparison of 1000-h Rupture Strengths

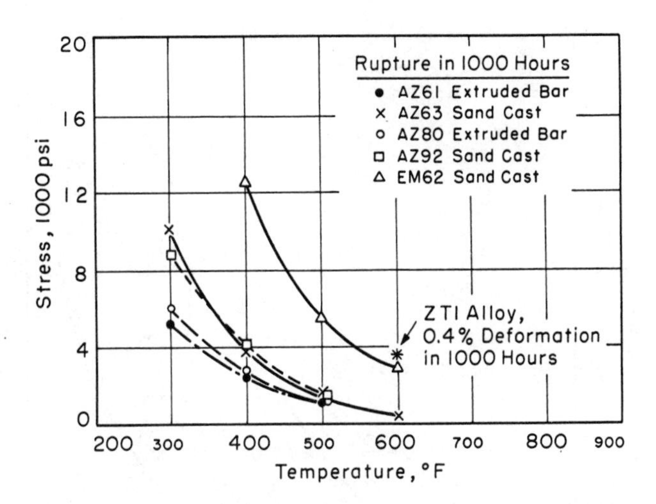

Rupture-strength curves for magnesium alloys.

Considerable attention has been given to the development of sand-cast Mg-RE and thorium-containing alloys. It has been determined that all rare-earth metals investigated improved the strength, hardness, and creep resistance of magnesium at room and elevated temperatures. One of the first alloys of this type was EM-62 alloy (6% Ce and 2% Mn), which had considerably better high-temperature-strength properties than the Mg-Al-Zn alloys. The figure above shows the stresses to produce rupture in 1000 h for EM-62 and several Mg-Al-Zn alloys. Corresponding 1000-h rupture properties of the more recent Mg-Th-Zr and Mg-Th-Zn-Zr alloys are not available for comparison. However, the asterisk in the figure shows the stress to produce 0.4% creep in 1000 h for the Mg-Zn-Zr-Th alloy (ZT1) at 600 °F. This indicates that the stress to produce rupture in 1000 h must be well above 3360 psi.

Source: H.C. Cross and W.F. Simmons, "Alloys and Their Properties for Elevated Temperature Service," in Utilization of Heat Resistant Alloys, American Society for Metals, Cleveland, OH, 1954, p 57-58

Mg-Mn Alloys: Comparison of Creep Resistance for Extruded and Solution-Heat-Treated Material

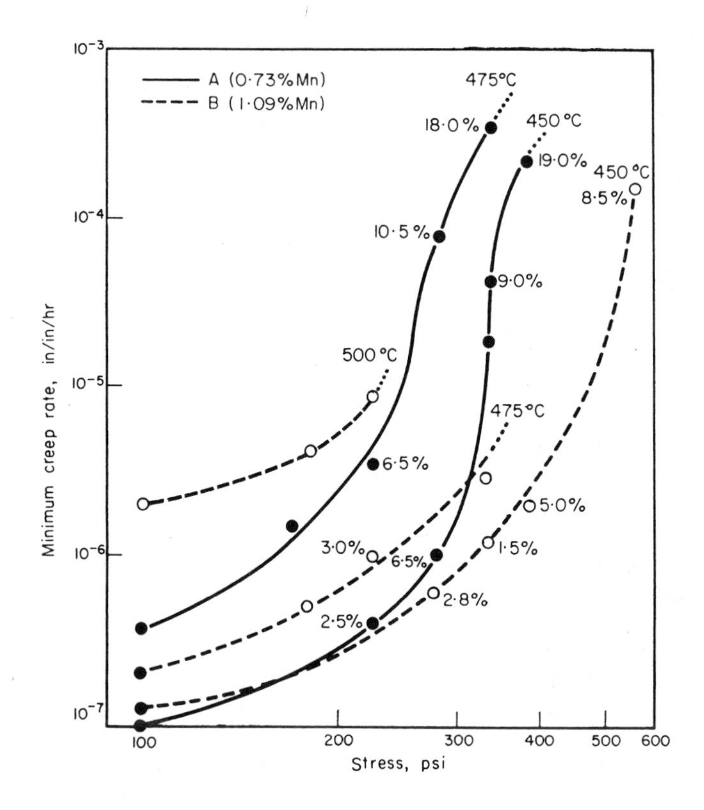

Creep resistance of extruded and solution-heat-treated magnesium-manganese alloys as a function of stress at 450 to 500 °C. The transition between lower and higher minimum creep rates corresponds with a change from intercrystalline to transcrystalline failure.

Source: E.F. Emley, Principles of Magnesium Technology, Pergamon, Oxford, 1966, p 868

Mg-RE Alloys: Comparison of 0.1% Creep Strains in 100 h, as Influenced by RE Content

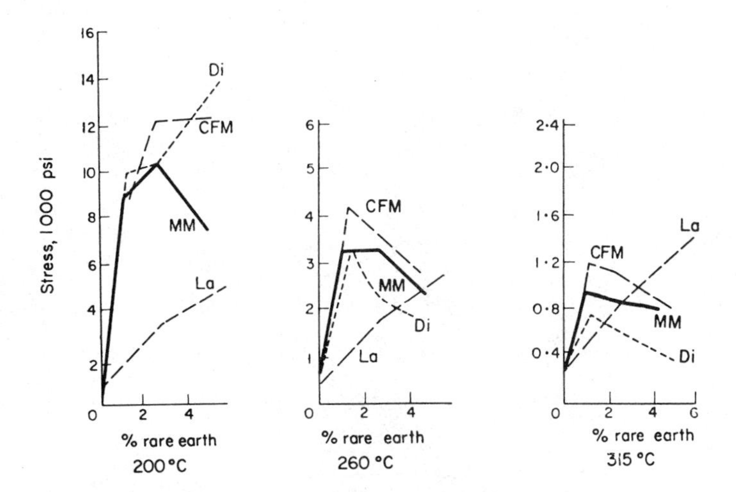

Effect of rare earth (RE) content on the stress for 0.1% creep strain in 100 h shown by extruded and fully heat treated Mg-RE alloys. MM = cerium mischmetal; CFM = cerium-free mischmetal; Di = didymium mischmetal.

Source: E.F. Emley, Principles of Magnesium Technology, Pergamon, Oxford, 1966, p 524

21.10 Mg-Th-Zr Alloys: Comparison of 0.1% Creep Strains in 100 h, as Influenced by Thorium Content and Heat Treatment

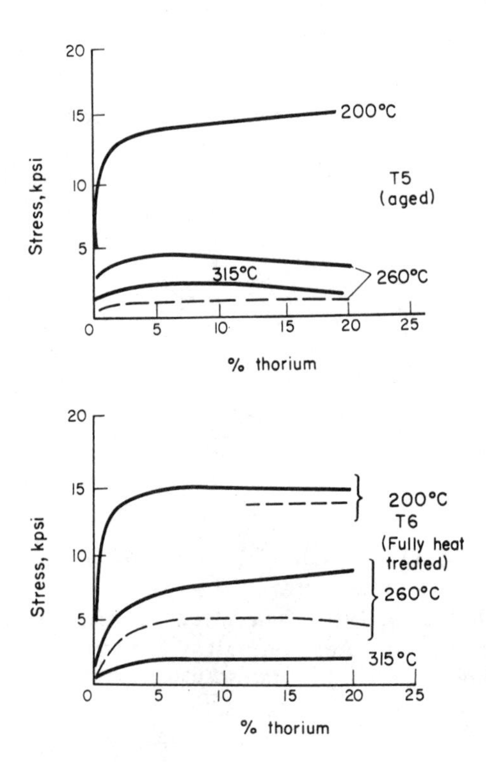

Effects of thorium content and heat treatment on the stress for 0.1% creep strain in 100 h shown by aged and fully heat treated extruded Mg-Th-Zr alloys, with and without zirconium. Dashed line represents specimens without zirconium.

Source: E.F. Emley, Principles of Magnesium Technology, Pergamon, Oxford, 1966, p 527

Magnesium Alloys: Comparison of 0.2% Total Extensions in 100 h for Sand-Cast Alloys

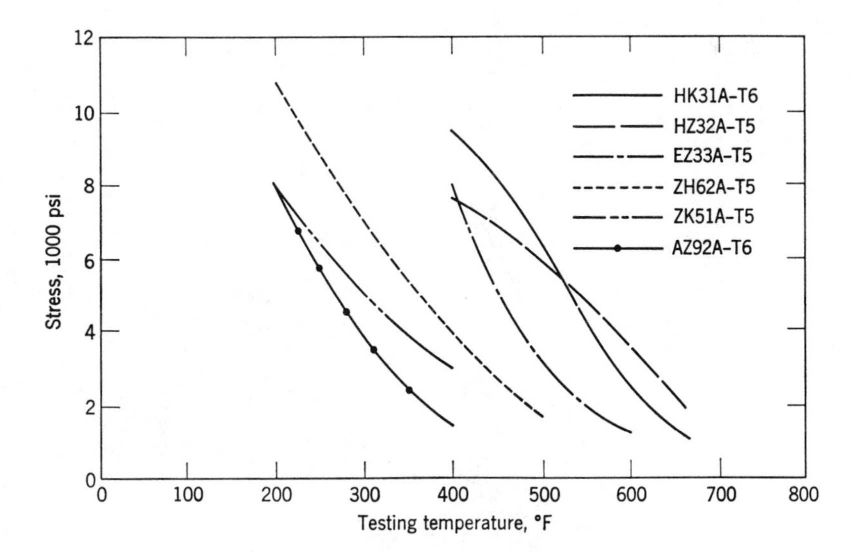

Temperature dependence of creep limit for 0.2% total extension in 100 h for sand-cast alloys using separately cast test bars.

The stresses in the figure above are commercially useful parameters that represent complete creep curves by one value of both strain and time. These data were obtained from separately cast test bars. Such test bars are more representative of sections of actual castings than are most other casting metals. Little or no change in mechanical properties is observed when the skin is machined from a Mg alloy casting.

Source: C.S. Roberts, Magnesium and Its Alloys, John Wiley & Sons, New York, 1960, p 132-133

21.12 Magnesium Alloys: Comparison of Creep Resistance for Several Alloys in Dry Carbon Dioxide

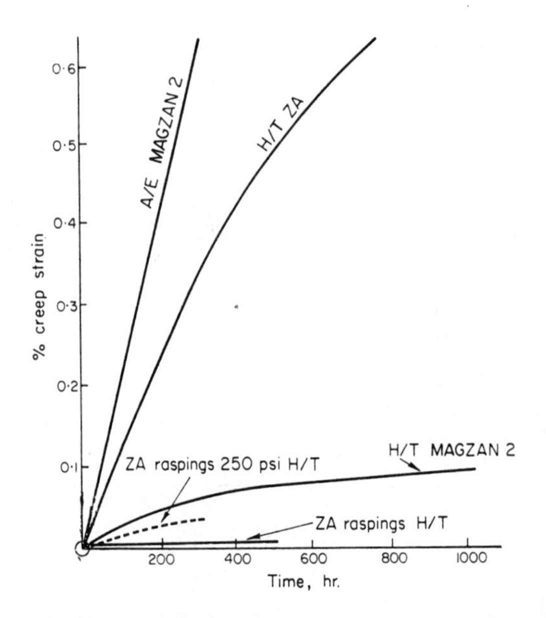

Creep resistance of various magnesium alloys tested at 500 °C and 100 psi in dry carbon dioxide. Under these conditions, heat treated ZA raspings alloy does not move, and the dotted curve shows its resistance at much higher stress. H/T = heat treated: ZA, 100 h at 600 °C; H/T = heat treated: Magzan, 4 h at 565 °C; A/C = as extruded.

Source: E.F. Emley, Principles of Magnesium Technology, Pergamon, Oxford, 1966, p 871

Magnesium Alloys: Comparison of Long-Term Creep Properties for Casting Alloys

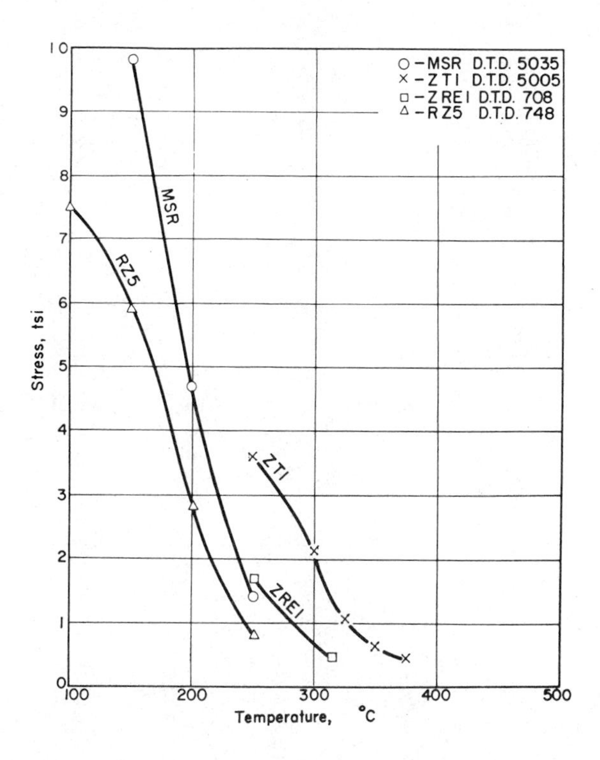

Comparative long-term creep performance of some cast alloys: effect of temperature on the stress for 0.5% creep strain in 1000 h.

Source: E.F. Emley, Principles of Magnesium Technology, Pergamon, Oxford, 1966, p 317

Magnesium Alloys: Comparison of Short-Time Creep Resistance for Casting Alloys

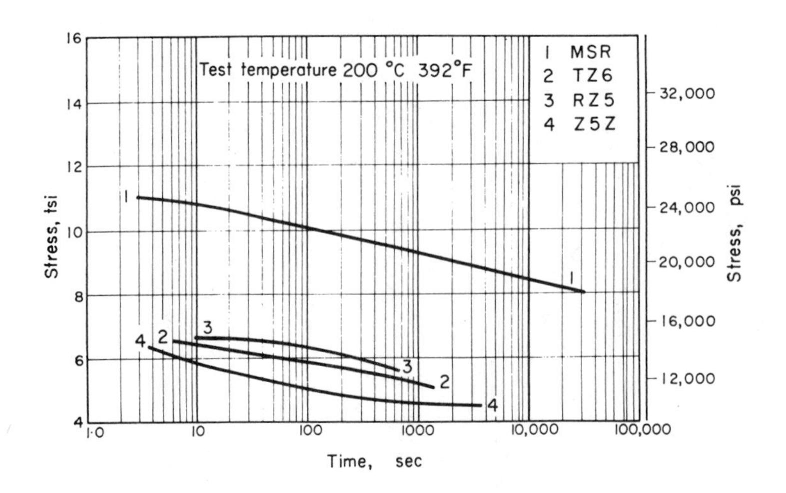

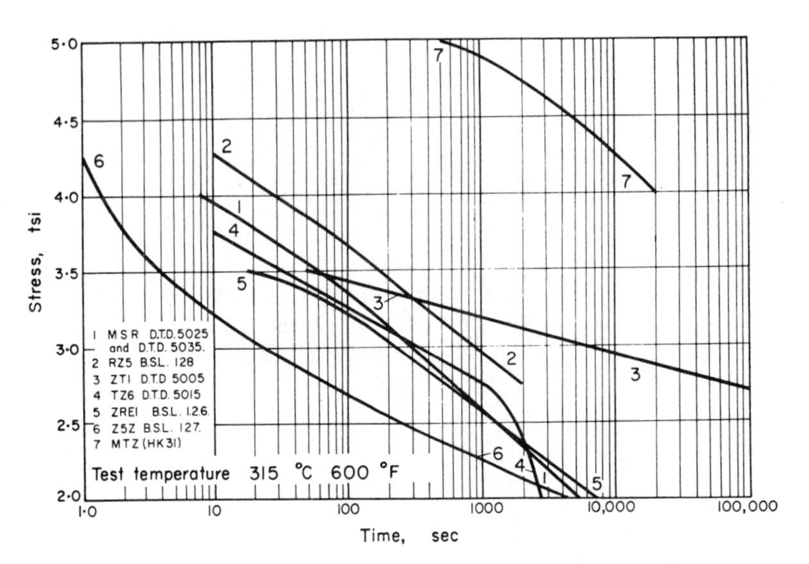

Comparative short-time creep resistance of cast alloys: stress-time relationships for 0.5% total strain at 200 °C (top) and 315 °C (bottom).

Source: E.F. Emley, Principles of Magnesium Technology, Pergamon, Oxford, 1966, p 317-318

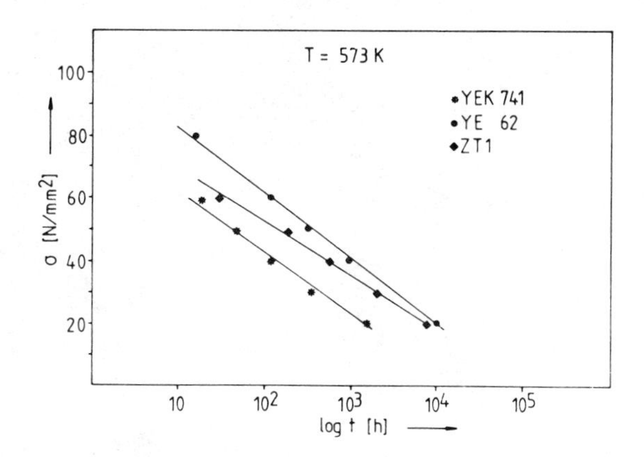

Time-stress relationship for 1% creep strain at 573 K for ZT1, YEK 741, and YE 62.

The figure above compares Zr-containing alloys YEK 741 and ZT1 and the Zr-free alloy YE 62 (Mg-6Y-2Nd). The superior creep properties of YE 62 can be attributed to the coarse grain size.

Source: B.L. Mordike and W. Henning, "Creep and High Temperature Properties of Magnesium Based Alloys," in Magnesium Technology (Proceedings of a Conference in London), The Institute of Metals, London, 1987, p 55, 59

22

Titanium and Titanium Alloys

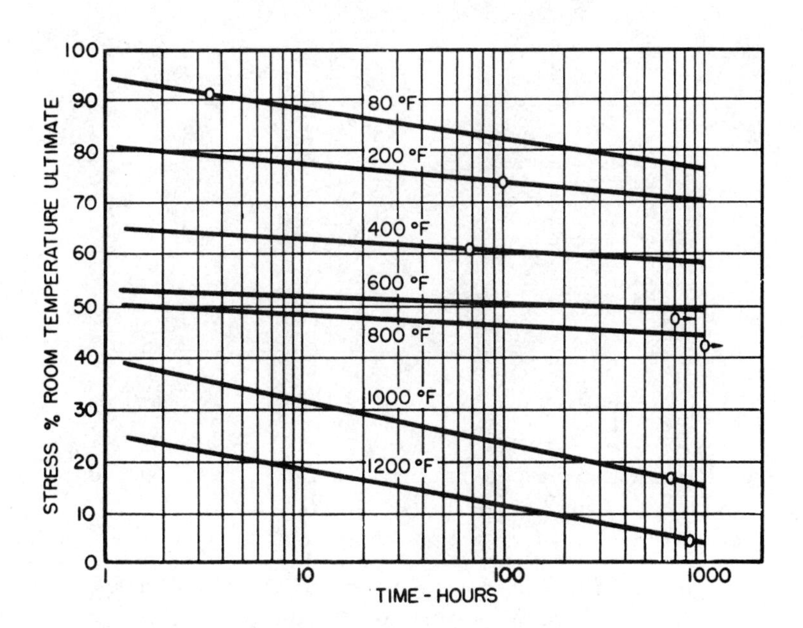

Stress-rupture curves for Ti-5Al-2.5Sn alloy from 80 to 1200 °F.

Source: Crucible A-110AT Titanium Base Alloy, Data Sheet, Crucible Steel Company of America, Pittsburgh, May 1958, p 5

Ti-5Al-2Sn-2Zr-4Mo-4Cr Alloy

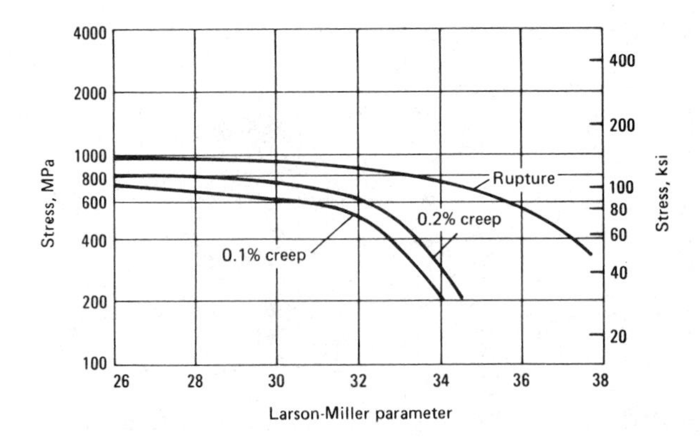

Parametric creep-rupture curves for forged Ti-5Al-2Sn-2Zr-4Mo-4Cr alloy.

Larson-Miller parameter equals $10^{-3}T$ (25 + log t), where T is temperature in °R and t is time in hours.

Source: Metals Handbook, Ninth Edition, Volume 3, Properties and Selection: Stainless Steels, Tool Materials and Special-Purpose Metals, American Society for Metals, Metals Park, OH, 1980, p 397

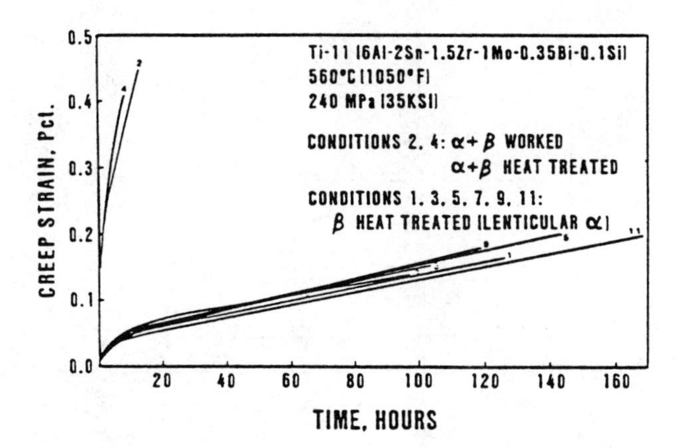

Creep strain vs time curves for two processing conditions of Ti-6Al-2Sn-1.5Zr-1Mo-0.35Bi-0.1Si at 1050 °F and 35 ksi.

Beta heat treated alloy with lenticular transformed α phase is stronger in creep than α–β worked and heat treated material.

Source: D. Eylon, S. Fujishiro, and F.H. Froes, "Titanium Alloys for High Temperature Applications: A Review," High Temperature Materials and Processes, Vol. 6, No. 1 and 2, 1984, p 87

Ti-6Al-2Sn-2Zr-2Cr-2Mo-0.25Si Alloy

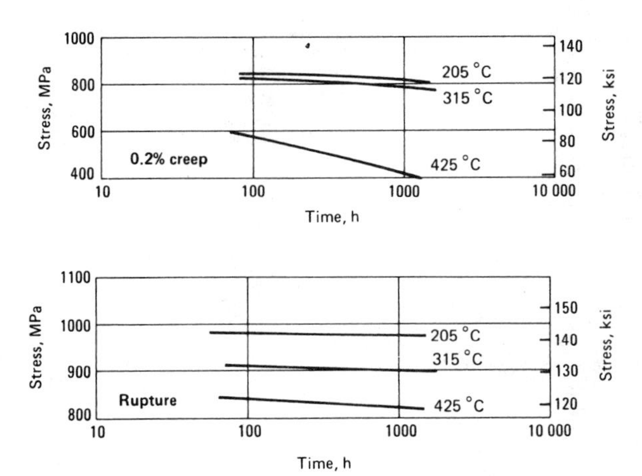

Creep and rupture properties of Ti-6Al-2Sn-2Zr-2Cr-2Mo-0.25Si alloy from 205 to 425 °C.

Source: Metals Handbook, Ninth Edition, Volume 3, Properties and Selection: Stainless Steels, Tool Materials and Special-Purpose Metals, American Society for Metals, Metals Park, OH, 1980, p 397

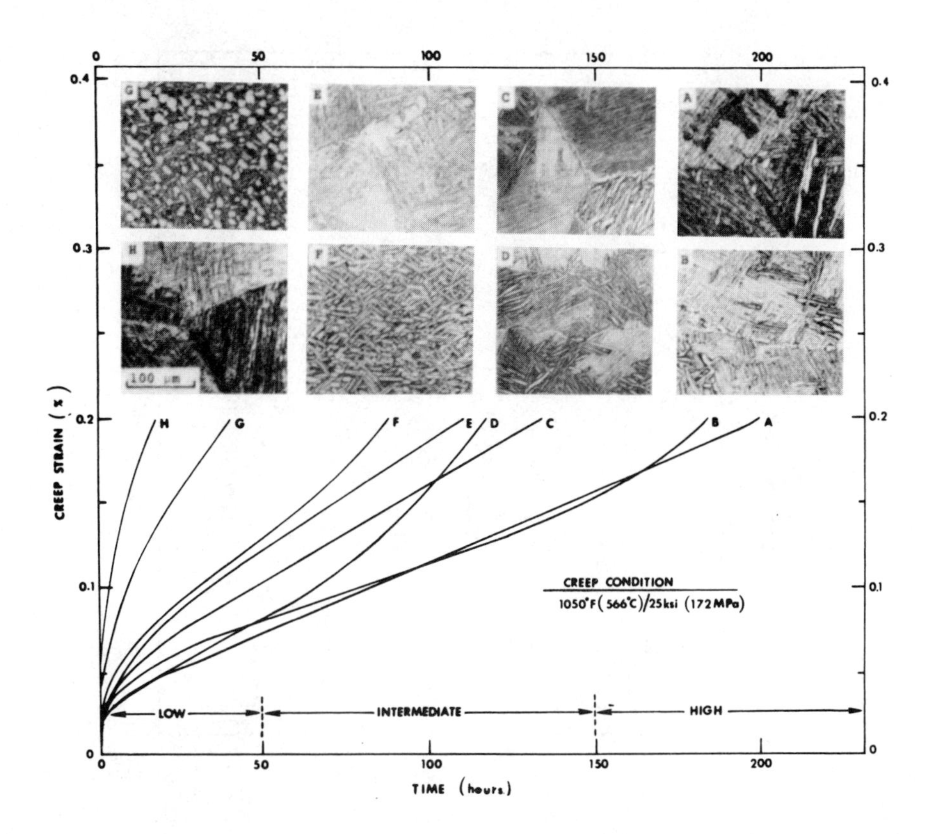

Creep strain vs time curves for eight microstructural conditions of Ti-6Al-2Sn-4Zr-2Mo-0.1Si pancake forgings at 1050 °F and 25 ksi.

The figure above shows that β or pseudo-β structures (Curve A) have the highest strength.

Source: C.C. Chen and J.E. Coyne, "Relationships between Microstructure and Mechanical Properties in Ti-6Al-2Sn-4Zr-2Mo-0.1Si Alloy Forgings," in Titanium/80, Volume 2 (Proceedings of the Fourth International Conference on Titanium, Kyoto), H. Kimura and O. Izumi, Eds., American Institute of Mining, Metallurgical, and Petroleum Engineers, Warrendale, PA, 1980, p 1198

Ti-6Al-2Sn-4Zr-2Mo Alloy: Effect of Silicon

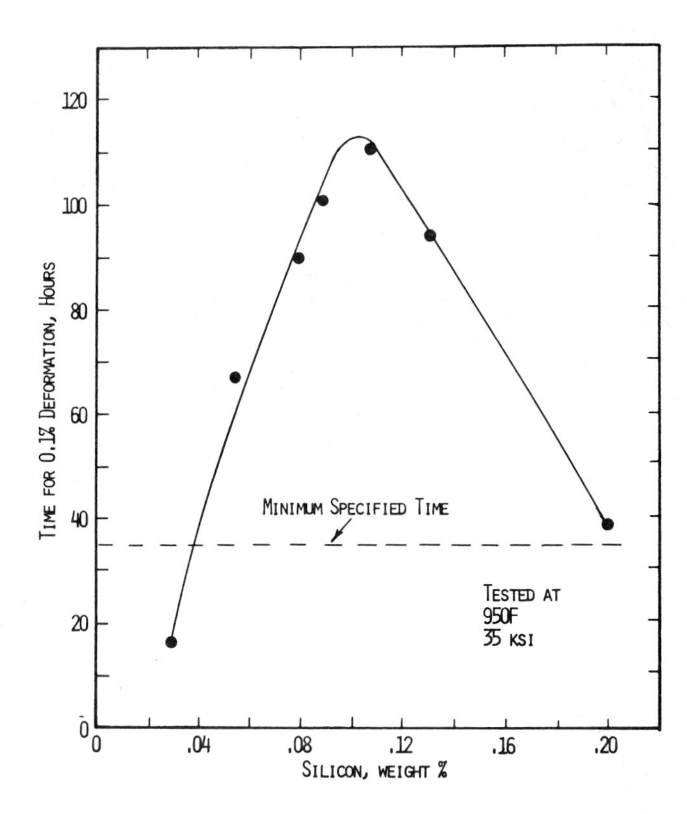

Effect of silicon on time to 1% creep for Ti-6Al-2Sn-4Zr-2Mo alloy bar stock at 950 °F and 35 ksi.

The specification's minimum 1% creep time is shown for the alloy. Note that a small amount of Si is needed to meet the creep requirement.

Source: S.R. Seagle, G.S. Hall, and H.B. Bomberger, "High Temperature Properties of Ti-6Al-2Sn-4Zr-2Mo-0.09Si," in Source Book on Materials Selection, Volume II, R.B. Gunia, Ed., American Society for Metals, Metals Park, OH, 1977, p 183

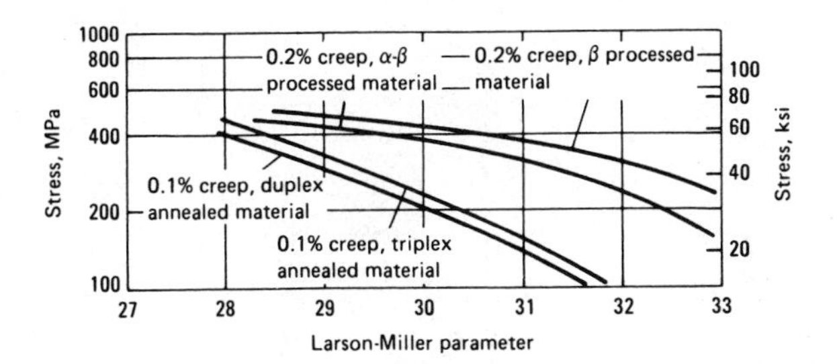

Parametric creep curves for Ti-6Al-2Sn-4Zr-2Mo alloy.

Larson-Miller parameter equals $10^{-3}T$ (20 + log t), where T is temperature in °R and t is time in hours. Note that β-processed material has the highest creep strength.

Source: Metals Handbook, Ninth Edition, Volume 3, Properties and Selection: Stainless Steels, Tool Materials and Special-Purpose Metals, American Society for Metals, Metals Park, OH, 1980, p 385

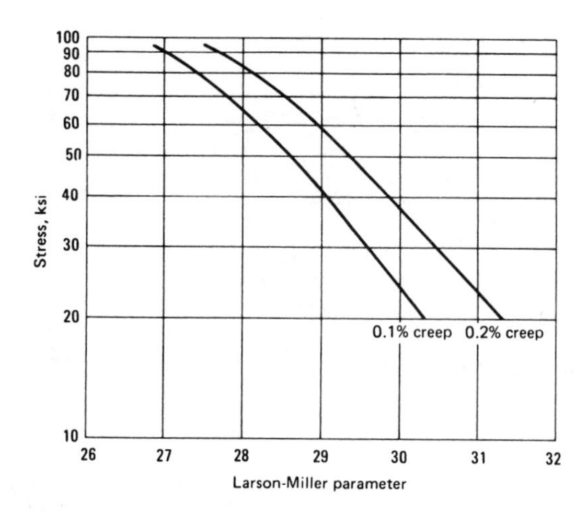

Typical parametric creep curves for Ti-6Al-2Sn-4Zr-6Mo.

Larson-Miller parameter equals $10^{-3}T$ (20 + log t), where T is temperature in °R and t is time in hours.

Source: Metals Handbook, Ninth Edition, Volume 3, Properties and Selection: Stainless Steels, Tool Materials and Special-Purpose Metals, American Society for Metals, Metals Park, OH, 1980, p 396

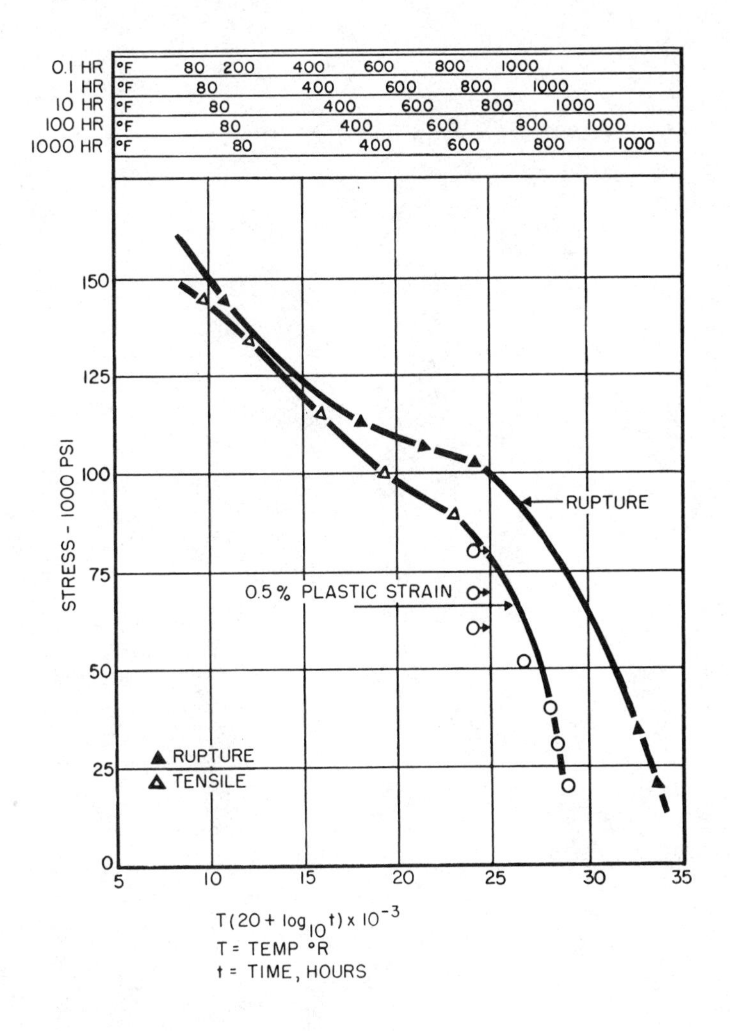

Parametric creep and rupture curves for wrought Ti-6Al-4V alloy.

Source: Crucible C-120AV Titanium Base Alloy, Data Sheet, Crucible Steel Company of America, Pittsburgh, May 1958, p 5

Ti-6Al-4V Alloy: Creep-Rupture Properties

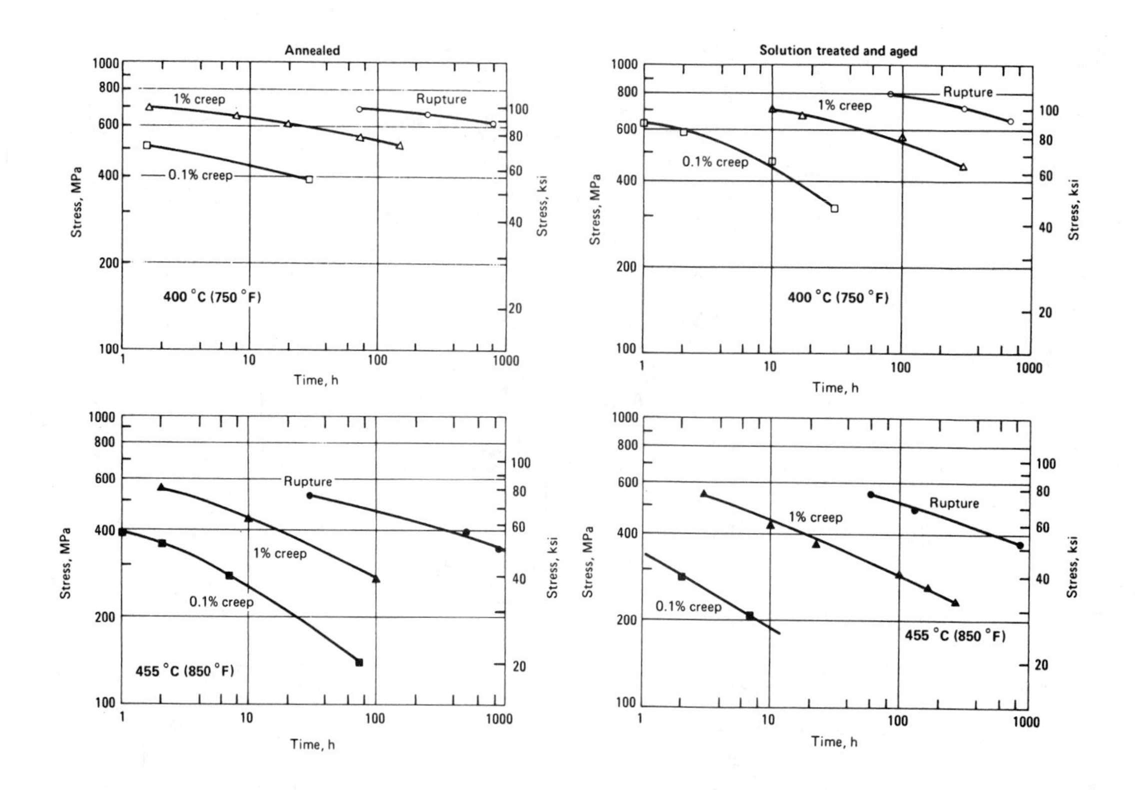

Creep-rupture properties of wrought Ti-6Al-4V bar at 750 and 850 °F for two heat treating conditions.

Source: Metals Handbook, Ninth Edition, Volume 3, Properties and Selection: Stainless Steels, Tool Materials and Special-Purpose Metals, American Society for Metals, Metals Park, OH, 1980, p 390

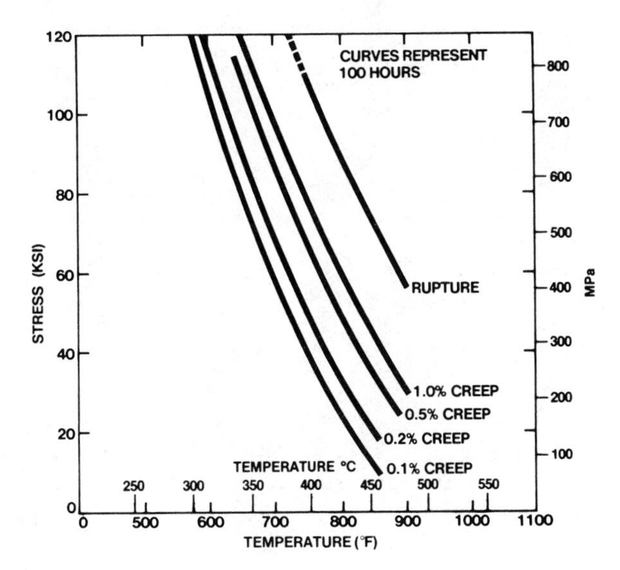

Typical isochronal 100-h creep and stress-rupture capability vs temperature for Ti-6Al-4V alloy wrought bar in the solution treated and aged condition.

Source: Properties and Processing Ti-6Al-4V, TIMET Division, Titanium Metals Corporation of America, Pittsburgh, April, 1980, p 19

Ti-7Al-4Mo Alloy

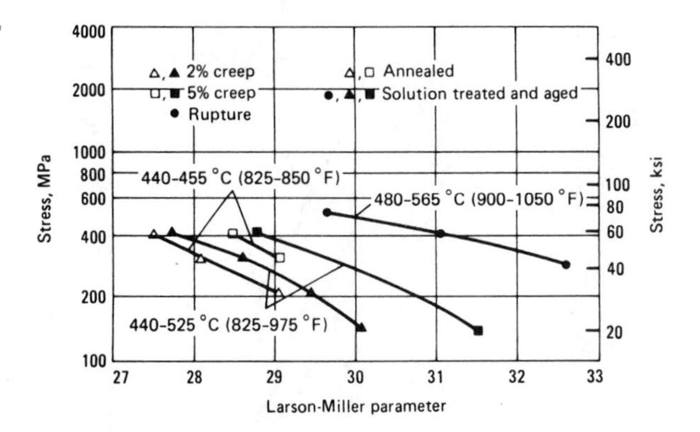

Parametric creep and stress-rupture curves for Ti-7Al-4Mo alloy in several heat treated conditions.

Specimens were taken from compressor wheel forgings forged below the beta transus. Open symbols refer to material annealed 1 h at 790 °C (1450 °F), then furnace cooled to about 565 °C (1050 °F), and air cooled to room temperature. Solid symbols refer to material annealed as above, then solution treated 1/2 h at 850 °C (1560 °F), air cooled, aged 24 h at 550 °C (1020 °F), and air cooled.

Source: Metals Handbook, Ninth Edition, Volume 3, Properties and Selection: Stainless Steels, Tool Materials and Special-Purpose Metals, American Society for Metals, Metals Park, OH, 1980, p 395

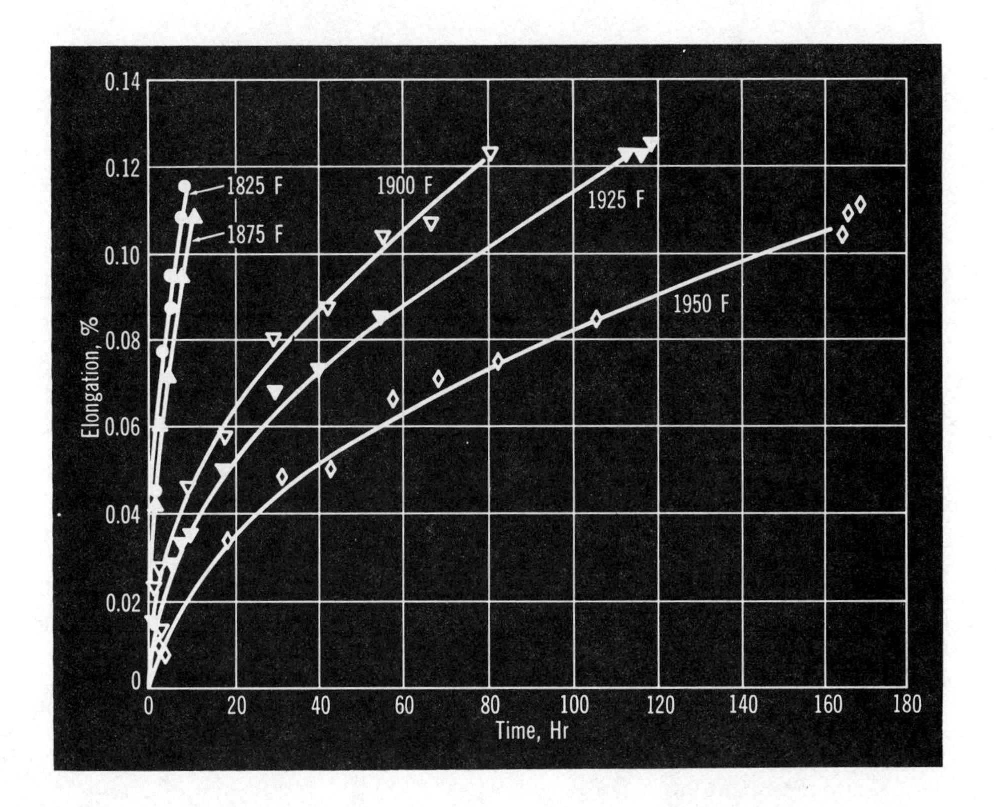

Creep strain vs time for Ti-8Al-1Mo-1V alloy at 50 ksi and 850 °F as a function of prior solution annealing temperature.

Tests show that resistance of Ti-8Al-1Mo-1V to creep at 850 °F improves greatly in specimens that are solution annealed above 1875 °F.

Source: D.B. Warmuth and H.D. Kessler, "How to Heat Treat Ti-8Al-1Mo-1V," Metal Progress, Vol. 89, March 1966, p 94

Ti-8Al-1Mo-1V Alloy: Creep Strain vs Time Curves

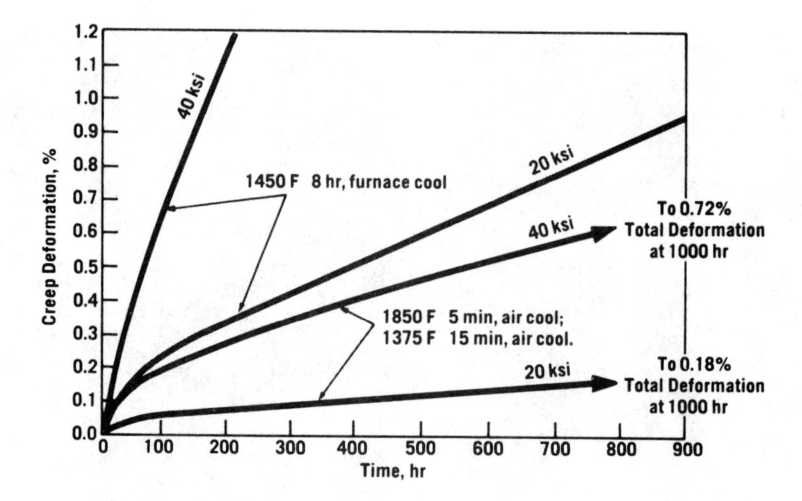

Comparison of 900 °F creep strain vs time curves for simple and duplex annealed Ti-8Al-1Mo-1V sheet at two stress levels.

Source: M.G. Manzone, Titanium Alloys with Molybdenum – the 1966 Picture, Climax Molybdenum Co., New York, 1966, p 18

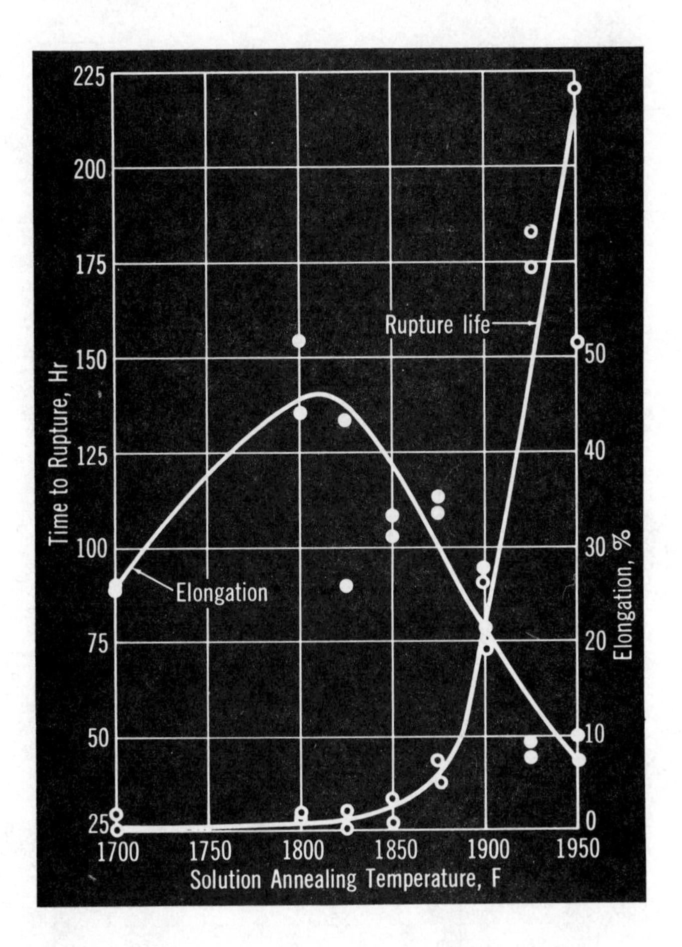

Stress-rupture life and elongation of Ti-8Al-1Mo-1V alloy at 57.5 ksi and 1000 °F as a function of prior solution annealing temperature.

Elongation is highest in Ti-8Al-1Mo-1V solution annealed at 1800 °F, but rupture life rises sharply in material solution annealed at 1850 °F and above.

Source: D.B. Warmuth and H.D. Kessler, "How to Heat Treat Ti-8Al-1Mo-1V," Metal Progress, Vol. 89, March 1966, p 93

Ti-8Mn Alloy

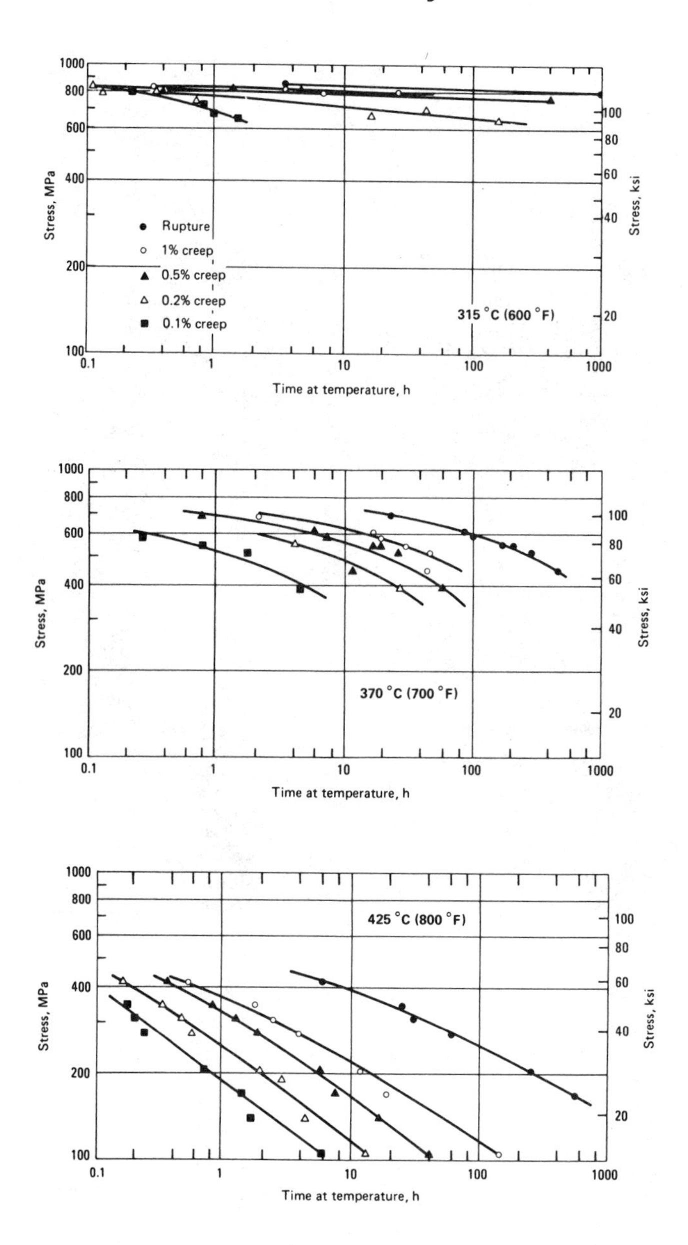

Creep-rupture properties of annealed Ti-8Mn alloy sheet at 600, 700, and 800 °F.

Source: Metals Handbook, Ninth Edition, Volume 3, Properties and Selection: Stainless Steels, Tool Materials and Special-Purpose Metals, American Society for Metals, Metals Park, OH, 1980, p 392

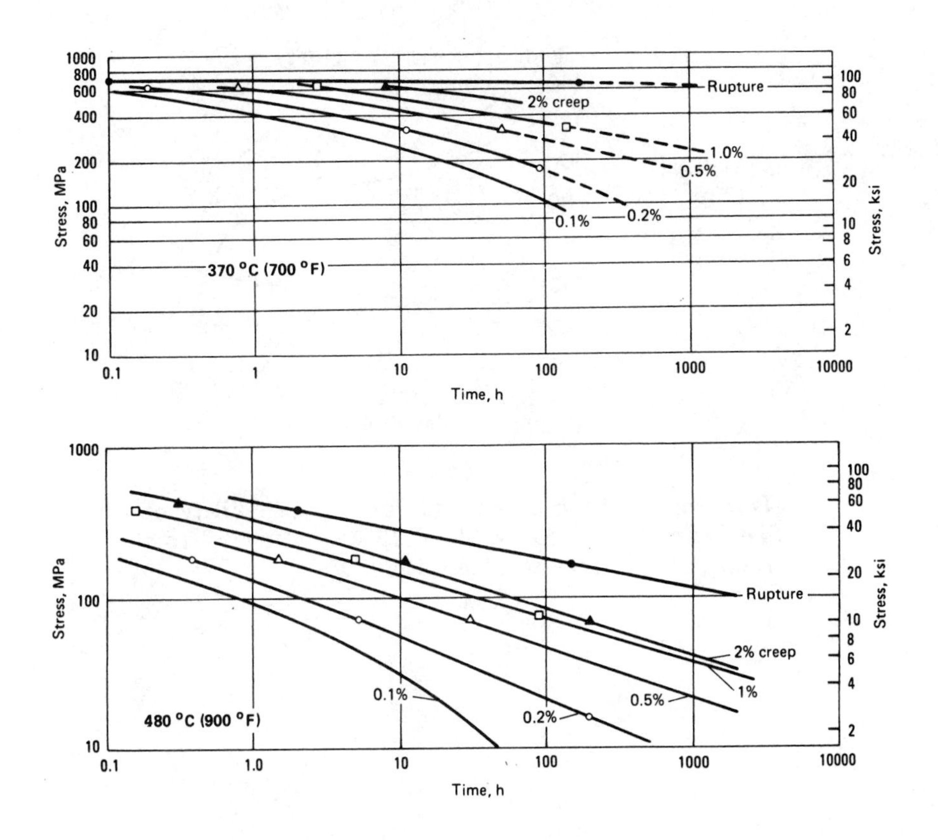

Creep and stress-rupture properties of Ti-10V-2Fe-3Al alloy solution treated and aged at 700 and 900 °F.

Source: Metals Handbook, Ninth Edition, Volume 3, Properties and Selection: Stainless Steels, Tool Materials and Special-Purpose Metals, American Society for Metals, Metals Park, OH, 1980, p 400

Ti-11Sn-2.25Al-5Zr-1Mo-0.25Si Alloy (IMI 679)

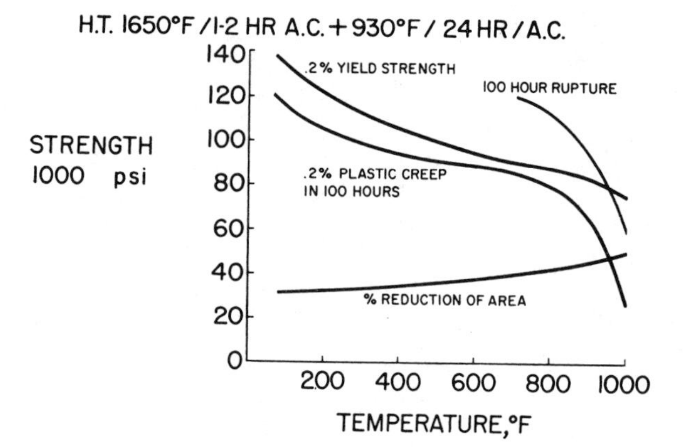

Isochronal 100-h creep and rupture capability of Ti-11Sn-2.25Al-5Zr-1Mo-0.25Si alloy from room temperature to about 1000 °F.

Source: V.J. Erdeman and F.W. Ross, "Long Time Stability of Ti-679 after Creep Exposure for Times to 15,000 Hours," in The Science, Technology and Application of Titanium (Proceedings of an International Conference in London), R.I. Jaffee and N.E. Promisel, Eds., Pergamon, Oxford, 1970, p 831

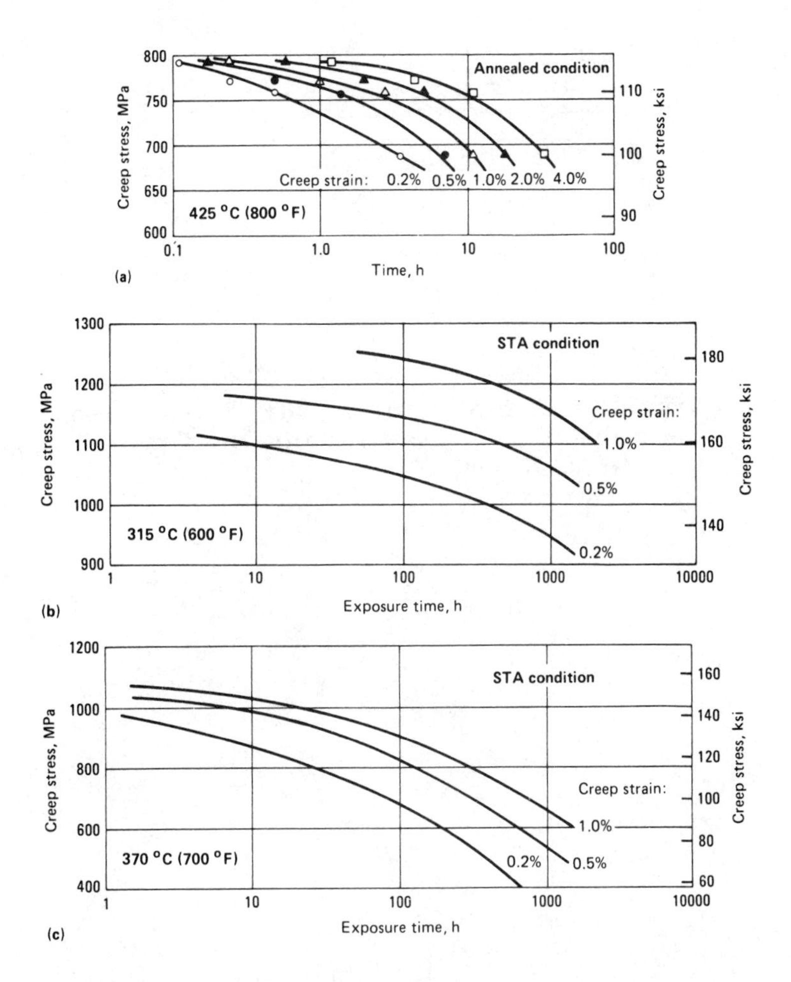

Creep properties of Ti-13V-11Cr-3Al alloy sheet at several temperatures and process conditions.

(a) Annealed 1.09-mm (0.043-in.) sheet. (b and c) 1.63-mm (0.064-in.) sheet aged 50 h at 470 °C (875 °F).

Source: Metals Handbook, Ninth Edition, Volume 3, Properties and Selection: Stainless Steels, Tool Materials and Special-Purpose Metals, American Society for Metals, Metals Park, OH, 1980, p 402

Titanium: Commercially Pure, Unalloyed, Grades 3 and 4

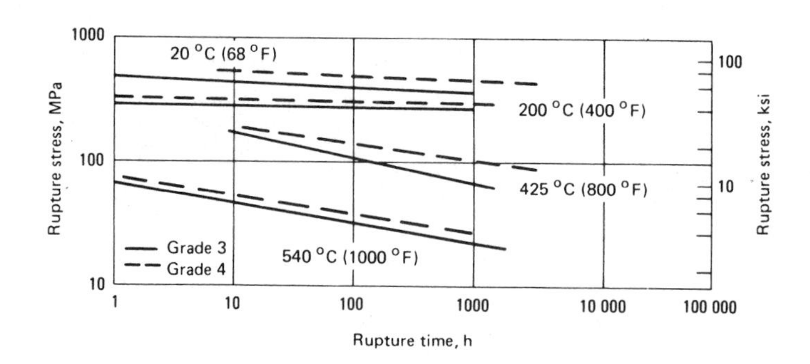

Typical isothermal stress-rupture behavior for commercially pure, unalloyed titanium Grades 3 and 4 at 68 to 1000 °F.

Source: Metals Handbook, Ninth Edition, Volume 3, Properties and Selection: Stainless Steels, Tool Materials and Special-Purpose Metals, American Society for Metals, Metals Park, OH, 1980, p 376

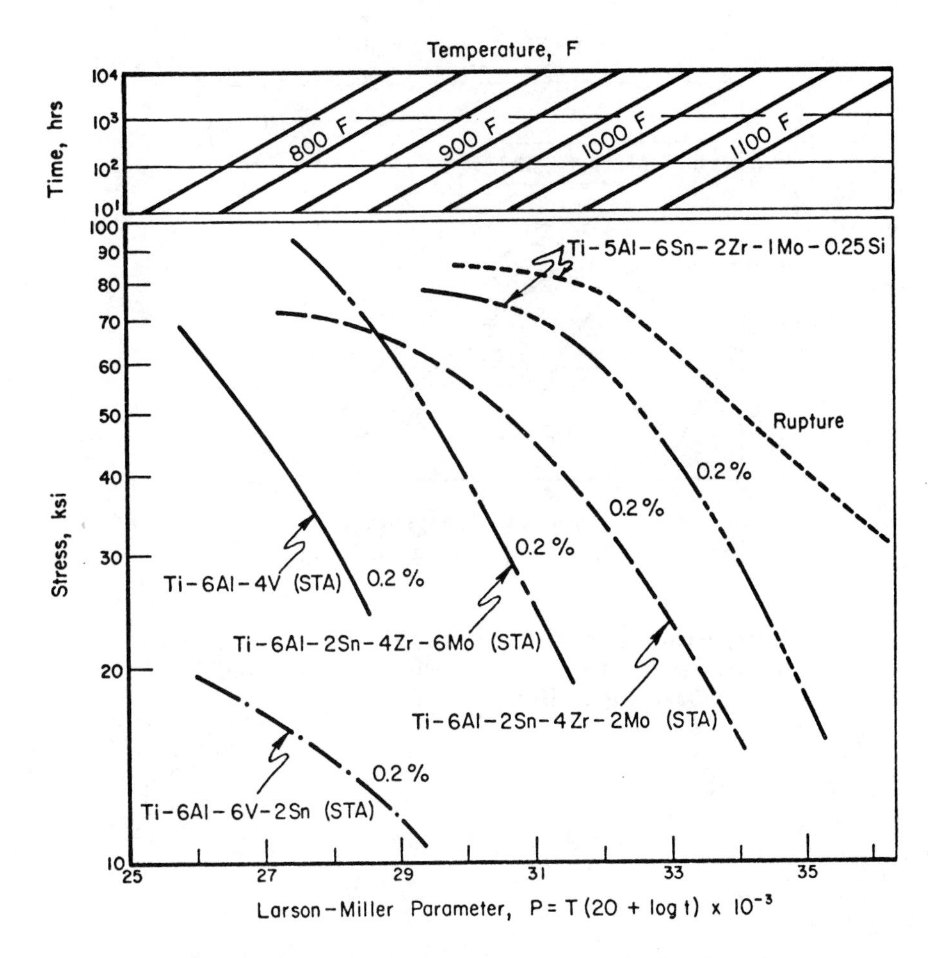

Typical creep and stress-rupture behavior for selected titanium alloys.

The figure above compares three materials on the basis of creep stress versus the Larson-Miller parameter, where time and temperature (for elevated-temperature exposure) are combined. As shown, the Ti-6Al-6V-2Sn alloy is not as creep resistant as Ti-6Al-4V alloy, which in turn is not as resistant as any of the alloys whose curves (depicting the creep conditions of time, temperature, and stress to result in 0.2% plastic strain) plot to the right of the curve for Ti-6Al-4V. These are typical curves for the alloys illustrated, and, like tensile and other properties, variations can occur with composition, microstructure, heat treatment, and testing variables. The rupture curve shown in the figure above for the Ti-5Al-6Sn-2Zr-1Mo-0.25Si alloy is displaced to the right of the 0.2% plastic creep curve, because, obviously, longer times, higher temperatures, or higher stress levels are required to produce the rupture end point of the deformation process defined as creep. Ti-8Al-1Mo-1V, Ti-2.5Al-11Sn-5Zr-1Mo-0.2Si, and Ti-6Al-2Sn-1.5Zr-1Mo-0.35Bi-0.1Si are also well known for their excellent creep resistance.

Source: "Nonspecification Mechanical Properties," in Titanium and Its Alloys, MIL-HDBK-697A, 1 June 74, Department of Defense, Washington, D.C., p 37

Titanium Alloys vs Inconel 718: A Comparison

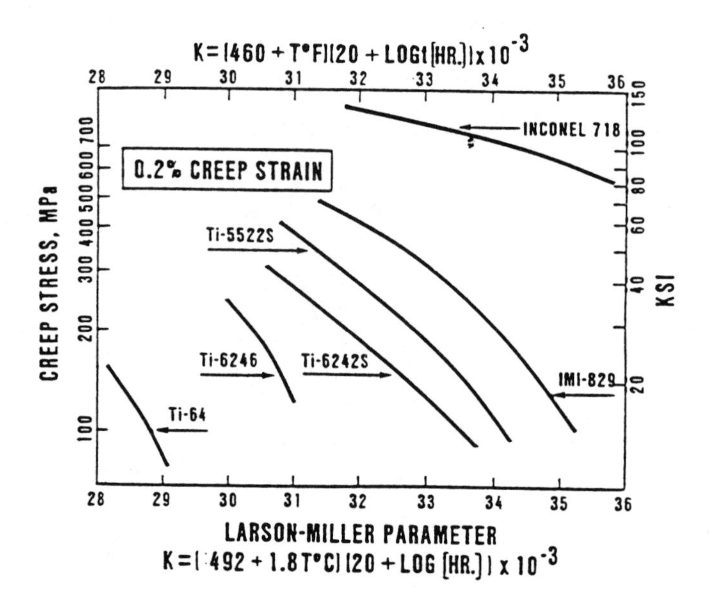

Comparison parametric plot of the 0.2% creep properties of five titanium alloys and Inconel 718 nickel-base superalloy.

Source: D. Eylon, S. Fujishiro, and F.H. Froes, "Titanium Alloys for High Temperature Applications: A Review," High Temperature Materials and Processes, Vol. 6, No. 1 and 2, 1984, p 86

Titanium: Comparison of the 0.1% Creep Properties of Various Alloys

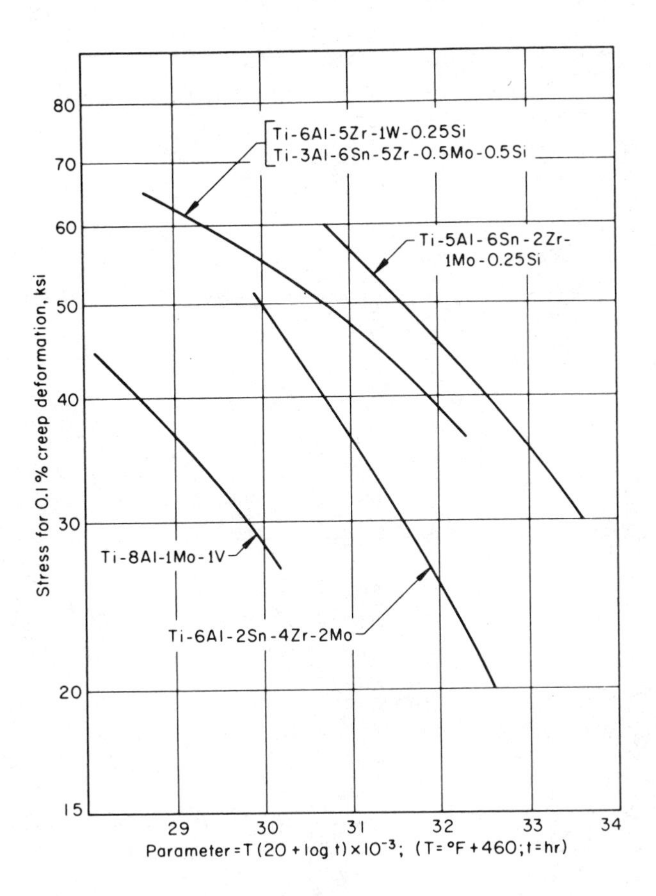

Comparison parametric plot of the 0.1% creep properties of five titanium alloys.

Source: E.F. Bradley and M.J. Donachie, "Materials for Aerospace Forgings," in Source Book on Materials Selection, Volume 1, R.B. Gunia, Ed., American Society for Metals, Metals Park, OH, 1977, p 15

22.24 Titanium: Comparison of the 0.2% Creep Properties of Various Alloys

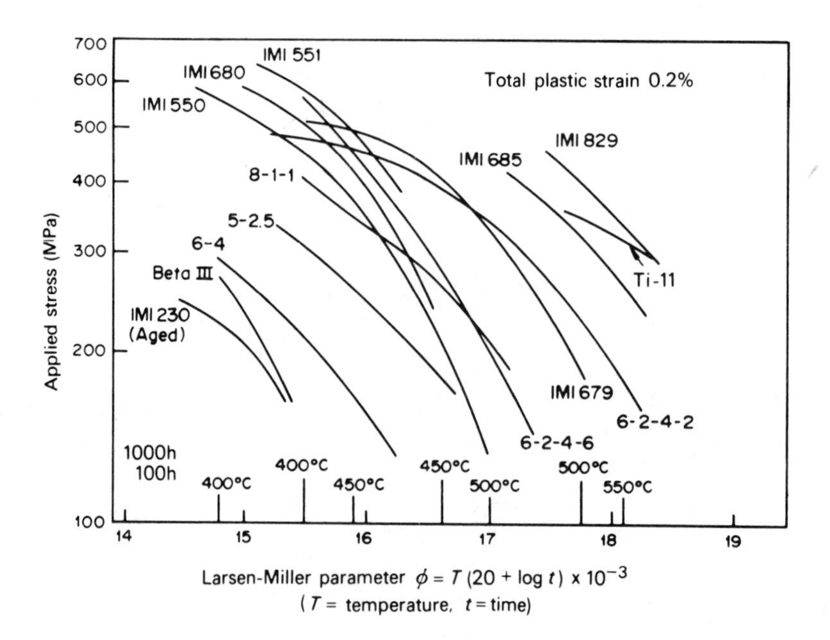

Comparison parametric plot of the 0.2% creep properties of fourteen titanium alloys.

A number of European (UK) titanium alloys are available in this comparison.

Source: I.J. Polmear, Light Alloys, American Society for Metals, Metals Park, OH, 1963, p 164

23

Appendix

Appendix

A

axial strain. Increase (or decrease) in length resulting from a stress acting parallel to the longitudinal axis of the specimen.

B

breaking load. The maximum load (or force) applied to a test specimen or structural member loaded to rupture.

breaking stress. (See *rupture stress.*)

C

Charpy test. An impact test in which a V-notched, keyhole-notched, or U-notched specimen, supported at both ends, is struck behind the notch by a striker mounted at the lower end of a bar that can swing as a pendulum. The energy that is absorbed in fracture is calculated from the height to which the striker would have risen had there been no specimen and the height to which it actually rises after fracture of the specimen.

creep. Time-dependent strain occurring under stress. The *creep strain* occurring at a diminishing rate is called primary or transient creep; that occurring at a minimum and almost constant rate, secondary or steady-rate creep; that occurring at an accelerating rate, tertiary creep.

creep rate. The slope of the creep-time curve at a given time determined from a Cartesian plot.

creep recovery. The time-dependent decrease in strain in a solid, following the removal of force. Recovery is usually determined at constant temperature.

creep-rupture strength. The stress that will cause fracture in a creep test at a given time in a specified constant environment. Also known as stress-rupture strength.

creep-rupture test. A test in which progressive specimen deformation and the time for rupture are both measured. In general, deformation is much greater than that developed during a creep test.

creep strain. The time-dependent total strain (extension plus initial gage length) produced by applied stress during a creep test.

creep strength. The stress that will cause a given creep strain in a creep test at a given time in a specified constant environment.

creep stress. The constant load divided by the original cross-sectional area of the specimen.

creep test. A method of determining the extension of metals under a given load at a given temperature. The determination usually involves the plotting of time-elongation curves under constant load. A single test may extend over many months. The results are often expressed as the elongation (in millimeters or inches) per hour on a given gage length (e.g., 25 mm or 1 in.).

D

Decarburization. Loss of carbon from the surface layer of a carbon-containing alloy due to reaction with one or more chemical substances in a medium that contacts the surface.

deformation. A change in the form of a body due to stress, thermal change, change in moisture, or other causes. Measured in units of length.

displacement. The distance that a chosen measurement point on the specimen displaces normal to the crack plane.

distortion. Any deviation from an original size, shape, or contour that occurs because of the application of stress or the release of residual stress.

ductile crack propagation. Slow crack propagation that is accompanied by noticeable plastic deformation and requires energy to be supplied from outside the body.

ductility. The ability of a material to deform plastically before fracturing. Measured by elongation or reduction in area in a tensile test, by height of cupping in a cupping test, or by the radius or angle of bend in a bend test.

E

elastic deformation. A change in dimensions directly proportional to and in phase with an increase or decrease in applied force.

elastic limit. The maximum stress which a material is capable of sustaining without any permanent strain (deformation) remaining upon complete release of the stress.

elongation. A term used in mechanical testing to describe the amount of extension of a test piece when stressed. (See also *elongation, percent.*)

elongation, percent. The extension of a uniform section of a specimen expressed as percentage of the original gage length.

F

flow. When essentially parallel planes within an element of a material move (slip or slide) in parallel directions; occurs under the action of shearing stress. Continuous action in this manner, at constant volume and without disintegration of the material, is termed yield, creep, or plastic deformation.

I

initial recovery. The decrease in strain in a solid during the removal of force before any creep recovery takes place, usually determined at constant temperature. Sometimes referred to as instantaneous recovery.

initial strain. The strain in a solid immediately upon achieving the given loading conditions in a creep test. Sometimes referred to as instantaneous strain.

initial stress. The stress in a specimen immediately upon achieving the given constraint conditions in a stress-relaxation test. Sometimes referred to as instantaneous stress.

L

Larson-Miller. This theory contends that for each combination of materials and stress level there exists a unique value or a parameter P which is related to temperature and rupture life by the following equation:

$$P = (T + 460)(C + \log t_r$$

where

P = parameter value for given material and stress
T = temperature, °F
C = a constant, usually assumed to be 20
t_r = time to rupture, hr

With this equation it is a simple matter to calculate a short-time combination of temperature and time which is equivalent to any desired long-time service operating requirement. According to the above noted equation, for example, following combinations should have equivalent rupture stresses:

Operating Conditions	Test Conditions
10,000 hr at 1000 °F	13 hr at 1200 °F
1,000 hr at 1200 °F	12 hr at 1350 °F
1,000 hr at 1350 °F	17 hr at 1500 °F
1,000 hr at 300 °F	2.2 hr at 400 °F

load. In the case of testing machines, a force applied to a test piece that is measured in units such as pound-force, newton, or kilogram-force.

M

mechanical properties. The properties of a material that reveal its elastic and inelastic behavior when force is applied or that involve the relationship between the intensity of the applied stress and the strain produced. The properties included under this heading are those that can be recorded by mechanical testing -- for example, modulus of elasticity, tensile strength, elongation, hardness, and fatigue limit. Compare with physical properties.

mechanical testing. The methods by which the mechanical properties of a metal are determined.

modulus of rupture. Nominal stress at fracture in a bend test or torsion test. In bending, modulus of rupture is the bending moment at fracture divided by the section modulus. In torsion, modulus of rupture is the torque at fracture divided by the polar section modulus. See also *modulus of rupture in bending* and *modulus of rupture in torsion.*

modulus of rupture in bending, *Sb.* The value of maximum tensile or compressive stress (whichever causes failure) in the extreme fiber of a beam loaded to failure in bending computed from the flexure equation:

$$S_b = \frac{Mc}{I}$$

modulus of rupture in torsion, *Ss.* The value of maximum shear stress in the extreme fiber of a member of circular cross section loaded to failure in torsion computed from the equation:

$$S_s = \frac{Tr}{J}$$

N

necking. (1) Reducing the cross-sectional area of metal in a localized area by stretching. (2) Reducing the diameter of a portion of the length of a cylindrical shell or tube.

O

Offset. The distance along the strain coordinate between the initial portion of a stress-strain curve and a parallel line that intersects the stress-strain curve at a value of stress (commonly 0.2%) that is used as a measure of the yield strength. Used for materials that have no obvious yield point.

P

parameter. In statistics, a constant (usually unknown) defining some property of the frequency distribution of a population, such as a population median or a population standard deviation.

plastic strain. Dimensional change that does not disappear when the initiating stress is removed. Usually accompanied by some elastic deformation.

R

rate of creep. (See *creep rate.*)

relaxation curve. A plot of either the remaining or relaxed stress as a function of time. (See also *relaxation rate.*)

relaxation rate. The absolute value of the slope of a relaxation curve at a given time.

relaxed stress. The initial stress minus the remaining stress at a given time during a stress-relaxation test. (See also *stress relaxation.*)

remaining stress. The stress remaining at a given time during a stress-relaxation test. (See also *stress relaxation.*)

rupture stress. The stress at failure. Also known as *breaking stress* or fracture stress.

S

shear stress. (1) A stress that exists when parallel planes in metal crystals slide across each other. (2) The stress component tangential to the plane on which the forces act. Also known as tangential stress.

slip. Plastic deformation by the irreversible shear displacement (translation) of one part of a crystal relative to another in a definite crystallographic direction and usually on a specific crystallographic plane. Sometimes called glide. (See also *flow.*)

specimen. A test object, often of standard dimensions or configuration, that is used for destructive or nondestructive testing. One or more specimens may be cut from each unit of a sample.

springback. The degree to which metal tends to return to its original shape or contour after undergoing a forming operation.

steady loads. Loads that do not change in intensity, or change so slowly that they may be regarded as steady.

strain. The unit of change in the size or shape of a body due to force. Also known as nominal strain.

strength. The maximum nominal stress a material can sustain. Always qualified by the type of stress (tensile, compressive, or shear).

stress. The intensity of the internally distributed forces or components of forces that resist a change in the volume or shape of a material that is or has been subjected to external forces. Stress is expressed in force per unit area and is calculated on the basis of the original dimension of the cross section of the specimen. Stress can be either direct (tension or compression) or shear.

stress relaxation. The time-dependent decrease in stress in a solid under constant constraint at constant temperature. The stress-relaxation behavior of a metal is usually shown in a *stress-relaxation curve.*

stress-relaxation curve. A plot of the remaining or relaxed stress as a function of time. The relaxed stress equals the initial stress minus the remaining stress. Also known as stress-time curve.

stress-rupture strength. (See *creep-rupture strength.*)

T

time-temperature parameters (TTP). A TTP is basically correlating the creep-rupture test variables of stress (load), temperature and time, which normally are recorded for every standard, uniaxial, isothermal, constant-load test performed throughout the world.

TTP. (See *time-temperature parameters.*)

U

ultimate strength. The maximum stress (tensile, compressive, or shear) a material can sustain without fracture, determined by dividing maximum load by the original cross-sectional area of the specimen. Also known as nominal strength or maximum strength.

unaxial strain. (See *axial strain.*)

uniform elongation. The elongation at maximum load and immediately preceding the onset of necking in a tensile test.

uniform strain. The strain occurring prior to the beginning of localization of strain (necking); the strain to maximum load in the tension test.

Z

zero time. The time when the given loading or constraint conditions are initially obtained in creep or stress-relaxation tests, respectively.